Unmanned Ground and Aerial Vehicles (UGVs-UAVs) for Digital Farming

Unmanned Ground and Aerial Vehicles (UGVs-UAVs) for Digital Farming

Editors

Monica Herrero-Huerta
Jose A. Jiménez-Berni
Shangpeng Sun
Ittai Herrmann
Diego González-Aguilera

MDPI

Basel • Beijing • Wuhan • Barcelona • Belgrade • Novi Sad • Cluj • Manchester

Editors

Monica Herrero-Huerta
University of Padova
Padua
Italy

Jose A. Jiménez-Berni
Spanish National Research Council (CSIC)
Madrid
Spain

Shangpeng Sun
McGill University
Ste-Anne-de-Bellevue, QC
Canada

Ittai Herrmann
The Hebrew University of Jerusalem
Rehovot
Israel

Diego González-Aguilera
University of Salamanca
Avila
Spain

Editorial Office
MDPI AG
Grosspeteranlage 5
4052 Basel, Switzerland

This is a reprint of articles from the Topic published online in the open access journals *Drones* (ISSN 2504-446X), *Agriculture* (ISSN 2077-0472), *Applied Sciences* (ISSN 2076-3417), *Remote Sensing* (ISSN 2072-4292), and *Sensors* (ISSN 1424-8220) (available at: https://www.mdpi.com/topics/UGVs_UAVs_digital_farming).

For citation purposes, cite each article independently as indicated on the article page online and as indicated below:

Lastname, A.A.; Lastname, B.B. Article Title. *Journal Name* **Year**, *Volume Number*, Page Range.

ISBN 978-3-7258-2095-5 (Hbk)
ISBN 978-3-7258-2096-2 (PDF)
doi.org/10.3390/books978-3-7258-2096-2

Contents

drones

MDPI

Article

Evaluation of Mosaic Image Quality and Analysis of Influencing Factors Based on UAVs

Xiaoyue Du [1,2], Liyuan Zheng [1,2], Jiangpeng Zhu [1,2], Haiyan Cen [1,2] and Yong He [1,2,*]

[1] College of Biosystems Engineering and Food Science, Zhejiang University, Hangzhou 310058, China
[2] Key Laboratory of Spectroscopy Sensing, Ministry of Agriculture and Rural Affairs, Hangzhou 310058, China
* Correspondence: yhe@zju.edu.cn; Tel.: +86-0571-88982143

Abstract: With the growing prominence of UAV-based low-altitude remote sensing in agriculture, the acquisition and processing of high-quality UAV remote sensing images is paramount. The purpose of this study is to investigate the impact of various parameter settings on image quality and optimize these parameters for UAV operations to enhance efficiency and image quality. The study examined the effects of three parameter settings (exposure time, flight altitudes and forward overlap (OF)) on image quality and assessed images obtained under various conditions using signal-to-noise ratio (SNR) and BRISQUE algorithms. The results indicate that the setting of exposure time during UAV image acquisition directly affects image quality, with shorter exposure times resulting in lower SNR. The optimal exposure times for the RGB and MS cameras have been determined as 0.8 ms to 1.1 ms and 4 ms to 16 ms, respectively. Additionally, the best image quality is observed at flight altitudes between 15 and 35 m. The setting of UAV OF complements exposure time and flight altitude; to ensure the completeness of image acquisition, it is suggested that the flight OF is set to approximately 75% at a flight altitude of 25 m. Finally, the proposed image redundancy removal method has been demonstrated as a feasible approach for reducing image mosaicking time (by 84%) and enhancing the quality of stitched images (by 14%). This research has the potential to reduce flight costs, improve image quality, and significantly enhance agricultural production efficiency.

Keywords: remote sensing; UAV; multispectral image; flight optimization; rice

Citation: Du, X.; Zheng, L.; Zhu, J.; Cen, H.; He, Y. Evaluation of Mosaic Image Quality and Analysis of Influencing Factors Based on UAVs. *Drones* **2024**, *8*, 143. https://doi.org/10.3390/drones8040143

Academic Editor: Pablo Rodríguez-Gonzálvez

Received: 27 January 2024
Revised: 25 March 2024
Accepted: 2 April 2024
Published: 4 April 2024

1. Introduction

With its ability to provide a high and wide monitoring range in real time that causes minimal crop damage [1], unmanned aerial vehicle (UAV) remote sensing technology has been extensively used in crop phenotype monitoring [2–6]. Currently, the combined crop information acquisition method using multiple imaging sensors is gradually gaining more attention from researchers [7]. However, as the variety of imaging sensors carried by UAVs continues to increase, with each sensor having a different resolution and field of view, setting flight parameters is becoming a challenge.

The operational workflow of UAVs is delineated in Figure 1. Initially, flight mission planning is conducted based on the dimensions of the experimental area, entailing the determination of crucial parameters such as flight altitude (H), flight speed (V), camera exposure time (ET), and image overlap [8]. Subsequently, multiple images are captured, and through the utilization of image stitching techniques, a comprehensive orthomosaic image (ortho-image) of the experimental area is generated [9]. Finally, a complete operational prescription map is generated for subsequent analysis and processing [10]. However, in this process, the configuration of flight parameters predominantly relies on empirical rules. The setting of flight altitude is generally manually determined to ensure flight safety, obstacle-free operation, and minimal disturbance to crops from UAV-generated wind. Camera exposure time is usually randomly set to avoid the overexposure or underexposure of images. Similarly, the setting of image overlap can be used for successful image stitching.

Notably, a higher overlap setting necessitates a shorter shot interval for the camera under fixed flight speed and altitude conditions, resulting in an increased number of images captured in a specific target area to complete the image stitching. However, this implies a lengthier image mosaicking process, requiring a processor with higher computational performance or a longer processing time, substantially elevating time and labor costs. Meanwhile, the quality of the ortho-image obtained by image stitching technology directly determines the amount of crop information, which is also affected by the environment (illumination change) and the parameter setting of spectral imaging equipment (aperture focal length, exposure time, etc.). Parameter selection through empirical methods frequently falls short in optimizing sensor utilization efficiency, leading to escalated experimental costs, resource consumption, and an inability to guarantee the quality and integrity of the acquired images.

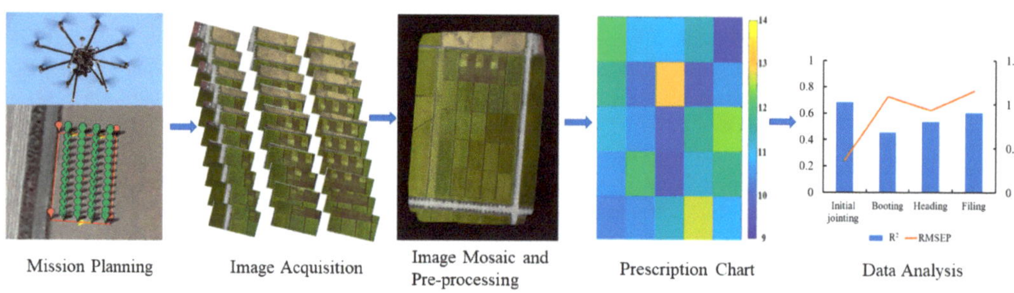

Figure 1. Flow chart from image acquisition to data analysis based on UAV.

Several researchers have acknowledged the significance of establishing UAV flight parameters and have explored the impact of these parameters on operational efficacy. Torres et al. [11] employed a UAV remote sensing platform to capture RGB color images and multispectral images at various flight altitudes, investigating the differentiation effects of vegetation indices on bare soil, crops, and weeds. Their findings revealed that NDVI could reliably distinguish between vegetation and bare soil; however, its ability to differentiate between weeds and crops was influenced by flight altitude, resulting in a higher misjudgment rate. In a similar vein, Faiçal et al. [12] integrated ground sensors with UAV spraying, dynamically adjusting flight parameters based on measured environmental parameters to achieve more uniform pesticide spraying. Song et al. [13] optimized the distribution of fertilizer particle deposition under different flight altitudes and speeds of multi-rotor UAVs during fertilization, ensuring a reasonable and effective deposition amount during actual fertilization processes. Furthermore, Gu et al. [14] explored the influence of different flight parameters on point cloud data quality, revealing a positive correlation between the root mean square error of the airborne lidar flight trajectory and flight altitude and speed. Lastly, Hu et al. [15] investigated the impact of different flight speeds and altitudes of plant protection UAVs on the distribution of fog droplets for pollinating oil tea and its fruit setting rate. He et al. [16] explored the variation of visible and multispectral-based vegetation indices in UAV imagery at different flight altitudes and the impact on estimating vegetation cover. These studies collectively emphasize the intricate relationship between flight parameters and the operational outcomes of UAV-based applications. However, the bulk of research on flight parameters has predominantly concentrated on facets such as pesticide spraying and fertilization in agricultural UAVs. Conversely, there has been comparatively limited explorations of the influence of parameter settings during the crop image capture process using UAVs, despite the pivotal role that these settings play in the precise acquisition and monitoring of crop information. Consequently, it is imperative to systematically investigate UAV flight parameters to ensure the acquisition of high-quality crop images, a critical prerequisite for gathering agricultural information in the context of smart farming.

The main contributions of this study include the following: (1) discussing the influence of different exposure times of cameras on the quality of crop images; (2) investigating the impact of different flight altitudes of UAVs on the quality of crop images; (3) exploring the effect of UAV flights with different overlap rates on the quality of ortho-images; and (4) proposing a method to improve the efficiency of image mosaicking and the quality of ortho-images to reduce flight costs.

2. Materials and Methods

2.1. Image Acquisition

This study employs a combined approach of outdoor and indoor experiments to investigate the impact of varying flight parameters on image quality. The outdoor experiments were conducted in 2019 in Anhua Town, Zhuji City, Zhejiang Province, China (29°31′5.35″ N, 120°6′6.12″ E). The experimental area consisted of 100 plots, each measuring 9×5 m^2, with a 1 m protection lane surrounding the plots. To acquire images of the rice crops, an octo-rotor UAV equipped with red–green–blue (RGB) and multispectral (MS) cameras was used. The UAV had a diameter of 1.1 m, a height of 0.35 m, and a maximum payload of 8 kg. The RGB camera used was the Sony A6000 micro single camera (Sony, Dugang District, Tokyo, Japan) equipped with a 16 mm fixed focus lens, and a resolution of 6000 pixels × 4000 pixels. The field of view of the lens is 83 degrees. The MS camera used was the MQ022MG-CM by XIMEA (Munster, Germany), equipped with a 16 mm fixed focus lens and a resolution of 409 pixels × 216 pixels. The field of view of the lens is 43.6 degrees. Throughout all experiments, manual exposure was employed, maintaining a consistent flight speed of 2.5 m/s, while setting the aperture of both cameras to 2.8. Further parameters regarding the RGB and MS cameras are detailed in Table 1.

Table 1. Parameters of RGB and MS cameras.

Camera	Name	Parameter
RGB	Weight	358 g
	Sensor size	23.4 mm × 15.6 mm
	Resolution	6000 pixels × 4000 pixels
	Focus lens	16 mm/fixed
	Field of view	83
MS	Weight	123 g
	Sensor size	11.27 mm × 6 mm
	Resolution	409 pixels × 216 pixels
	Focus lens	16 mm/fixed
	Field of view	43.6
	Bands	600–1000 nm

Note: The field of view is calculated diagonally.

Images were acquired under cloudless and windless weather conditions. The flight altitude varied between 15 m, 25 m, 30 m, 35 m, 40 m, 45 m, 50 m, 55 m, 100 m and 150 m, corresponding to ground sampling distances (GSDs) of 0.37 cm, 0.61 cm, 0.73 cm, 0.85 cm, 0.98 cm, 1.01 cm, 1.22 cm, 1.34 cm, 2.44 cm, and 3.65 cm for RGB cameras, respectively. Additionally, the GSDs of MS cameras were 2.58 cm, 4.29 cm, 5.16 cm, 6.02 cm, 6.88 cm, 7.74 cm, 8.59 cm, 9.46 cm, 17.18 cm, and 25.76 cm, respectively. Five images were taken at each flight altitude in the hovering state of the UAV. The same region of interest (ROI) was selected using MATLAB (2018a, 9.4.0.813654, MathWorks, Natick, Boulder, CO, USA) software and evaluated to represent the level of image quality at that altitude.

For a fixed altitude of 25 m, different combinations of exposure time and image overlap settings were explored. The flight speed of UAV is fixed at 2.5 m/s. The apertures of both cameras were set at 2.8. The exposure time of the RGB camera was set at two gradients 1.25 ms and 1 ms. For the MS camera, exposure times were set at five gradients: 5, 6, 7, 16 and 20 ms. Due to the disparate field of view of the two cameras, the flight overlap was set based on the camera with the smaller field of view (MS). The designed gradients for the forward overlap were 65%, 75%, and 80%, while for the side overlap, three gradients of 55% 60%, and 65% were employed. The specific experimental design of the flight parameter combinations is outlined in Table 2.

Table 2. Summary of flight campaigns and parameter settings at a flight altitude of 25 m.

Camera	Experiments	Exposure Time (ms)	Forward Overlap (%)	Side Overlap (%)	Number of Images
RGB	Exp. 1	1	65	55	221
	Exp. 2	1	80	65	577
	Exp. 3	1	75	60	387
	Exp. 4	1.25	75	60	388
MS	Exp. 5	5	75	60	387
	Exp. 6	6	65	55	211
	Exp. 7	7	80	65	577
	Exp. 8	16	75	60	387
	Exp. 9	20	75	60	388

The MS images from the same flight mission were radiometrically calibrated and then stitched together to produce orthomosaic images using Agisoft Photoscan (Version 1.2.5 Agisoft LLC, St. Petersburg, Russia) software. This process involved the input of images and geographic coordinates, image alignment, mesh generation, texture generation, DEM creation and orthomosaic image generation. Geographic coordinates were acquired using GPS through a trigger signal on the UAV synchronized with the image. The mesh was computed based on the sparse point cloud due to the flat terrain. As the field of view of the RGB lens is greater than that of the MS, the acquired wide-angle images (RGB) were initially proportionally cropped to match the field of view of the MS images using MATLAB. Subsequently, these processed images were input into Agisoft Photoscan to obtain ortho-images following the same procedure. The MS and RGB ortho-images of the rice crops in the experimental field at a flight altitude of 25 m are shown in Figure 2.

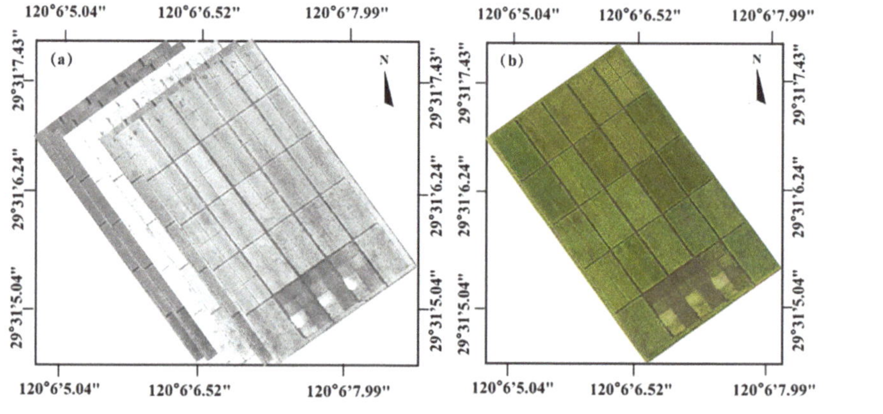

Figure 2. Ortho-images of multispectral camera (**a**) and RGB camera (**b**) at a flight altitude of 25 m.

To mitigate additional influences on image quality due to environmental factors such as surrounding wind speed and lighting changes during UAV flights, exposure time experiments were also conducted indoors. Using the same sensors as the UAV, the focal length and altitude were fixed to capture standardized reflectance gradient boards (with reflectance values of 12%, 25%, 50%, and 99%). For the RGB camera, exposure times were set at 24 gradients in a darkroom environment, ranging from 0.25 to 100 ms, corresponding to shutter speeds of 1/4000 to 1/10 s. The MS camera was set at 18 exposure time gradients ranging from 0.25 to 110 ms. It is worth noting that the aperture of both cameras remained constant during all experiments, both at 2.8.

2.2. Overlap Calculation

During the UAV flight, the overlap between two adjacent images on the same flight strip is referred to as forward overlap (OF), while the overlap between two pictures on adjacent strips is known as side overlap (OS). There are generally two methods for calculating the degree of overlap. One is to represent the degree of OF and OS according to the length ratio of overlapping areas in different directions (Equations (1) and (2)). One is to represent the degree of image overlap according to the percentage of overlapping regions in the captured image area (Equations (3) and (4)). In this paper, Equation (3) is used as the standard of image overlap degree. Typically, an OF from 60% to 80% is required, with a minimum of 53%, while the OS should be at least 8%, with an optimal range from 15% to 60%. However, there is no consensus on how to select the overlap degree to obtain higher quality images. Figure 3 shows the schematic diagram of the overlap calculations method.

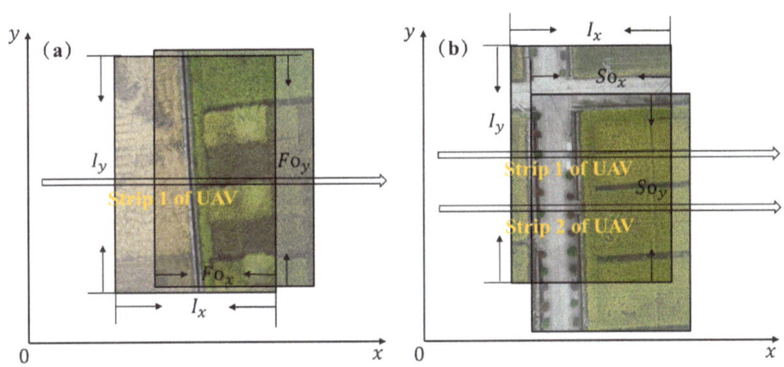

Figure 3. Diagram of the degree of image forward overlap (**a**) and side overlap (**b**).

Based on the definition of overlap, OF and OS can be described as Equations (1) and (2). The formula for calculating the degree of image overlap based on the overlap area is shown in Equation (3).

$$\mathrm{OF} = \frac{Fo_x}{I_x} \times 100\% \tag{1}$$

$$\mathrm{OS} = \frac{So_y}{I_y} \times 100\% \tag{2}$$

$$\mathrm{OF} = \frac{Fo_x \times Fo_y}{I_x \times I_y} \times 100\% \tag{3}$$

$$\mathrm{OS} = \frac{So_x \times So_y}{I_x \times I_y} \times 100\% \tag{4}$$

where Fo_x and Fo_y represent the size of the overlap area in the flight direction. So_x and So_y represent the overlap of two pictures on two flight strips, and I_x and I_y indicates the size of the UAV image.

2.3. Image Evaluation

2.3.1. Conventional Image Quality Evaluation

Signal-to-noise ratio (SNR) is a commonly used metric to evaluate the noise level in an image. It measures the ratio of the signal strength to the noise level in the image. Higher SNR values indicate less noise and better image quality. This indicator is used to assess image quality when the exposure time is different and other conditions remain unchanged.

$$\mathrm{SNR} = 10Lg\frac{P_s}{P_n}, \tag{5}$$

where P_s is the power of the signal, P_n is the power of the noise, and Lg denotes the logarithm of the base 10.

2.3.2. BRISQUE Algorithm

As remote sensing images lack original images for reference, a no-reference image quality assessment algorithm (NR-IQA) is selected to evaluate the quality of the obtained remote sensing images. Among the existing NR-IQA models for natural images, the Blind/Referenceless Image Spatial Quality Evaluator (BRISQUE) [10] is considered to be one of the most advanced and optimal models. BRISQUE not only considers image luminance [17], contrast [18], distortion [19], the complex statistic of an image [20], texture statistic [21], and Natural Scene Statistics (NSS) [22], but also maintains a relatively low computational complexity.

The algorithm can be summarized as follows. For a given (possibly distorted) image, locally normalized luminances are first computed via local mean subtraction and divisive normalization. Such an operation may be applied to a given intensity image *I(i, j)* to produce:

$$I(i,j) = \frac{I(i,j) - \mu(i,j)}{\sigma(i,j) + C}, \tag{6}$$

where, $i \in 1, 2, \ldots, \mathrm{M}$, $j \in 1, 2, \ldots, \mathrm{N}$ are spatial indices; M and N are the image height and width, respectively; $C = 1$ is a constant that prevents instabilities from occurring when the denominator tends to zero and

$$\sigma(i,j) = \sqrt{\sum\sum w_{k.l}\left(I_{k.l}(i,j) - \mu(i,j)^2\right)}, \tag{7}$$

$$\mu(i,j) = \sum\sum w_{k.l} I_{k.l}, \tag{8}$$

where $w = \{\, w_{k.l} \mid k = -K, \ldots, K, l = -L, \ldots, L\}$ is a 2D circularly symmetric Gaussian weighting function sampled out to three standard deviations and rescaled to unit volume. In our implementation, $K = L = 3$. Meanwhile, the pre-processing model (6) is developed and refers to the transformed luminances *I(i, j)* as mean subtracted contrast normalized (MSCN) coefficients. Then, an asymmetric generalized Gaussian distribution (AGGD) can be used to effectively capture a broader spectrum of distorted image statistics. Additionally, Pearson linear correlation coefficients (PLCC) of MSCN adjacent coefficients in four directions of horizontal, vertical, main diagonal, and sub-diagonal are added to describe the characteristics of the overall structural distortion of the image. Finally, a support vector machine regression (SVM) [23] is used to learn from the feature space to quality scores, so it is worth noting that the better the image quality, the lower the score.

However, the algorithm only evaluates the quality of RGB images and is not applicable to MS images obtained using remote sensing. Notably, the research team adapted the BRISQUE algorithm to MS images by changing the image input to single band input and analyzing all bands to use the GGD fit of the MSCN coefficients. The average value of all bands is used as the quality evaluation score of the MS images. The algorithm flow chart is shown in Figure 4.

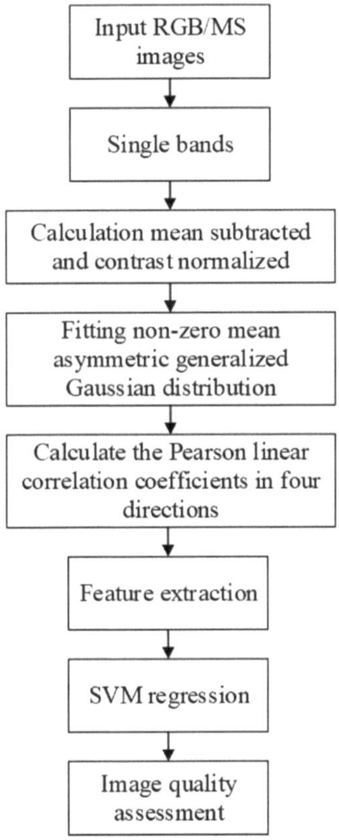

Figure 4. Algorithm flow chart of the improved BRISQUE algorithm.

2.4. Methods for Removing Redundancy

The large amount of image data causes an increase in the time cost of the image stitching process. To solve this problem, a redundancy removal approach is proposed to streamline the UAV image data. The specific operation process of this method is shown in Figure 5. Step 1: input the UAV images with their respective BRISQUE scores. Step 2: Determine the redundancy interval, denoted as i, with a value range of integers greater than or equal to 0. The criterion for selection is the removal of redundancy to enable image mosaicking, as excessively low overlap between images can lead to mosaic failure. Step 3: Retain the image with the lowest quality score within the range of (i + 1) and remove the others. Then, proceed to the next redundancy interval. Step 4: repeat the above steps until all input images have been filtered. Step 5: Output all retained high-quality images. It is important to note that we need to perform image mosaicking on the retained images to obtain the final ortho-image. If the mosaicking process encounters difficulties, the redundancy interval should be reduced.

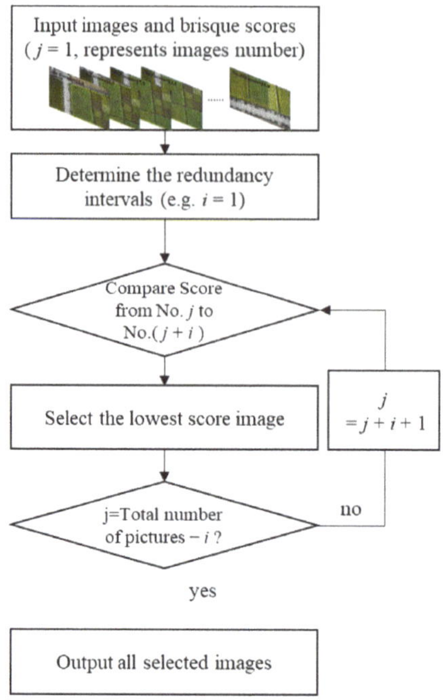

Figure 5. Image selection method for removing redundancy.

3. Results and Discussion

3.1. Calculation of Actual Overlap

Due to the complexity of the external environment, this study used the area method to calculate the actual overlap between two adjacent images, which is shown in Equation (3) in order to ensure the accuracy. According to the calculation method of overlap in the previous chapter, the data of all flights are analyzed. Taking the second experiment as an example, the results are illustrated in Figure 6. It can be observed that the overlap of the actual adjacent images acquired during the flight along the same strip fluctuates around a set value, rather than remaining at a fixed value. This fluctuation occurs due to the triggering signal method for UAV-controlled image capture being distance-based. Additionally, the autonomously developed UAV utilizes GPS for positioning, which may lead to positioning and drift errors during flight, resulting in this fluctuation phenomenon.

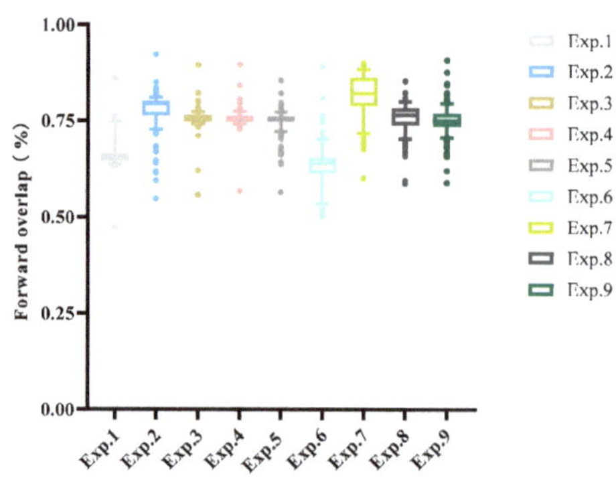

Figure 6. Forward overlap (OF) of all experiments.

We also observed in Figure 6 that in Experiments 2 and 7, although the flight overlap was set at 80%, some actual values were below 60%. This is due to the fact that, given the determined UAV flight altitude and speed, increasing the image overlap requires shortening the capture interval. However, the exposure time during camera capture and the image storage time impose constraints on the capture interval. When the capture interval cannot meet the image storage requirements, some images may be missed. Therefore, in the scenario of a UAV flying at a speed of 2.5 m per second and at a flight altitude of 25 m, it is recommended to set the OF to not exceed 75% to ensure data integrity. When a higher flight speed is attainable, adjusting the flight altitude to maintain the capture interval in accordance with image acquisition requirements is feasible. Alternatively, a lower altitude can be employed in conjunction with reduced flight speed. Substituting the imaging sensor with a faster storage speed can also fulfill the demand for increased image overlap.

3.2. UAV Image Quality Evaluation

3.2.1. Influence of Exposure Time on Image Quality Evaluation

The linearity of the ratio of DN to exposure time for the four reflectivity targets is demonstrated in Figure 7. Notably, as the exposure time increases, the linearity of the targets becomes more stable. This observation holds true even when keeping the light intensity, camera height, and focal length constant.

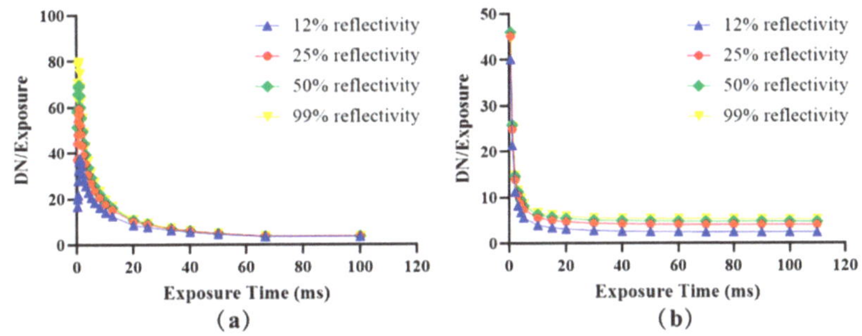

Figure 7. Ratio of digital number (DN) to exposure time for four reflectivity targets with an RGB camera (**a**) and MS camera (**b**).

Figure 8 depicts the SNR values obtained for different exposure times and three different types of reflection. It can be observed that the image SNR values increase as the exposure time increases. However, the SNR values reach a plateau after a certain level due to the saturation of image DNs. The maximum SNR levels for different channels are dependent on the dark noise level of each channel. In practical scenarios, it is recommended to select integration time settings that optimize image SNRs while avoiding image saturation. For instance, in low irradiance conditions and low reflectivity land covers, the optimal integration time coincides with the maximum integration time. On the other hand, for high-reflectivity targets, exposure times with better linearity and lower sensitivity should be chosen.

In the presence of constant light intensity, the relationship between the image quality score and exposure time for both the RGB and MS cameras is illustrated in Figure 9. Evidently, the image quality scores obtained by both cameras exhibit a trend of initially decreasing and then increasing with longer exposure times. Notably, the RGB camera is capable of achieving a higher image quality (QS < 10) between 0.8 ms and 5 ms, whereas the MS images demonstrate this between 4 ms and 50 ms.

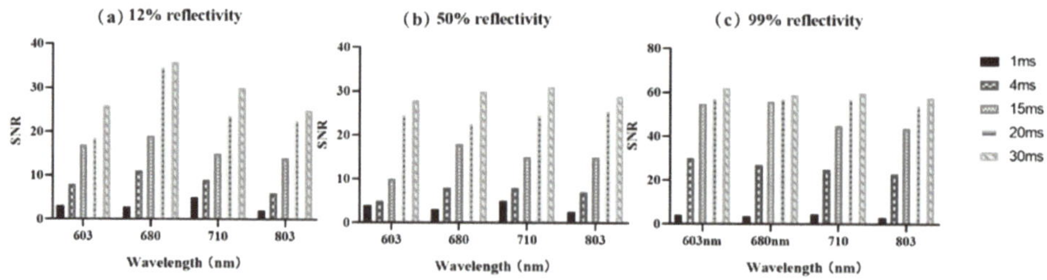

Figure 8. Signal-to-noise ratio (SNR) for different exposure time settings and targets with an MS camera: (**a**) 12% reflectivity, (**b**) 50% reflectivity, and (**c**) 99% reflectivity under constant illumination. The x-axis represents the different channels, and the y-axis is the image SNR.

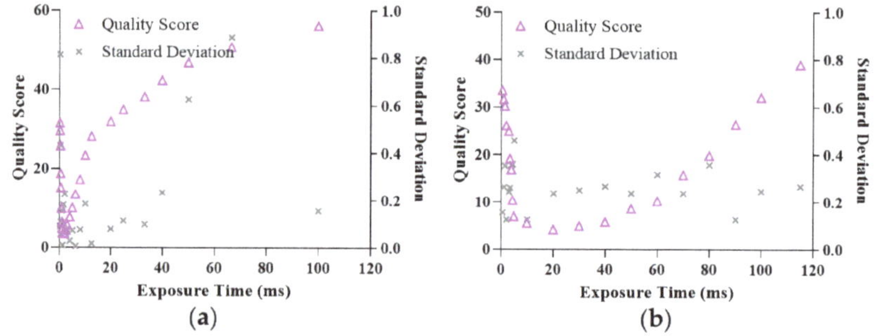

Figure 9. Relationship between quality score (QS) and exposure time with an RGB camera (**a**) and MS camera (**b**).

In the outdoor environment, the BRISQUE algorithm is used to evaluate the image quality of five different exposure time setting experiments of MS camera. The probability analysis of five flights is shown in Figure 10. We can find that the probability distribution of image quality is different under different exposure time settings. Overall, during a single UAV flight experiment, fluctuations in external environmental conditions lead to variations in the quality of image acquisition.

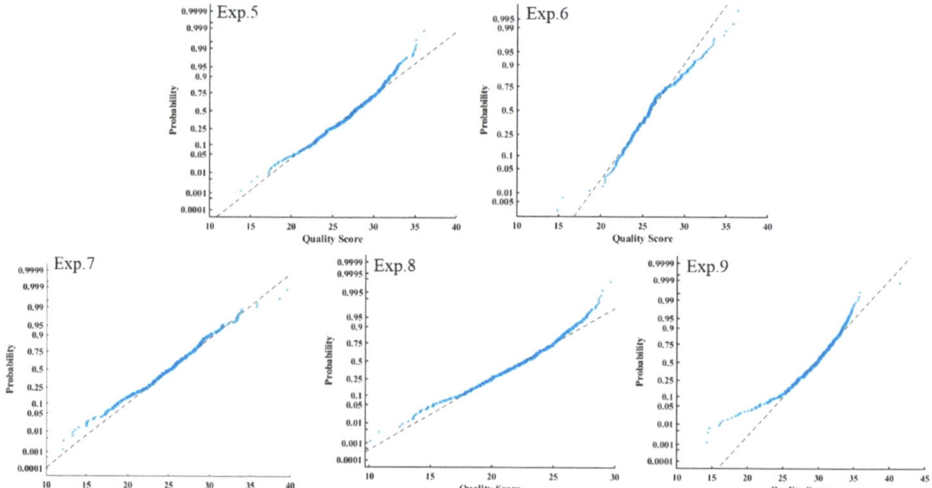

Figure 10. Probability analysis of quality evaluation distribution of five flights with an MS camera.

3.2.2. Image Quality Evaluation of Single Experiment

In this paper, the improved BRISQUE algorithm is used to evaluate the image quality of MS images and RGB images obtained using a UAV. The quality score is shown in Figure 11. The lower the score that we calculated, the better the required image quality since support vector machine regression is used in this algorithm. It can be seen from the figure that the quality score of remote sensing images in the same experience is in the form of a normal distribution. The position of the expected value depends on the matching degree between the exposure time set by the experimenter at takeoff and the light intensity at that time. In the process of UAV operation, due to the change in illumination intensity, the image quality obtained in one experiment is not invariable. Therefore, we can try to explore whether removing lower-quality remote sensing images can obtain higher-quality mosaic images. The setting of high overlap makes it possible to reduce image redundancy.

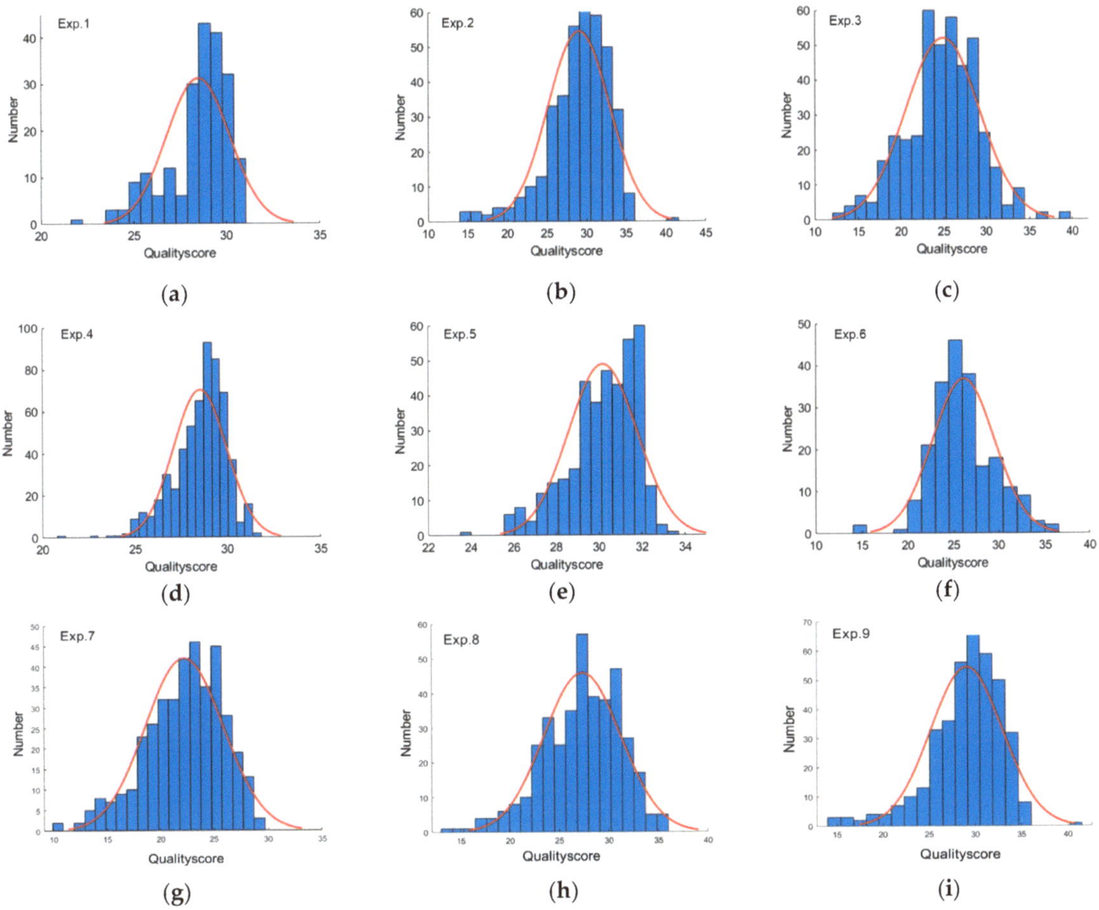

Figure 11. Quality evaluation distribution of nine flight experiments (**a–i**) based on the BRISQUE algorithm.

3.2.3. Image Quality Evaluation of Different Flight Altitude

The UAV flies at different altitudes to obtain five crop RGB and MS remote sensing images. The obtained image quality is evaluated using the BRISQUE algorithm. The box chart (Figure 12) shows the image quality scores at different altitudes (15 m, 25 m, 30 m, 35 m, 40 m, 45 m, 50 m, 100 m, 150 m). Comparing the average scores of different altitudes, we can find that the image quality scores of RGB images obtained between 15 m and 150 m show a fluctuating increase, but there is a large difference at the same altitude. This is because the factors affecting the image quality score at lower flight altitudes depend more on the suitability between the variation in light intensity and the exposure time setting.

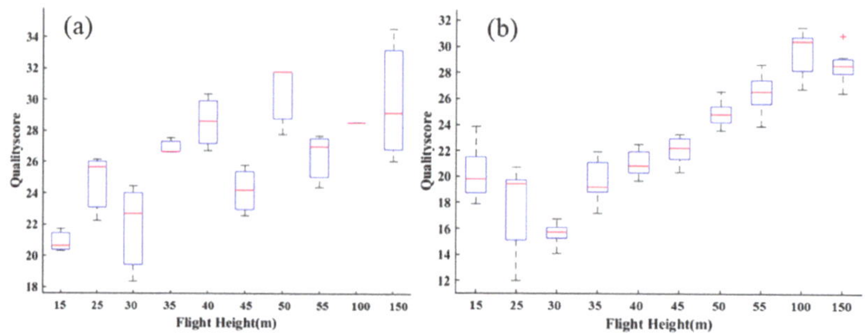

Figure 12. Quality evaluation of different flight altitudes based on BRISQUE algorithm with RGB camera (**a**) and MS camera (**b**).

The MS image quality score exhibits a trend of decreasing and then increasing, with an average optimal score achieved at a flight altitude of 30 m. In general, higher flight altitudes lead to lower image quality. The higher image quality at 30 m may be attributed to the camera acquiring more crop information under relatively stable external environmental conditions, thereby improving the quality score to a certain extent. However, due to the low resolution of the MS camera, exceeding a certain flight altitude results in a low ground resolution that erases surface texture information, ultimately affecting the comprehensive score. Therefore, when using the two cameras described in this paper to obtain remote sensing images of crops, it is recommended to maintain a flight altitude between 15 and 35 m.

3.3. *Image Mosaic and Redundancy Reduction*

Initially, all images based on the same flight were input into the software to obtain an orthomosaic image, and the time from the image input to final ortho-image was recorded. A fixed redundancy interval was selected based on the OF, and the images were stitched again. This process was repeated until stitching failed. The number of images and splicing time of the last successful splicing were recorded as the final result of redundant processing. The image quality of the acquired ortho-images was evaluated using the BRISQUE algorithm.

Figure 13 illustrates the variation in image overlap after three rounds of redundancy removal in the second experiment, decreasing from the original 80% to fluctuate at around 45%. As the decrease in overlap makes image mosaicking challenging, the images after three rounds of redundancy removal cannot be stitched to obtain ortho-images. Therefore, the final retained high-quality images are those processed for redundancy twice. According to the results, it is recommended that during the flight, the forward overlap factor should reach at least 60% to ensure a proper image alignment. However, considering the inherent errors associated with the use of consumer-grade small-format digital cameras (such as the SONY camera we utilized) in small UAV photogrammetry, as well as the importance of larger overlaps in mitigating variations due to slight topographic height differences and enhancing the overall robustness of the image block adjustments, the setting of overlap parameters can be increased as much as possible while ensuring complete image coverage.

Figure 14 illustrates changes in the number of images and mosaic image quality scores before and after redundancy removal in all experiments. Additionally, Table 3 displays the completion time of obtaining the mosaic image. The image mosaic efficiency and the mosaic image quality of experiments were improved to varying degrees after reducing redundancy, except in the case of Experiment 6. Fortunately, the maximum improvement in image stitching efficiency was 84% (Experiment 7), while the image quality score improved by 7%. Meanwhile, Experiment 2, with the largest improvement in image quality scores (13%), had a 30% reduction in completion time. The reason for the absence of redundancy reduction in Experiment 6 was that the resolution of the MS camera is lower than that of RGB, and there are fewer feature points in the process of image stitching, which makes the stitching more difficult. In addition, the overlap is too low after redundancy reduction, which leads to stitching failure. Therefore, it is not necessary to reduce the redundancy of low-overlap MS images. The higher the overlap degree is set, the more images of the target area are obtained, and the larger the amount of redundancy data. Therefore, for a large number of images obtained using a high overlap, image redundancy reduction can not only improve the efficiency of image mosaicking, but also improve the quality of image stitching to a certain extent. Simultaneously, it is possible to appropriately increase the fixed redundancy interval based on the overlap to enhance image processing efficiency.

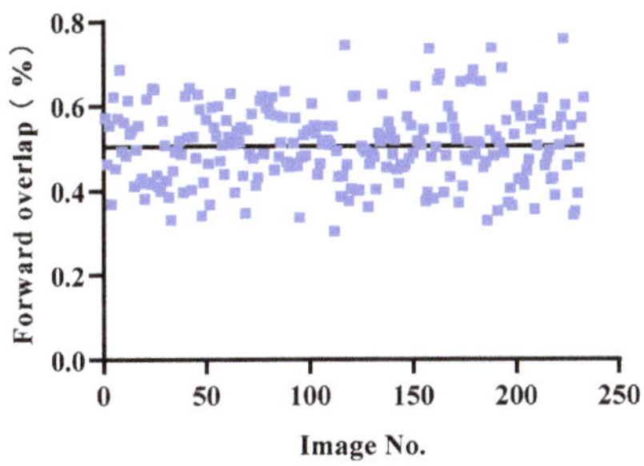

Figure 13. Forward overlap (OF) of the second experiment after two redundancy removals.

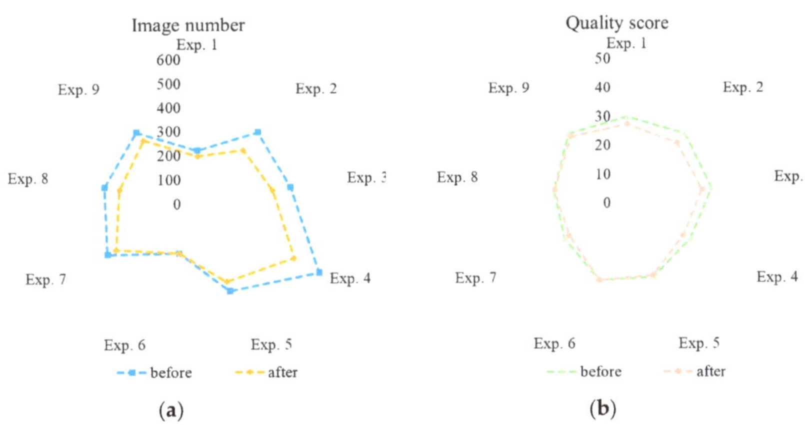

Figure 14. The changes in the number of images (**a**) and quality evaluation of the mosaic image (**b**) before and after the removal redundancy in each flight experiment.

Table 3. Quality evaluation of mosaic image before and after redundancy reduction.

Camera	Experiments	Completion Time (h)		
		Before	After	Improved
RGB	Exp. 1	1.02	0.75	26%
	Exp. 2	2.16	1.5	30%
	Exp. 3	2.4	1.75	27%
	Exp. 4	19.4	10	48%

Table 3. *Cont.*

Camera	Experiments	Completion Time (h)		
		Before	After	Improved
MS	Exp. 5	0.08	0.07	13%
	Exp. 6	0.05	0.05	0
	Exp. 7	0.5	0.08	84%
	Exp. 8	0.09	0.08	11%
	Exp. 9	0.09	0.08	11%

4. Discussion of Flight Strategy

When considering the utilization of UAVs for image acquisition, a comprehensive assessment of various factors such as flight altitude, exposure time, flight speed, and flight overlap is essential to obtain comprehensive crop information. It is important to note that the configuration of these parameters directly impacts the quality and efficiency of image acquisition, and their interrelationships play a crucial role in devising effective flight strategies. In the realm of camera optics, to obtain clear images without causing motion blur during the process of capturing dynamic images, the exposure time and flight speed must satisfy the condition outlined in Equation (9). The calculation of GSD is directly related to the flight altitude (Equation (10)). As a result, the interplay between flight altitude, flight speed, and exposure time is elucidated through the relationships depicted in Equation (9) and Equation (10). It is noteworthy that the configuration of flight speed primarily influences the capture of clear images. When GSD remains constant, setting the exposure time requires careful consideration of factors such as camera performance solar radiation intensity, and white balance, thereby adding complexity to the decision making process. These factors directly impact the imaging quality of the acquired images underscoring the importance of prioritizing exposure time requirements when formulating flight strategies. In contrast, considerations for flight speed are relatively straightforward once exposure time requirements are met. Consequently, this study focuses on exploring methods for setting exposure time, while maintaining a consistent flight speed.

$$ET \times V \leq \text{GSD}, \quad (9)$$

$$\text{GSD} = \frac{S \times H}{f}, \quad (10)$$

where ET is exposure time, V is flight speed, GSD is ground sample distance, S is camera single pixel size, H is flight altitude above ground, and f is focal length.

By utilizing this formula, the longest exposure times required during UAV-mounted camera flights can be determined. Consequently, in this study, under the conditions of a flight speed of 2.5 m/s and a flight altitude of 25 m, the respective maximum exposure times for the RGB and MS cameras are within 1.1 ms and 16 ms. Figures 7–10 demonstrate that exposure time directly influences image quality, with shorter exposure times resulting in lower signal-to-noise ratios. The results indicate that maintaining the exposure time for the RGB camera between 0.8 ms and 5 ms, as well as selecting an exposure time between 4 ms and 50 ms for the MS camera, contributes to achieving higher signal-to-noise ratios and better image quality. Therefore, considering the aforementioned requirements, the exposure time for the RGB camera should be set between 0.8 ms and 1.1 ms, and for the MS camera, it should be set between 4 and 16 ms. Furthermore, adjustments to aperture and ISO can maintain a consistent white balance in the images, even amidst changes in external environmental conditions.

The setting of the UAV's flight forward overlap is closely related to the field of view of the mounted camera, as expressed by Equation (5).

$$OF = \frac{2 \times H \times \tan\left(\frac{\theta}{2}\right)}{V}, \tag{11}$$

where OF is forward overlap, H is flight altitude, V is flight speed, θ is field of view of lens along the short frame side.

This indicates that the forward overlap is inversely proportional to the flight speed and directly proportional to the flight height and the camera's field of view, as demonstrated by Equation (11). A higher OF can be achieved by reducing the flight speed, increasing the flight altitude, or employing a camera with a larger field of view while maintaining a consistent shooting interval. Figure 12 confirms that the appropriate GSD leads to a high image quality, with an optimal image quality obtained within the flight altitude range of 15–35 m. Therefore, priority should be given to reducing the flight speed or adjusting the camera's field of view to enhance the overlap. Additionally, under constant flight altitude, speed, and camera specifications, increasing the overlap can also be achieved by reducing the photo capture interval. However, this method requires consideration of the camera's exposure time, as outlined in the aforementioned Equation (9).

In the post-data collection processing and analysis workflow, greater emphasis is placed on the quality and efficiency of image stitching. Figure 11 demonstrates the uneven quality of images captured during the same flight mission. Figure 14 illustrates that by selecting high-quality images and reducing redundancy, not only can the efficiency of image stitching be improved, but the stitching quality can also be enhanced. Therefore, during UAV image acquisition, setting a higher overlap can contribute to obtaining higher-quality stitched images, while ensuring exposure time.

Based on the above analysis, when using both RGB and MS cameras for crop image acquisition in this study, it is recommended to maintain a flight altitude of between 15 and 35 m. Furthermore, it is recommended to set the exposure time for the RGB camera between 0.8 ms and 1.1 ms, and for the MS camera within the range of 4–16 ms, when both camera apertures are set to 2.8. This approach is more likely to yield higher-quality images. Regarding the flight OF, a value of approximately 75% is suggested, as it adequately meets the requirements for both flight altitude and exposure time. Certainly, it is possible to increase the overlap while ensuring the complete acquisition of images.

In addition, UAV flight parameters should be varied accordingly for different terrains [24]. This is because factors such as different terrain types, elevation, vegetation density and reflective surfaces have a significant impact on the planning and execution of UAV survey or mapping missions. For instance, areas with tall vegetation may introduce shadows and varying light levels, mandating careful adjustments to exposure settings to ensure optimal image quality. Similarly, snow-covered areas, water bodies, and urban environments may require exposure adjustments to prevent over- or underexposure. Moreover, higher elevations may demand adjustments to the flight altitude to maintain the desired GSD and uphold image quality, while complex terrains, such as mountainous regions or areas with significant elevation changes, may require a higher overlap between images to capture comprehensive data effectively. Additionally, dense vegetation can obstruct the view of the ground, potentially requiring a higher overlap to capture sufficient data for accurate image stitching and subsequent analysis. Considering these aspects enhances the robustness of the image acquisition strategy, ensuring that the set parameters align with the specific characteristics of the UAV-mapped area and thereby contribute to the acquisition of high-quality and comprehensive image data for subsequent analysis.

5. Conclusions

In this study, we achieved UAV flight parameter optimization and obtained high-quality stitched images by investigating the effects of different exposure times, flight altitudes, and OF settings on the image quality of rice phenology. We used an eight-rotor UAV flight platform equipped with two image acquisition sensors, including MS and RGB cameras, to capture images of rice crops. The experimental design included a combination of three different variables such as exposure time, flight altitude and overlap. All experimental flights were conducted at the same speed. The SNR was used to evaluate the noise level of the images, and an improved BRISQUE algorithm was proposed to assess the quality of the MS and RGB remote sensing images. The results indicate that exposure time directly influences the quality of low-altitude remote sensing images obtained by UAVs, with shorter exposure times leading to lower signal-to-noise ratios. Additionally obtaining clear images during UAV motion requires specific conditions. Based on our study when both camera apertures are set to 2.8, the optimal exposure time for the RGB camera should be between 0.8 ms and 1.1 ms, and for the MS camera, the exposure time should range between 4 and 16 ms. Based on these experiments, it is recommended to capture images at flight altitudes between 15 m and 35 m. Meanwhile, the OF setting of the UAV needs to take into account the flight altitude and exposure time, and a high OF will lead to image loss. Therefore, choosing the appropriate OF is crucial for image acquisition, and the results of this study show that a flight OF of about 75% can satisfy the needs of image acquisition. Finally, in the subsequent image processing stage, a method for removing image redundancy proposed in this study can effectively improve the quality (13%) and efficiency (84%) of image stitching, which provides an idea for the effective utilization of data and reduction in time cost.

6. Patents

The US invention patents resulting from the work reported in this manuscript have been authorized—patent number: US 11636582B1.

Author Contributions: Conceptualization, Methodology, Software, Formal Analysis, X.D.; Investigation, X.D. and L.Z.; Data Curation, X.D., L.Z. and J.Z.; Writing—Original Draft Preparation X.D.; Methodology, Writing—Review and Editing, H.C.; Writing—Review and Editing, Funding Acquisition, Y.H. All authors have read and agreed to the published version of the manuscript.

Funding: This research was funded by Key R & D projects in Zhejiang Province, grant number 2021C02023.

Data Availability Statement: All data used to support the findings of this study are included within the article.

Conflicts of Interest: The authors declare no conflicts of interest.

References

1. Lan, Y.; Thomson, S.J.; Huang, Y.; Hoffmann, W.C.; Zhang, H. Current Status and Future Directions of Precision Aerial Application for Site-Specific Crop Management in the USA. *Comput. Electron. Agric.* **2010**, *74*, 34–38. [CrossRef]
2. Lu, W.; Okayama, T.; Komatsuzaki, M. Rice Height Monitoring between Different Estimation Models Using UAV Photogrammetry and Multispectral Technology. *Remote Sens.* **2022**, *14*, 78. [CrossRef]
3. Yue, J.; Yang, G.; Tian, Q.; Feng, H.; Xu, K.; Zhou, C. Estimate of Winter-Wheat above-Ground Biomass Based on UAV Ultrahigh-Ground-Resolution Image Textures and Vegetation Indices. *ISPRS J. Photogramm. Remote Sens.* **2019**, *150*, 226–244. [CrossRef]
4. Zhang, X.; Zhang, K.; Wu, S.; Shi, H.; Sun, Y.; Zhao, Y.; Fu, E.; Chen, S.; Bian, C.; Ban, W. An Investigation of Winter Wheat Leaf Area Index Fitting Model Using Spectral and Canopy Height Model Data from Unmanned Aerial Vehicle Imagery. *Remote Sens* **2022**, *14*, 5087. [CrossRef]
5. Ma, Y.; Zhang, Q.; Yi, X.; Ma, L.; Zhang, L.; Huang, C.; Zhang, Z.; Lv, X. Estimation of Cotton Leaf Area Index (LAI) Based on Spectral Transformation and Vegetation Index. *Remote Sens.* **2021**, *14*, 136. [CrossRef]
6. Xu, X.Q.; Lu, J.S.; Zhang, N.; Yang, T.C.; He, J.Y.; Yao, X.; Cheng, T.; Zhu, Y.; Cao, W.X.; Tian, Y.C. Inversion of Rice Canopy Chlorophyll Content and Leaf Area Index Based on Coupling of Radiative Transfer and Bayesian Network Models. *ISPRS J Photogramm. Remote Sens.* **2019**, *150*, 185–196. [CrossRef]
7. Kefauver, S.C.; Vicente, R.; Vergara-Díaz, O.; Fernandez-Gallego, J.A.; Kerfal, S.; Lopez, A.; Melichar, J.P.E.; Serret Molins, M.D.; Araus, J.L. Comparative UAV and Field Phenotyping to Assess Yield and Nitrogen Use Efficiency in Hybrid and Conventional Barley. *Front. Plant Sci.* **2017**, *8*, 1733. [CrossRef] [PubMed]
8. Wan, L.; Cen, H.; Zhu, J.; Zhang, J.; Zhu, Y.; Sun, D.; Du, X.; Zhai, L.; Weng, H.; Li, Y.; et al. Grain Yield Prediction of Rice Using Multi-Temporal UAV-Based RGB and Multispectral Images and Model Transfer—A Case Study of Small Farmlands in the South of China. *Agric. For. Meteorol.* **2020**, *291*, 108096. [CrossRef]

9. Fu, Z.; Jiang, J.; Gao, Y.; Krienke, B.; Wang, M.; Zhong, K.; Cao, Q.; Tian, Y.; Zhu, Y.; Cao, W.; et al. Wheat Growth Monitoring and Yield Estimation Based on Multi-Rotor Unmanned Aerial Vehicle. *Remote Sens.* **2020**, *12*, 508. [CrossRef]
10. Liu, Y.; An, L.; Wang, N.; Tang, W.; Liu, M.; Liu, G.; Sun, H.; Li, M.; Ma, Y. Leaf Area Index Estimation under Wheat Powdery Mildew Stress by Integrating UAV-based Spectral, Textural and Structural Features. *Comput. Electron. Agric.* **2023**, *213*, 108169. [CrossRef]
11. Torres-Sánchez, J.; Peña-Barragán, J.M.; Gómez-Candón, D.; De Castro, A.I.; López-Granados, F. Imagery from Unmanned Aerial Vehicles for Early Site Specific Weed Management. In Proceedings of the Precision Agriculture, Lleida, Spain, 7–11 July 2013; pp. 193–199.
12. Faiçal, B.S.; Freitas, H.; Gomes, P.H.; Mano, L.Y.; Pessin, G.; de Carvalho, A.C.P.L.F.; Krishnamachari, B.; Ueyama, J. An Adaptive Approach for UAV-Based Pesticide Spraying in Dynamic Environments. *Comput. Electron. Agric.* **2017**, *138*, 210–223. [CrossRef]
13. Song, C.; Liu, L.; Wang, G.; Han, J.; Zhang, T.; Lan, Y. Particle Deposition Distribution of Multi-Rotor UAV-Based Fertilizer Spreader under Different Height and Speed Parameters. *Drones* **2023**, *7*, 425. [CrossRef]
14. Gu, J.; Zhang, R.; Dai, X.; Han, X.; Lan, Y.; Kong, F. Research on Setting Method of UAV Flight Parameters Based on SLAM. *J. Chin. Agric. Mech.* **2022**, *43*, 2095–5553.
15. Hu, S.; Cao, X.; Deng, Y.; Lai, Q.; Wang, G.; Hu, D.; Zhang, L.; Liu, M.; Chen, X.; Xiao, B.; et al. Effects of the Flight Parameters of Plant Protection Drone on the Distribution of Pollination Droplets and the Fruit Setting Rate of Camellia. *Trans. Chin. Soc. Agric. Eng. (Trans. CSAE)* **2023**, *39*, 92–100. [CrossRef]
16. He, Y.; Du, X.; Zheng, L.; Zhu, J.; Cen, H.; Xu, L. Effects of UAV Flight Height on Estimated Fractional Vegetation Cover and Vegetation Index. *Trans. Chin. Soc. Agric. Eng. (Trans. CSAE)* **2022**, *38*, 63–72.
17. Mittal, A.; Moorthy, A.K.; Bovik, A.C. No-Reference Image Quality Assessment in the Spatial Domain. *IEEE Trans. Image Process.* **2012**, *21*, 4695–4708. [CrossRef] [PubMed]
18. Fang, Y.; Ma, K.; Wang, Z.; Lin, W.; Fang, Z.; Zhai, G. No-Reference Quality Assessment of Contrast-Distorted Images Based on Natural Scene Statistics. *IEEE Signal Process. Lett.* **2015**, *22*, 838–842. [CrossRef]
19. Moorthy, A.K.; Bovik, A.C. Blind Image Quality Assessment: From Natural Scene Statistics to Perceptual Quality. *IEEE Trans. Image Process.* **2011**, *20*, 3350–3364. [CrossRef]
20. Ye, P.; Doermann, D. No-Reference Image Quality Assessment Using Visual Codebooks. *IEEE Trans. Image Process.* **2012**, *21*, 3129–3181. [CrossRef]
21. Tang, H.; Joshi, N.; Kapoor, A. Learning a Blind Measure of Perceptual Image Quality. In Proceedings of the IEEE Computer Society Conference on Computer Vision and Pattern Recognition, Colorado Springs, CO, USA, 20–25 June 2011; pp. 305–312.
22. Saad, M.A.; Bovik, A.C.; Charrier, C. Blind Image Quality Assessment: A Natural Scene Statistics Approach in the DCT Domain. *IEEE Trans. Image Process.* **2012**, *21*, 3339–3352. [CrossRef]
23. Schölkopf, B.; Smola, A.J.; Williamson, R.C.; Bartlett, P.L. New Support Vector Algorithms. *Neural Comput.* **2000**, *12*, 1207–1245. [CrossRef] [PubMed]
24. Storch, M.; Jarmer, T.; Adam, M.; de Lange, N. Systematic Approach for Remote Sensing of Historical Conflict Landscapes with UAV-Based Laserscanning. *Sensors* **2022**, *22*, 217. [CrossRef] [PubMed]

agriculture

MDPI

Article

Research on Path Tracking of Articulated Steering Tractor Based on Modified Model Predictive Control

Baocheng Zhou [1], Xin Su [2], Hongjun Yu [2], Wentian Guo [2] and Qing Zhang [1,*]

1 College of Engineering, China Agricultural University, Beijing 100083, China; 17356549377@139.com
2 Beijing Institute of Space Launch Technology, Beijing 100076, China
* Correspondence: zhangqingbit@cau.edu.cn

Abstract: With the development of agricultural mechanization and information technology, automatic navigation tractors are becoming a more common piece of farm equipment. The accuracy of automatic navigation tractor path tracking has become critical for maximizing efficiency and crop yield. Aiming at improving path tracking control accuracy and the real-time performance of the traditional model predictive control (MPC) algorithm, the study proposed an adaptive time-domain parameter with MPC in the path tracking control of the articulated steering tractor. Firstly, the kinematics model of the articulated steering tractor was established, as well as the multi-body dynamics model by RecurDyn. Secondly, the genetic algorithm was combined with MPC. The genetic algorithm was used to calculate the optimal time domain parameters under real-time vehicle speed, vehicle posture and road conditions, and the adaptive MPC was realized. Then, path tracking simulations were conducted by combining RecurDyn and Simulink under different path types. Compared with the traditional MPC algorithm under the three paths of U-shaped, figure-eight-shaped and complex curves, the maximum lateral deviations of the modified MPC algorithm were reduced by 59.0%, 24.9% and 13.2%, respectively. At the same time, the average lateral deviation was reduced by 72%, 43.5% and 20.3%, respectively. Finally, the real path tracking tests of the articulated steering tractor were performed. The test results indicated that under the three path tracking conditions of straight line, front wheel steering and articulated steering, the maximum lateral deviation of the modified MPC algorithm was reduced by 67.8%, 44.7% and 45.1% compared with the traditional MPC. The simulation analysis and real tractor tests verified the proposed MPC algorithm, considering the adaptive time-domain parameter has a smaller deviation and can quickly eliminate the deviation and maintain tracking stability.

Keywords: articulated steering tractor; path tracking; genetic algorithm; adaptive MPC; algorithm optimization

Citation: Zhou, B.; Su, X.; Yu, H.; Guo, W.; Zhang, Q. Research on Path Tracking of Articulated Steering Tractor Based on Modified Model Predictive Control. *Agriculture* **2023**, *13*, 871. https://doi.org/10.3390/agriculture13040871

Academic Editors: Monica Herrero-Huerta, Jose A. Jiménez-Berni, Shangpeng Sun, Ittai Herrmann and Diego González-Aguilera

Received: 21 March 2023
Revised: 11 April 2023
Accepted: 13 April 2023
Published: 15 April 2023

1. Introduction

The front and rear body of the articulated steering tractor can be relatively deflected, resulting in an excellent passing ability, small turning radius and convenient operation, which has been widely utilized in orchards and small-pitch farmland [1,2]. In recent years, with the rapid development of satellite navigation, sensors and control technology, the research on autonomous navigation agricultural equipment has also grown rapidly [3]. The automatic navigation control technology of agricultural machinery has become an important factor in liberating productivity and realizing agricultural automation [4,5]. It is of great significance to investigate the automatic navigation of the articulated steering tractor for improving the orchard intellectualization and unmanned technology [6,7].

The path tracking control algorithm is the key to automatic navigation technology of agricultural equipment [8]. With a complex structure and strong nonlinearity, the path tracking control of articulated steering vehicles is more difficult than that of ordinary vehicles. At present, the commonly used path tracking control algorithms include the pure tracking control algorithm, linear feedback control algorithm, Stanley control algorithm, model predictive control (MPC) algorithm and so on [9,10]. The MPC algorithm has the advantage of predicting future trajectories and handling multiple constraints, which has been widely employed in path tracking control. In the application, there are three essential steps in MPC, i.e., the model prediction, rolling optimization and feedback correction. The most obvious advantage of MPC is that it can add multiple constraints in the control process, since these constraints play an influential role in the planning and control of vehicle motion [11–15]. MPC solves an optimal control problem (OCP) to get a sequence of control commands over a finite receding horizon that optimizes a certain control metric (objective); then, the first portion of the resulting sequence is applied to the system. The main advantage of using MPC for path following in comparison with the non-predictive controllers presented above is the ability of MPC to handle constrained and nonlinear systems and it has been widely adopted in path tracking.

Beal et al. used the model prediction algorithm to design the path tracking controller. The stability boundary was determined according to the maximum available tire force to ensure the driving stability of the vehicle in the process of tracking the path [16]. Arun et al. established a path tracking control model based on MPC and vehicle dynamics, and it was implemented in a simulation with a car-sim model [17]. Zhang et al. applied the state lattice method to the upper trajectory planning controller and designed an MPC controller for path tracking based on the kinematic model [18]. Ji et al. used a 3D virtual dangerous potential field and designed the path tracking controller using the multi-constrained MPC method [19]. These studies that applied MPC have achieved excellent results; however, these studies all applied MPC to road vehicles with a good working environment, and the effect would be worse when they were applied to agricultural vehicles with a bad working environment. Therefore, the improved application of the MPC control scheme has also been favored by researchers. Considering the lateral and heading deviation to the reference trajectory, Mata et al. presented a tube-based robust MPC approach [20]. Wei et al. designed nonlinear MPC based on corridors to realize smooth and comfortable track control [21]. Based on the steering geometric constraints, Liu proposed a path planning algorithm based on local deviation correction, which contains a new following vehicle distance solving algorithm to improve the accuracy of seismic vehicle path tracking [22–25]. Based on the MPC algorithm, Bai et al. proposed two optimization schemes to reduce the number of control steps or reduce the control frequency. The results indicated that the path control accuracy was higher by reducing the number of control steps. Meanwhile, the control frequency was also reduced to meet the real-time requirements, while the error was slightly larger than that of the reduced control step scheme [26]. Furthermore, Meng et al. constructed the MPC controller based on preview distance. The simulation tests proved the enhanced accuracy and stability of path tracking [27]. For path tracking control of articulated vehicles, Li et al. designed an MPC controller based on the dynamic model by considering the multi-point preview error of the path [28], which can effectively improve the path tracking accuracy of articulated vehicles. Joseph et al. considered a model predictive path following control (MPFC). The closed-loop asymptotic stability under MPFC without terminal constraints or costs is rigorously proven and a stabilizing-horizon length is calculated. The analysis is based on verifying the cost-controllability assumption by deriving an upper bound of the MPFC value function with a finite prediction horizon [29]. Yue et al. proposed a model free predictive control (MFAPC) strategy using particle swarm optimization (PSO) to overcome structural and unstructured uncertainties. The control scheme of MFAPC is improved by integrating vehicle state parameters. The experimental results show that the proposed scheme does not require an accurate mathematical model and can quickly track the reference path [30]. Nevertheless, the real-time problem of time-domain parameters in MPC was rarely considered in these research studies. The setting of time-domain parameters is mostly fixed and not updated with the real-time status of vehicles, which diminishes the path tracking accuracy and the applicability.

In this paper, in order to improve the real-time performance of MPC, the study proposed an adaptive time-domain parameter with MPC. We modified the MPC by combining the genetic algorithm. The genetic algorithm was employed to calculate the optimal time domain parameters under real-time vehicle speed, vehicle attitude and road conditions. The effectiveness and superiority of the new algorithm was verified through co-simulation of path tracking and real tractor tests.

2. Materials and Methods

2.1. Kinematics Model of the Articulated Steering Tractor

In the study, the articulated steering tractor has two steering modes, i.e., front wheel steering and articulated steering. When the steering terrain can satisfy the steering requirements, then only front wheel steering is adopted. Conversely, the articulated steering method is involved. In the following parts, the front wheel steering kinematic model and the articulated steering kinematic model are constructed separately.

2.1.1. Front Wheel Steering Kinematic Model

It is assumed that the articulated steering tractor does not have lateral slip and roll as well as the interaction between the tires and the ground. The kinematic model of front wheel steering of the articulated steering tractor can be simplified into a two-wheel vehicle model with two degrees of freedom. The kinematic relationship of front wheel steering is shown in Figure 1.

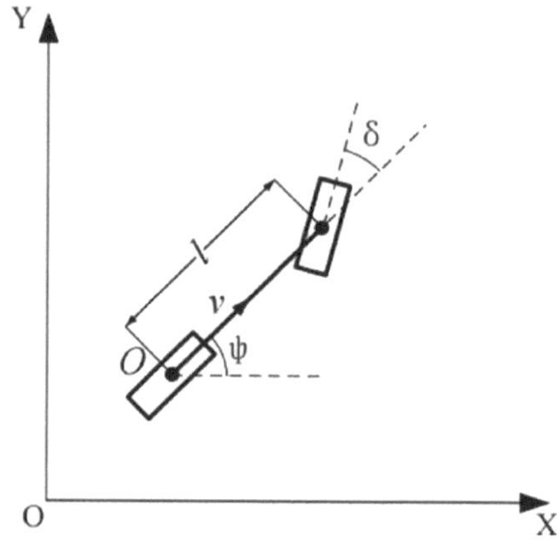

Figure 1. Front wheel steering kinematic relationship of the articulated steering tractor.

The rear axle center of the articulated steering tractor is depicted in coordinates (x, y). The components of velocity v in the X and Y axes can be calculated as follows.

$$\dot{x} = v\cos\psi \tag{1}$$

$$\dot{y} = v\sin\psi \tag{2}$$

The change rate of the heading angle can be calculated by,

$$\dot{\psi} = \frac{v\tan\delta}{l} \tag{3}$$

The motion equation of the articulated steering tractor with the front wheel steering mode is shown as follows.

$$\begin{bmatrix} \dot{x} \\ \dot{y} \\ \dot{\psi} \end{bmatrix} = v \begin{bmatrix} cos\psi \\ sin\psi \\ \frac{\tan\delta}{l} \end{bmatrix} \tag{4}$$

where v is speed of rear wheel (m/s), $\dot{x}$ is the component of v in the X-axis direction (m/s), $\dot{y}$ is the component of v in the Y-axis direction (m/s), ψ is heading angle (°), $\dot{\psi}$ is rate of change of heading angle (rad/s^2), δ is front wheel turning angle (°) and l is distance from the center of the front wheel to the center of the rear wheel (m).

2.1.2. Kinematic Model of Articulated Steering

In the mode of articulated steering, the steering process of the tractor is to first turn the front wheel angle to the limit in the shortest time, and then begin articulated steering. In order to obtain the kinematic model of articulated steering, the following assumptions are adopted: The articulated angle φ remains constant under small displacement. Moreover, it is assumed that the track length does not change during the driving process. Furthermore, there is no slip between the track and the track wheel. According to the structure of the articulated steering tractor, the kinematic relationship of the articulated steering process is constructed, as shown in Figure 2.

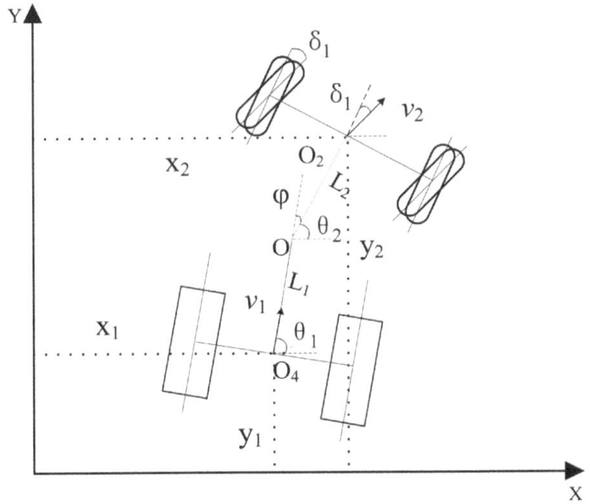

Figure 2. Articulated steering kinematic relationship of the articulated steering tractor.

The symbols in Figure 2 are listed as follows.

O_1 is the center of the rear axle, O_2 is the center of the front axle, O is the articulated steering hinge point, L_1 is the distance from the hinge point to the rear axle (m), L_2 is the distance from the hinge point to the front axle (m), δ_1 is the maximum value of the front wheel turning angle (°), φ is the articulated steering angle (°), θ_1 is the azimuth angle of the rear body (°), θ_2 is the azimuth angle of the front body (°), v_1 is the center speed of the rear axle (m/s) and v_2 is the center speed of the front axle (m/s).

The kinematic constraints of the articulated steering tractor can be expressed as,

$$\begin{cases} \dot{x}_1 sin\theta_1 - \dot{y}_1 cos\theta_1 = 0 \\ \dot{x}_2 sin\theta_2 - \dot{y}_2 cos\theta_2 = 0 \end{cases} \tag{5}$$

where $\dot{x}_1$ is the component of v_1 in the X-axis direction (m/s), $\dot{y}_1$ is the component of v_1 in the Y-axis direction (m/s), $\dot{x}_2$ is the component of v_2 in the X-axis direction (m/s) and $\dot{y}_2$ is the component of v_2 in the Y-axis direction (m/s).

The relationship between the rate of the articulated steering angle and the rate of front and rear body azimuth is shown in the following equation.

$$\dot{\varphi} = \dot{\theta}_1 - \dot{\theta}_2 \tag{6}$$

where $\dot{\varphi}$ is rate of change of articulated steering angle (rad/s^2), $\dot{\theta}_1$ is rate of change of the rear body azimuth (rad/s^2) and $\dot{\theta}_2$ is rate of change of the front body azimuth (rad/s^2).

Therefore, the relative velocity equations of the front and rear bodies can be shown as follows.

$$\begin{cases} v_1 cos\varphi = v_2 cos\delta_1 + \dot{\theta}_1 L_1 sin\varphi \\ v_1 sin\varphi = \dot{\theta}_2 L_2 + \dot{\theta}_1 L_1 cos\varphi + v_2 sin\delta_1 \end{cases} \tag{7}$$

The rate of the rear body azimuth is calculated by,

$$\dot{\theta}_1 = \frac{(\sin\varphi - \tan\delta_1 \cos\varphi)v_1 + \dot{\varphi}L_2}{L_1(\cos\varphi + \tan\delta_1 \sin\varphi) + L_2} \tag{8}$$

In the proposed method, the variable is the rate of the articulated angle denoted by ω. The articulated steering kinematic model can be expressed as Equation (9).

$$\begin{bmatrix} \dot{x}_1 \\ \dot{y}_1 \\ \dot{\theta}_1 \\ \dot{\varphi} \end{bmatrix} = \begin{bmatrix} \cos\theta_1 & 0 \\ \sin\theta_1 & 0 \\ \frac{(\sin\varphi - \tan\delta_1 \cos\varphi)}{L_1(\cos\varphi + \tan\delta_1 \sin\varphi) + L_2} & \frac{L_2}{L_1(\cos\varphi + \tan\delta_1 \sin\varphi) + L_2} \\ 0 & 1 \end{bmatrix} \begin{bmatrix} v_1 \\ \omega \end{bmatrix} \tag{9}$$

Based on the aforementioned analysis, the kinematic model of the articulated steering tractor with the center of the rear axle as the control point is established.

2.2. Multi-Body Dynamics Model of the Articulated Steering Tractor

In order to verify the performance of the proposed adaptive MPC path tracking controller, a multi-body dynamics model of the articulated steering tractor was established in software RecurDyn, as shown in Figure 3. The model parameters are shown in Table 1.

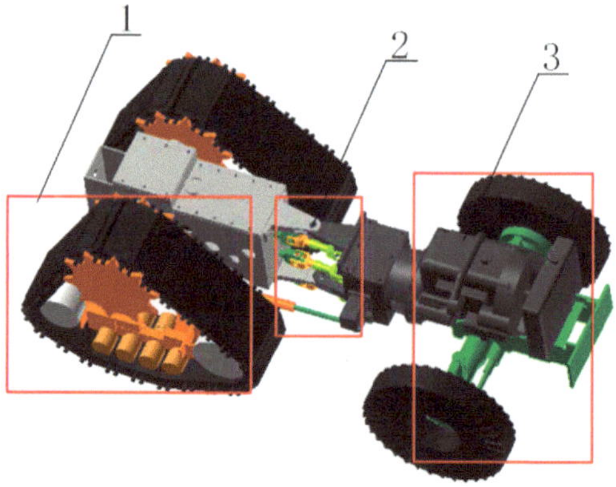

Figure 3. The multi-body dynamics model of the articulated steering tractor, where 1 represents the track system, 2 represents the articulated steering system and 3 represents the front wheel steering system.

Table 1. Model parameters of the articulated steering tractor.

Parameters	Value	Parameters	Value
Overall vehicle mass (kg)	1992.2	Maximum climbing degree (°)	30
Front body mass (kg)	632.8	Front wheel spacing (mm)	930
Rear body mass (kg)	1359.4	Rear wheel spacing (mm)	1080
Length (mm)	3100	Wheelbase (mm)	1850
Width (mm)	1230	Maximum articulated angle (°)	34
Height (mm)	1640	Crawler grounding length (mm)	460
Minimum radius of front wheel steering (m)	4.0	Minimum radius of articulated steering (m)	2.2

2.3. Parametric Adaptive Path Tracking Controller Design

2.3.1. MPC Path Tracking Algorithm Based on Kinematic Model

It can be seen from Equations (4) and (9) that when the front wheel steers, the articulated steering tractor can be regarded as a control system with input $\boldsymbol{u}(v,\delta)$ and state quantity $\chi(x,y,\psi)$, while as a control system with input $u(v,\omega)$ and state quantity $\chi(x_1, y_1, \theta_1, \varphi)$ at the time of articulated steering. The following is an analytical calculation of the articulated steering process. The kinematic model of the articulated steering tractor is presented in Equation (10) [31,32].

$$\dot{\chi} = f(\chi, \boldsymbol{u}) \tag{10}$$

For a planned target path, each reference point satisfies the above equation. Using r to represent the reference quantity, Equation (10) can be rewritten as,

$$\dot{\chi}_{\mathrm{r}} = f(\chi_{\mathrm{r}}, \boldsymbol{u}_{\mathrm{r}}) \tag{11}$$

where $\chi_{\mathrm{r}} = [\chi_{\mathrm{r}}\ y_{\mathrm{r}}\ \theta_{\mathrm{r}}\ \varphi_{\mathrm{r}}]$, $\boldsymbol{u}_{\mathrm{r}} = [v_{\mathrm{r}}\ \omega_{\mathrm{r}}]$.

After expanding Equation (10) with the Taylor series at the reference point and ignoring the higher order terms, Equation (12) can be found.

$$\dot{\chi} = f(x_{\mathrm{r}}, \boldsymbol{u}_{\mathrm{r}}) + \frac{\partial f(\chi, \boldsymbol{u})}{\partial \chi}(\chi - \chi_{\mathrm{r}}) + \frac{\partial f(\chi, \boldsymbol{u})}{\partial \boldsymbol{u}}(\boldsymbol{u} - \boldsymbol{u}_{\mathrm{r}}) \tag{12}$$

Equation (12) minus Equation (11) obtains the linear error model of the articulated steering tractor.

$$\dot{\tilde{\chi}} = \dot{\chi} - \dot{\chi}_{\mathrm{r}} = \boldsymbol{A}\tilde{\chi} + \boldsymbol{B}\tilde{\boldsymbol{u}} \tag{13}$$

where $A = \frac{\partial f(\chi,\boldsymbol{u})}{\partial \chi}$, $B = \frac{\partial f(\chi,\boldsymbol{u})}{\partial u}$, $\tilde{\chi}(k) = \chi(k) - \chi_r(k)$, $\tilde{\boldsymbol{u}}(k) = u(k) - u_r(k)$, $\tilde{\boldsymbol{u}}(k)$ is the control volume increment, k is the current sampling moment and $k + 1$ is the next sampling moment.

At any reference point, Equation (14) can be obtained by linear discretization of Equation (13).

$$\dot{\tilde{\chi}} = \frac{\tilde{\chi}(k+1) - \tilde{\chi}(k)}{T} = A\tilde{\chi} + B\tilde{u} \tag{14}$$

where T = the control period.

The discrete state space equations of the kinematic model of the articulated steering tractor can be obtained after rectification.

$$\tilde{\chi}(k+1) = (TA + E)\tilde{\chi}(k) + T\boldsymbol{B}\tilde{\boldsymbol{u}}(k). \tag{15}$$

Transform the above model to build a new state vector.

$$\xi(k|k) = \begin{bmatrix} \tilde{\chi}(k|k) \\ \tilde{\boldsymbol{u}}(k-1|k) \end{bmatrix} \tag{16}$$

Then the new state space expression can be obtained as follows.

$$\xi(k+1|k) = \begin{bmatrix} \tilde{\chi}(k+1|k) \\ \tilde{\boldsymbol{u}}(k|k) \end{bmatrix} = \begin{bmatrix} \tilde{\boldsymbol{A}} & \tilde{\boldsymbol{B}} \\ 0 & I_1 \end{bmatrix} \xi(k|k) + \begin{bmatrix} \tilde{\boldsymbol{B}} \\ I_1 \end{bmatrix} \Delta\tilde{\boldsymbol{u}}(k|k) = \boldsymbol{a}\xi(k|k)\boldsymbol{b}\Delta\tilde{\boldsymbol{u}}(k|k) \tag{17}$$

The output equation is by,

$$\eta(k|k) = \begin{bmatrix} I_2 & 0 \end{bmatrix} \begin{bmatrix} \tilde{\chi}(k|k) \\ \tilde{\boldsymbol{u}}(k-1|k) \end{bmatrix} = c\xi(k|k) \tag{18}$$

where I_1, I_2 = unit matrices.

The output of the system in the predicted future time-domain N_p can be predicted as Equation (19).

$$Y(t) = \Psi\xi(k|k) + \Theta\Delta U(k) \quad (19)$$

where $Y = \begin{bmatrix} \eta(k+1|k) \\ \eta(k+2|k) \\ \cdots \\ \eta(k+N_c|k) \\ \cdots \\ \eta(k+N_p|k) \end{bmatrix}$, $\Psi = \begin{bmatrix} ca \\ ca^2 \\ \cdots \\ ca^{N_c} \\ \cdots \\ ca^{N_p} \end{bmatrix}$, $\Theta = \begin{bmatrix} cb & 0 & \cdots & 0 \\ cab & cb & \cdots & 0 \\ \vdots & \vdots & \ddots & 0 \\ ca^{N_p-1}b & ca^{N_p-2}b & \cdots & ca^{N_p-N_c}b \end{bmatrix}$

$\Delta U = \begin{bmatrix} \Delta\tilde{u}(k|k) \\ \Delta\tilde{u}(k+1|k) \\ \Delta\tilde{u}(k+2|k) \\ \cdots \\ \Delta\tilde{u}(k+N_c-1|k) \end{bmatrix}$, N_p = predicted time-domain, N_c = control time-domain.

In order to ensure that the tractor can track the target trajectory rapidly and stably, the increment of the articulated steering angle is used as the control quantity of the objective function. Therefore, the optimized objective function of the path tracking model can be drawn as follows [33,34].

$$J(\xi(k), u(k-1), \Delta U(k)) = min\sum_{i=1}^{N_p} \Delta\eta(k+i \mid k)_Q^2 + \sum_{i=1}^{N_c-1} \Delta u(k+i \mid k)_R^2 + \rho\varepsilon^2 \quad (20)$$

where $\Delta\eta(k+i \mid k) = \eta(k+i \mid k) - \eta_r(k+i \mid k)$, $\Delta\eta(k+i \mid k)$ = Difference between actual output and reference output, I = 1,2, . . . , *Np*.

Furthermore, Q, R and ρ are the weight matrices and ε represents the relaxation factor.

2.3.2. Genetic Algorithm to Optimize Time-Domain Parameters

In Equation (20), the prediction time-domain N_p determines the length of the rolling optimization solution process. The control time-domain N_c affects the tractor's tracking performance as well as the control speed. Therefore, the values of N_p and N_c make a great impact on the path tracking performance of the unmanned tractors. However, the time-domain parameters of the traditional MPC controller are fixed at different speeds and different road conditions, which make it difficult to adapt to different road conditions. In the study, in order to obtain the optimal time-domain parameters in real time, firstly the whole path tracking process is segmented according to the system sampling frequency, and then the MPC time-domain parameters within each sampling frequency are optimized by genetic algorithm.

Furthermore, the optimization of the time-domain parameters is made by the genetic algorithm. The genetic algorithm was introduced by Professor Holland in 1975 according to the phenomena of reproduction, hybridization and mutation in nature [35]. The genetic algorithm is a stochastic global search and optimization method that imitates the mechanism of biological evolution [36,37]. By selecting high-quality individuals and eliminating inferior individuals, the law of survival of the fittest in the natural world is simulated. Reproduction, hybridization and mutation are carried out among the selected high-quality individuals. Then, the individuals with better qualities that may be produced and iterated repeatedly are selected to make the population better and better until the expected fitness value is met. The genetic algorithm uses a probabilistic mechanism for iteration with the purpose of avoiding traps in a local optimum. The genetic algorithm is not constrained by the search space and has no continuous, derivable or single-peaked requirements for the objective function. Therefore, genetic algorithms are suitable for solving multi-objective optimization problems while being scalable and convenient to combine with other algorithms [38,39].

The optimization principle of the time-domain parameters is shown in Figure 4. The steps to optimize the time-domain parameters by the genetic algorithm are as follows. Firstly, the population is initialized, in which each individual of the population will be assigned a value and rounded according to the range of time-domain parameters. Secondly, the adaptation degree of each individual is calculated by combining the information of tractor speed and position through the adaptation degree function. Finally, the optimal parameters are obtained when the termination condition is satisfied. If the termination condition is not satisfied, then selection, crossover and mutation are performed to obtain a new population and the fitness function value is calculated again.

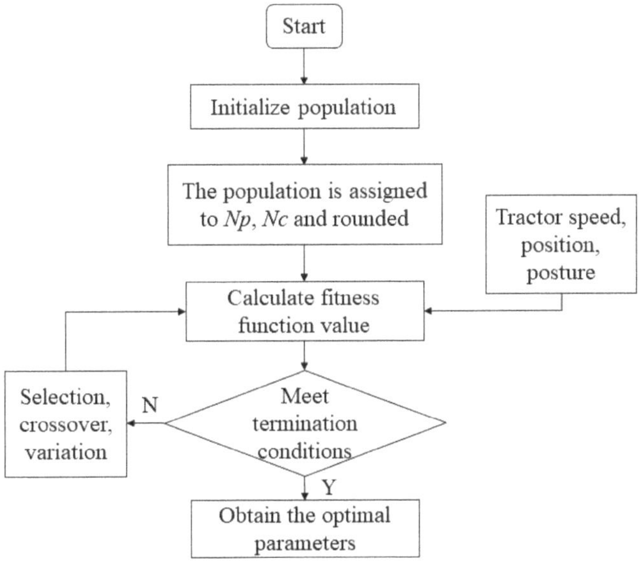

Figure 4. Time-domain parameter optimization principle.

The fitness function of the genetic algorithm is presented as follows.

$$L = \frac{1}{J(\xi(k), \boldsymbol{u}(k-1), \Delta \boldsymbol{U}(k))} \tag{21}$$

The population size is set as 200. Furthermore, the probabilities of crossover and variation are set as 0.6 and 0.1 separately. Terminated evolutions are set at 20. To improve the optimization efficiency, the prediction time-domain is taken in the range (0, 60), and the control time-domain is taken in the range (0, 30). Finally, by optimizing the time-domain parameters within each sampling frequency, the adaptive time-domain parameters of the whole section of the path tracking can be realized.

3. Results and Discussion

3.1. Simulation Test

3.1.1. Construction of the Simulation System

To verify the performance of the proposed adaptive time-domain parametric model, the co-simulation model of RecurDyn and Simulink is established, as shown in Figure 5. The model can be divided into four parts, i.e., the tractor model to search for nearest target point, the adaptive MPC controller, and the genetic algorithm module.

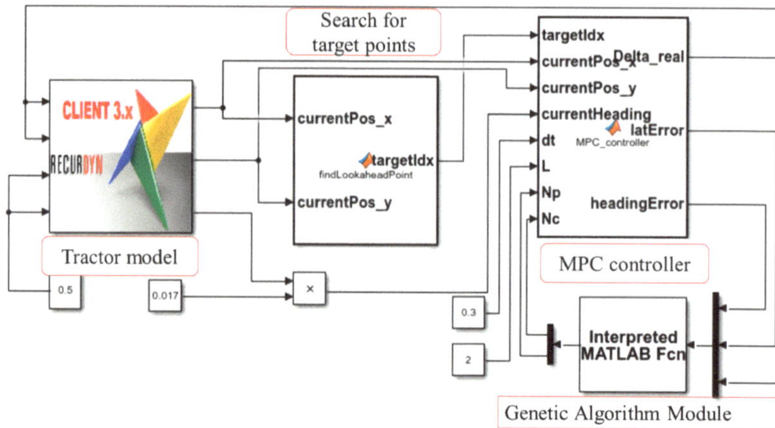

Figure 5. Co-simulation model of RecurDyn and Matlab/Simulink.

The co-simulation principle is shown in Figure 6. The control performance of the adaptive MPC algorithm on the straight driving, front wheel steering and articulated steering process is verified through three paths of U-shaped, figure-eight-shaped and complex curves. These three path conditions include all the conditions of the articulated steering tractor in the actual working process of the orchard, i.e., the straight-line conditions the front wheel steering conditions and the articulated steering conditions. The system sampling frequency is set to 0.5 s, and the tractor driving speed is set to 0.5 m/s.

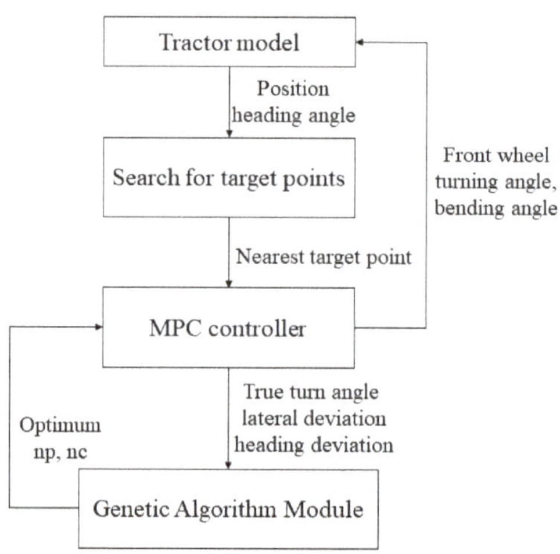

Figure 6. Co-simulation principle of RecurDyn and Matlab/Simulink.

3.1.2. U-Shaped Curve Path Tracking Simulation

The comparison of the tracking performance of the adaptive MPC and the traditional MPC on a U-shaped curve path is shown in Figure 7a. It can be seen that the tracking path of the adaptive MPC is more stable and smoother, and the tracking effect is also better. From Figure 7b,c, it can be seen that the maximum values of lateral deviation and heading deviation occur at the articulation of straight and curved lines. Adaptive MPC has a smaller maximum lateral deviation and maximum heading deviation than MPC. The lateral deviation and heading deviation of adaptive MPC fluctuate greatly during turning which shows that adaptive MPC can quickly adjust the tractor to prevent the deviation from changing too much when there is a deviation. When there is a deviation in the whole path tracking process, the adaptive MPC can adjust the deviation to zero more quickly.

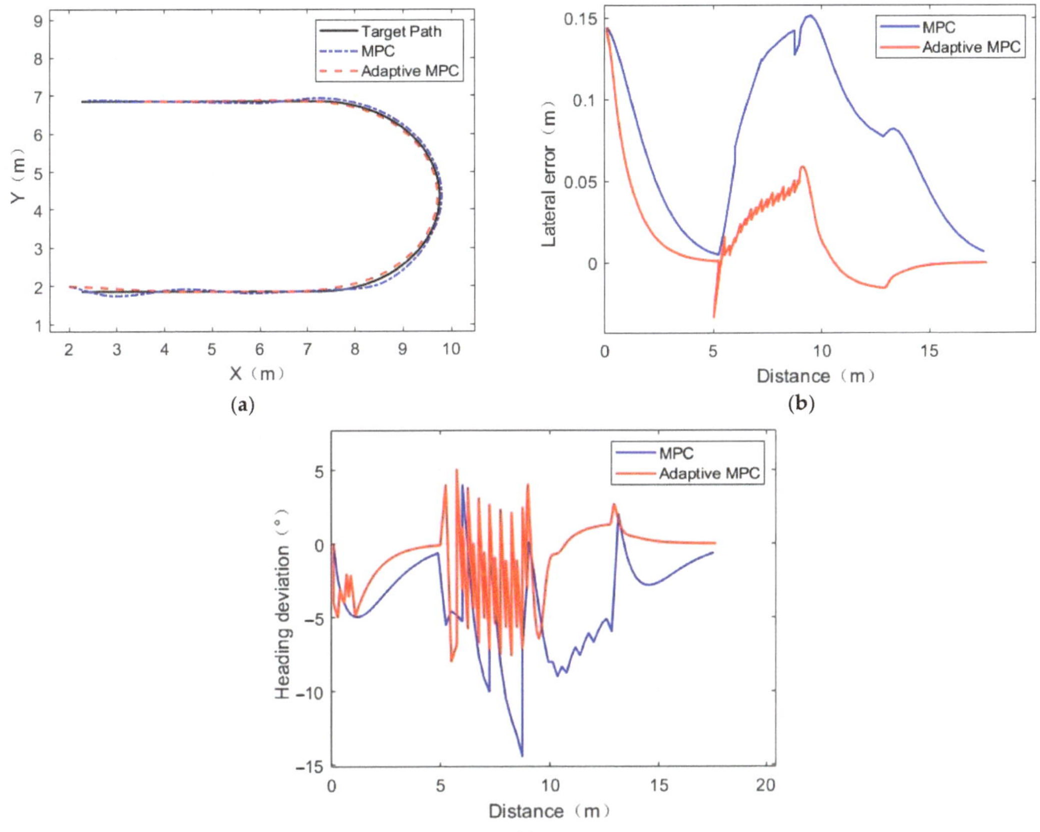

Figure 7. U-shaped curve simulation results. (**a**) Path tracking performance comparison, (**b**) lateral deviation and (**c**) heading deviation.

The deviation statistics results are shown in Table 2. It can be seen that the maximum value, average value and standard deviation of the lateral deviation of the adaptive MPC are reduced by 59.0%, 72% and 39.7% compared with the traditional MPC. At the same time, the maximum, average and standard deviations of the heading deviation decreased by 44.6%, 58.7% and 36.3%, respectively. The maximum values of lateral deviation and heading deviation occur at the articulation of straight and curved lines. From the average value and standard deviation, it can be seen that the accuracy and stability of adaptive MPC are better. Adaptive MPC significantly reduces the maximum and average values of deviation.

Table 2. The deviation statistics results.

Category	Category	MPC	Adaptive MPC
lateral deviation (cm)	Maximum	15.13	6.21
	Average	7.53	2.10
	SD	4.71	2.84
heading deviation (°)	Maximum	14.45	8.00
	Average	4.19	1.73
	SD	3.14	2.00

3.1.3. Figure-Eight-Shaped Curve Path Tracking Simulation

The comparisons of the tracking performance of the adaptive MPC and the traditional MPC on a figure-eight-shaped curve are shown in Figure 8a. It can be seen that the adaptive time-domain parameter MPC achieves a better tracking effect. From Figure 8b,c, it can be distinguished that the deviation is large at the beginning of tracking and at the junction of two circles. At this time, the adaptive MPC controller reduces the heading deviation significantly. MPC only keeps the deviation stable and does not reduce the deviation when there is a deviation. However, the adaptive MPC can quickly adjust the tractor to reduce the deviation, which further proves the superiority and accuracy of the adaptive MPC.

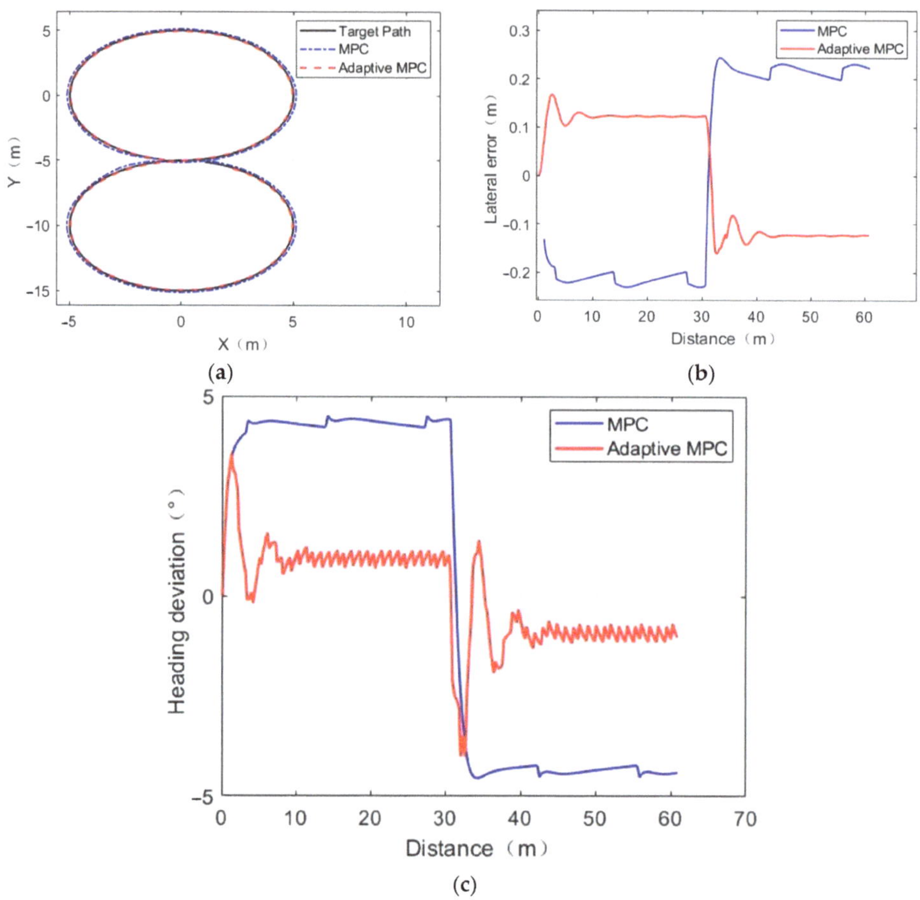

Figure 8. Figure-eight-shaped curve simulation results. (**a**) Path tracking performance comparison, (**b**) lateral deviation and (**c**) heading deviation.

The deviation statistics results are shown in Table 3. It can be seen that the maximum value, average value and standard deviation of the lateral deviation of the adaptive MPC are reduced by 24.9%, 43.5% and 16.9% compared with the traditional MPC. The maximum, average and standard deviation of heading deviation decreased by 11.9%, 74.9% and 25.0%, respectively. Adaptive MPC significantly reduces the average values of lateral deviation and heading deviation. From the maximum value and Figure 8, adaptive MPC can reduce the maximum deviation more obviously when turning.

Table 3. The deviation statistics results.

Category	Category	MPC	Adaptive MPC
lateral deviation (cm)	Maximum	22.41	16.83
	Average	21.31	12.05
	SD	2.37	1.97
heading deviation (°)	Maximum	4.54	4.00
	Average	4.26	1.07
	SD	0.60	0.45

3.1.4. Complex Curve Path Tracking Simulation

The comparison of the tracking performance of the adaptive MPC and the traditional MPC on the complex curve path is shown in Figure 9a. It can be seen that the adaptive time-domain parameter MPC controller has a better tracking effect. According to Figure 9b,c, it can be seen that the maximum deviation occurs at the junction of the curve. The lateral deviation and heading deviation of articulated steering are smaller than front wheel steering. The lateral deviation and heading deviation of adaptive MPC fluctuate greatly, which shows that adaptive MPC adjusts the vehicle more times. The adaptive MPC can reduce the deviation more quickly and maintain stability when there is a deviation.

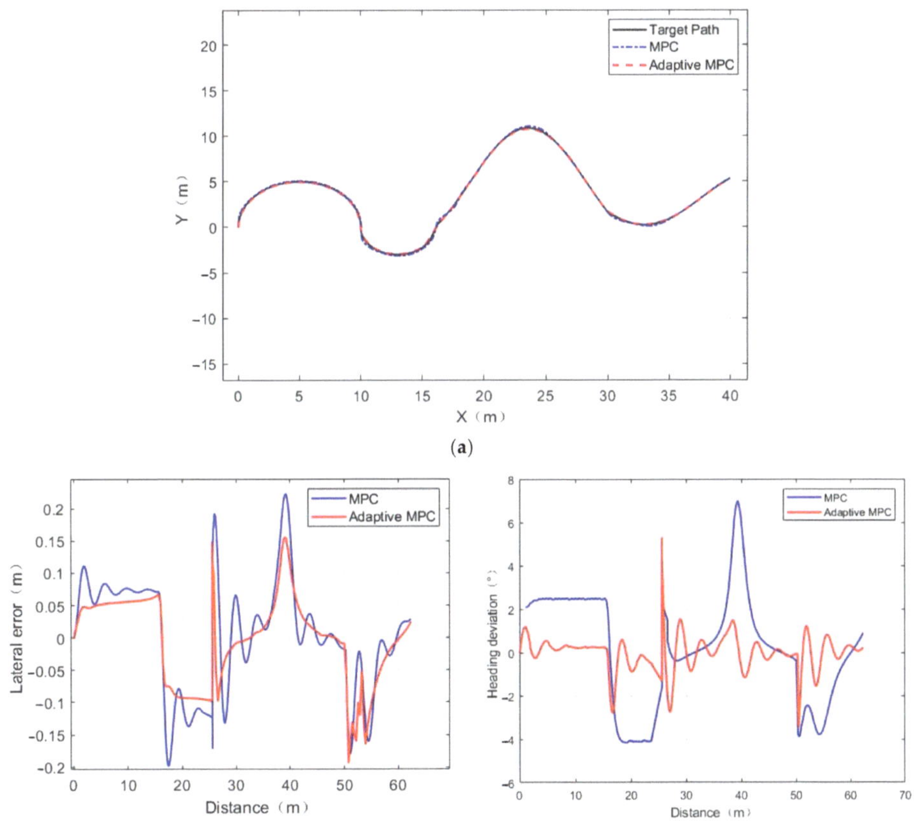

Figure 9. Complex curve simulation results. (**a**) Path tracking performance comparison, (**b**) lateral deviation and (**c**) heading deviation.

The deviation statistics results are shown in Table 4. It can be seen that the maximum value, average value and standard deviation of the lateral deviation of the adaptive MPC are reduced by 13.2%, 20.3% and 19.2% compared with the traditional MPC. The maximum average and standard deviation of heading deviation decreased by 24.1%, 68.5% and 48.7%, respectively.

Table 4. The deviation statistics results.

Category	Category	MPC	Adaptive MPC
lateral deviation (cm)	Maximum	22.34	19.38
	Average	6.99	5.57
	SD	5.37	4.34
heading deviation (°)	Maximum	7.01	5.32
	Average	1.97	0.62
	SD	1.56	0.80

3.2. Test Verification

In order to further verify the effectiveness of the adaptive MPC controller, real tractor path tracking tests were conducted in the study. By taking the articulated steering tractor as the test platform, the test equipment consists of an industrial personal computer (IPC) display screen, satellite positioning equipment, steering controller, angle sensor and so on. The equipment utilized on the test platform are shown in Figure 10.

The path tracking test of the tractor was carried out in the standardized demonstration orchard of Shijiazhuang Xinnong Machinery Co., Ltd. (Shijiazhuang, China). The test road is well-maintained and flat, and the orchard has 10 rows of fruit trees, with a single row being about 70 m long, the average height of the fruit trees being 2.5 m and the average row spacing being 4 m. The position information and heading angle information obtained during the test were saved in the IPC. Each experiment was carried out three times. The data were exported for statistical processing after the test. The maximum, minimum average value and standard deviation of the lateral deviation and heading deviation were obtained by data analysis software.

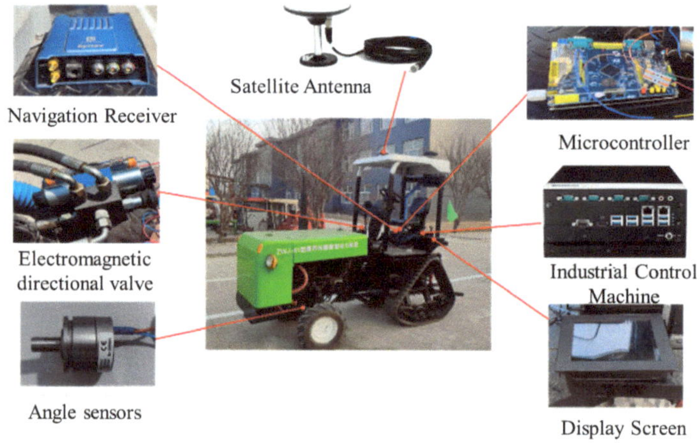

Figure 10. Test platform hardware arrangement of articulated steering tractor.

The principle of path tracking control is shown in Figure 11. The program of path tracking is written and compiled in the industrial control machine by Python. The position and heading angle information are obtained in real time by satellite positioning equipment, steering controller, angle sensor, etc. The front wheel and articulated angle are calculated in the industrial control computer. The control information can be obtained through the communication subprogram between the microcontroller controller and the industrial control computer, so as to control each motor to control the angle in real time.

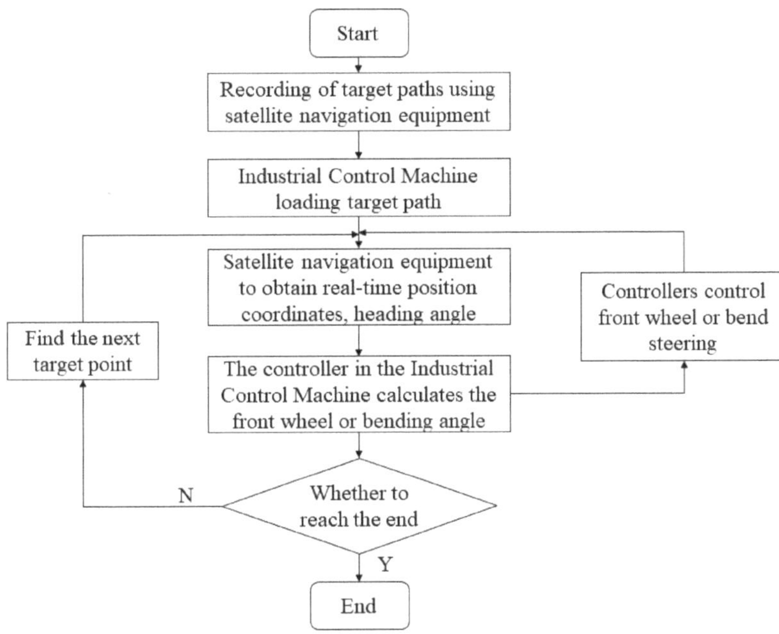

Figure 11. Test platform hardware arrangement.

Articulated steering tractors are mainly utilized in orchard environments. The driving path in the orchard can be divided into straight-line operating sections and headland turning sections. The effectiveness and accuracy of adaptive MPC in straight ahead, front wheel steering and articulated steering conditions need to be tested and verified. Therefore, the straight-line path, front wheel steering path and articulated steering path tracking tests were conducted in the study. The tractor driving speed was set to 0.5 m/s.

3.2.1. Straight-Line Path

The straight-line path tracking test circumstances are shown in Figure 12a. Figure 12b depicts the path tracking comparison between adaptive MPC and traditional MPC, wherein it can be seen that the adaptive MPC has a better tracking effect. Figure 12c,d presents the lateral and heading deviation comparisons. It can be seen that MPC will have a large lateral deviation and heading deviation, and the adaptive MPC deviation fluctuation is smaller. The adaptive MPC proposed in the study achieved significant reduction in the lateral and heading deviations in straight-line path tracking compared with the traditional MPC.

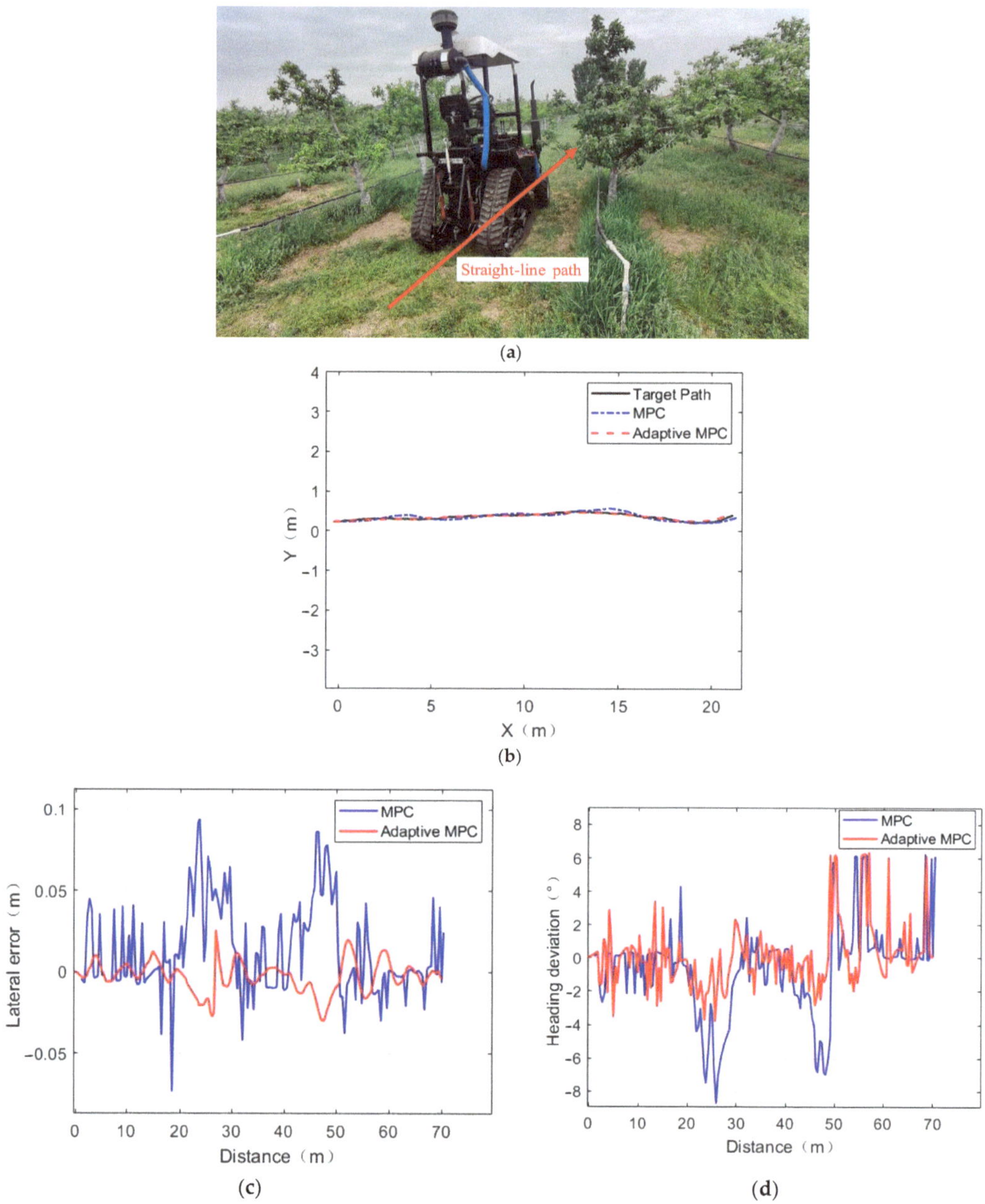

Figure 12. Straight-line path tracking test results. (**a**) Test site, (**b**) path tracking performance comparison, (**c**) lateral deviation and (**d**) heading deviation.

Table 5 presents the deviation statistics results. It can be seen that the maximum value average value and standard deviation of the lateral deviation of the proposed adaptive MPC are reduced by 67.8%, 65.3% and 68.8%, respectively. At the same time, in comparison with the traditional MPC, the maximum, average and standard deviation of heading deviation are also decreased by 26.8%, 28.1% and 35.7%, respectively. Adaptive MPC can significantly improve lateral deviation.

Table 5. The deviation statistics results.

Category	Category	MPC	Adaptive MPC
lateral deviation (cm)	Maximum	9.39	3.02
	Average	2.19	0.76
	SD	2.21	0.69
heading deviation (°)	Maximum	8.73	6.39
	Average	1.85	1.33
	SD	2.21	1.42

3.2.2. Front Wheel Steering Path

The front wheel steering path tracking tests were conducted to verify the control effect of adaptive MPC. The front wheel steering path tracking test environment is shown in Figure 13a. Figure 13b shows the comparisons of path tracking between adaptive MPC and traditional MPC. It can be seen from Figure 13b that the adaptive MPC tracks better. Figure 13c,d presents that the lateral deviation and heading deviation of the adaptive MPC proposed in the study are significantly lower than those of the traditional MPC in the curve paths. In the latter part of the path tracking, the heading angle fluctuates greatly, which may be caused by the vibration of the tractor body. The tractor made a big deviation during the two turns. However, the deviations produced by adaptive MPC are both smaller than MPC. Adaptive MPC can adjust tractors more times, faster and better.

(a)

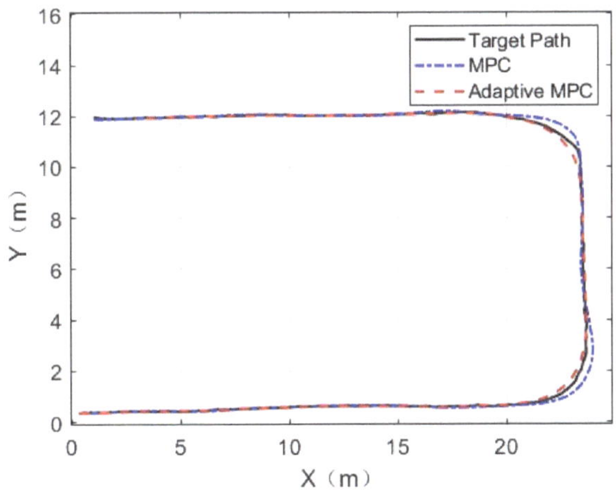

(b)

Figure 13. *Cont.*

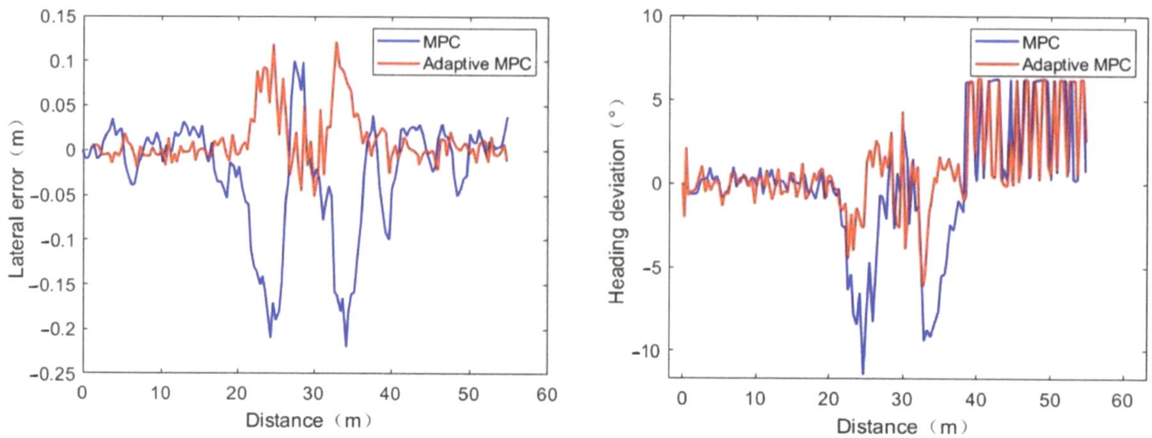

Figure 13. Front wheel steering path tracking test results. (**a**) Test site, (**b**) path tracking performance comparison, (**c**) lateral deviation and (**d**) heading deviation.

Table 6 depicts the deviation statistics results. It can be seen that the maximum value average value and standard deviation of the lateral deviation of the adaptive MPC are reduced by 44.7%, 57.4% and 53.2% compared with the traditional MPC. Furthermore, the maximum, average and standard deviation of heading deviation are also decreased by 44.9%, 31.4% and 28.6%, respectively.

Table 6. The deviation statistics results.

Category	Category	MPC	Adaptive MPC
lateral deviation (cm)	Maximum	22.00	12.17
	Average	4.69	2.00
	SD	5.38	2.52
heading deviation (°)	Maximum	11.44	6.30
	Average	2.74	1.88
	SD	2.94	2.10

3.2.3. Articulated Steering Path

The articulated steering path tracking tests were conducted in an orchard to verify the control effect of adaptive MPC. The articulated steering path tracking environment is shown in Figure 14a. Figure 14b depicts the path tracking comparisons. From Figure 14b, it can be seen that the adaptive MPC tracks better. Figure 14c,d indicates that the adaptive MPC reduces the lateral deviation and heading deviation over the whole path compared with the traditional MPC, in which a significant reduction can be easily distinguished. In the whole path tracking process, the MPC deviation fluctuates greatly, and the adjustment speed is slow. Furthermore, after adjusting the deviation, the adaptive MPC can keep the deviation at a lower level. In the latter part of path tracking, the heading angle fluctuates greatly but the frequency of adaptive MPC adjustment is faster and the deviation is lower. The test results further prove that traditional MPC has greater errors when used in agricultural tractors, and it needs to be improved. The adaptive MPC proposed in this study improves the path tracking accuracy when used in agricultural tractors.

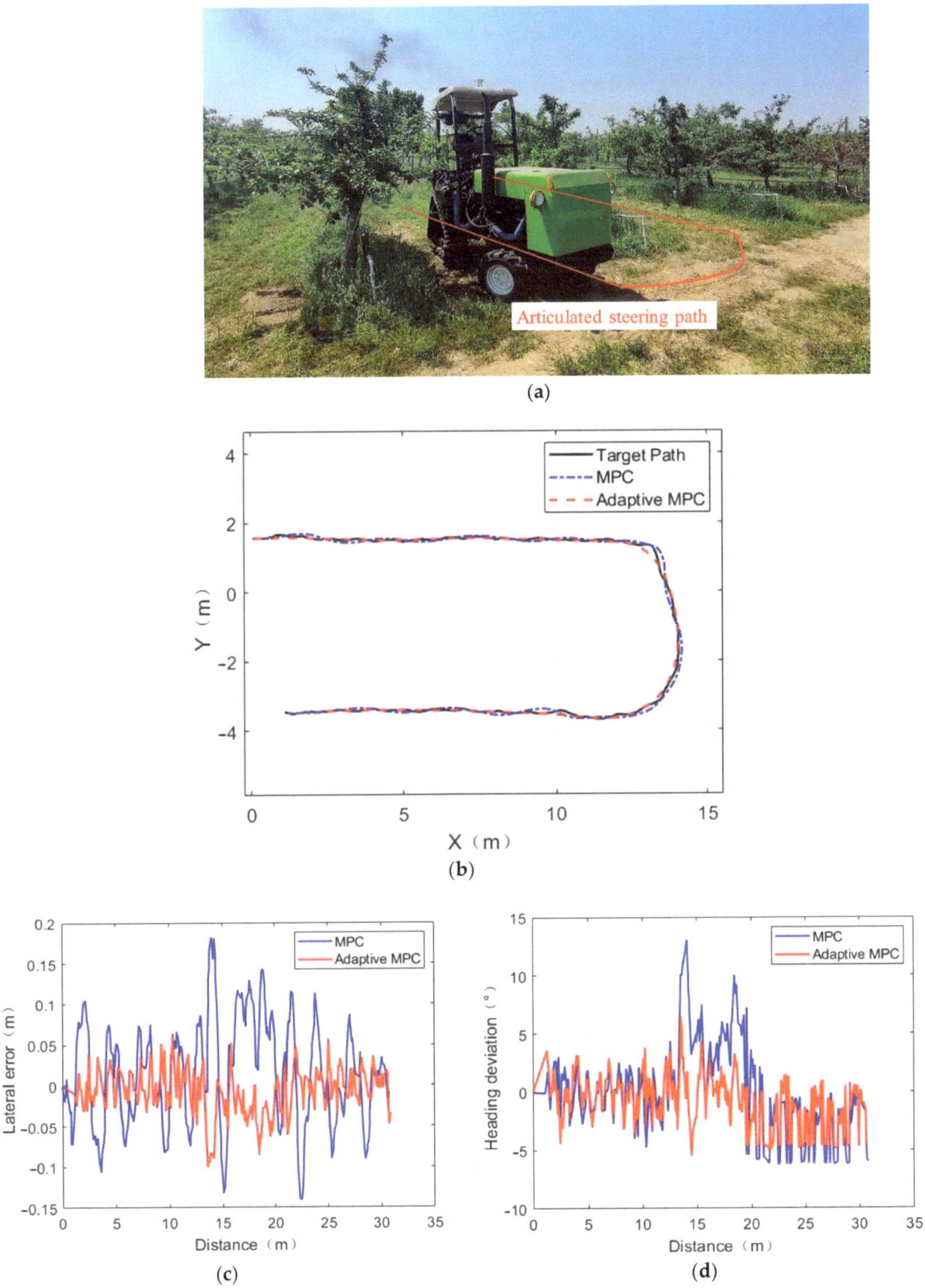

Figure 14. Articulated steering path tracking test results. (**a**) Test site, (**b**) path tracking performance comparison, (**c**) lateral deviation and (**d**) heading deviation.

Table 7 presents the specific results in path tracking of the articulated steering tractor. It can be seen that compared with the traditional MPC, the maximum value, average value and standard deviation of the lateral deviation of the adaptive MPC are reduced by 45.1%, 60.9% and 51.6%, respectively. The maximum, average and standard deviation of heading deviation are decreased by 50.2%, 34.7% and 39.2%, respectively. The average and standard deviation of lateral deviation are reduced more, which ensures the stability of tracking deviation. The heading deviation is also greatly improved, which ensures that the car body will not vibrate too violently.

Table 7. The deviation statistics results.

Category	Category	MPC	Adaptive MPC
lateral deviation (cm)	Maximum	18.24	10.01
	Average	5.45	2.22
	SD	3.84	1.86
heading deviation (°)	Maximum	13.07	6.51
	Average	2.91	1.90
	SD	2.60	1.58

4. Conclusions

In order to improve the real-time performance of MPC, the present study proposed an adaptive time-domain parameter with traditional MPC in path tracking control of the articulated steering tractor. The genetic algorithm was adopted to calculate the optimal time-domain parameters under real-time tractor speed, tractor attitude and road conditions. The conclusions are as follows.

- The kinematics model and multi-body dynamics model of the articulated steering tractor were established. Then, the co-simulations by RecurDyn and Simulink were conducted under a U-shaped, figure-eight-shaped and complex curves path. The maximum lateral deviations of the adaptive MPC were reduced by 59.0%, 24.9% and 13.2%, respectively. At the same time, the average lateral deviations were reduced by 72%, 43.5% and 20.3% compared with the traditional MPC. The maximum heading deviations of the adaptive MPC were reduced by 44.6%, 11.9% and 24.1%, respectively. The average lateral deviations were reduced by 58.7%, 74.9% and 68.5%.
- Taking the articulated steering tractor as the test platform, the performance of adaptive MPC was tested in real tractors through a straight-line path, front wheel steering path and articulated steering path. The results indicated that the maximum lateral deviations of the adaptive MPC were reduced by 67.8%, 44.7% and 45.1%, respectively. Compared with the traditional MPC, the average lateral deviations of the adaptive MPC were reduced by 65.3%, 57.4% and 60.9%, respectively. The maximum heading deviations of the adaptive MPC were reduced by 26.8%, 44.9% and 50.2%, respectively. The average lateral deviations were reduced by 28.1%, 31.4% and 34.7%.
- The results of simulations and real tractor tests show that the real-time and path tracking performance of the proposed adaptive MPC is superior to the traditional MPC. Adaptive MPC can adjust the tractor faster when deviations occur, and the adjustment frequency of adaptive MPC is faster and the effect is better. The adaptive MPC can effectively enhance the path tracking accuracy of the articulated steering tractor.

Author Contributions: Conceptualization, B.Z. and Q.Z.; methodology, B.Z. and Q.Z.; software, B.Z. and X.S.; validation, B.Z. and Q.Z.; formal analysis, B.Z. and Q.Z.; investigation, W.G. and X.S.; resources, W.G. and H.Y.; data curation, B.Z. and Q.Z.; writing—original draft preparation, B.Z. and Q.Z.; writing—review and editing, X.S., W.G. and H.Y.; visualization, B.Z. and Q.Z.; supervision, H.Y.; project administration, Q.Z.; funding acquisition, Q.Z. All authors have read and agreed to the published version of the manuscript.

Funding: This research was funded by the National Key R&D Program of China (2022YFD2001902).

Institutional Review Board Statement: Not applicable.

Data Availability Statement: The data presented in this study are available on request from the corresponding author.

Conflicts of Interest: The authors declare no conflict of interest.

References

1. Franceschetti, B.; Rondelli, V.; Capacci, E. Lateral Stability Performance of Articulated Narrow-Track Tractors. *Agronomy* **2021**, *11*, 2512. [CrossRef]
2. Liang, C.; Pan, K.; Zhao, M.; Lu, M. Multi-Node Path Planning of Electric Tractor Based on Improved Whale Optimization Algorithm and Ant Colony Algorithm. *Agriculture* **2023**, *13*, 586. [CrossRef]
3. Innam, L.; Kyunghyun, L.; Jeehyong, L.; Kwanho, Y. Autonomous Greenhouse Sprayer Navigation Using Automatic Tracking Algorithm. *Appl. Eng. Agric.* **2015**, *31*, 17–21. [CrossRef]
4. Shutske, J.M. Agricultural automation & autonomy: Safety and Risk Assessment Must be at the Forefront. *J. Agromed.* **2022**, *28*, 5–10. [CrossRef]
5. Md-Tahir, H.; Zhang, J.; Zhou, Y.; Sultan, M.; Ahmad, F.; Du, J.; Ullah, A.; Hussain, Z.; Xia, J. Engineering Design, Kinematic and Dynamic Analysis of High Lugs Rigid Driving Wheel, a Traction Device for Conventional Agricultural Wheeled Tractors. *Agriculture* **2023**, *13*, 493. [CrossRef]
6. Huang, Y.; Hoffmann, W.C.; Lan, Y.; Wu, W.; Fritz, B.K. Development of a Spray System for an Unmanned Aerial Vehicle Platform. *Appl. Eng. Agric.* **2009**, *25*, 803–809. [CrossRef]
7. Wang, Z.; Xin, P.; Sun, H.T. Path Tracking of Unmanned Vehicles Based on Contraction Constraint Model Predictive Control. *Control. Decis.* **2022**, *37*, 625–634.
8. Zheng, J.; Wang, L.; Wang, X.; Shi, Y.; Yang, Z. Parameter Calibration of Cabbages (*Brassica oleracea* L.) Based on the Discrete Element Method. *Agriculture* **2023**, *13*, 555. [CrossRef]
9. DiazdelRio, F.; SanchezCuevas, P.; IñigoBlasco, P.; SevillanoRamos, J.L. Improving Tracking of Trajectories through Tracking Rate Regulation: Application to UAVs. *Sensors* **2022**, *22*, 9795. [CrossRef] [PubMed]
10. Wu, H.C.; Zhang, H.H.; Feng, Y.X. MPC-Based Obstacle Avoidance Path Tracking Control for Distributed Drive Electric Vehicles. *World Electr. Veh. J.* **2022**, *13*, 221. [CrossRef]
11. Zheng, K.; Zhao, X.; Han, C.; He, Y.; Zhai, C.; Zhao, C. Design and Experiment of an Automatic Row-Oriented Spraying System Based on Machine Vision for Early-Stage Maize Corps. *Agriculture* **2023**, *13*, 691. [CrossRef]
12. Huang, Y.R.; Fu, J.H.; Xu, S.Y.; Han, T.L.; Yu, W. Research on Integrated Navigation System of Agricultural Machinery Based on RTK-BDS/INS. *Agriculture* **2022**, *12*, 1169. [CrossRef]
13. Qiang, S.; Mao, H.P.; Guan, X.P. Numerical Simulation and Experimental Verification of the Deposition Concentration of an Unmanned Aerial Vehicle. *Appl. Eng. Agric.* **2019**, *35*, 367–376. [CrossRef]
14. Zhang, S.; Guo, C.; Gao, Z.; Sugirbay, A.; Chen, J.; Chen, Y. Research on 2D Laser Automatic Navigation Control for Standardized Orchard. *Appl. Sci.* **2020**, *10*, 2763. [CrossRef]
15. Kang, N.; Han, Y.; Wang, B.Y. Linear quadratic regulator based on extended state observer–based active disturbance rejection control of autonomous vehicle path following control. *J. Syst. Control. Eng.* **2023**, *273*, 102–120. [CrossRef]
16. Beal, C.E.; Gerdes, J.C. Model Predictive Control for Vehicle Stabilization at the Limits of Handling. *IEEE Trans. Control. Syst. Technol.* **2013**, *21*, 1258–1269. [CrossRef]
17. Arun, M.; Hiroyuki, O.; Tatsuya, S. Path tracking control using model predictive control with on GPU implementation for autonomous driving. *J. Arid. Land Stud.* **2018**, *28*, 163–167. [CrossRef]
18. Zhang, C.; Chu, D.; Liu, S. Trajectory planning and tracking for autonomous vehicle based on state lattice and model predictive control. *IEEE Intell. Transp. Syst. Mag.* **2019**, *11*, 29–40. [CrossRef]
19. Ji, J.; Khajepour, A.; Melek, W.W. Path planning and tracking for vehicle collision avoidance based on model predictive control with multiconstraints. *IEEE Trans. Veh. Technol.* **2017**, *66*, 52–64. [CrossRef]
20. Mata, S.; Zubizarreta, A.; Pinto, C. Robust tube-based model predictive control for lateral path tracking. *IEEE Trans. Intell. Veh.* **2019**, *4*, 569–577. [CrossRef]
21. Wei, C.; Romano, R.; Merat, N. Risk-based autonomous vehicle motion control with considering human driver's behaviour. *Transp. Res. C* **2019**, *107*, 1–14. [CrossRef]

22. Liu, L.X.; Guo, X.; Fang, Y.C. A Reinforcement Learning-Based Strategy of Path Following for Snake Robots with an Onboard Camera. *Sensors* **2022**, *22*, 9867. [CrossRef] [PubMed]
23. Bai, G.X.; Meng, Y.; Liu, L.; Gu, Q.; Huang, J.; Liang, G.D. Path Tracking for Car-like Robots Based on Neural Networks with NMPC as Learning Samples. *Electronics* **2022**, *24*, 4232. [CrossRef]
24. Hua, Z.X. Overview of path tracking control algorithms for autonomous vehicles. *Equip. Manuf. Technol.* **2021**, *6*, 100–103.
25. Liu, P.T.; Wang, S.K.; Wang, J.Z. Design and application of unmanned control algorithm for seismic vehicle. *J. Mech. Eng.* **2022**, *58*, 211–220
26. Bai, G.X.; Liu, L.; Meng, Y.; Liu, S.Y. Real-time path tracking of mobile robot based on nonlinear model predictive control. *J. Agric Mach.* **2020**, *51*, 47–52+60.
27. Meng, Y.; Gan, X.; Bai, G.X. Path tracking predictive control of articulated vehicle for underground mine based on preview distaNce. *J. Eng.* **2019**, *41*, 662–671.
28. Li, S.X.; Xu, B.; Hu, M.J. Multi-point preview path tracking method for articulated vehicles based on dynamic model predictive control. *Automot. Eng.* **2021**, *43*, 1187–1194.
29. Joseph, C.; Mehrez, M.W.; Han, J.; Jeon, S.; Melek, W. Model Predictive Path Following Control without terminal constraints for holonomic mobile robots. *Control. Eng. Pract.* **2023**, *132*, 105406. [CrossRef]
30. Yue, J.; Xu, X.; Zhang, L.; Zou, T. Model free predictive path tracking control of variable-configuration unmanned ground vehicle *ISA Trans.* **2022**, *129*, 485–494. [CrossRef]
31. Nie, Y.X.; Hua, Y.D.; Zhang, M.L.; Zhang, X.J. Intelligent Vehicle Trajectory Tracking Control Based on VFF-RLS Road Friction Coefficient Estimation. *Electronics* **2022**, *11*, 3119. [CrossRef]
32. Fu, Z.Q.; Xiong, L.; Qian, Z.X.; Leng, B.; Zeng, D.Q.; Huang, Y.J. Model Predictive Trajectory Optimization and Tracking in Highly Constrained Environments. *Int. J. Automot. Technol.* **2022**, *23*, 927–938. [CrossRef]
33. Chen, X.; Peng, D.; Hu, J.B.; Li, C.; Zheng, S.L.; Zhang, W.H. Adaptive torsional vibration active control for hybrid electric powertrains during start-up based on model prediction. *Proc. Inst. Mech. Eng. Part D J. Automob. Eng.* **2022**, *236*, 2219–2229 [CrossRef]
34. Vu, T.M.; Moezzi, R.; Cyrus, J.; Hlava, J.P. Parallel Hybrid Electric Vehicle Modelling and Model Predictive Control. *Appl. Sci.* **2021**, *11*, 10668. [CrossRef]
35. Holland, J.H. *Adaptation in Natural Artificial Systems*; MIT Press: Cambridge, UK, 1975.
36. Mohammadi, S.; Hejazi, S.R. Using particle swarm optimization and genetic algorithms for optimal control of non-linear fractional-order chaotic system of cancer cells. *Math. Comput. Simul.* **2022**, *206*, 538–560. [CrossRef]
37. Dong, T.S.; Chen, S.Y.; Huang, H.; Han, C.; Dai, Z.Q.; Yang, Z.H. Large-Scale Truss Topology and Sizing Optimization by an Improved Genetic Algorithm with Multipoint Approximation. *Appl. Sci.* **2021**, *21*, 407. [CrossRef]
38. Ochelska, M.J.; Poniszewska, M.; Aneta, M.W. Selected Genetic Algorithms for Vehicle Routing Problem Solving. *Electronics* **2021**, *10*, 3147. [CrossRef]
39. Zhang, Z.F.; Xie, D.Q.; Lv, F.; Liu, R.K.; Yang, Y.W. Intelligent geometry compensation for additive manufactured oral maxillary stent by genetic algorithm and backpropagation network. *Comput. Biol. Med.* **2023**, *157*, 106716. [CrossRef] [PubMed]

sensors

MDPI

Article

Procapra Przewalskii Tracking Autonomous Unmanned Aerial Vehicle Based on Improved Long and Short-Term Memory Kalman Filters

Wei Luo [1,2,3,4,5], Yongxiang Zhao [1], Quanqin Shao [2,6], Xiaoliang Li [1], Dongliang Wang [2,*], Tongzuo Zhang [6,7], Fei Liu [8], Longfang Duan [1,3,4], Yuejun He [1,3,4], Yancang Wang [1,3,4], Guoqing Zhang [1], Xinghui Wang [1,3,4] and Zhongde Yu [1]

1 North China Institute of Aerospace Engineering, Langfang 065000, China
2 Key Laboratory of Land Surface Pattern and Simulation, Institute of Geographic Sciences and Natural Resources Research, Chinese Academy of Sciences, Beijing 100101, China
3 Aerospace Remote Sensing Information Processing and Application Collaborative Innovation Center of Hebei Province, Langfang 065000, China
4 National Joint Engineering Research Center of Space Remote Sensing Information Application Technology, Langfang 065000, China
5 Agricultural Information Institute of Chinese Academy of Agricultural Sciences, Key Laboratory of Agricultural Monitoring and Early Warning Technology, Ministry of Agriculture and Rural Affairs, Beijing 100081, China
6 University of Chinese Academy of Sciences, Beijing 101407, China
7 Key Laboratory of Adaptation and Evolution of Plateau Biota, Northwest Institute of Plateau Biology, Chinese Academy of Sciences, Xining 810001, China
8 Intelligent Garden and Ecohealth Laboratory (iGE), College of Biosystems Engineering and Food Science, Zhejiang University, 866 Yuhangtang Road, Hangzhou 310058, China
* Correspondence: wangdongliang@igsnrr.ac.cn

Citation: Luo, W.; Zhao, Y.; Shao, Q.; Li, X.; Wang, D.; Zhang, T.; Liu, F.; Duan, L.; He, Y.; Wang, Y.; et al. Procapra Przewalskii Tracking Autonomous Unmanned Aerial Vehicle Based on Improved Long and Short-Term Memory Kalman Filters. *Sensors* **2023**, *23*, 3948. https://doi.org/10.3390/s23083948

Academic Editors: Diego González-Aguilera, Jose A. Jiménez-Berni, Ittai Herrmann, Shangpeng Sun and Monica Herrero-Huerta

Received: 20 March 2023
Revised: 2 April 2023
Accepted: 10 April 2023
Published: 13 April 2023

Abstract: This paper presents an autonomous unmanned-aerial-vehicle (UAV) tracking system based on an improved long and short-term memory (LSTM) Kalman filter (KF) model. The system can estimate the three-dimensional (3D) attitude and precisely track the target object without manual intervention. Specifically, the YOLOX algorithm is employed to track and recognize the target object, which is then combined with the improved KF model for precise tracking and recognition. In the LSTM-KF model, three different LSTM networks (f, Q, and R) are adopted to model a nonlinear transfer function to enable the model to learn rich and dynamic Kalman components from the data. The experimental results disclose that the improved LSTM-KF model exhibits higher recognition accuracy than the standard LSTM and the independent KF model. It verifies the robustness, effectiveness, and reliability of the autonomous UAV tracking system based on the improved LSTM-KF model in object recognition and tracking and 3D attitude estimation.

Keywords: Procapra przewalskii protection; autonomous unmanned aerial vehicle; object tracking; Kalman filter; long and short-term memory

1. Introduction

Procapra przewalskii is an endangered ungulate endemic to the Qinghai-Tibet Plateau. Its type specimen was collected by Nikolai M. Przewalski in the Ordos Plateau of Inner Mongolia in 1875 [1]. It is listed as a national key protected wild animal by the *List of National Key Protected Wild Animals* issued in 1988. In 2001, it was listed as one of the 15 species in urgent need of rescue in the *National Wildlife Conservation and Nature Reserve Construction Project Master Plan*. In December 2002, the State Forestry Administration formulated the *Overall Plan of the National Proctor* for the gazelle protection project. Meanwhile, it was determined to be a critically endangered (CE) species by the Red List of the International Union for Conservation of Nature (IUCN) in 1996 and 2003. Due to the recovery of local populations and the discovery of new populations, it was adjusted as endangered (EN) from CE by the Red List of the IUCN, but it was still listed as CE in the assessment results of the Red List of Biodiversity in China [2]. Perennial cold and oxygen deficiency in plateau areas pose more difficulty in continuous artificial monitoring of animal behaviors.

As a non-contact data acquisition method, unmanned aerial vehicles (UAVs) exhibit unique advantages in wildlife monitoring. Thanks to the significant progress in computer vision technologies, autonomous UAVs equipped with edge reasoning units have been applied in preliminary real-time monitoring [3,4]. UAVs are limited by size, weight, and power (SWaP), and a single camera is usually equipped as the best sensor. However when the single-shot attitude estimation is applied to video data, it has to estimate the extraordinary noise and confuse the visually similar but spatially different image features if the time information is ignored. Therefore, a time filter is an effective way to increase the accuracy of attitude estimation. The Kalman filter (KF) [5] is a very wide choice due to its simplicity and versatility. In addition, the extended KF (EKF) [6] can address the nonlinear systems of measurement and transition models.

However, these measurement and transition models cannot be specified a priori, severely restricting the application of KFs. Therefore, they can try to directly learn the motion models from training data with support vector machines [7] or with long and short-term memory (LSTM) [8] to overcome the above limitations. It can release the modeler from time-consuming selection and optimization of the KF by learning the motion models simultaneously enriching the underlying models. However, sufficient training data should be ensured to cover all possible motion paths of the tracked object when learned motion models are applied to enforce temporal consistency in pose estimation.

This work mainly contributes to an autonomous UAV target-tracking system based on the improved LSTM-KF model. (1) A three-dimensional (3D) attitude tracking algorithm integrating a learning-based real-time target detection algorithm with a UAV embedded system based on target detection, stereo reconstruction technology, and the KF model, is implemented in the low-cost UAV system to automatically identify, locate, and track the Procapra przewalskii. (2) In the LSTM-KF model, the calculation formulas of the output gate Q_t of three different standard LSTM networks are modified by connecting them with the input gate to solve overfitting during the training of many datasets about the Procapra przewalskii video sequences. (3) During the training iteration of the LSTM-KF model, the Adam optimizer is modified by extending the update rules based on the L^2 norm in Adam optimization to those based on the L^p norm. The improved model is able to accurately track the behavioral attitudes of the LSTM-KF model to identify the original model and reduce the modeler's a priori burden to specify motion and noise models.

This paper is structured as follows. In the "Introduction" section, the background and definitions of Procapra przewalskii, autonomous UAV object-tracking system, KF, and LSTM model are introduced. In the "Related Works" section, the background of the CNN-based animal-monitoring algorithm and the learning-based KF structure are described. The "Materials and Methods" section describes the hardware system architecture used for UAV object tracking, the study area and dataset, the definition and principles of animal 3D pose estimation, the YOLOX model, and the structure of the improved LSTM-KF model. Next, the experimental results are introduced and analyzed. The performances of the YOLOX model at different resolutions are compared, and the tracking results and average errors of the improved LSTM-KF model are verified. In addition, the tracking effects of Procapra przewalskii based on simulation systems and actual flight scenarios are presented. In the "Conclusions" section, the results and discussions are summarized, and future research directions are prospected.

2. Related Works

2.1. Animal-Monitoring Algorithm Based on Convolutional Neural Network (CNN)

CNN-based methods that can achieve accurate result-oriented image feature extraction and representation have been increasingly accepted and applied in vision and image processing [9–11]. Meanwhile, they have exhibited extensive applications in animal identification without pre-specifying any features [12] but have limited usages in monitoring the activities of farming animals [13]. The two-stream network proposed by [14] is one of the tracking models to track moving objects. Multi-layers and the optical flow of convolutional networks enable the capturing the frames of relevant information on an object and tracking the object movement among various frames. Shortly after the proposal of two-stream networks, long-term recurrent convolutional networks (LRCNs) were developed. LRCNs [15] generally comprise several CNNs, namely Inception modules, ResNet, VGG, and Xception, enabling the extraction of spatial and temporal features.

LRCN was the most applied tracking model thanks to its reasonable architecture in object tracking. Generic object tracking using regression network (GOTURN) [16] is another lightweight network that achieves 100 frames per second (fps) for object tracking. GOTURN was initially trained with the generic-objects-filled datasets. The regions of interest (ROIs) on the frames are undertaken as input data for the trained network, providing the possibility of continuously predicting the location of the target. The Slow Fast network [17], on the other hand, tracks objects using two streams of frames, namely slow and high pathways. Many other algorithms can be introduced for animal monitoring, including, but not limited to, simple online and real-time tracking (SORT), the Hungarian algorithm (HA), the Munkres variant of the Hungarian assignment algorithm (MVHAA), the spatial-aware temporal response filter (STRF), and the channel and spatial reliability discriminative correlation filter (CSRDCF) [18–21].

These algorithms were utilized by object-detection models, such as Faster R-CNN, FCN, SSD, VGG, and YOLO, to detect and track animals in images using their geometric features in continuous frames [13]. To provide a reliable and efficient method to monitor the behavioral activity in cows, [22] presented a tracking system embedded with ultra-wideband technology. Similarly, [23] employed a computer vision module to analyze and detect the positive and negative social interactions in feeding behavior among cows. The system was implemented and tested on seven dairy cows in the feeding area, realized an accuracy with a mean error of 0.39 m and standard deviation of 0.62 m, and achieved a detection accuracy of social interactions of 93.2%. However, the real-time locating system (RTLS) exhibits poor accuracy in identifying individual cows if they are in close body contact.

The CNN-based algorithm presents limited applications in monitoring agricultural animals because it does not pre-specify any features of any target. In this case, the YOLOX model, a lightweight network with an anchorless frame at the head of the network, is applied in this paper, which is equipped with several high-performance detectors and a faster network convergence, so that animals with specified features can be monitored. LRCN is the most widely used tracking model due to its reasonable structure in target tracking. However, it fails to effectively monitor the pose behavior of animals and consider the variations in measurement noise. Therefore, the LSTM-KF model with good robustness, reliability, and validity in target tracking and 3D pose estimation is proposed in this paper, which can alleviate the effect of measurement noise and modify the measurement results.

2.2. Learning-Based KF Architecture

In the current work, machine learning and KF models are combined for temporal regularization. The approaches can be classified into those that learn the static parameters of the KF and those that actively regress the parameters during filtering. The noise covariance matrices (NCMs) were optimized statically by [24] to replace the manual fine-tuning of noise parameters in a robotic navigation. Additionally, a coordinate ascent algorithm was employed and each element of the NCM was optimized. However, this approach is only applicable for noisy but time-invariant systems. As opposed to the dynamic model adopted in this study, a change in measurement noise cannot be considered and will therefore lower the accuracy in state estimates.

Reference [8] learned the underlying state transition function that controlled the dynamics of a hidden process state. However, only the state space equations of the KF are used instead of the prediction and update scheme with good performance under linear state transitions and additive Gaussian noise [6]. The neural network models that jointly learn to propagate the state, incorporate the measurement updates, and react to control inputs were trained. In addition, the covariances were designed as constants during the entire estimation. This approach can estimate the state better than a distinct prediction and update model, especially when large-scale training data are insufficient, as demonstrated in the experiment section in the present study.

The dynamic regression of KF parameters was put forward by [7] who adopted support vector regression (SVR) to estimate a linear state transition function, jointly with which the predicted NCM was estimated. The SVR-based system can deal with time-variant systems and outperforms manually tuned KF models in object tracking. As opposed to the model adopted here, the measurement noise covariances (MNCs) are kept constant and the transition function is modeled as a matrix multiplication. In this case, it can only estimate the linear motion models, while the model employed in the present study can estimate the nonlinear transition functions based on all previous state observations.

Reference [25] focused on integrating a one-shot estimation as a measurement into a KF model, which required a prediction of the MNC. They demonstrated that the integrated model exhibited superior performance by comparing it with two other models. In contrast the model designed in the present work undertakes the measurement updates as a black box system and automatically estimates the MNC, so that they can be combined with the current one-shot estimators.

Previous work has extensively investigated the temporal regularization for bit-pose estimation, and priority attention has been given to works that focus on implicit regularization schemes and that explicitly use a learning-based KF structure to infer temporal coherence. In contrast to other models, the proposed model introduces the LSTM-KF, which mitigates the modeler-induced influence on specifying motion and noise models a priori while allowing rich models to learn from data that are extremely difficult to write down explicitly. An extensive series of experiments reveals that the LSTM-KF outperforms both the stand-alone KF and LSTM in terms of temporal regularization.

3. Materials and Methods

3.1. Overall Technical Architecture

The videos in the research area were acquired using Prometheus 230 intelligent UAVs (Chengdu Bobei Technology Co., Ltd., Chengdu, China), as shown in Figure 1. An Intel RealSense D435i stereo camera was selected for the system to acquire sensing and depth data because its features include light weight, wide field of view (FoV), high depth accuracy and good stability. Furthermore, a powerful graphics processing unit (GPU) was employed for the embedded systems, and the NVIDIA Jetson AGX Xavier embedded platform was selected to process the deep-learning-based algorithms. A flight controller (Pixhawk 4 (PX4)) was deployed and communicated with the MAVROS package, which was connected to the planner node.

As illustrated in Figure 2, the designed system consists of (1) a perception module (2) an object-tracking algorithm, (3) a UAV maneuver, and (4) a ground station visualization module. In brief, the UAV system perceives the red-green-blue (RGB) images and the depth data first, and the drone recognizes the Procapra przewalskii with YOLOX (a deep-learning based detector). Next, the 2D bounding boxes are fused with the depth measurement to estimate the 3D pose of the Procapra przewalskii. Finally, the improved LSTM-KF model proposed here is integrated to assist in predicting the motion of the Procapra przewalskii. During this, the visualization user interface is included.

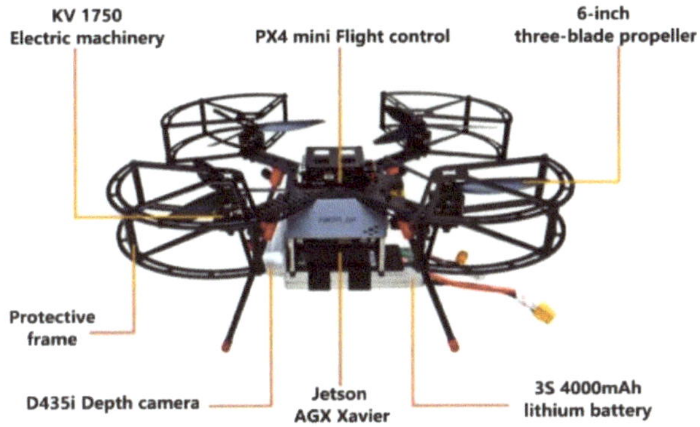

Figure 1. Prototype of the proposed autonomous object-tracking system.

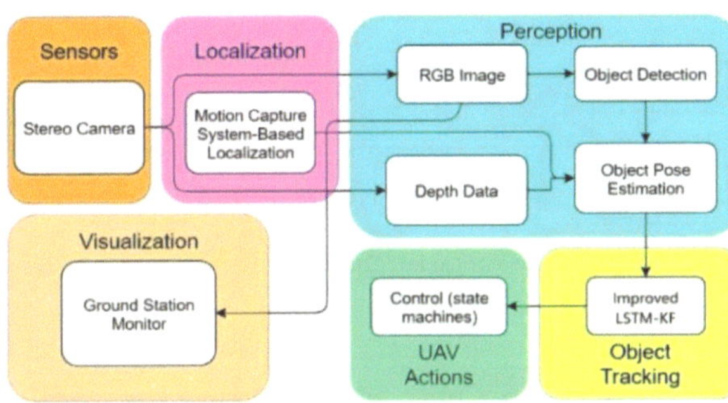

Figure 2. Software architecture of the system.

3.2. Dataset Establishment and Training

Precepting an object in a 3D world is essential for detecting and tracking an object. A deep-learning-based detector is employed here to generate related 2D information and perform 3D stereo reconstruction. This is very challenging because the object may move fast (e.g., running), the training data are low, the detection accuracy is not high, and the position of the object is changing continuously. As mentioned above, the YOLOX algorithm is selected as the baseline model to more accurately detect and track the Procapra przewalskii. Its structural framework is illustrated in Figure 3.

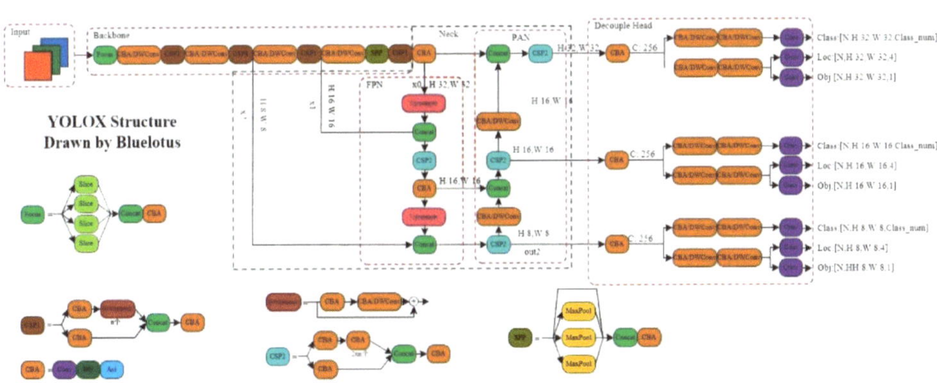

Figure 3. Structural framework of the YOLOX algorithm.

By referring to YOLOV3 and Darknet53, the YOLOX model adopts the structural architecture and spatial pyramid poling (SPP) layer of the latter. In addition, the model is equipped with several high-performance detectors.

In August 2022, we went to the research area (Qinghai Lake) (Figure 4a) to acquire the UAV data and verify the actual flight. There were 40 flights in five days, and each flight lasted about half an hour. The average flight height was about 100 m, and the aerial photography coverage area reached 9744 km^2. Figure 4b shows the flight landing sites. With the captured videos, an object-tracking database was established to identify the moving Procapra przewalskii, match them in different frames, and track their motions.

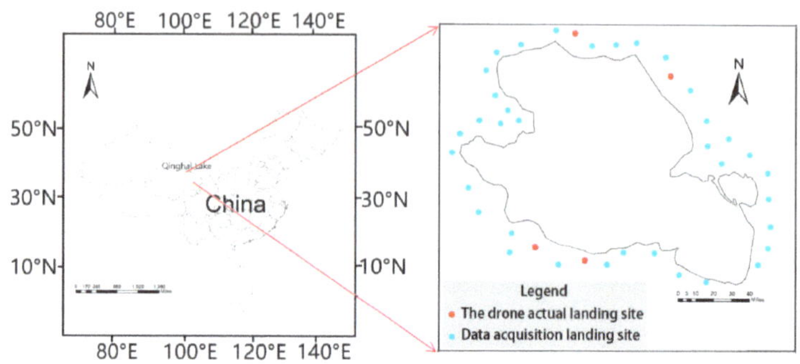

(a) Geographical location of Qinghai Lake (b)Distribution of UAV landing sites in August

Figure 4. Research area: (**a**) geographical location of Qinghai Lake; (**b**) distribution of UAV landing sites in August.

A total of 6 video sequence databases, which were composed of 3 training databases and 3 test databases, are marked. The data were divided into a training set, a test set, and a verification set at a ratio of 3:2:1 (The Supplementary Data Set is available at link https://pan.baidu.com/s/1vEYdFFTKUE9Z9cC67lCH_Q?pwd=56vx). There were three major motions for Procapra przewalskii, which are standing, walking, and running, as displayed in Figure 5a–c (male), Figure 5d–f (female) and Figure 5g–i (young). The database was trained based on the YOLOX model by adjusting the weight ratio, confidence threshold intersection over union (IoU) threshold of nms, and activation function. In this way, a stable and accurate model was obtained.

(a) Male procapra przewalskii standing (b) Male procapra przewalskii walking (c) Male procapra przewalskii running
(d) Female procapra przewalskii standing (e) Female procapra przewalskii walking (f) Female procapra przewalskii running
(g) Young procapra przewalskii standing (h) Young procapra przewalskii walking (i) Young procapra przewalskii running

Figure 5. Different motions of male, female, and young Procapra przewalskii.

3.3. 3D Pose Estimation

Herein, the predicted bounding box was saved as S_{ROI}, and the 3D pose of the object was recovered and dynamically tracked by the object coordinates on the 2D frame based on the depth information obtained from the stereo camera. In addition, an interior rectangle Si was firstly generated by contracting S_{ROI} with a scaling factor θ, as computed in Equations (1) and (2). In the equations below, S_i is the predicted bounding box; θ refers to the scaling factor to adjust the size of the bounding box; c_x and c_y are the coordinates of the bounding box; and w and h represent the width and height of the bounding box, respectively.

$$S_{ROI} = [c_x\ c_y\ w\ h] \tag{1}$$

$$S_{\text{i}} = [c_x\ c_y\ \theta w\ \theta h] \tag{2}$$

S_i, as displayed in Figure 6b, serves as the *ROI* to obtain the depth information. The unfilled pixels are filtered out from the depth image captured by the stereo camera, and the remaining depth data in S_i are averaged as S, which is assumed as the distance between the observer and the target object.

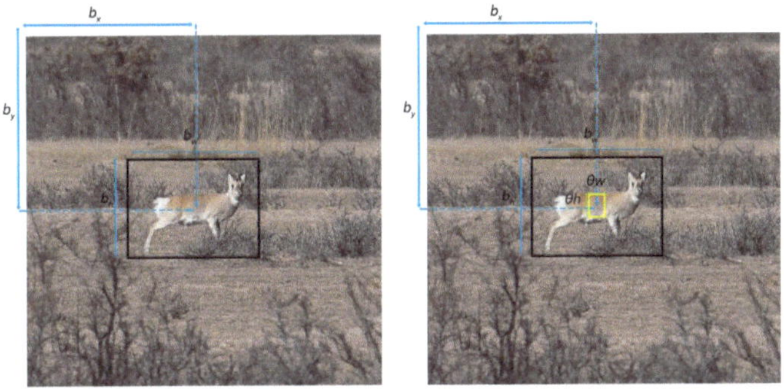

Figure 6. (**a**) Bounding box coordinates; (**b**) *ROI* for depth.

Then, with the boundary box coordination, coordination transformation was performed to obtain the relative attitude of the camera and the global attitude in the world frame. Frame transformation was carried out according to Equations (3) and (4) below:

$$S[uv]^T = K \cdot \begin{bmatrix} X_{\text{i}}^C \\ 1 \end{bmatrix} \tag{3}$$

$$\begin{bmatrix} X_{\text{i}}^W \\ 1 \end{bmatrix} = T_B^W T_C^B \begin{bmatrix} X_i^C \\ 1 \end{bmatrix}, T_C^W T_C^B \in SO(3) \tag{4}$$

In the above equations, u and v are the pixel coordinates of S_t; K is the inherent matrix of the local camera and X_i^C is the object pose vector in camera frame; and X_i^W refers to the object pose vector in the world frame. Specifically, the transformation matrix can be calculated using Equations (5) and (6).

$$T_C^B = \begin{bmatrix} 0 & 0 & 1 & 0 \\ -1 & 0 & 0 & 0 \\ 0 & -1 & 0 & 0 \\ 0 & 0 & 0 & 1 \end{bmatrix} \tag{5}$$

$$T_B^W = \begin{bmatrix} r_{11} & r_{12} & r_{13} & o_x \\ r_{12} & r_{22} & r_{23} & o_y \\ r_{31} & r_{32} & r_{33} & o_z \\ 0 & 0 & 0 & 1 \end{bmatrix} \quad (6)$$

where $r_{i\,j}$ is an element in the observer pose rotation matrix; o_x, o_y, and o_z denote the position of the observer (UAV) relative to the world frame; and T_C^B and T_B^W are the pose transformation matrices. Rotation of the coordinate system is usually represented by a rotation matrix or a quaternion representation.

3.4. Tracking Based on the Improved LSTM-KF Model

In this study, the YOLOX algorithm is employed because it can balance speed and accuracy. Dynamic states of the target Procapra przewalskii and the quadrotor reduce the robustness of the pose estimation based on the descriptions in Section 3.3. The target Procapra przewalskii could not always be captured in the FoV during a surveillance as false positive or negative results may be found. In addition, partial or full occlusion might occur but not often. To address the above issues, the KF model is utilized to enhance tracking, but it requires the specification of a motion model and a measurement model in advance, which increases the burden on the modeler.

3.4.1. Model Structure and Prediction Steps

As introduced above, an improved LSTM-KF model is proposed in the current study, which is a time regularization model for attitude estimators. Its main idea is to use the KFs without specifying a linear conversion function or fixed process and measuring the covariance matrixes Q and R.

The network of the standard LSTM (Figure 7) exhibits memory units, forgetting gates (f_t), input gates (i_t), and output gates (O_t). Some information of cell state C_{t-1} is retained in the current cell state C_t, and the amount of retained information is determined by f_t as given in Equation (7). Meanwhile, i_t and O_t can be calculated with Equations (8) and (9), respectively.

$$f_t = \sigma(W_f \cdot [h_{t-1}, x_t] + b_f) \quad (7)$$

$$\begin{aligned} i_t &= \sigma(W_i \cdot [h_{t-1}, x_t] + b_i) \\ C_t &= \tanh(W_c \cdot [h_{t-1}, x_t] + b_c) \end{aligned} \quad (8)$$

$$\begin{aligned} O_t &= \sigma(W_o \cdot [h_{t-1}, x_{t-1}] + b_o) \\ h_t &= O_t \cdot \tanh(C_t) \end{aligned} \quad (9)$$

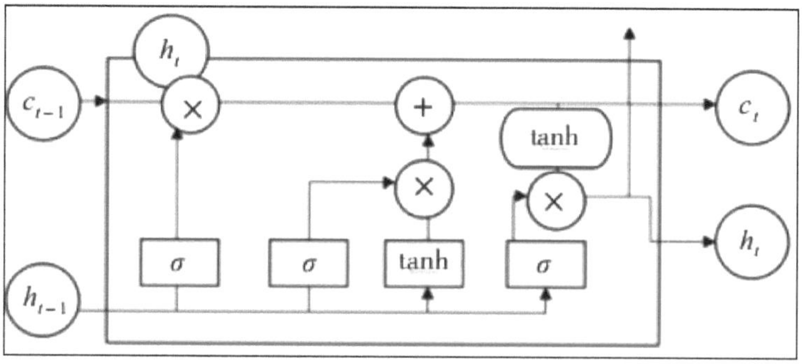

Figure 7. Network structure of the standard LSTM.

Here, O_t on the standard LSTM network is modified as follows:

$$Y_t = V_t C_t + \sum_{n=0}^{t-1} W_{nt} X_n \quad (10)$$

In Equation (10) above, $\{X_0, X_1, \ldots, X_{t-1}\}$ and $\{Y_t, Y_{t+1}, \ldots, Y_{t+n}\}$ are the input and output of the LSTM network, respectively; $\{W_{0(t)}, W_t, \ldots, W_{(t-1)t}\}$ and $\{W_{0(t+1)}, W_{t(t+1)}, \ldots W_{(t-1)(t+1)}\}$ represent the direct weights of the input and output, respectively; C refers to the current state of the LSTM network; and V is the coefficient.

KF is an optimal state estimator under the assumptions of linear and Gaussian noise. Specifically, if the state and the measurement state are expressed as y_t and z_t, respectively, the hypothetical model here can be expressed in Equations (11) and (12).

$$y_t = Ay_{t-1} + \mathrm{w}, w \to N(0, R) \quad (11)$$

$$Z_t = Hy_t + v, v \to N(0, R) \quad (12)$$

Because the incoming measurements are noisy estimates of potential states and $H = I$ in Equation (11), Equations (11) and (12) can be modified to Equations (13) and (14), respectively, which are the basic models of LSTM-KF. In the equations below, Z_t denotes the model measurement state; W_t is the weight at moment t; Q_t and R_t are covariance matrices; and f is a nonlinear transfer function.

$$y_t = f(y_{t-1}) + w_t, w_t \to N(0, Q_t) \quad (13)$$

$$Z_t = y_t + v_t, v \to N(0, R_t) \quad (14)$$

The prediction step can be defined by Equations (15) and (16).

$$\hat{y}'_t = f(\hat{y}_{t-1}) \quad (15)$$

$$\hat{P}' = F\hat{P}_{t-1}F^T + \hat{Q}_t \quad (16)$$

where f is modeled by an LSTM module, F is the Jacobian matrix of f relative to $\hat{y}_{t-1}$, and $\hat{Q}_t$ is the output of the second LSTM model. Thus, the updating steps are specified in Equations (17)–(19).

$$K_t = \hat{P}_t'(\hat{P}_t + \hat{R}_t)^{-1} \quad (17)$$

$$\hat{y}_t = \hat{y}'_t + K_t(\hat{z}_t - \hat{y}'_t) \quad (18)$$

$$\hat{P}_t = (I - K_t)\hat{P}_t' \quad (19)$$

where $\hat{R}_t$ is the output of the third LSTM module and $\hat{z}_t$ refers to the observed measurement at time t. Next, these LSTM modules are described in detail.

3.4.2. Architecture and Loss Function

In this paper, $LSTM_f$, $LSTM_Q$, and $LSTM_R$ are selected to represent the three LSTM modules of f, $\hat{Q}_t$, and $\hat{R}_t$, respectively. $LSTM_f$ is composed of three stacked layers (1024 hidden cells in each) and three fully connected (FC) layers (with 1024, 1024, and 48 hidden cells). Similarly, standard LSTM is built as $LSTM_f$, but O_t of the $LSTM_f$ is modified to connect to i_t. In addition, ReLU nonlinearity is introduced to all FC layer activations except the last, and each LSTM layer is followed by a lost layer with a retention probability of 0.7. $LSTM_Q$ and $LSTM_R$ follow with 256 hidden unit monolayer frameworks and 48 hidden units. Meanwhile, O_t connects with i_t to prevent the video time sequence of the agency and the Procapra przewalskii training to avoid overfitting. Figure 8 shows each module of the LSTM-KF model and Figure 9 displays an overview of the system.

Figure 8. Structure of the LSTM-KF model.

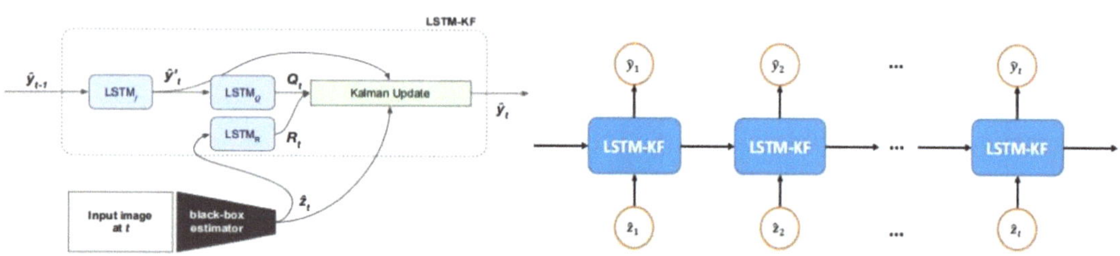

Figure 9. Overview of the LSTM-KF model: (**a**) a high-level depiction of the architecture which uses three LSTM modules to predict the internals of the KF; (**b**) LSTM-KF unrolled over time, which can be trained end to end with backpropagation.

At each t, taking $\hat{y}_{t-1}$ as an input, $LSTM_f$ generates the intermediate state $\hat{y}'_t$ without depending on the current measurement; $LSTM_Q$ takes $\hat{y}'_t$ and $\hat{Q}_t$ as the input and output respectively, and estimates the process covariance; and taking Z_t and $\hat{R}_t$ as an input and out put, respectively, $LSTM_R$ only estimates the measured covariance. Finally, $\hat{y}'_t$ and Z_t, along with the covariance estimates made here, are fed into a standard KF (Equations (16)–(19)) ultimately yielding a new prediction $\hat{y}_t$. Moreover, Q and R are restricted to diagonal and positive definite by indexing the output of the $LSTM_Q$ and $LSTM_R$ modules in this study.

Preliminarily, the standard Euclidean loss summation is applied throughout the entire process, but the $LSTM_f$ module fails to learn reasonable mapping. Therefore, the loss function is introduced with a term to enhance the gradient flow to the $LSTM_f$ module in the current study. Equation (20) expresses the specific loss function.

$$L(\theta) = \frac{1}{T}\sum_{t=1}^{T} ||y_t - \hat{y}_t(\theta)||^2 + \lambda||y_t - \hat{y}'_t(\theta)||^2 \quad (20)$$

3.4.3. Optimization of Parameters

All parameters θ in the loss function are optimized to minimize the loss given by Equation (20) regarding all free parameters in the model applied in this work, which is the connection of the ownership weight matrix and bias from all three LSTM modules, which are a combination of the LSTM layer and the linear layer (Figure 8).

The LSTM-KF model can achieve end-to-end training, and the gradient can be obtained through the time backpropagation algorithm [26]. All the computations and states in the model are presented by a single data flow graph, so that communication between the sub-computations can be displayed, which is conductive to the parallel execution of independent computations to obtain the gradient as soon as possible.

In addition, the Adam-based optimizer [27] is introduced for training iteration to update the gradient, ensuring high stability and high predictability of the model. Meanwhile, the update rules are extended based on the L^2 norm to those based on the L^p norm. A large p results in numerical instability of these variants. However, in a special case of $p \to \infty$, the algorithm is simple but stable, which can be calculated with Equation (21).

$$\mathrm{v}_t = \beta_2^p v_{t-1} + (1 - \beta_2^p)|g_t|^p \tag{21}$$

With the L^p norm, the step size at time t is supposed to be inversely proportional to $v_t{}^{1/p}$, and then Equation (21) can be modified to Equation (22).

$$\mathrm{v}_t = (1 - \beta_2^p)\sum_{i=1}^{t} \beta_2^{p(t-i)} \cdot |g_i|^p \tag{22}$$

It should be noted that the attenuation term is equivalently parameterized here to β_2^p instead of β^2. If $p \to \infty$ and $v_t = lim_{p \to \infty}(v_t)^{1/p}$ are defined, Equations (23)–(26) can be obtained. This corresponds to a very simple recursive equation (Equation (27)).

$$\mathrm{v}_t = \lim_{p\to\infty}(v_t)^{1/p} = \lim_{p\to\infty}\left((1 - \beta_2^p)\sum_{i=1}^{t} \beta_2^{p(t-i)} \cdot |g_i|^p\right)^{1/p} \tag{23}$$

$$= \lim_{p\to\infty}(1 - \beta_2^p)^{1/p}\left(\sum_{i=1}^{t} \beta_2^{p(t-i)} \cdot |g_i|^p\right)^{1/p} \tag{24}$$

$$= \lim_{p\to\infty}\left(\sum_{i=1}^{t} (\beta_2^{(t-i)} \cdot |g_i|^p)\right)^{1/p} \tag{25}$$

$$= \max(\beta_2^{t-1}|g_1|, \beta_2^{t-2}|g_2|, \ldots, \beta_2|g_{t-1}|, |g_t|) \tag{26}$$

$$\mathrm{v}_t = \max(\beta_2 \cdot v_{t-1}, |g_t|) \tag{27}$$

The initial value is $v_0 = 0$. Note that conveniently, the initialization bias does not necessarily need to be corrected. The improved Adam-based optimizer is simpler than the original and is easier for gradient updating.

4. Result

The performance of the trained model was assessed on a Jetson AGX Xavier onboard computer, where the coupled detection head of the original YOLO model was replaced with the decoupled head, which greatly accelerated training convergence. As mentioned, the robustness of the proposed model was observed on a streaming video with various techniques for quantitative analysis Lastly, several intensive flight tests were performed on a self-assembled quadrotor platform to evaluate the overall performance.

4.1. Detection Effect of the YOLOX Model

The YOLOX model outputs a predictive bounding box that classifies detected objects and marks their locations, which plays a key role in subsequent pose estimation using the UAV. The model training lasted for 1000 iterations, during which time the loss did not degrade. Because the accuracy of the model is affected by different neural network resolutions, the YOLOX model was trained with different resolutions (i.e., 416 × 416,

512 × 512, 640 × 640) to evaluate the best performance. The four input resolutions are compared in Table 1.

Table 1. Performances of YOLOv4-Tiny and YOLOX with respect to different resolutions.

Method	Backbone	Size	mAP@0.5 (AP50)	FPS
YOLOv4	CSPDarknet-53	416 × 416	83.19%	6.12
		512 × 512	85.87%	6.24
		640 × 640	88.67%	6.02
YOLOX	Darknet-53	640 × 640	93.25%	13.56

Furthermore, the surveillance performed by UAVs was realized based on real-time perception solutions, which focus more on object detection and tracking. In this case the detection speed and accuracy had to be balanced to ensure consistent detection and tracking, in which delay can be neglected and accuracy is high enough. Thus, the YOLOX and YOYLOv4 models of the same network resolution were compared to examine their accuracy and speeds.

After training, the model performance in detecting target Procapra przewalskii on real-time videos captured was evaluated on an Intel RealSense D435i stereo camera. The trained model was proven to be robust under various environments and exhibits low false positives and negatives. Procapra przewalskii tracking was then successively assessed after assuring the validity of the model.

4.2. Tracking Performance on Target

In this study, the LSTM-KF model was employed to track and identify the behaviors and gestures of the video sequence dataset of Procapra przewalskii. Six object-tracking sequences were comprehensively generated from the Procapra przewalskii dataset, and the 6-DOF ground realistic attitude was available. The LSTM-KF model was trained at 2×10^{-5}, decaying by 0.95 from the second period. Before training, gradients of 100 time steps were propagated using a truncated backpropagation time.

However, for a single-layer LSTM with 16 hidden units, batch size is set to 2 and the learning rate is designed as 5×10^{-4}. After the model is trained for 120 periods, the gradient is propagated again for 10-time steps in the same way. It is the same case for the standard LSTM method evaluated in this work.

The tracking algorithm can be evaluated by employing the successive frames of depth frame sequence tracking through the 3D CAD model of 3D pose. Therefore, all the task methods are compared here to obtain a target-tracking method which is superior to the existing methods. Table 2 displays the results of tracking recognition under the scenario.

Table 2. Effects of temporal regularization on object-tracking estimations of Procapra przewalskii. The errors in translation are denoted as [mm].

	Male Trans	Female Trans	Young Trans	Mean Trans
D.J. Tan et al.	1.58	2.55	3.85	2.66
+Kalm an Vel.al	1.49	2.42	3.34	2.42
+Kalm an Acc.	1.49	2.41	3.32	2.41
+EMA	1.59	2.56	3.87	2.67
+Std. LSTM	40.98	45.98	50.15	45.7
LSTM-KF (ous)	0.58	0.67	1.22	0.82

4.3. Verification of Field Tracking Flight

To integrate a perception to reaction and evaluate the surveillance system, four sites (the red points in Figure 4b) were selected to verify the tracking effect of the UAV on Procapra przewalskii. The parameter settings for the flight test are shown in Table 3.

Table 3. Defined parameters for flight test.

Parameters	Value
θ	12 deg
$V_{\theta max}$	45 deg/s
V_{Zmax}	2 m/s
V_{Xmax}	2 m/s
R_{safe}	5 m
R_{sur}	30 m

Due to the complexity of the algorithm and the uncertainty of the research site, simulation verification and parameter adjustment of the algorithm are necessary before actual flight. Prometheus is an open-source autonomous drone software platform with seamless switching from simulation to real operation (https://wiki.amovlab.com/public/prometheus-wiki/Prometheus-%E8%87%AA%E4%B8%BB%E6%97%A0%E4%BA%BA%E6%9C%BA%E5%BC%80%E6%BA%90%E9%A1%B9%E7%9B%AE/Prometheus-%E8%87%AA%E4%B8%BB%E6%97%A0%E4%BA%BA%E6%9C%BA%E5%BC%80%E6%BA%90%E9%A1%B9%E7%9B%AE.html, accessed on 12 August 2021), emergency safety protection mechanisms, interactive ground stations, unified interfaces, and code specifications. Figure 10 shows the validation scenario of the Prometheus simulation system, which simulated the LSTM-KF model proposed in this paper to track Procapra przewalskii. The field verification results are displayed in Figure 11.

Additionally, the estimated dynamic position is compared with the ground truth of the tracked Procapra przewalskii. Figures 12 and 13 demonstrate that the system can basically track the poses of Procapra przewalskii in 3D space. Regardless of jittering and occasional drift, the Procapra przewalskii can be relocated accurately after several frames. In addition, the figure shows that error is basically within 0.5 m in all axes of the world frame.

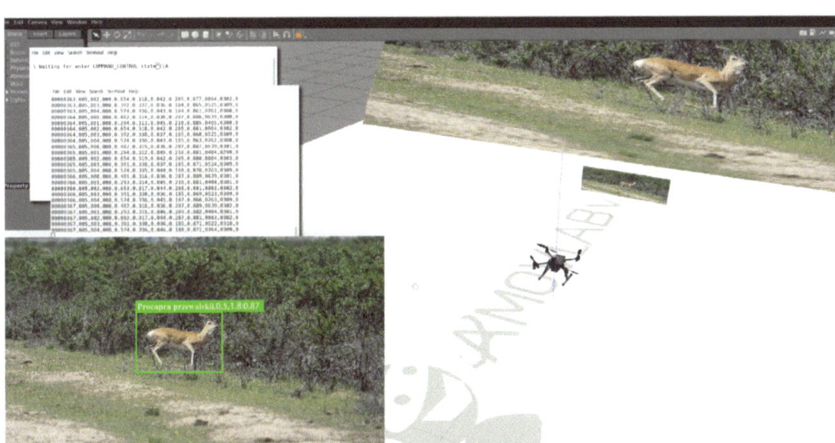

Figure 10. Verification scenario for the tracking algorithm of Procapra przewalskii based on the Prometheus simulation system.

Figure 11. First-person views of the UAV system during the Procapra przewalskii track (the yellow and the red bounding boxes are the detected and predicted states of the object, respectively).

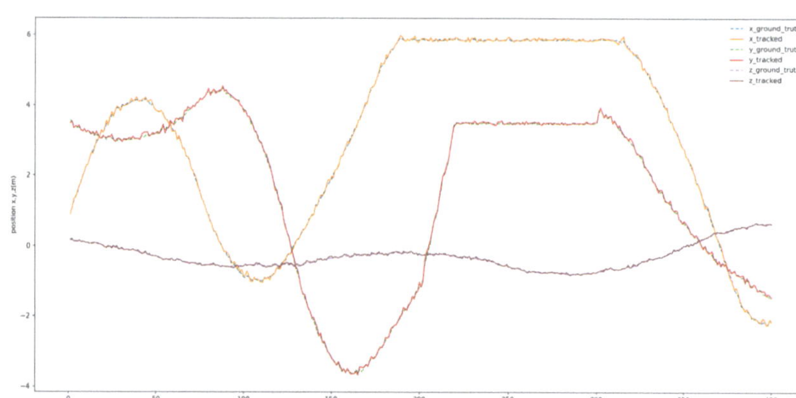

Figure 12. Comparison between the estimated position and ground truth of Procapra przewalskii.

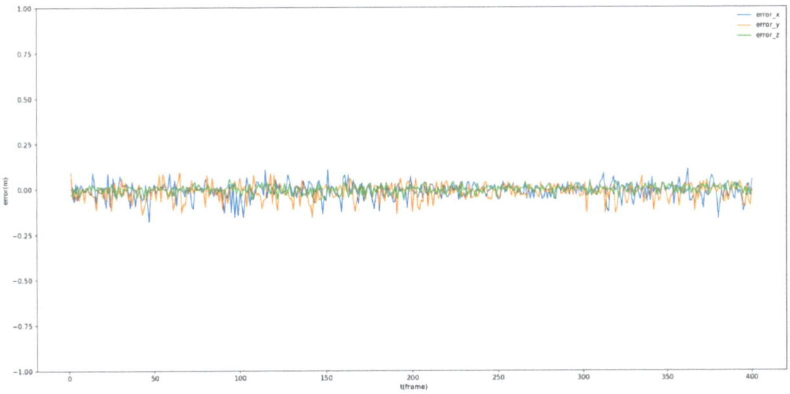

Figure 13. Error throughout the mission time.

In addition, the root-mean-square error (*RMSE*) and mean absolute error (*MAE*) (defined in Equations (28) and (29), respectively) were calculated, as presented in Table 4. Here, *RMSE* indicates the degree of prediction error generated by the model, and a large error results in heavier weights. *MAE* reflects the error between the predicted and actual values, and a smaller *MAE* corresponds to a better model performance. y_i and $\hat{y}_i$ in Equations (28) and (29) are the true value and the predicted value, respectively.

$$RMSE = \sqrt{\frac{1}{n}\sum_{i=1}^{n}(y_i - \overline{y_i})^2} \tag{28}$$

$$MAE = \frac{1}{n}\sum_{i=1}^{n}|y_i - \overline{y_i}| \tag{29}$$

Table 4. Calculated *RSME* and *MAE* for position estimation of the dynamic Procapra przewalskii.

Error Evaluation	X(m)	Y(m)	Z(m)
RMSE	0.0477	0.0478	0.0221
MSE	0.0022	0.0023	0.0005

In addition, during monitoring, the distance between the UAV and the target Procapra przewalskii constantly changed, so it is necessary to further analyze the accuracy at different distances. As shown in Table 5, the performance of the proposed method remained basically stable regardless of the distance between the UAV and the target Procapra przewalskii.

Table 5. Calculated *RSME* and *MAE* of position estimation of dynamic Procapra przewalskii with different object distances.

Object Distance		10–30			30–50	
Error Evaluation	X(m)	Y(m)	Z(m)	X(m)	Y(m)	Z(m)
RMSEMSE	0.0477 0.0022	0.0478 0.0023	0.0221 0.0005	0.0773 0.0059	0.0805 0.0648	0.0656 0.0431

5. Discussion

It should be highlighted that if the input resolution is too large, the best possible mAP increases, but the training and detection speed are indeed negatively affected. Therefore, the higher input resolution is not trained in this work, because the currently obtained speed and accuracy at 640 × 640 are acceptable, and the mAP and the union threshold intersection are 88.67% and 0.50 (AP50), respectively. Unfortunately, the fps is not as fast as the resolution of 512 × 512. Meanwhile, it was found that the mAP of the YOLOX model was lower in contrast to the YOLOv4 model, but the fps was higher. In consideration of the greater significance of fps in real-time predictions, the YOLOX model with a higher fps was selected to balance the accuracy and speed.

Table 2 clearly displays that the motion models that do not investigate the training data, i.e., Kalman Vel Kalman Acc and EMA, are not meaningfully improved for translational estimation and rotation. However, the improved LSTM-KF model proposed in this paper performs better in predicting the target position (0.82 mm) with a mean error of 61.26%, which is higher than the original estimate and better than the results in [28] using the KF algorithm alone for target tracking, and exhibits a lower average error. In addition, the LSTM-KF model greatly improves the original measurements with all actions outperforming the standard LSTM by an average of 14% compared to the state-of-the-art method. In contrast, the standard LSTM method estimates the position and rotation with such a large error that they fail to meet the requirements.

As shown in Tables 3 and 4, the 3D object pose-estimation systems focused on by other scholars [29,30] highlight the objects in static states, while the model applied in this paper exhibited higher errors in estimating the dynamic position of Przewalski's Tibetan antelope in real time. However, it possesses better robustness and is also more accurate than the model in [28] that uses the KF algorithm alone for 3D pose estimation of the target. In addition, it mitigates the influence of the modeler on the a priori specified motion and noise models. The model for 3D pose estimation of dynamic targets using an improved spatio-temporal context algorithm in comparison to that in [31] exhibits higher accuracy, a fast network convergence and fewer impacts from measurement noise. Overall, the proposed LSTM-KF possesses a better performance in pose estimation than the sole use of the KF algorithm and improved spatio-temporal context algorithm. If the target animal, Przewalski's Tibetan antelope, moves suddenly, redundant overshot periods follow, slightly affecting the overall performance of the model. Thus, it further proves that the proposed model can be better applied to autonomous UAV monitoring systems in a real-time and manipulable manner.

6. Conclusions

In this paper, a deep-learning-based model was employed to build an autonomous UAV tracking system to help monitor Procapra przewalskii. The LSTM-KF model is proposed and applied to track the target, and the YOLOX model is employed to identify the target. Meanwhile, they were combined to estimate the pose of the protozoa in 3D images, thus improving the performance of object tracking. In addition, the three different standard LSTM networks modeled in the LSTM-KF model are optimized by modifying and connecting the computation of Q_t. During the training iterations of the LSTM-KF model, Adam as an optimizer is improved by extending the L^2 criterion-based update rule to an L^p criterion-based rule. The results show that the improved LSTM-KF model can achieve the best result (0.82 mm) in predicting the target position with an average error of 61.26%, which is higher than the original estimate, significantly improving the accuracy of the measurement results. In addition, the YOLOX model exhibits an mAP of 93.25% and an FPS of 13.56 on images with 640 × 640 resolution, which are higher than those of the YOLOv4 model. Overall, the proposed improved LSTM-KF model is robust, valid, and reliable for animal tracking, recognition, and pose estimation.

However, this paper is subject to several shortcomings for Procapra przewalskii tracking. For example, when UAVs are applied to track dense herds of Procapra przewalskii, accuracy is decreased if Procapra przewalskii individuals occlude each other. Based on this, we are trying to solve this problem in future research using algorithms based on depth sorting. In addition, it is believed that a visual servo controller can be designed to control the UAV to explore the environment or avoid obstacles using only one camera, ensuring that the tracked object is always in the field of view of the camera.

Supplementary Materials: The following supporting information can be downloaded at: https://pan.baidu.com/s/1vEYdFFTKUE9Z9cC67lCH_Q?pwd=56vx, accessed on 19 March 2023.

Author Contributions: Conceptualization, X.L.; Methodology, Y.W. and X.W.; Validation, Z.Y.; Formal analysis, Y.Z.; Investigation, Y.H.; Data curation, T.Z.; Writing–original draft, W.L.; Writing–review & editing, D.W.; Visualization, G.Z.; Supervision, L.D.; Funding acquisition, Q.S. and F.L. All authors have read and agreed to the published version of the manuscript.

Funding: This research was funded by the National Natural Science Foundation of China (No.: 42071289); the Open Fund of Key Laboratory of Agricultural Monitoring and Early Warning Technology, Ministry of Agriculture and Rural Affairs (JCYJKFKT2204); the Open Fund of Key Laboratory of Spectroscopy Sensing, Ministry of Agriculture and Rural Affairs, P.R. China (No.: 2022ZJUGP004); the Innovation Fund of Production, Study and Research in Chinese Universities (No.: 2021ZYA08001); the Central Government Guided Local Science and Technology Development Fund Project (No.: 216Z0303G); the National Key Research and Development Plan "Establishment of Spectral Earth with Medium Spatial Resolution and Its Application Research" (No.: 2019YFE0127300); Hebei Province Full-time Introduction of Top Talent Research Project (No.: 2020HBQZYC002); the National Science and Technology Major Project "Application and Demonstration of High Resolution Remote Sensing Monitoring Platform for Ecological Environment in Xiong'an New Area" (No.: 67-Y50G04-9001-22/23); the High Resolution Earth Observation System National Science and Technology Major Project; the National Basic Research Plan Project (No.: 2019YFE0126600); and the Doctoral Research Startup Fund Project (No.: BKY-2021-32).

Institutional Review Board Statement: This study does not require "ethics committee" approval.

Informed Consent Statement: Informed consent was obtained from all members who participated in this study.

Data Availability Statement: The supplemental data set for this study is available in Supplementary Materials.

Conflicts of Interest: The authors declare that they have no known competing financial interests or personal relationships that could have appeared to influence the work reported in this paper.

References

1. Ellerman, J.R.; Morrison-Scott, T.C. *Checklist of Palaearctic and Indian Mammals, 1758 to 1946*; British Museum: London, UK, 1951.
2. Jiang, Z.G.; Jiang, J.P.; Wang, Y.Z.; Zhang, Y.Y.; Li, L.L.; Xie, F.; Cai, B.; Cao, L.; Zheng, G.M.; Dong, L.; et al. Red List of China's Vertebrates. *Biodivers. Sci.* **2016**, *24*, 500–551.
3. Luo, W.; Zhang, Z.; Fu, P.; Wei, G.; Wang, D.; Li, X.; Shao, Q.; He, Y.; Wang, H.; Zhao, Z.; et al. Intelligent Grazing UAV Based on Airborne Depth Reasoning. *Remote Sens.* **2022**, *14*, 4188. [CrossRef]
4. Luo, W.; Li, X.; Zhang, G.; Shao, Q.; Zhao, Y.; Li, D.; Zhao, Y.; Li, X.; Zhao, Z.; Liu, Y.; et al. High-Accuracy and Low-Latency Tracker for UAVs Monitoring Tibetan Antelopes. *Remote Sens.* **2023**, *15*, 417. [CrossRef]
5. Kalman, R.E. A new approach to linear filtering and prediction problems. *J. Basic Eng.* **1960**, *82*, 35–45. [CrossRef]
6. Welch, G.; Bishop, G. *An Introduction to the Kalman Filter*; Technical Report 1; University of North Carolina: Chapel Hill, NC, USA, 2006.
7. Salti, S.; Di Stefano, L. Online support vector regression of the transition model for the kalman filter. *Image Vis. Comput.* **2012**, *31*, 487–501. [CrossRef]
8. Krishnan, R.G.; Shalit, U.; Sontag, D. Deep Kalman filters. In Proceedings of the NIPS Workshop on Advances in Approximate Bayesian Inference and Black Box Inference, Montreal, QC, Canada, 11 December 2015.
9. Bello, R.W.; Mohamed, A.S.A.; Talib, A.Z. Contour extraction of individual cattle from an image using enhanced mask R-CNN instance segmentation method. *IEEE Access* **2021**, *9*, 56984–57000. [CrossRef]
10. Bello, R.W.; Mohamed, A.S.A.; Talib, A.Z. Enhanced mask R-CNN for herd segmentation. *Int. J. Agric. Biol. Eng.* **2021**, *14*, 238–244. [CrossRef]
11. Bello, R.W.; Mohamed, A.S.A.; Talib, A.Z.; Olubummo, D.A.; Enuma, O.C. Enhanced deep learning framework for cow image segmentation. *IAENG Int. J. Comput. Sci.* **2021**, *48*, 1182–1191.
12. Kumar, S.; Pandey, A.; Satwik, K.S.R.; Kumar, S.; Singh, S.K.; Singh, A.K.; Mohan, A. Deep learning framework for recognition of cattle using muzzle point image pattern. *Measurement* **2018**, *116*, 1–17. [CrossRef]
13. Li, G.; Huang, Y.; Chen, Z.; Chesser, G.D.; Purswell, J.L.; Linhoss, J.; Zhao, Y. Practices and applications of convolutional neural network-based computer vision systems in animal farming: A review. *Sensors* **2021**, *21*, 1492. [CrossRef] [PubMed]
14. Simonyan, K.; Zisserman, A. Two-stream convolutional networks for action recognition in videos. *arXiv* **2014**, arXiv:1406.2199.

15. Donahue, J.; Anne Hendricks, L.; Guadarrama, S.; Rohrbach, M.; Venugopalan, S.; Saenko, K.; Darrell, T. Long-term recurrent convolutional networks for visual recognition and description. In Proceedings of the IEEE Conference on Computer Vision and Pattern Recognition, Boston, MA, USA, 7–12 June 2015; pp. 2625–2634.
16. Held, D.; Thrun, S.; Savarese, S. Learning to track at 100 fps with deep regression networks. In Computer Vision—ECCV 2016, Proceedings of the European Conference on Computer Vision, Amsterdam, The Netherlands, 11–14 October 2016; Springer: Cham, Switzerland, 2016; pp. 749–765.
17. Feichtenhofer, C.; Fan, H.; Malik, J.; He, K. Slowfast networks for video recognition. In Proceedings of the 2019 IEEE/CVF International Conference on Computer Vision, Seoul, Republic of Korea, 27 October–2 November 2019; pp. 6202–6211.
18. Alameer, A.; Kyriazakis, I.; Bacardit, J. Automated recognition of postures and drinking behaviour for the detection of compromised health in pigs. *Sci. Rep.* **2020**, *10*, 13665. [CrossRef] [PubMed]
19. Chen, G.; Shen, S.; Wen, L.; Luo, S.; Bo, L. Efficient pig counting in crowds with keypoints tracking and spatial-aware temporal response filtering. In Proceedings of the 2020 IEEE International Conference on Robotics and Automation (ICRA), Paris, France, 31 May–31 August 2020.
20. Cowton, J.; Kyriazakis, I.; Bacardit, J. Automated individual pig localisation, tracking and behaviour metric extraction using deep learning. *IEEE Access* **2019**, *7*, 108049–108060. [CrossRef]
21. Zhang, L.; Gray, H.; Ye, X.; Collins, L.; Allinson, N. Automatic individual pig detection and tracking in pig farms. *Sensors* **2019**, *19*, 1188. [CrossRef] [PubMed]
22. Ren, K.; Bernes, G.; Hetta, M.; Karlsson, J. Tracking and analysing social interactions in dairy cattle with real-time locating system and machine learning. *J. Syst. Archit.* **2021**, *116*, 102139. [CrossRef]
23. Salau, J.; Lamp, O.; Krieter, J. Dairy cows' contact networks derived from videos of eight cameras. *Biosyst. Eng.* **2019**, *188*, 106–113. [CrossRef]
24. Abbeel, P.; Coates, A.; Montemerlo, M.; Ng, A.Y.; Thrun, S. Discriminative Training of Kalman Filters. In *Robotics: Science and Systems I, Proceedings of the Robotics: Science and Systems Conference, Cambridge, MA, USA, 8–11 June 2005*; MIT Press: Cambridge, MA, USA, 2005; pp. 289–296.
25. Haarnoja, T.; Ajay, A.; Levine, S.; Abbeel, P. Backprop KF: Learning Discriminative Deterministic State Estimators. In Proceedings of the International Conference on Neural Information Processing Systems (NIPS), Barcelona, Spain, 5–10 December 2016.
26. Werbos, P.J. Backpropagation through time: What it does and how to do it. *Proc. IEEE* **1990**, *78*, 1550–1560. [CrossRef]
27. Kingma, D.; Ba, J. Adam: A method for stochastic optimization. *arXiv* **2014**, arXiv:1412.6980.
28. Lo, L.-Y.; Yiu, C.H.; Tang, Y.; Yang, A.-S.; Li, B.; Wen, C.-Y. Dynamic Object Tracking on Autonomous UAV System for Surveillance Applications. *Sensors* **2021**, *21*, 7888. [CrossRef] [PubMed]
29. Feng, Y.; Tse, K.; Chen, S.; Wen, C.-Y.; Li, B. Learning-Based Autonomous UAV System for Electrical and Mechanical (E&M) Device Inspection. *Sensors* **2021**, *21*, 1385. [CrossRef] [PubMed]
30. Steich, K.; Kamel, M.; Beardsley, P.; Obrist, M.K.; Siegwart, R.; Lachat, T. Tree cavity inspection using aerial robots. In Proceedings of the IEEE/RSJ International Conference on Intelligent Robots and Systems (IROS), Daejeon, Republic of Korea, 9–14 October 2016.
31. Zhang, K.; Zhang, L.; Liu, Q.; Zhang, D.; Yang, M.H. Fast Visual Tracking via Dense Spatio-temporal Context Learning. In *Computer Vision—ECCV 2014, Part V, Proceedings of the 2014 European Conference on Computer Vision—ECCV, Zurich, Switzerland, 6–12 September*; LNCS; Springer: Cham, Switzerland, 2014; Volume 8693, pp. 127–141.

applied sciences

MDPI

Article

Centralized Mission Planning for Multiple Robots Minimizing Total Mission Completion Time

Nam Eung Hwang *, Hyung Jun Kim and Jae Gwan Kim

Hanwha Systems Co., Seongnam-si 13524, Gyeonggi-do, Republic of Korea
* Correspondence: skadnd144@hanwha.com

Abstract: Most mission planning algorithms solve multi-robot-multi-mission problems based on mixed integer linear programming. In these algorithms, the rewards (or costs) of missions for each robot are calculated according to the purpose of the user. Then, the (robot-mission) pair that has maximum rewards (or minimum costs) is found in the rewards (or costs) table and the mission is allocated to the robot. However, it is hard to design the reward for minimizing total mission completion time because not only a robot, but also the whole robots' mission plans must be considered to achieve the purpose. In this paper, we propose centralized mission planning for multi-robot-multi-mission problems, minimizing total mission completion time. First, mission planning for single-robot-multi-mission problems is proposed because it is easy to solve. Then, this method is applied for multi-robot-multi-mission problems, adding a mission-plan-adjustment step. To show the excellent performance of the suggested algorithm in diverse situations, we demonstrate simulations for 3 representative cases: a simple case, which is composed of 3 robots and 8 missions, a medium case, which is composed of 4 robots and 30 missions, and a huge case, which is composed of 6 robots and 50 missions. The total mission completion time of the proposed algorithm for each case is lower than the results of the existing algorithm.

Keywords: mission planning; task allocation; task alignment; task planning; multi-robot systems; automation; robot control

Citation: Hwang, N.E.; Kim, H.J.; Kim, J.G. Centralized Mission Planning for Multiple Robots Minimizing Total Mission Completion Time. *Appl. Sci.* **2023**, *13*, 3737. https://doi.org/10.3390/app13063737

Academic Editor: Diego González-Aguilera

Received: 21 February 2023
Revised: 10 March 2023
Accepted: 13 March 2023
Published: 15 March 2023

1. Introduction

As automation technology highly advances, concepts of swarming, Manned and Unmanned Teaming (MUM-T), etc., occurred [1–9]. These concepts focused on overcoming a variety of missions which are hard for single robot by using multiple robots [10–14]. For example, multiple robots can complete surveillance and reconnaissance missions of a specific area within a short time, while a single robot takes much more time. If there are multiple missions to solve, making a mission plan for each robot considering the purpose of the user is important for efficiency. Although there are various purposes, such as conducting important missions first, minimizing fuel consumption (or minimizing total moving distances), etc., most users want to complete whole missions as soon as possible, which means minimizing total mission completion time.

Mission planning technologies have been developed in various ways [15–33]. One of the general mission planning methods is Consensus-based Bundle Algorithm (CBBA) which is based on scores, i.e., rewards or costs of missions for each robot [25]. The CBBA algorithm is separated into two steps: the bundle construction step and conflict resolution step. In the bundle construction step, each robot calculates the sequence of missions based on the scores under the assumption that the robot accomplishes all missions. The scoring scheme is designed using the expected start/finish time of tasks, distances between robots and missions, etc., based on the user's purpose. In the conflict resolution step, each robot decides which missions it will do based on scores calculated in the bundle construction step. Usually, the robot who has the highest score will take the mission that has the maximum score. CBBA is a deterministic allocation algorithm, so results are always same if the inputs, the information of the robots and missions, are same. In addition, the computation cost of CBBA is much lower than for other heuristic algorithms.

The modified task allocation that can handle some constraints based on CBBA is Coupled-Constraint CBBA (CCBBA) [26]. The missions that have a constraint relationship are grouped as 'Activity' and each mission in the activity is set as 'Element'. Such settings enable easy judgement of whether a mission can be allocated to a robot or not. In addition, it uses optimistic and pessimistic bidding strategies. Each robot bids to the missions based on the strategy for obeying the constraints. However, there are some problems regarding convergence and calculation time for mission planning. Hence, our previous work proposed the Task Allocation and Alignment based on Constraint Table and Alignment Rules (TACTAR), which is modified based on CBBA for centralized mission planning systems [33]. It handles some spatial and temporal constraints using a constraint table and four alignment rules for mission planning. These methods commonly use only the robot's own information to score the missions, but the plans of all robots should be considered to minimize the total task completion time. Hence, CBBA, CCBBA, and TACTAR cannot be used for mission allocation, minimizing total mission completion time without any modification.

Qin et al. proposed path planning algorithms for multiple Unmanned Aerial Vehicles (UAVs) to minimize mission completion time [24]. The problem in this paper is defined as a collection of sparse multiple sensors. The sensor points which have distances less than the communication distance of UAVs are grouped, and the center point of each group is defined as hovering point. Then, an ant colony algorithm for solving Traveling Salesman Problem (TSP) is used to visit whole hovering points by one UAV, and a cycle is made as the result. The k-cycles algorithm is used for decomposing the cycle into the number of UAVs. However, this algorithm is based on basic TSP, which means that UAVs must come back to home, and the home positions for all UAVs must be the same. In addition, the ant colony algorithm is used for solving TSP, so the calculation cost of the algorithm will be fatal when the number of hovering points becomes larger.

Wang et al. also proposed path planning algorithms for multiple fixed-wing UAVs for minimizing mission completion time [34]. The algorithm focused on the path planning of fixed-wing UAVs for UAV-to-UAV (U2U) communication with the aim of minimum information transmission time in the presence of ground transmitters. The constraints contain the communication throughput requirement, interference from ground transmitters, the speed and acceleration range of each UAV, and the minimum communication distance of UAVs. However, they only consider only one mission: communicating two UAVs with each other. Hence, the algorithm calculates the local path planning for each UAV to minimize information transmission time, not the sequence of missions for each UAV.

In this paper, we propose a centralized mission planning algorithm for multi-robot multi-mission problems, minimizing total mission completion time. The proposed algorithm is designed based on CBBA to overcome some limitations such as computation time and necessity to set the same home for all robots in the above algorithms. First, we introduced the algorithm for single-robot-multi-mission problems, minimizing total mission completion time based on the method using the bundle construction step in CBBA. It focused on how to decide the sequence of missions that the robot has to work on. Then, the algorithm was expanded for multi-robot-multi-mission problems. To overcome limitations of CBBA, CCBBA, and TACTAR, we proposed bundle adjustment, which adjusts the mission plan only if the total mission completion time could be lower by the adjustment. In order to apply the status of each robot in real time, the proposed algorithm should be run iteratively. However, if the positions of robots are changed, the mission plan could fall into another local optimal solution, so the mission plan could be changed frequently. To prevent this phenomenon, the initial bundle construction, which is based on the arrival times to missions for robots, is performed only at the first run, and the previous mission plan is used from the second iteration instead of the initial bundle construction. Then, some simulation results using the proposed algorithm will be provided.

2. Preliminaries

The mission planning problem for minimizing total mission completion time is to find a sequence of missions for each robot that has the minimum total mission completion time as seen below:

$$\mathrm{Min}\left(\mathrm{Max}\left(t_e(j,\ p_i)x_{ij}\right)\right) \tag{1}$$

subject to

$$\sum_{i=1}^{N_r} x_{ij} \leq 1,\ \forall j \in J \tag{2}$$

$$\sum_{j=1}^{N_m} x_{ij} \leq L_m,\ \forall i \in I \tag{3}$$

$$\sum_{i=1}^{N_r}\sum_{j=1}^{N_m} x_{ij} = N_{min} \tag{4}$$

$$x_{ij} \in \{0,\ 1\} \tag{5}$$

where $t_e(j,\ p_i)$ is finish time of mission j if robot i conduct mission j along the path p_i; x_{ij} is the flag that is 1 if mission j is taken to robot i and 0, or otherwise, N_r and N_m are the number of robots and missions, respectively, L_m is the maximum number of missions that the robot can take, and N_{min} is the minimum number between N_m and $N_r L_m$. I and J are robot and task group, respectively, so the number of elements in I is N_r and the number of elements in J is N_m. Note that one mission is taken to only one robot as described in Equation (2).

The mission planning is composed of two parts: mission allocation and mission alignment. The mission allocation is to decide which robot will take which mission. The mission alignment is to calculate the start time and the finish time of each mission based on the result of mission allocation. Some mission planning algorithms conduct mission allocation at first for all missions; then, the mission alignment is performed once based on the mission allocation result. The other mission planning algorithms conduct mission allocation for one mission, and then the mission alignment is performed for one mission. This procedure repeats for all missions. This sequence is used when the mission allocation uses the start time or the finish time of each mission for the results of mission alignment.

3. Related Works

This section reviews the most common mission planning works related to the proposed algorithm; one is CBBA and the other is TACTAR [25,33].

3.1. Consensus-Based Bundle Algorithm (CBBA)

CBBA is one of the mission planning algorithms for the multi-robot-multi-mission problem. It is inspired by a decentralized auction process. CBBA is composed of two steps: the bundle construction step and the conflict resolution phase. In bundle construction step, robots make their mission lists, called a bundle, and decide their own sequence of missions assuming that each robot takes all missions. For this step, each agent calculates the rewards of the possible mission. The reward calculation scheme depends on the purpose of users. The procedure of bundle construction is as in Algorithm 1.

In lines 1–4, the results of the previous iteration are copied into the results of the current iteration. $\boldsymbol{y}_i$, z_i, $\boldsymbol{b}_i$, and $\boldsymbol{p}_i$ are winning bids (the highest rewards of each mission) that robot i knows, winning robots (who take each mission) that robot i knows, the bundle (which mission that robot to do) of robot i, and the sequence of the missions of the robot i, respectively. Note that t is the number of iterations (maybe L_m or N_m times), not time. If the number of bundles for robot i is lower than L_m, the rewards of missions that are not included in the bundle is calculated. $\boldsymbol{S}_i^{\boldsymbol{p}_i}$ means the total reward of robot i if it completes along the sequence $\boldsymbol{p}_i$. $\boldsymbol{p}_i(+)_n\{j\}$ means adding mission j into the nth position of sequence

$\boldsymbol{p}_i$, so the reward of the mission is the highest score difference $S_i^{\boldsymbol{p}_i(+)_n\{j\}} - S_i^{\boldsymbol{p}_i}$ when the mission is added to the sequence. In line 7, the reward of the mission is compared to the winning bid of the mission. $\boldsymbol{h}_{ij}$ is 1 if the reward is higher than the winning bid, and 0 otherwise. If the rewards of all possible missions are calculated, then the robot i chooses the mission that has the highest rewards. Finally, the chosen mission is added to the results of the current iteration.

Algorithm 1 CBBA Phase 1: Bundle construction

1: $\boldsymbol{y}_i(t) = \boldsymbol{y}_i(t-1)$
2: $\boldsymbol{z}_i(t) = \boldsymbol{z}_i(t-1)$
3: $\boldsymbol{b}_i(t) = \boldsymbol{b}_i(t-1)$
4: $\boldsymbol{p}_i(t) = \boldsymbol{p}_i(t-1)$
5: **while** $|\boldsymbol{b}_i| < L_t$, **do**
6: $\quad c_{ij} = max_{n \le |\boldsymbol{p}_i|} S_i^{\boldsymbol{p}_i(+)_n\{j\}} - S_i^{\boldsymbol{p}_i}, \forall j \in \boldsymbol{J} \setminus \boldsymbol{b}_i$
7: $\quad h_{ij} = \mathrm{II}\left(c_{ij} > y_{ij}\right), \forall j \in \boldsymbol{J}$
8: $\quad JJ_i = argmax_j c_{ij} \cdot h_{ij}$
9: $\quad n_{i,JJ_i} = argmax_n \; S_i^{\boldsymbol{p}_i(+)_n\{JJ_i\}}$
10: $\quad \boldsymbol{b}_i = \boldsymbol{b}_i(+)_{end}\{JJ_i\}$
11: $\quad \boldsymbol{p}_i = \boldsymbol{p}_i(+)_{n_{i,JJ_i}}\{JJ_i\}$
12: $\quad y_{i,JJ_i}(t) = c_{i,JJ_i}$
13: $\quad z_{i,JJ_i}(t) = i$
14: **end while**

The next step of CBBA is the conflict resolution step, which is making a decision on who will take the mission by comparing the rewards of each robot. Basically, CBBA is for a decentralized mission planning system, so two communicable robots decide the owner of missions following the rules described in Table 1.

Table 1. Conflict resolution rules for mission j of robot i which communicates with robot k.

Robot k's (Sender) z_{kj} Is	Robot k's (Sender) z_{kj} Is	Receiver's Action (Default: Leave)
k	i	update if $(y_{kj} > y_{ij})$
	k	update
	$m \notin \{i, k\}$	update if $(s_{km} > s_{im})$ | | $(y_{kj} > y_{ij})$
	none	update
i	i	leave
	k	reset
	$m \notin \{i, k\}$	reset if $(s_{km} > s_{im})$
	none	leave
$m \notin \{i, k\}$	i	update if $(s_{km} > s_{im})$ & $(y_{kj} > y_{ij})$
	k	update if $(s_{km} > s_{im})$, reset else
	m	update if $(s_{km} > s_{im})$
	$n \notin \{i, k, m\}$	update if $(s_{km} > s_{im})$ & $((s_{kn} > s_{in})$ | | $(y_{kj} > y_{ij}))$ reset if $(s_{kn} > s_{in})$ & $(s_{im} > s_{km})$
	none	update if $(s_{km} > s_{im})$
none	i	leave
	k	update
	$m \notin \{i, k\}$	update if $(s_{km} > s_{im})$
	none	leave

If CBBA is used in centralized mission planning, the conflict resolution step based on the rules in Table 1 is replaced into the Sequential Greedy Algorithm (SGA). When the bundle construction step is performed, the rewards of missions for each robot are calculated and the central computer makes a reward table based on the rewards of each robot. Then, the SGA just finds the (robot-mission) pair that has the maximum reward and conducts mission allocation and alignment.

3.2. Centralized Task Allocation and Alignment Based on Constraint Table and Alignment Rules (TACTAR)

CBBA cannot solve the problem with several constraints. The constraints are separated into two types: temporal and spatial constraints. The temporal constraints are constraints about the start/finish time of missions. For instance, 'mission A ends before mission B starts', 'mission A starts simultaneously with mission B', etc., are temporal constraints. The spatial constraints are constraints about mission allocation. For example, 'mission A is local MUTEX with mission B' means mission A cannot be allocated to the robot who takes mission B and vice versa. TACTAR is a centralized mission planning algorithm that can handle spatial and temporal constraints based on CBBA. Supported constraints in TACTAR are depicted in Table 2.

Table 2. Supported type of constraints in TACTAR.

Name	Description
Simultaneous	Mission A and B must start at the same time
After	Mission A must start after mission B's finish time
Start During	Mission A must start between mission B's start time and finish time
End During	Mission A must end between mission B's start time and finish time
Local MUTEX	Mission A and B are locally mutually exclusive
Global MUTEX	Mission A and B are globally mutually exclusive

The purpose of TACTAR is to calculate the mission plan considering spatial and temporal constraints quickly. TACTAR does not do as described in line 6 of Algorithm 1, just calculate rewards of missions, assuming missions are performed right after the robot's last mission. The optimization performance may suffer slightly, but calculation speed is boosted much faster than CBBA. In addition, some filtering process is added to TACTAR before the beginning of the mission allocation and alignment to not violate constraints. In the mission alignment of TACTAR, some delay or renewal of the mission time (arrival time to mission area, start/finish time of mission) are performed based on constraint table and alignment rules. Further details can be found in [33].

4. Mission Planning Scheme

In this section, we propose a centralized mission planning algorithm for minimizing the total mission completion time. It is based on minimizing total mission completion time for a single-robot-multi-mission problem, modified from the bundle construction in CBBA. So, the algorithm for single-robot is introduced first, and it expands to the algorithm for the multi-robot problems. Note that only mission alignment is needed in single-robot-single-mission problems. In addition, note that this section is focused on the mission planning, not path planning, for minimizing total mission completion time.

4.1. Single-Robot-Multi-Mission Problem

In Single-Robot-Multi-Mission (SRMM) problem, one robot has to conduct all missions, so there is no need to select missions to conduct. All the mission planning has to achieve is to decide the sequence of missions and to calculate start/finish time of each mission. To find the optimal sequence of missions, we propose a process similar to line 6 of Algorithm 1 as described in Algorithm 2.

Algorithm 2 Mission planning: SRMM

```
1: while |p_i| < L_m, do
2:     t_ij = min_{n≤|p_i|} t_i^e(p_i(+)_n{j}), ∀j ∈ J\p_i
3:     JJ_i = argmin_j(t_ij)
4:     n_{i,JJ_i} = argmin_n(t_i^e(p_i(+)_n{JJ_i}))
5:     p_i = p_i(+)_{n_{i,JJ_i}}{JJ_i}
6: end while
```

1: **while** $|\boldsymbol{p}_i| < L_m$, **do**
2: $t_{ij} = min_{n\le|\boldsymbol{p}_i|} t_i^e(\boldsymbol{p}_i(+)_n\{j\}), \forall j \in \boldsymbol{J}\backslash \boldsymbol{p}_i$
3: $JJ_i = argmin_j\left(t_{ij}\right)$
4: $n_{i,\,JJ_i} = argmin_n\left(t_i^e(\boldsymbol{p}_i(+)_n\{JJ_i\})\right)$
5: $\boldsymbol{p}_i = \boldsymbol{p}_i(+)_{n_{i,\,JJ_i}}\{JJ_i\}$
6: **end while**

In line 2, the total mission completion time is calculated when a mission is added to nth space of sequence using the speed of the robot, the distances between the robot and mission areas, and duration of missions. It repeats for all missions and lengths of sequence. Then, it finds the (mission-space) pair that has the minimum total mission completion time. Finally, the found mission is added to the found space of sequence. Lines 2–5 of Algorithm 2 repeat until all missions (or L_m missions) are allocated. After the final mission is allocated the start/finish times of each mission are calculated.

4.2. Multi-Robot-Multi-Mission Problem

Unlike the SRMM problems, each robot must select which missions it will perform in Multi-Robot-Multi-Mission (MRMM) problems, considering the equipment status of robots. For example, the robot that does not have camera cannot conduct surveillance and reconnaissance missions. However, it is hard to conduct a mission allocation because not only a robot but also the whole robots' planning results must be considered to achieve total mission completion time minimization. So, we propose initial bundle construction and bundle adjustment methods as described in Algorithm 3.

Algorithm 3 Mission planning: MRMM

if first time
    Construct initial bundle
else
    bundle = result of adjustment
end if
Adjust bundle

If it is first time for the mission planning, initial bundle construction step runs. If not, the result of the previous adjustment is copied into the bundle for preventing to fall into local optimization solutions during mission planning iterations. Then, the bundle adjustment for total mission completion time minimization is run.

In initial bundle construction, as described in Algorithm 4, each robot calculates the expected finish time of missions based on the positions of the robot and missions, the speed of the robot, and the duration of each mission for the robot. Then, the table of finish time of missions for robots will be made. The (robot-mission) pair that has the minimum finish time of the mission is found in the table. If the pair has an infinity finish time of mission (initial value of the table), all possible missions are already added to the bundles, so it breaks out of the iterations. If not, it checks the selected robot that can conduct the selected mission. If the robot cannot conduct the mission, $t_{i_{sel}j_{sel}}$ in the table is set to infinity (or initial value) and another pair in the table is found. If the robot can conduct the mission the mission is added to the bundle of the robot. The column (or row) about the mission is set to infinity (or initial value) in the table, and the procedure is repeated until all missions are added to the bundle.

Algorithm 4 Construct initial bundle

$t_{ij} = d_{ij}/v_i + dur_{ij}, \forall i \in \boldsymbol{I}, \forall j \in \boldsymbol{J}$
while 1
  $i_{sel} = argmin_i\left(t_{ij}\right)$
  $j_{sel} = argmin_j\left(t_{i_{sel}j}\right)$
  if $t_{i_{sel}j_{sel}} = \infty$
    break
  else
    if i_{sel} cannot do j_{sel}
      $t_{i_{sel}j_{sel}} = \infty$
    else
      $\boldsymbol{b}_{i_{select}} = \boldsymbol{b}_{i_{sel}}(+)_{end}\{j_{sel}\}$
      $\boldsymbol{t}_{j_{sel}} = \infty$
    end if
  end if
end while

In the bundle adjustment, line 6 of Algorithm 3, the robot that has the maximum total mission completion time is found. Then, one of the missions in the bundle of that robot is moved to the bundles of the others who can conduct the mission. The total mission completion time for each robot can be calculated based on the bundles following Algorithm 2, because fixing the bundles of each robot means MRMM problems are simplified into SRMM problems for each robot. The (i_s, j_s) pair that has the minimum total mission completion time is searched. If there are multiple (i_s, j_s) pairs that have the same minimum total completion time, a (i_s, j_s) pair in which the larger value is the smallest among the total mission completion time of i_{max} and i_s is selected. Finally, the total mission completion times for the results of previous and current iterations are compared. If the current one is lower than the previous one, the current bundle and total mission completion time are copied into the previous results. Then, it goes to line 4 of Algorithm 5 and the procedures are repeated. If not, it means the previous results are the optimal mission plans for minimizing the total mission completion time, so it outputs the previous results.

Algorithm 5 Adjust bundle

1: $t^e_{max,prev} = max(\boldsymbol{t}^e)$
2: $\boldsymbol{b}_{prev} = \boldsymbol{b}$
3: **while** 1
4:   $i_{max} = argmax_i\left(t^e_i\right)$
5:   **for** $n = 1 : size\left(\boldsymbol{b}_{i_{max}}\right)$
6:     $\boldsymbol{b}_{i_{max}} = \boldsymbol{b}_{i_{max},prev}$
7:     $j_{sel}(n) = \boldsymbol{b}_{i_{max}}(n)$
8:     $\boldsymbol{b}_{i_{max}} = \boldsymbol{b}_{i_{max}}(-)_n$
9:     $\boldsymbol{b}_i = \boldsymbol{b}_i(+)_{end}\{j_{sel}(n)\}, \forall i \in \boldsymbol{I} \backslash i_{max}$ if i can do $j_{sel}(n)$
10:     **calculate** $max\left(\boldsymbol{t}^e_{in}\right)$ for $\boldsymbol{b}$
11:   **end for**
12:   $t^e_{max} = min\left(max\left(\boldsymbol{t}^e_{in}\right)\right)$
13:   $i_s = argmin_i\left(max\left(\boldsymbol{t}^e_{in}\right)\right), \forall i \in \boldsymbol{I} \backslash i_{max}$
14:   $j_s = j_{sel}\left(argmin_n\left(max\left(\boldsymbol{t}^e_{in}\right)\right)\right)$
15:   **if** multiple (i_s, j_s) that has same t^e_{max}
16:     **choose** that has $min\left(\max\left(t^e_{i_{max}}, t^e_{i_s}\right)\right)$
17:   **end if**
18:   **if** $t^e_{max,prev} > t^e_{max}$
19:     $t^e_{max,prev} = t^e_{max}$
20:     $\boldsymbol{p}_{prev} = \boldsymbol{p}$
21:   **else**
22:     **break**
23:   **end if**
24: **end while**

5. Result and Discussion

In this section, some simulation results are shown to assess the algorithm that we have proposed. We simulated three cases, a simple case, medium case, and huge case, to prove the performance of proposed algorithm. The simple case is composed of 3 robots and 8 missions, the medium case is composed of 4 robots and 30 missions, and the huge case is composed of 6 robots and 50 missions. The speed of all robots and the duration of all missions are equally set to 2 m/s and 5 s, respectively, to simplify the problem in the simple case. In the medium and huge cases, the speed of each robot and the duration time of each mission are set differently to represent the general situations. The detailed scenarios of simulations are described in Table 3.

Table 3. The scenarios of simulations.

Contents	Specification Simple	 Medium	 Huge
# of robots	3	4	6
# of missions	8	30	50
Pos. of robots	[1, 0], [2, 0], [3, 0]	[1, 0], [1, 1], [3, 0], [3, 1]	[1, 0], [1, 1] [2, 0], [2, 1] [3, 0], [3, 1]
Pos. of missions	[−2, 3], [−4, 3] [−4, 5.1], [−2, 5.2] [7, 7], [9, 7] [9, 9.1], [7, 9.2]	20·rand(30, 2)-10	20·rand(50, 2)-10
Spd. of robots	2 [m/s] for all	2.0 [m/s], 1.5 [m/s], 1.0 [m/s], 0.5 [m/s]	2.0 [m/s], 1.8 [m/s], 1.6 [m/s], 1.4 [m/s], 1.2 [m/s], 1.0 [m/s]
Dur. of missions	5 [s] for all	2.1:0.1:5 [s]	2.1:0.1:7 [s]

CBBA and TACTAR are used to show that the proposed algorithm is improved over the existing algorithm. The total mission completion time and computational cost are set as performance indicators. We used OpenMP parallel computing to run each algorithm, and the computational cost of each algorithm is measured based on the result of the parallel computation. The environment of simulations is described in Table 4.

Table 4. The environment of simulations.

Contents		Specification
CPU		Intel® Core™ i5-10500 3.10 GHz (6 cores, 12 threads)
GPU		Intel® UHD Graphics 630
RAM		16 GB
Storage		256 GB SSD
Program	Calculation	Visual Studio Professional 2013 C++
	Plotting	MATLAB 2018b

5.1. The Simple Case

The results of the simple case are shown in Figures 1–3. Figure 1 depicts the results of CBBA, Figure 2 depicts the results of TACTAR, and Figure 3 depicts the results of the proposed algorithm. Panel (a) in Figures 1–3 shows the 2D situation map to easily check the mission plan result of each algorithm. The 'R' and 'M' means robot and mission, respectively. The following number means the index of robots and missions. The red lines mean the sequence of missions for each robot. For example, R01 will conduct M01, M03 and M07 sequentially, as seen in Figure 1a. As seen in Figures 1 and 2, the mission plan result of CBBA is absolutely the same as with the result of TACTAR. However, the mission plan result of the proposed algorithm is totally different to them. R01 and R03 will conduct

(M01-M02-M03) and (M05-M06-M07), which are the closest missions to the R01 and R03, respectively. R02 will conduct M04 and M08, which were the last missions for R01 and R03, so the total mission completion time is minimized.

Panel (b) in Figures 1–3 shows a timetable of each algorithm to easily check the mission completion time of each robot. The blue boxes mean the duration times of missions. The 'R', 'M', and the number after them have the same meanings as in the panel (a). In the panel (b) of Figures 1 and 2, R01 will complete the last mission, M07, at 25.37 s. So, the total mission completion times of CBBA and TACTAR are 25.37 s. In the panel (b) of Figure 3, R03 will complete the last mission, M07, at 21.08 s, which is about 4.3 s lower than the results of CBBA and TACTAR.

The calculation times of CBBA, TACTAR, and the proposed algorithm for the simple case are 0.0055 s, 0.0068 s, and 0.0155 s, respectively. The computational cost of the proposed algorithm is about 2~3 times larger than the others, but it is still fast enough to run it in real-time.

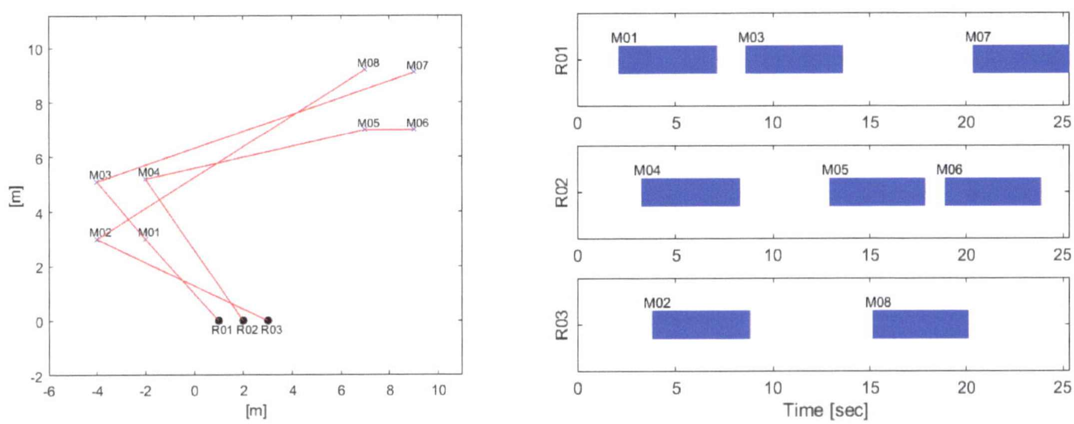

(a) (b)

Figure 1. The results for simple case using CBBA: (**a**) 2D situation map; (**b**) timetable.

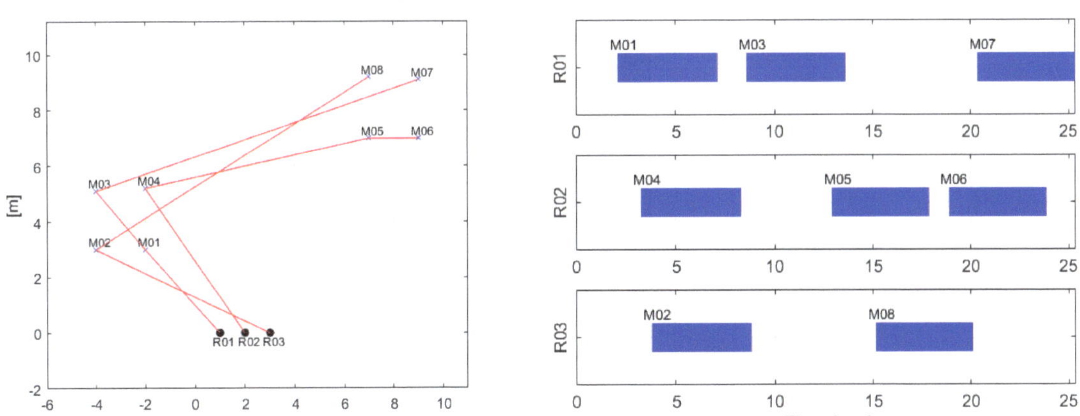

(a) (b)

Figure 2. The results for simple case using TACTAR: (**a**) 2D situation map; (**b**) timetable.

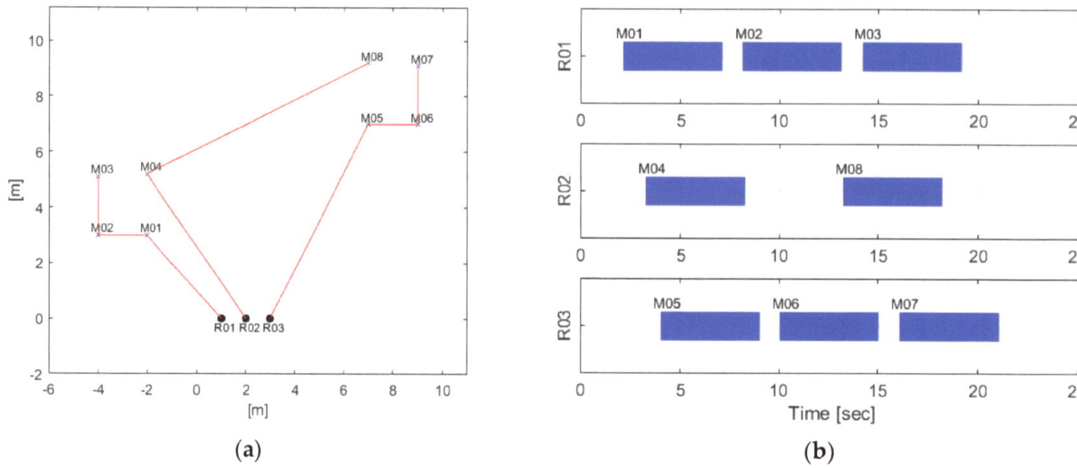

Figure 3. The results for simple case using proposed algorithm: (**a**) 2D situation map; (**b**) timetable.

5.2. The Medium Case

The results of the medium case are shown in Figures 4–6. Figure 4 depicts the results of CBBA, Figure 5 depicts the results of TACTAR, and Figure 6 depicts the results of the proposed algorithm. Unlike the simple case, the mission plan result of CBBA is different to the result of TACTAR, as seen in Figures 4 and 5. R01 will conduct M13 and go to M08 in Figure 4a, while it will conduct M13 and go to M02 in Figure 5a. In Figure 6a, R01 will initially conduct the same mission as the result of CBBA, M13-M08-M10. However, R01 will conduct M01 after finishing M10 in Figure 6a, while it will conduct M26 in Figure 4a. The mission plans of R02~04 calculated by the algorithms are very different.

Panel (b) in Figures 4–6 shows a timetable of each algorithm for the medium case. In the panel (b) of Figure 4, R02 will complete the last mission, M24, at 67.30 s. In the panel (b) of Figure 5, R01 will complete the last mission, M22, at 47.07 s, which is about 20.3 s lower than the results of CBBA. In the panel (b) of Figure 6, R01 will complete the last mission, M04, at 44.59 s, which is about 22.8 and 2.5 s lower than the results of CBBA and TACTAR, respectively. The main reason is the bundle adjustment step in the proposed algorithm. The number of missions each robot should conduct in Figure 4b are 11, 11, 5, and 3. However, in the number of missions each robot should conduct in Figure 6b are 8, 6, 7, and 9. It means that the missions of the robots that need to conduct more missions are distributed to the robots that have fewer missions by the bundle adjustment step.

The calculation times of CBBA, TACTAR, and the proposed algorithm for the simple case are 0.0436 s, 0.0199 s, and 0.0342 s, respectively. Unlike the results for the simple case, the proposed algorithm has a lower computational cost than CBBA for the medium case. The computational cost is slightly increased, but it is still fast to run it in real-time.

5.3. The Huge Case

The results of the huge case are shown in Figures 7–9. Figure 7 depicts the results of CBBA, Figure 8 depicts the results of TACTAR, and Figure 9 depicts the results of the proposed algorithm. Like the medium case, the mission plan results of CBBA, TACTAC and the proposed algorithm are totally different from each other. R01 will conduct M46, M27, M37, M12, M47, M20, M29, M01, M05, and M39 sequentially in Figure 7a. However, it will conduct M07-M13-M09-M35-M19-M45-M38-M46 and M06-M12-M28-M05-M21-M20-M01-M26-M29-M27 in Figures 8a and 9a, respectively.

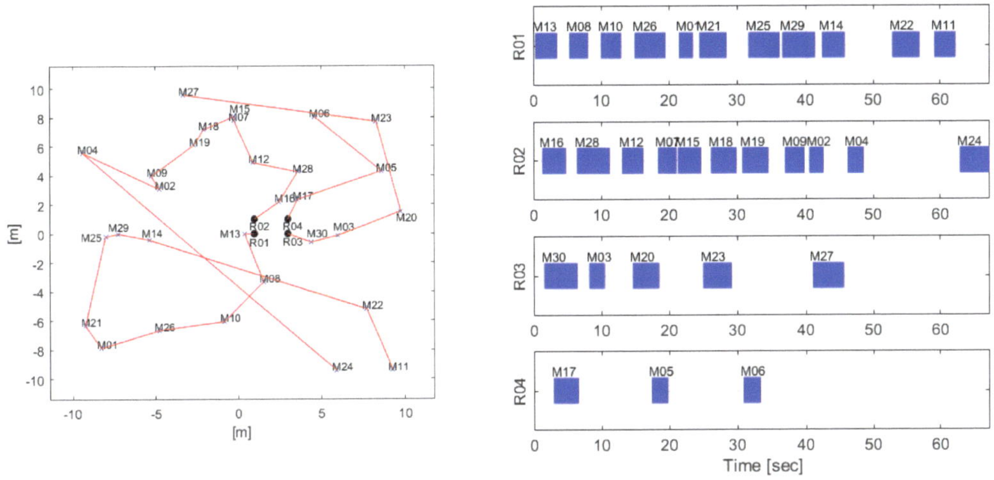

(a) (b)

Figure 4. The results for medium case using CBBA: (**a**) 2D situation map; (**b**) timetable.

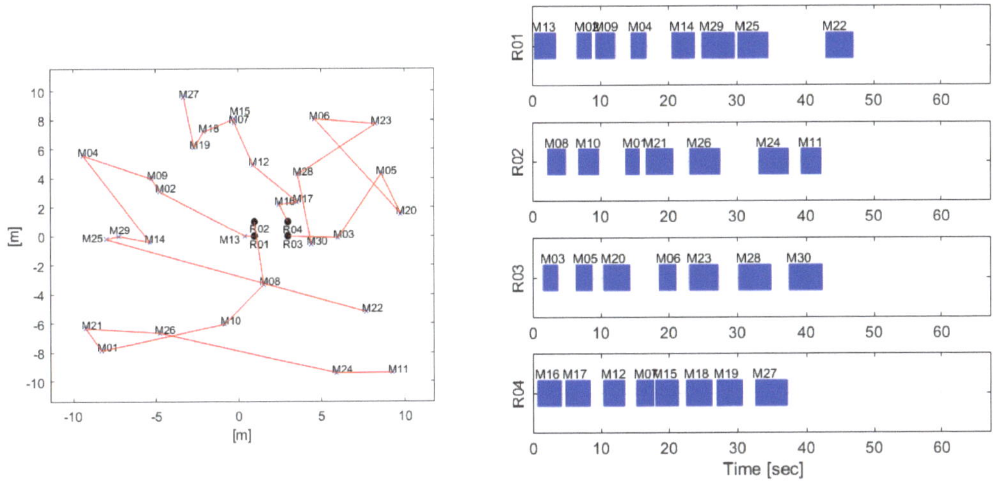

(a) (b)

Figure 5. The results for medium case using TACTAR: (**a**) 2D situation map; (**b**) timetable.

Panel (b) in Figures 7–9 shows a timetable of each algorithm for the huge case. In the panel (b) of Figure 7, R06 will complete the last mission, M22, at 63.83 s. In the panel (b) of Figure 8, R05 will complete the last mission, M48, at 59.27 s, which is about 4.6 s lower than the results of CBBA. In the panel (b) of Figure 9, R05 will complete the last mission, M49, at 56.76 s, which is about 7.1 and 2.5 s lower than the results of CBBA and TACTAR, respectively. The performance of the bundle adjustment step in the proposed algorithm can also be seen in the huge case. However, unlike the medium case, the number of missions each robot should conduct in Figures 7b and 9b are (10, 8, 8, 10, 7, 7) and (10, 10, 7, 7, 8, 8), respectively, which are similar to each other. It means that the bundle adjustment step in the proposed algorithm considers not only the number of missions but also the total mission time of each robot.

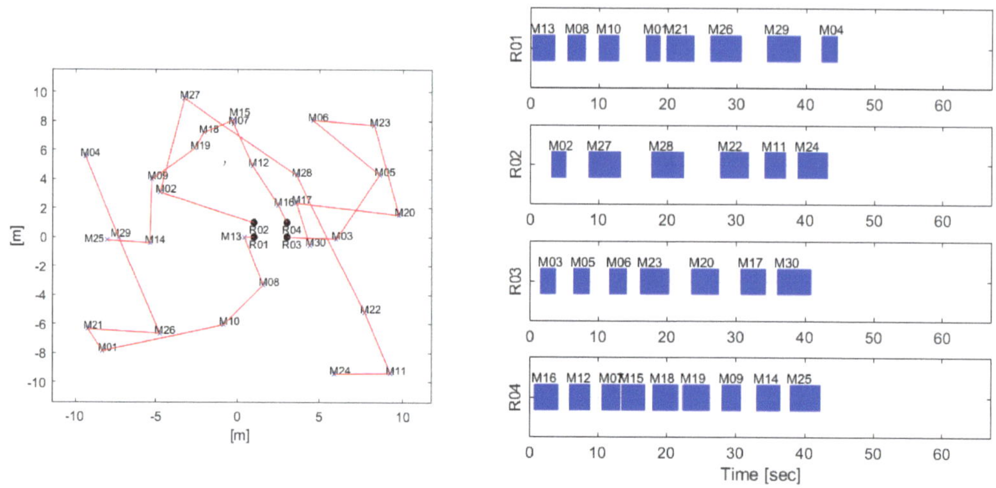

(**a**) (**b**)

Figure 6. The results for medium case using proposed algorithm: (**a**) 2D situation map; (**b**) timetable.

The calculation times of CBBA, TACTAR, and the proposed algorithm for the simple case are 0.1815 s, 0.0754 s, and 0.1636 s, respectively. Like the results for medium, the proposed algorithm has a lower computational cost than CBBA for the huge case. The computational cost is highly increased compared to the simple case, but it is still fast to run it in real-time.

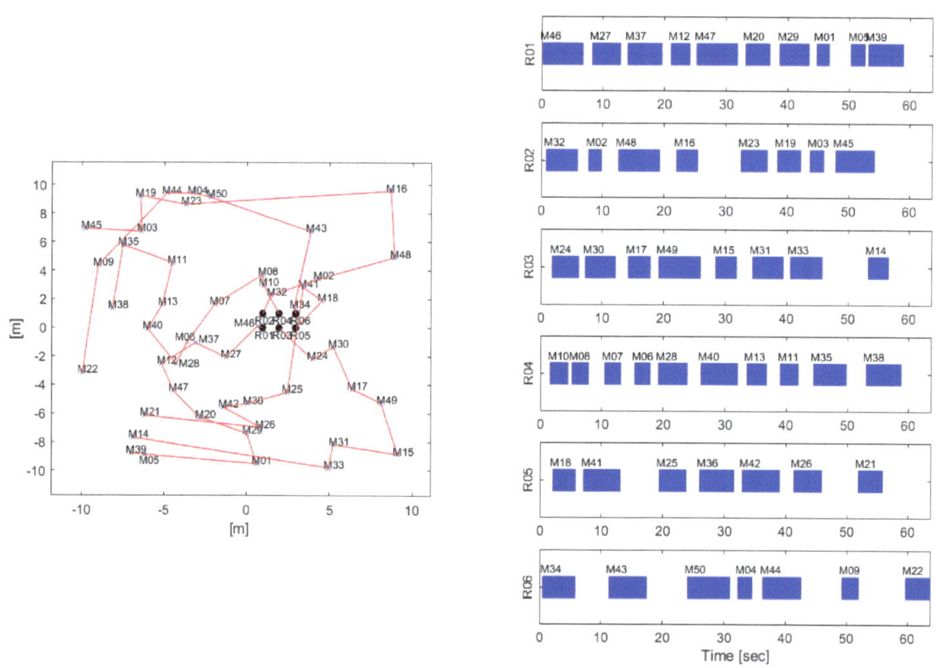

(**a**) (**b**)

Figure 7. The results for huge case using CBBA: (**a**) 2D situation map; (**b**) timetable.

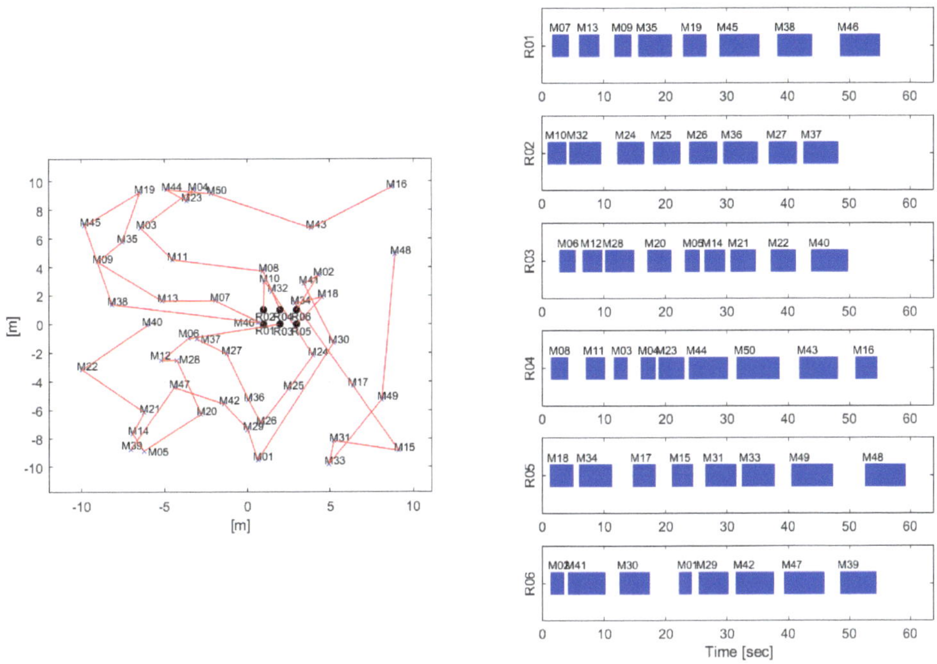

(a) (b)

Figure 8. The results for huge case using TACTAR: (**a**) 2D situation map; (**b**) timetable.

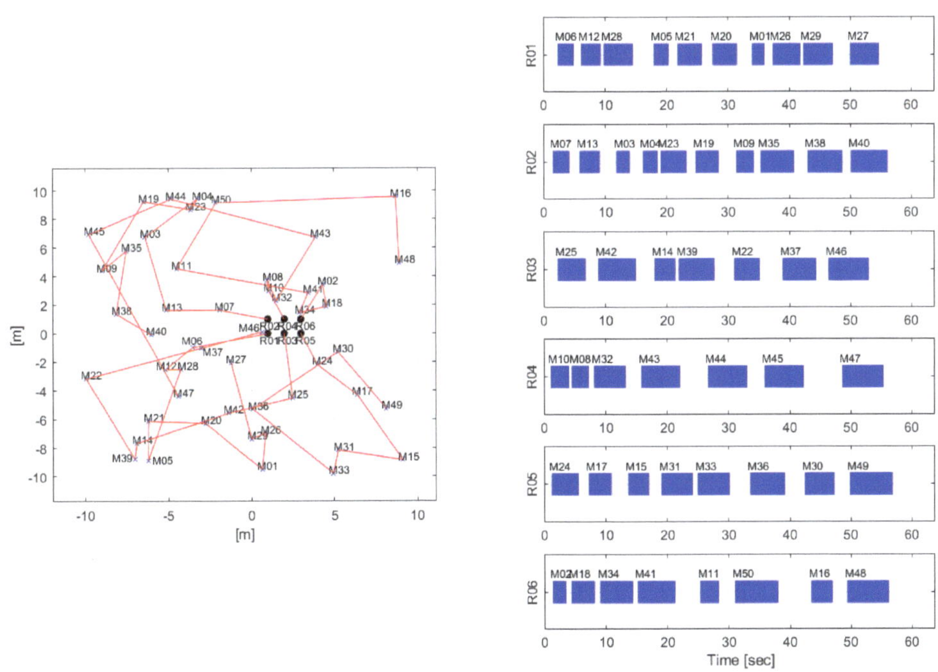

(a) (b)

Figure 9. The results for huge case using proposed algorithm: (**a**) 2D situation map; (**b**) timetable.

6. Conclusions

In this paper, we suggested a centralized mission planning algorithm of single/multiple robots for minimizing total mission completion time. First, we introduced the algorithm for single-robot-multi-mission (SRMM) problems, minimizing total mission completion time based on the bundle construction step in CBBA. In SRMM problems, which robot will take the mission was already decided, so it focused on how to decide the sequence of missions that the robot has to work on. Then, the algorithm was expanded for multi-robot-multi mission (MRMM) problems. Unlike SRMM problems, it had to decide which robot will take the mission. However, it is hard to decide at once because the whole situation needs to be considered to minimize the total mission completion time. So, the initial bundles were constructed based on the expected finish time of each mission for robots and bundles and were adjusted iteratively. Finally, we demonstrated simulation results comparing them with CBBA and TACTAR, the existing mission planning algorithm, and the proposed algorithm. To verify the performance of these algorithms for various situations, we selected 3 cases: the simple case, which was composed of 3 robots and 8 missions, the medium case, which was composed of 4 robots and 30 missions, and the huge case, which was composed of 6 robots and 50 missions. The results showed that the proposed algorithm always outputs the mission plan that has the minimum total mission completion time. In addition, the computational cost of the proposed algorithm is lower than CBBA, and larger than TACTAR. However, the calculation time of our algorithm is fast enough to use in real-time applications.

One of the limitations in our algorithm is that it cannot handle constraints like TACTAR. It is important because most users want to set constraints for various applications. The constraints are separated into two types: the spatial ones and the temporal ones. The spatial constraints are related to mission allocation. A typical spatial constraint is MUTual Exclusive (MUTEX). The temporal constraints are related to the mission alignment, e.g., mission time. The typical spatial constraints are 'Before', 'After', 'Simultaneous', etc. The modification of our algorithm to handle those constraints is the focus of our next work.

Author Contributions: Conceptualization, N.E.H.; methodology, N.E.H.; software, N.E.H.; validation, N.E.H.; formal analysis, N.E.H.; investigation, N.E.H.; resources, N.E.H.; data curation, N.E.H.; writing—original draft preparation, N.E.H.; writing—review and editing, N.E.H.; visualization, N.E.H.; supervision, H.J.K. and J.G.K.; project administration, H.J.K. and J.G.K.; funding acquisition, H.J.K. and J.G.K. All authors have read and agreed to the published version of the manuscript.

Funding: This work was supported by the Korea Research Institute for Defense Technology planning and advancement (KRIT) grant funded by the Korea government (DAPA (Defense Acquisition Program Administration)) (No. KRIT-CT-21-009, Development of Realtime Automatic Mission Execution and Correction Technology based on Battlefield Information, 2022).

Data Availability Statement: The presented data in this paper can be provided if the request is reasonable.

Conflicts of Interest: The authors declare no conflict of interest.

References

1. Rubenstein, M.; Comejo, A.; Nagpal, R. Programmable self-assembly in a thousand-robot swarm. *Science* **2014**, *345*, 795–799. [CrossRef] [PubMed]
2. Nouryan, S.; Campo, A.; Dorigo, M. Path formation in a robot swarm. *Swarm Intell.* **2007**, *2*, 1–23. [CrossRef]
3. Penders, J.; Alboul, L.; Witkowski, U.; Naghsh, A.; Saez-Pons, J.; Herbrechtsmeier, S.; El-Habbai, M. A robot swarm assisting a human fire-fighter. *Adv. Robot.* **2012**, *25*, 93–117. [CrossRef]
4. Chen, W.; Liu, J.; Guo, H.; Kato, N. Toward robust and intelligent drones swarm: Challenges and future directions. *IEEE Netw.* **2020**, *34*, 278–283. [CrossRef]
5. Asaamoning, G.; Mendes, P.; Rosario, D.; Cerqueira, E. Drone swarms as networked control systems by integration of networking and computing. *Sensors* **2021**, *21*, 2642. [CrossRef] [PubMed]
6. Taylor, G.; Turpin, T. Army Aviation Manned-Unmanned Teaming (MUM-T): Past, Present, and Future. In Proceedings of the 18th International Symposium on Aviation Psychology, Dayton, OH, USA, 4–7 May 2015.

7. Das, A.N.; Doelling, K.; Lundberg, C.; Sevil, H.E.; Lewis, F. A Mixed reality based on hybrid swarm control architecture for manned-unmanned teaming (MUM-T). In Proceedings of the ASME 2017 International Mechanical Engineering Congress and Exposition, Tampa, FL, USA, 3–9 November 2017.
8. Frey, M.; Schulte, A. TacARA: Tactical Analysis and Reconnaissance Assistant to Support Pilots in MUM-T Scenarios. In Proceedings of the AIAA SCITECH 2022 Forum, San Diego, CA, USA, 3–7 January 2022.
9. Uhrmann, J.; Stenzke, R.; Schulte, A. Task-based guidance of multiple detached unmanned sensor platforms in military helicopter operations. In Proceedings of the Cognitive Systems with Interactive Sensors, Crawley, UK, 22 November 2010.
10. Goetz, J.; Kiesler, S.; Powers, A. Matching robot appearance and behavior to tasks to improve human-robot cooperation. In Proceedings of the 12th IEEE International Workshop on Robot and Human Interactive Communication, Millbrae, CA, USA, 2 November 2003.
11. Parker, L.E. Heterogeneous Multi-Robot Cooperation. Ph.D. Thesis, Massachusetts Institute of Technology, Cambridge, MA, USA, 1994.
12. Khan, A.; Rinner, B.; Cavallaro, A. Cooperative robots to observe moving targets: Review. *IEEE Trans. Cybern.* **2018**, *48*, 187–198. [CrossRef] [PubMed]
13. Xiang, C.; Zhou, Y.; Dai, H.; Qu, Y.; He, S.; Chen, C.; Yang, P. Reusing Delivery Drones for Urban Crowdsensing. *IEEE Trans. Mob. Comput.* **2021**, 1. [CrossRef]
14. Xiang, C.; Li, Y.; Zhou, Y.; He, S.; Qu, Y.; Li, Z.; Gong, L.; Chen, C. A Comparative Approach to Resurrecting the Market of MOD Vehicular Crowdsensing. In Proceedings of the IEEE INFOCOM 2022-IEEE Conference on Computer Communications, London, UK, 2–5 May 2022.
15. Billonnet, A.; Costa, M.; Sutter, A. An Efficient Algorithm for a Task Allocation Problem. *J. ACM* **1992**, *39*, 502–518. [CrossRef]
16. Moon, S.; Kim, H.J. Cooperation with Ground and Ariel Vehicles for Multiple Tasks: Decentralized Task Assignment and Graph Connectivity Control. *J. Inst. Contr. Robot Syst.* **2012**, *18*, 218–223. [CrossRef]
17. Lim, M.C.; Choi, H.L. Improving Computational Efficiency in Crowded Task Allocation Games with Coupled Constraints. *Appl. Sci.* **2019**, *9*, 2117. [CrossRef]
18. Oh, K.T.; Kim, W.D. Task Assignment Algorithm for Rendezvous of Multiple UAVs. In Proceedings of the Korean Society for Aeronautical and Space Sciences Fall Conference, Jeju, Republic of Korea, 14–16 November 2012.
19. Marcarthur, K.S.; Stranders, R.; Ramchum, S.D.; Jennings, N.R. A Distributed Anytime Algorithm for Dynamic Task Allocation in Multi-Agent Systems. In Proceedings of the AAAI Conference on Artificial Intelligence, San Francisco, CA, USA, 7–11 August 2011.
20. Jin, Y.; Minai, A.A.; Polycarpou, M.M. Cooperative Real-Time Search and Task Allocation in UAV Teams. In Proceedings of the 42nd IEEE International Conference on Decision and Control, Maui, HI, USA, 9–12 December 2003.
21. Thenepalle, J.K.; Singamsetty, P. An open close multiple traveling salesman problem with single depot. *Decis. Sci. Lett.* **2019**, *8*, 121–136. [CrossRef]
22. Ahmeda, Z.H.; Al-Dayelb, I. An Exact Algorithm for the Single-Depot Multiple Traveling Salesman Problem. *Int. J. Netw. Secur.* **2020**, *20*, 65–74.
23. Levchuk, G.M.; Levchuk, Y.N.; Luo, J.; Pattipatu, K.R.; Kleinman, D.L. Normative design of organizations. I. Mission planning. *IEEE Trans. Syst. Man Cybern. Part A Syst. Hum.* **2022**, *32*, 346–359. [CrossRef]
24. Qin, A.; Li, A.; Dong, C.; Dai, H.; Xu, Z. Completion time minimization for multi-UAV information collection via trajectory planning. *Sensors* **2019**, *19*, 4032. [CrossRef] [PubMed]
25. Choi, H.L.; Brunet, L.; How, J.P. Consensus-Based Decentralized Auctions for Robust Task Allocation. *IEEE Trans. Robot.* **2009**, *25*, 912–926. [CrossRef]
26. Whitten, A.K.; Choi, H.L.; Johnson, L.B.; How, J.P. Decentralized Task Allocation with Coupled Constraints in Complex Missions. In Proceedings of the 2011 American Control Conference, San Francisco, CA, USA, 29 June–1 July 2011.
27. Whitten, A.K. Decentralized Planning for Autonomous Agents Cooperating in Complex Missions. Master's Thesis, Massachusetts Institute of Technology, Cambridge, MA, USA, 2010.
28. Wei, H.; Lv, Q.; Duo, N.; Wang, G.S.; Liang, B. Consensus Algorithms Based Multi-Robot Formation Control under Noise and Time Delay Conditions. *Appl. Sci.* **2019**, *9*, 1004. [CrossRef]
29. Lagoudakis, M.G.; Berhault, M.; Koenig, S.; Keskinocak, P.; Kleywegt, A.J. Simple Auctions with Performance Guarantees for Multi-Robot Task Allocation. In Proceedings of the 2004 IEEE/RSJ International Conference on Intelligent Robots and Systems, Sendai, Japan, 28 September–2 October 2004.
30. Lee, C.H.; Moon, G.H.; Yoo, D.W.; Tahk, M.J.; Lee, I.S. Distributed Task Assignment Algorithm for SEAD Missions of Heterogeneous UAVs Based on CBBA Algorithm. *J. Korea Soc. Aeronaut. Space Sci.* **2011**, *40*, 988–996.
31. Kim, K.S.; Kim, H.Y.; Choi, H.L. A bid-based grouping method for communication-efficient decentralized multi-UAV task allocation. *Int. J. Aeronaut. Space Sci.* **2019**, *21*, 290–302. [CrossRef]
32. Oh, G.; Kim, Y.; Ahn, J.; Choi, H.L. Task allocation of multiple UAVs for cooperative parcel delivery. In Proceedings of the 4th CEAS Specialist Conference on Guidance, Navigation and Control, Warsaw, Poland, 25–27 April 2017.

33. Hwang, N.E.; Kim, H.J.; Kim, J.G. Centralized Task Allocation and Alignment based on Constraint Table and Alignment Rules. *Appl. Sci.* **2022**, *12*, 6780. [CrossRef]
34. Wang, H.; Wang, J.; Ding, G.; Chen, J.; Gao, F.; Han, Z. Completion Time Minimization with Path Planning for Fixed-Wing UAV Communications. *IEEE Trans. Wirel. Commun.* **2019**, *18*, 3485–3499. [CrossRef]

drones

MDPI

Article

Structural Component Phenotypic Traits from Individual Maize Skeletonization by UAS-Based Structure-from-Motion Photogrammetry

Monica Herrero-Huerta [1,2,*], Diego Gonzalez-Aguilera [1] and Yang Yang [2]

1 Department of Cartographic and Land Engineering, Higher Polytechnic School of Avila, Universidad de Salamanca, Hornos Caleros 50, 05003 Avila, Spain
2 Institute for Plant Sciences, College of Agriculture, Purdue University, West Lafayette, IN 47906, USA
* Correspondence: monicaherrero@usal.es

Abstract: The bottleneck in plant breeding programs is to have cost-effective high-throughput phenotyping methodologies to efficiently describe the new lines and hybrids developed. In this paper, we propose a fully automatic approach to overcome not only the individual maize extraction but also the trait quantification challenge of structural components from unmanned aerial system (UAS) imagery. The experimental setup was carried out at the Indiana Corn and Soybean Innovation Center at the Agronomy Center for Research and Education (ACRE) in West Lafayette (IN, USA). On 27 July and 3 August 2021, two flights were performed over maize trials using a custom-designed UAS platform with a Sony Alpha ILCE-7R photogrammetric sensor onboard. RGB images were processed using a standard photogrammetric pipeline based on structure from motion (SfM) to obtain a final scaled 3D point cloud of the study field. Individual plants were extracted by, first, semantically segmenting the point cloud into ground and maize using 3D deep learning. Secondly, we employed a connected component algorithm to the maize end-members. Finally, once individual plants were accurately extracted, we robustly applied a Laplacian-based contraction skeleton algorithm to compute several structural component traits from each plant. The results from phenotypic traits such as height and number of leaves show a determination coefficient (R^2) with on-field and digital measurements, respectively, better than 90%. Our test trial reveals the viability of extracting several phenotypic traits of individual maize using a skeletonization approach on the basis of a UAS imagery-based point cloud. As a limitation of the methodology proposed, we highlight that the lack of plant occlusions in the UAS images obtains a more complete point cloud of the plant, giving more accuracy in the extracted traits.

Keywords: phenotyping; unmanned aerial vehicle (UAV); photogrammetry; skeleton; deep learning

Citation: Herrero-Huerta, M.; Gonzalez-Aguilera, D.; Yang, Y. Structural Component Phenotypic Traits from Individual Maize Skeletonization by UAS-Based Structure-from-Motion Photogrammetry. *Drones* **2023**, *7*, 108. https://doi.org/10.3390/drones7020108

Academic Editor: Görres Grenzdörffer

Received: 22 December 2022
Revised: 27 January 2023
Accepted: 3 February 2023
Published: 4 February 2023

1. Introduction

Nowadays, climate change and environmental degradation are increasing the risk of fiber, fuel and food insecurity; cost-effective phenotyping methods are needed to meet this challenge. Traits in plants serve as features that are able to highlight the associations between genetic or physiological characteristics [1] and are imperative to plant breeding programs, biomass and yield estimations [2,3] and growth simulations [4]. Recently, phenotypic data were manually measured in the field, which is time-consuming, labor intensive and error-prone, not to mention destructive. The demand for precise agriculture and the development of close-range remote sensing technology makes image-based methods the solution to the phenotypic trait extraction challenge regarding plant physiology and structure [2], yield-related traits [3], canopy over [5] or root architecture [6,7]. Another one of the current challenges for plant phenotyping is to, accurately and with high-throughput, extract the structural components, usually composed of the root, stem, leaf, flower, fruit and seed [8]. Structural component traits are directly connected to functional phenomics, an emerging discipline leading to an increased understanding of plant functioning by leveraging high-throughput phenotyping and data analytics [9].

Evaluating the information encoded in the shape of a plant is vital to understanding the function of plant organs [10]. A powerful shape descriptor of plant networks is the skeleton, easily computed from imaging data [11]. The skeleton opens a wide range of possibilities for quantitative phenotyping at a plant level, including describing hierarchies and branching plant networks. From the literature, there are several methods to extract the curve-skeleton from a solid, usually classified into two key types: volumetric and geometric [12]. This classification system relies on the solid's representation, depending on whether one is using an interior representation or a surface representation. Regarding volumetric approaches, they normally use a volumetric discrete representation, either a regularly partitioned voxelized representation or a discretized function demarcated in the 3D space. The potential loss of details within the solid and numerical instability due to inappropriate discretization resolution are the general disadvantages of this method [13]. On the other hand, geometric approaches directly work on the meshes or point sets. The most common used geometric methods are the Voronoi diagram [14] and medial axis [15]. Currently, Reeb graph-based methods have increased in popularity [16]. In addition there are another group of approaches based on 3D modeling: voxel approaches and parametric surface methods. It is worth mentioning that voxel-based approaches are limited in modelling irregular surfaces.

Recently, unmanned aerial systems (UAS) have positioned themselves as a basic tool for high-throughput plant phenotyping in precision agriculture [3]. The latest advances in technology and miniaturization of their components provides additional opportunities for UAS data collection platforms. As high-resolution imaging sensors, light detection and ranging (LiDAR) has the capacity to acquire 3D measurements of plants, even in the absence of light [17,18]. This technology relies on the reflection of laser beams from the surfaces [19,20]. Currently, there are several studies using the terrestrial LiDAR to perform organ stratification (even leaf labeling) and its angles from field maize [21–24]. However, the payload reducing and cost increasing nature of LiDAR onboard UAS are the main disadvantages. On the other hand, passive imaging technologies, such as visible cameras, are lighter and less expensive. In addition, SfM (structure from motion), defined as a photogrammetric range imaging technique, offers the opportunity to acquire point clouds on the basis of images taken from various viewpoints [3]. Point clouds as three dimensional, and massive data can be used for extracting complex structural information [25]. In addition, deep learning consists of methods which can deal with object detection, classification and segmentation tasks [8], based on voxels, octree, multi-surface, multi-view and directly on point clouds. The challenge of its high cost of computing memory means these networks are mainly used in small data applications. There are some approaches using UAS imagery-based point clouds to compute basic traits such as plant height or the leaf area index in maize [26–28].

Still, methodologies to fully exploit the potential of UAS-collected data in agriculture are urgently required. In this paper, we present a novel pipeline to automatically and accurately characterize several structural component phenotypic traits in maize trails. To the best of our knowledge, the skeletonization of maize from UAS imagery-based point clouds has not been performed before. RGB images using UAS is the input of the proposed workflow to acquire a georeferenced dense point cloud of the entire study field using SfM. Topological and deep learning-based algorithms were combined to extract individual plants from the point cloud. Once a surface reconstruction process from each individual plant was achieved, the skeleton extraction algorithm was applied. Finally, we were easily able to compute structural component traits highly demanded in phenotypic tasks, comparing them with on-field and digital measurements. The paper is structured as follows: after this brief introduction, the materials, including experimental setup, data acquisition and proposed methodology are described in detail. Next, the experimental results are described, validated and discussed. Finally, the more important conclusions reached with this study are addressed, along with future perspectives.

2. Materials and Methods

2.1. Experimental Setup and Data Acquisition

The research trial was located at the Indiana Corn and Soybean Innovation Center Manager at the Agronomy Center for Research and Education (ACRE) in West Lafayette (IN, USA) at Purdue University. Figure 1 illustrates the visualization of the workflow to follow.

DATA ACQUISITION

1. *Ground Control Points*
2. *RGB UAS flights*
3. *Ground measures*

METHODOLOGY

1. *Imagery-based Point Cloud*
2. *Filter*
3. *Individualization*
4. *Curve-skeleton*
5. *Phenotypic traits*

ACCURACY ASSESSMENT

✓ *Error metrics*

Figure 1. Proposed workflow.

The dates of planting (DOP) were June 6 and 17, 2021. The trail was designed with an arrangement of 18 ranges and 4 rows, planting at two different densities as Figure 2 shows: approximately 14 (DOP June 17) and 18 seeds*row-1 (DOP June 6); 3 ranges and 4 rows the first density and 15 ranges and 4 rows the second one. Four GCPs (ground control points) were placed on the ground and measured using a GNSS device for georeferencing. The material of these accuracy markers was highly reflected to be easily detected in the UAS imagery dataset. The flights were carried out on 27 July (flight 1) and 3 August (flight 2), 2021 around noon solar time on sunny and no-cloud days. A Sony Alpha ILCE-7R RGB camera with a Sony 35 mm lens was the photogrammetric sensor onboard a DJI Matrice 600 Pro (M600P) platform (Gryfn, West Lafayette, IN, USA). This platform is a rotocopter UAS with onboard GPS, IMU and magnetometer and a maximum payload of 6 kg. The photogrammetric flight configuration was set up with an along- and across-track overlap of 88% and a flight altitude of 22 m. A total of 530 and 518 images from flights 1 and 2, respectively, were captured with a dimension of 7952 × 5304 pixels, given the characteristic of the photogrammetric sensor as pixel size of 4.52 μm, focal length of 35 mm and size of 35.9 × 23.9 mm^2. The sensor configuration was ISO (the International Organization of Standardization) 200, an aperture with a F-stop of f/5.6 and a fixed exposure time of 1/1250 s.

Figure 2. Point cloud sample processed with the imagery dataset from two flights performed (27 July and 3 August 2021) at different densities: 14 seeds*row-1 (DOP June 17) (**a**) and 18 seeds*row-1 (DOP June 6) (**b**).

As for ground measurements, stem count and plant height for the full experiment were taken at the same date as the image acquisition from UAS. Notice that before the second flight, 30% of the plants were pulled over in order to avoid occlusions from the aerial images.

2.2. Imagery-Based Point Cloud

Pix4Dmapper software package (Pix4D SA, Lausanne, Switzerland) was used to process aerial images, which includes camera calibration, image orientation and dense point cloud extraction. In this way, the point cloud of the study field was obtained and accurately georeferenced to the earth reference system World Geodetic System 84. However point clouds automatically generated by SfM techniques probably englobe outlier points. To remove these points, a statistical outlier removal-based filter was applied. First, it computes the mean distance of each point to its neighbors (considering k nearest neighbors for each—k is the first parameter). Then, it rejects the points that are farther than the mean distance plus a number of times the standard deviation (second parameter). In other words, the process computes a threshold based on the Gaussian distribution of all pairwise distances in the neighborhood defined by a specific number of points (mean distance) and a number (k) to multiply the standard deviation (std. deviation), as Equation (1) shows. Points within a distance larger than the threshold are classified as outliers and removed from the point cloud [17].

$$threshold = \mu + k * \sigma \quad (1)$$

where μ is the mean distance, σ is the standard deviation, and k is a constant.

Figure 3 illustrates the low-cost photogrammetry result to 3D reconstruct a random plant from UAS imagery and the outlier removal process defined before.

(a) (b) (c)

Figure 3. Photogrammetric 3D reconstruction of a random plant: 2D manual picture from the ground (**a**) scaled 3D point cloud from UAS imagery (**b**), clean point cloud (outlier removal) (**c**).

2.3. Individual Maize Extraction

A 3D deep learning unified architecture named PointNet [29] was employed to automatically perform a semantic segmentation to extract the plants from the point cloud As the main advantage, PointNet directly runs on point clouds; that means the permutation invariance of points is not altered. Moreover, PointNet is highly robust, with little perturbation of the input points and in dealing with outliers and missing data. PointNet architecture works as follows: each point is represented by six values, its three coordinates (x, y, z) and its colors (R, G, B). The final fully connected layers of the network aggregate these optimal values into the global descriptor for the extraction. It is easy to independently apply rigid or affine transformations to each point due to the input format. Therefore, a data-dependent spatial transformer network was added, which attempted to standardize the data with the intention to further improve the results. In addition, we reduced over-fitting using a data augmentation procedure that works by creating a new dataset using label-preserving transformations [30]. The first stage in the data augmentation process generates n translations in the training dataset defined by manually extracting individual maize from the point cloud. The second stage proceeds to modify the RGB intensities. For this purpose, principal component analysis was computed on the RGB value set for each training point cloud. We added multiples of the found principal components m times, with magnitudes proportional to the corresponding eigenvalues times a random variable drawn from a Gaussian with μ (mean) of zero and σ (standard deviation) of 0.1. In that way, the training set was increased by a factor of n*m. In terms of geometry and the intensity and color of the illumination, the corn plant characteristics were mainly invariant.

Once the semantic segmentation of the plants was undertaken, we extracted individual mazes by connected component labeling and setting up an octree level to define the minimum gap between two components; this means the corresponding cell size of the 3D grid for extraction [31]. This processing consists of an octree decomposition, followed by a split-and-merge procedure. First, a decomposition of a point cloud into an octree based on point density is performed. Then, the points are split within each voxel into spatially connected components. Finally, a recursive merging of components across voxels is carried out, based on a connectivity criterion until the root node is reached. As a visual example, Figure 4 displays the outputs from the steps of our pipeline to extract individual maize from the point cloud within a random plot.

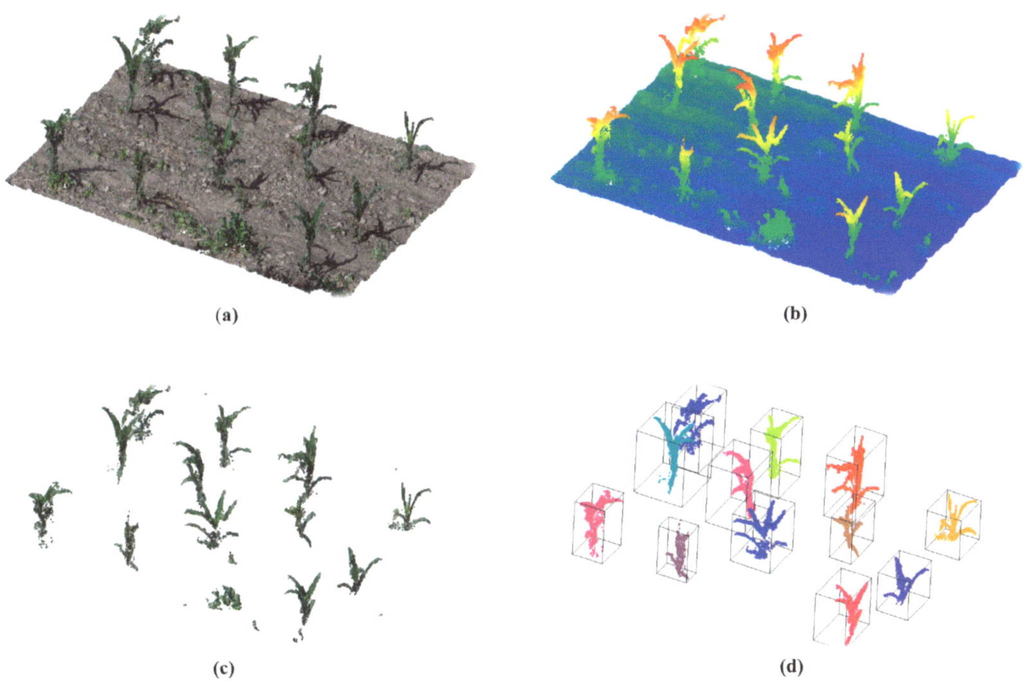

Figure 4. Partial and global outputs of the plant extraction pipeline within a random plot: RGB-based point cloud (**a**), height-based point cloud (**b**), vegetation-based semantic classification (**c**) and individual maize extraction and labeling (**d**).

2.4. Curve-Skeleton Extraction

Once the individual plants were extracted, the skeletonization process was applied to each point cloud. The skeleton structure is basically able to abstract the model volume and topological characteristics. In this case, a Laplacian-based contraction algorithm was used [13], which worked directly on the point cloud and operated on every point [32]. Advantageously, no resampled volumetric representation was required. Moreover, it was pose-insensitive and invariant to global rotation. We summarize the stages of the skeletonization process as follows: first, the mesh is contracted into a zero-volume skeletal shape, iteratively moving all the vertices along their curvature normal directions. After each iteration, all the collapsed faces from the degenerated mesh are removed until no triangles exist. During the contraction, the mesh connectivity is not altered, retaining all the key features using sufficient skeletal nodes. Lastly, the skeleton's geometric embedding is refined, moving each node to the center of mass of its local mesh region [32,33]. After these steps, we get the curve-skeleton of each individual maize.

2.5. Phenotypic Traits of Structural Components

The curve-skeleton is a structure that extracts the volume and topological characteristics of each individual plant represented by a point cloud and 3D line. In that way, we can easily define individual plant traits and different structural components of the plant: stem and leaves. As an individual plant phenotypic trait, we extracted the total height (difference between $z_{maximum}$ and $z_{minimum}$), crown diameter (difference between $x_{maximum}$ and $x_{minimum}$) and plant azimuth. The azimuth angle is defined as the angle between the maximum eigenvector of the plant skeleton and the north direction on the vertical projection plane. The origin of coordinate axes was selected as the leftmost point of the plant skeleton, with a value between 0 and 180°. The stem was defined as the most vertical line. The leaves originate from stem bifurcations and have a dead-end as a topological rule. In addition, leaves must have a minimum length to be considered a proper leaf. The stem lodging was calculated by computing the orientation between medium points from the beginning and end stretch (defined by a minimum distance) of the stem skeleton; from each leaf, we mathematically computed the length based on the length of the skeleton defined as leaf and the azimuth. The leaf azimuth is defined as the angle between the maximum eigenvector of a leaf skeleton and the north direction on the vertical projection plane. The origin of the coordinate axes was selected as the connection node between the proper leaf and the stem. The value of leaf azimuth is between 0 and 360° [34]. Figure 5 shows how the traits were extracted from the skeleton.

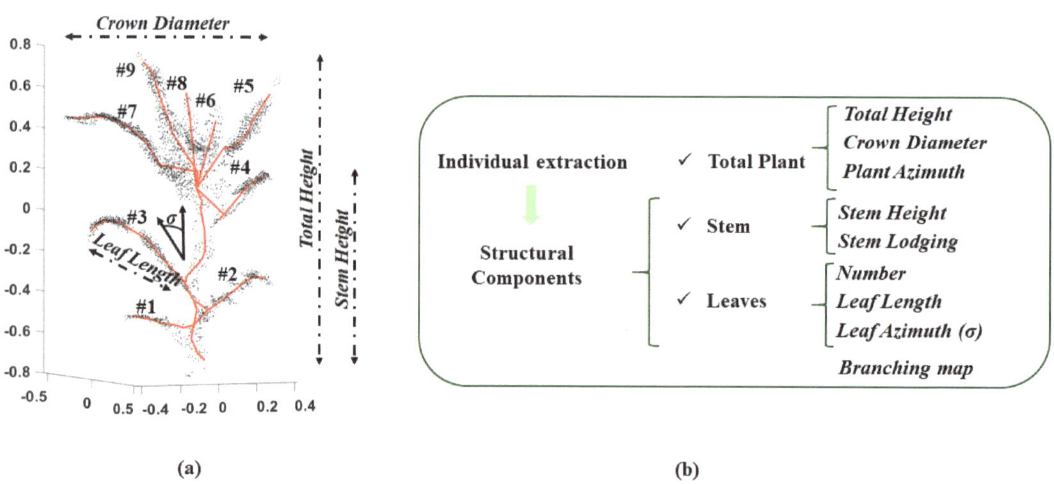

Figure 5. Skeleton-based structural component phenotypic traits: visual definition of the traits (**a**) and all the computed traits listed by type of the structural component (**b**).

Furthermore, this skeletal structure drives the registration process in temporal series. The registration process is critical to being able to automatically evaluate the growth of each individual plant. To register a temporal series, principal component analysis (PCA) was performed [35]. In general, the principal components are eigenvectors of the data's covariance matrix. More specifically, this statistical analysis uses the first and second moments of the curve-skeleton, resulting in three orthogonal vectors grouped on its center of gravity. The PCA summarizes the distribution of the lines along the three dimensions and models the principal directions and magnitudes of the curve-skeleton distribution around the center of gravity. Thereby, the registration of the temporal series was carried out by overlapping the principal component axes. After the registration, we can robustly monitor the growth as orientation and length variation.

2.6. Accuracy Assessment

The correlation between the plant height, stem count and number of leaves estimated by the skeleton and the on-field measurements or digital leaf counts was verified to evaluate the accuracy of the proposed methodology. Moreover, the rest of the skeleton algorithm-derived phenotypic traits (length and angles) were compared with manual and digital measurements from the point cloud of each individual plant. The leaf azimuth was manually measured by choosing the best suitable view direction which had the largest inclination. The determination coefficient (R^2), root mean square error (RMSE) and normalized root means square error (nRMSE) were calculated. The R^2 value was used to evaluate the coincidence between the computed and the measured value. The RMSE was used to measure the deviation between both values. The nRMSE represents the degree of difference between these both values (nRMSE < 10% indicates no difference, 10% ≤ nRMSE < 20% denotes a small difference, 20% ≤ nRMSE < 30% is moderate, and nRMSE ≤ 30% represents a large difference) [36]. Among them, a larger R^2 value indicates better data fit, and smaller RMSE and nRMSE values indicate higher estimation accuracy [37]. The calculation formulas of R^2, RMSE and nRMSE are shown in the following Formulas (2)–(4):

$$R^2 = 1 - \frac{\sum_{i=1}^{n}\left(x_{comp}^{i} - \underline{x_{act}}\right)^2}{\sum_{i=1}^{n}\left(x_{act}^{i} - \underline{x_{act}}\right)^2} \tag{2}$$

$$RMSE = \sqrt{\frac{\sum_{i=1}^{n}\left(x_{comp}^{i} - x_{act}^{i}\right)^2}{n}} \tag{3}$$

$$nRMSE = \frac{RMSE}{\underline{x_{act}}} \tag{4}$$

where x_{act}^{i} and $\underline{x_{act}}$ represent the actual value and the average of them, respectively (on-field measured in case of plant height and manually measured in the individual point cloud for the length and angles), x_{comp}^{i} represents the computed value of the trait, and n represents the number of samples (leaf, stem or individual plant).

Furthermore, the mean bias error (MBE), the absolute mean bias error (AMBE), the relative error (RE) and the absolute error (AE) were computed as follows (Equations (5)–(8)):

$$MBE = \frac{\sum_{i=1}^{n}\left(x_{comp}^{i} - x_{act}^{i}\right)}{n} \tag{5}$$

$$AMBE = \frac{\sum_{i=1}^{n}\left|\left(x_{comp}^{i} - x_{act}^{i}\right)\right|}{n} \tag{6}$$

$$RE = 100 * \frac{\sum_{i=1}^{n}\frac{\left(x_{comp}^{i} - x_{act}^{i}\right)}{x_{act}^{i}}}{n} \tag{7}$$

$$AE = 100 * \frac{\sum_{i=1}^{n} \frac{\left|\left(x_{pred}^{i} - x_{act}^{i}\right)\right|}{x_{act}^{i}}}{n} \quad (8)$$

In addition, the Nash and Sutcliffe index, η is also computed (Equation (9)) and used in modelling to characterize the error related to the spatial heterogeneity:

$$\eta = 1 - \frac{\sum_{i=1}^{n}\left(x_{pred}^{i} - x_{act}^{i}\right)^2}{\sum_{i=1}^{n}\left(x_{pred}^{i} - \underline{x_{act}}\right)^2} \quad (9)$$

Some of these evaluation metrics have been extensively used to analysis the power of regression models [38]. Smaller values of MBE, AMBE, RE and AE and larger values of η ($\infty < \eta \leq 1$) indicate better precision and accuracy of the prediction model.

3. Results

All the experimental results obtained below were run on a 3.6-GHz desktop computer with an Intel CORE I7 CPU and 32-GB RAM. We started the image processing using a Pix4D mapper software package (Pix4D SA, Lausanne, Switzerland) as a commercial solution for SfM. RGB imagery and ground control points were measured with terrestrial GPS works as inputs to finally reconstruct the study field into a scaled 3D point cloud. As an outcome, two point clouds from flights on different dates (27 July, first flight, and 3 August 2021 second flight) were computed and exactly georeferenced to EPSG 32616, WGS84 CRS. The point cloud was formed using a total of 35,983,365 points for the first flight and 34,851,008 points for the second flight, with a spatial resolution better than >24,800 points/m^2 in both cases. These values are valid within the limits of the study field (14 $\times$ 100 m^2). Due to the automated and massive character of the photogrammetric processing, an uncertainty quantity of outliers could be enclosed into these point clouds. A statistical analysis was carried out by supposing a Gaussian distribution of neighbors' distances to establish the threshold and determine outliers. The procedure has already been executed by [39]. We reached a spatial resolution better than 22,100 points/m^2 for both flights once the outlier detection approach was finished. A total of 257 plants for the first flight and 172 for the second flight were counted in the field, and all the plants were correctly and accurately extracted within the point clouds from both flights (30% of the plants were pulled over after the first flight and before the second in order to avoid occlusions from the aerial image). The average number of points per plant is 3968. Figure 6a represents the point clouds from the two dates and the corresponding individual plant extraction in top view. In Figure 6b, a zoomed window is shown with a 3D view. Figure 6c illustrates the growth of the individual plants within this zoomed area between the two dates precisely quantified in meters. We registered the point cloud of each individual plant from these two dates by computing PCA from the skeleton and overlapping the principal component axes. In this particular case, the maximum growth is 0.41 m. In addition, we can calculate the average maximum growth per plant, which is 0.22 m, and we can point out that the growth is always bigger at the upper part of the corn between these two dates.

Next, once the individual plants were extracted, the skeleton was computed from each point cloud using a Laplacian-based contraction algorithm, as Section 2.4 explains. Figure 7 graphically shows, in 3D, the individual point cloud in black overlapping the skeleton in red of the 16 plant cases: the maximum and minimum height, crown diameter, number of points and grown increment from the two flights (27 July 2011 and 3 August 2011). It is worth mentioning that the axes are in relative values. Below, Table 1 shows the plant data location and traits from the plant samples: the number of points, UTM coordinate of the point cloud center, and dimension of the bounding box from each individual point cloud, as well as traits computed by the individual point cloud skeletonization, such as the number of leaves, plant height, crow diameter, plant azimuth, lodging calculated as stem azimuth, stem height, mean leaf azimuth and mean leaf length.

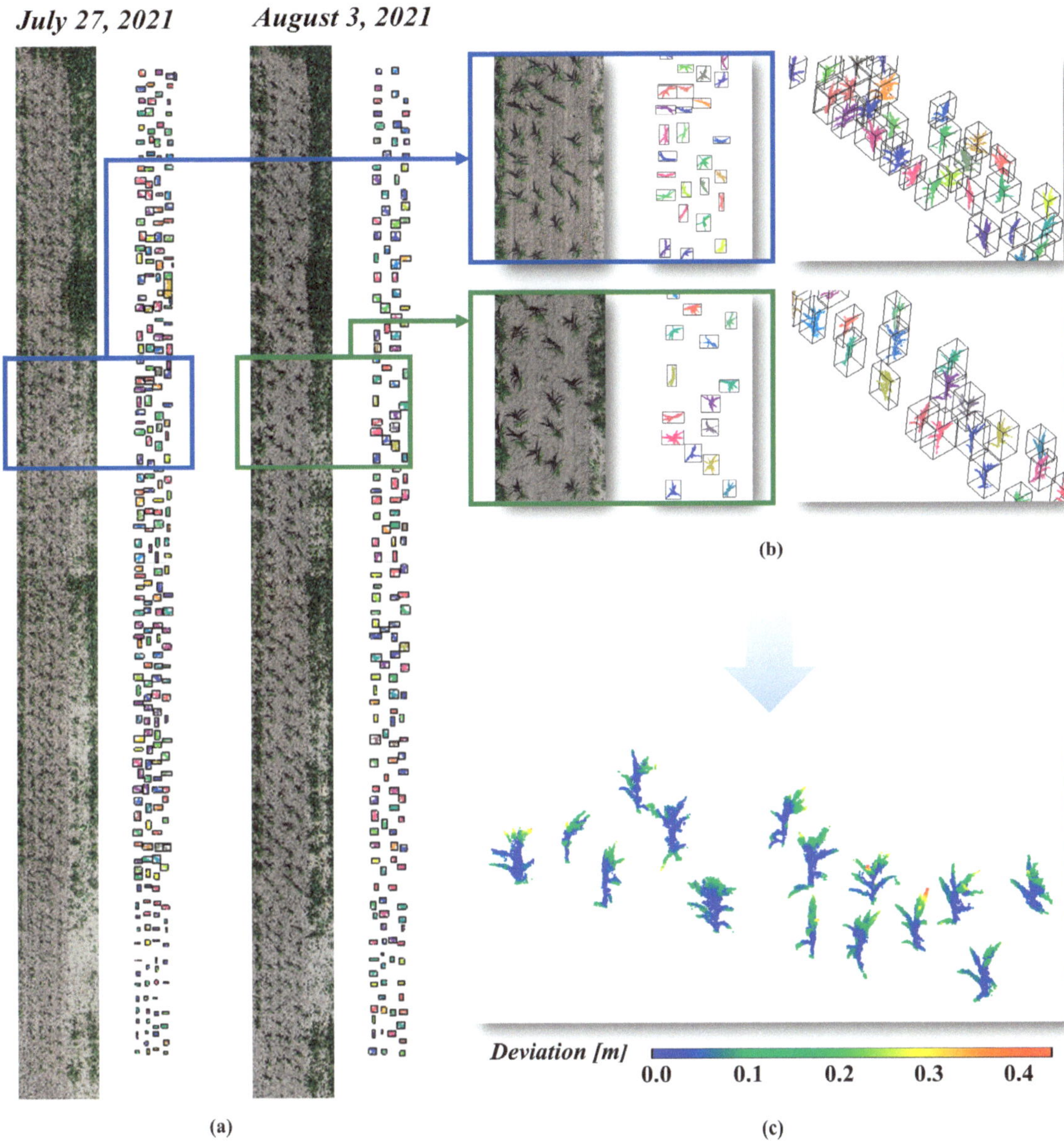

Figure 6. Point clouds from the two dates and individual plant extraction in different colors surrounded by a bounding box (**a**); zoomed window of a random area with a 3D view of the extracted plants (**b**); plant growth within this zoomed area from the two dates quantified in meters (**c**).

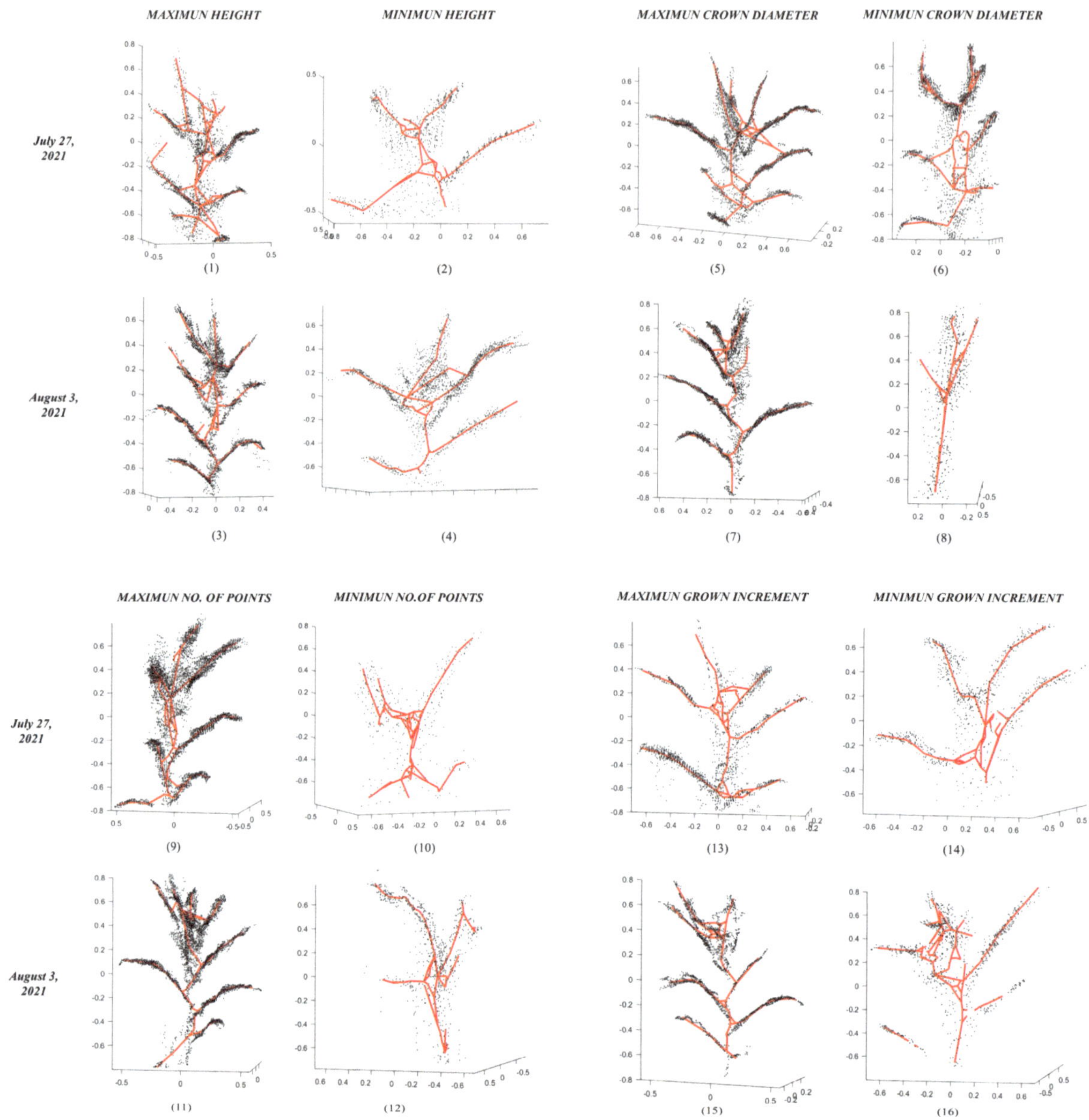

Figure 7. The 3D skeleton extraction in red overlapped the individual point cloud in black of 16 plant samples: maximum and minimum height, crown diameter, number of points and grown increment from the two flights (27 July 2011 and 3 August 2011).

Table 1. Plant data and traits from the 16 plant samples: number of points, UTM coordinate of the point cloud center and dimension of the bounding box from the point cloud; traits computed by the individual point cloud skeletonization, such as number of leaves, plant height, crow diameter, plant azimuth, lodging calculated as stem azimuth, stem height, mean leaf azimuth (LA) and mean leaf length (LL).

Plant ID	Point Cloud							Skeleton							
	Nº. Points	UTM Coord. Center			Bounding Box (m)			Trait Extraction							
		AX+500,000	AY+4,480,000	AZ+0	X	Y	Z	#Leaves	Height (m)	Crow Diam. (m)	Plant Azimuth (º)	Lodging (º)	Stem Height (m)	Mean LA (º)	Mean LL (m)
1	2429	224.21	202.40	181.93	0.66	0.30	1.04	9	0.97	0.69	339.4	351.0	0.58	344.4	0.21
2	956	225.36	156.35	180.90	0.38	0.26	0.25	4	0.22	0.37	338.2	341.1	0.16	340.7	0.13
3	3481	223.47	205.84	182.20	0.43	0.82	1.32	8	1.29	0.78	1.1	3.2	0.87	1.4	0.34
4	1729	223.11	156.49	181.19	0.58	0.34	0.32	5	0.29	0.52	27.8	19.4	0.24	24.3	0.20
5	3929	223.33	190.44	181.87	0.90	0.31	0.83	8	0.81	0.88	6.7	8.4	0.56	8.3	0.24
6	2259	225.88	221.57	181.49	0.16	0.23	0.81	7	0.79	0.24	358.1	5.5	0.61	1.7	0.12
7	3891	223.3	172.99	181.98	0.89	0.64	1.15	7	1.12	0.82	3.9	2.5	0.89	2.7	0.30
8	740	223.79	149.17	180.86	0.139	0.31	0.39	3	0.36	0.29	19.8	23.3	0.16	22.0	0.07
9	4641	224.96	199.35	181.90	0.59	0.59	0.87	7	0.85	0.57	18.8	8.1	0.60	15.6	0.21
10	369	223.85	149.24	180.68	0.44	0.40	0.26	5	0.23	0.47	9.4	12.4	0.12	10.8	0.06
11	5045	223.44	197.97	182.17	0.88	0.58	1.22	9	1.20	0.81	357.6	1.2	0.89	259.6	0.32
12	547	223.22	158.67	181.20	0.36	0.42	0.40	4	0.35	0.40	345.6	354.8	0.13	347.2	0.10
13	2115	225.82	212.08	181.68	0.86	0.45	1.00	6	0.97	0.79	17	4.0	0.49	4.9	0.23
14	947	223.87	153.23	180.80	0.40	0.43	0.45	4	0.39	0.41	2.9	6.2	0.16	5.8	0.15
15	1636	225.72	212.19	181.68	0.74	0.23	0.82	8	0.79	0.68	349.2	356.1	0.50	355.6	0.38
16	847	223.85	153.22	180.84	0.30	0.34	0.56	8	0.51	0.38	6.1	4.2	0.14	6.7	0.11

4. Discussion

In this section, the results are discussed and validated. The stem counts measured in the field with GPS were exactly the same as the digitally taken stem counts for both flights 257 plants for the first flight and 172 for the second one. The individual height of each plant was also measured in the field using a tape with centimetric precision. Comparing this on field-measurement with the digital height computed from the point cloud of each extracted plant, an R2 of more than 0.99 was achieved. No outliers were detected in this regression, guaranteeing accurate and precise height results. From Table 1, it is remarkable that the plants with a greater number of leaves are the ones with the maximum plant height and a greater number of points. It seems reasonable because when the point density is higher the plant has more detail to distinguish the leaves, and higher plants have more chance to have leaves. On the other hand, the plants with less recognizable leaves coincide with the minimum crown diameter plants. Another bit of information we can extract is that the more vertical plants are the highest ones, while the more inclined plants (lodging) coincide with the minimum crown diameter one. The following table, Table 2, illustrates statistics values from the computed traits of all the individual plant point clouds from both flights (27 July and 3 August 2011): mean, standard deviation (Std), median, normalized median absolute deviation (NMAD) (Equation (10)) and square root of the biweight midvariance (BwMv) (Equation (11)). It is worth mentioning that for the computation of the table, the outliers were discarded according to the studentized residuals for a significance level of 0.05 with two-tail distribution.

$$NMAD = 1.4826 \cdot MAD \tag{10}$$

$$BwMv = \frac{n\sum_{i=1}^{n} a_i (x_i - m)^2 (1 - U_i^2)^4}{\left(\sum_{i=1}^{n} a_i (1 - U_i^2)(1 - 5U_i^2)\right)^2} \tag{11}$$

$$a_i = \{1, if|U_i| < 1\; 0, if|U_i| \geq 1 \tag{12}$$

$$U = \frac{x_i - m}{9MAD} \tag{13}$$

being the median absolute deviation (MAD) and the median (m) of the absolute deviations from the data's median (mx).

Table 2. Statistics of the computed traits (mean, Std, median, NMAD and BwMv) and error metrics of all plants from both flights at 95% confidence interval (MBE, AMBE, RMSE, NMAD, RE, AE and η).

	#Leaf	Height (cm)	Crown Diam. (cm)	Azimuth (°)	Lodging (°)	H_{stem} (cm)	Mean LA (°)	Mean LL (cm)
Mean	5.98	70.16	54.66	1.18	4.56	42.76	−4.94	19.19
Std	1.40	34.84	21.43	15.61	8.48	28.12	26.82	9.61
Median	7	79.81	54.91	1.71	4.66	51.57	3.54	20.55
NMAD	2.43	44.84	28.45	14.45	13.2	35.80	23.13	11.44
BwMv	0.25	82.66	31.43	12.22	4.82	53.80	24.65	6.22
R^2 (%)	90.9	99.8	99.7	99.8	99.9	99.4	99.7	68.8
RMSE	0.661	1.769	1.137	8.456	4.650	2.341	11.054	8.231
nRMSE (%)	10.5	2.5	2.0	6.1	4.9	5.2	6.1	32.7
MBE	0.063	−0.431	−0.150	−3.375	−2.125	−0.544	−3.563	−5.000
AMBE	0.438	1.244	0.888	6.000	3.875	1.906	10.063	5.613
RE	0.781	−0.026	−0.052	−0.429	−0.122	−0.333	−0.294	−2.780
AE	0.781	0.026	0.104	0.429	0.122	0.333	0.294	14.053
η	0.879	0.997	0.997	0.997	0.999	0.993	0.996	0.267

Analyzing the computed traits, the plant height and the stem height have more dispersed values (in this position), as the larger values of NMAD and BwMv show. With regard to errors, the determination coefficient is superior to 0.9 for all the traits, except for the mean leaf length. The algorithm fails to recognize the short leaves, and it, therefore concludes that the mean of the computed leaf length is much larger. This is the same reason why the normalized root mean square error and the Nash and Sutcliffe index get the worst score in this computed trait. The rest of the errors show no difference between the actual and computed values at all.

5. Conclusions

As this study highlights, skeletons are powerful descriptors for analyzing plant networks by defining structural components and computing several phenotypic traits. Moreover, close-range platforms together with novel deep learning networks show a powerful combination to extract individual maize plants. Therefore, the approach proposed here is pretty rapid, accurate and cost-effective. It is worth mentioning that particular attention has been paid to the spatial resolution and completeness of the computed point cloud to effectively run our approach. These aspects are directly related to the plant spacing, which could generate shadows and to the variables coming from the flight (overlap, altitude and flight direction) to get a dense point cloud. In this study, the image acquisition strategy was only from nadir. However, oblique images will improve the completeness of the plant point cloud. Future analyses are needed to be able to apply our pipeline to different plant species and phenotypic growth stages, as well as to investigate the influence of environmental factors such as soil properties and light conditions. In addition, several platforms for high-throughput phenotyping, even terrestrial platforms or LiDAR-collected point clouds, are intended to be tested.

Author Contributions: M.H.-H. conceived the idea, developed the data analysis pipelines and software, performed the data analysis and visualization and wrote the manuscript; D.G.-A. supported the research and edited the manuscript; Y.Y. supervised the research. All authors have read and agreed to the published version of the manuscript.

Funding: MHH was supported by the Spanish Government under Maria Zambrano (Requalification of the Spanish university system for 2021–2023). This research was also funded by the European project H2020 CHAMELEON: A Holistic Approach to Sustainable, Digital EU Agriculture, Forestry, Livestock and Rural Development based on Reconfigurable Aerial Enablers and Edge Artificial Intelligence-on-Demand Systems. Ref: 101060529, Call: HORIZON-CL6-2021-GOVERNANCE-01-21.

Institutional Review Board Statement: Not applicable.

Informed Consent Statement: Not applicable.

Data Availability Statement: The data analyzed during the current study (the point cloud of the area of interest on the two dates and the point cloud of each individual plant and its skeleton) can be found as supplementary data at https://drive.google.com/file/d/1Yp-5ujfXRvtNqqmZ4xr-uRGKvXKJjpYM/view?usp=share_link (accessed on 9 January 2020).

Acknowledgments: The authors would like to thank the Institute for Plant Sciences (College of Agriculture), Jason Adams, Brian Dilkes and his lab and all at Purdue University (IN, USA) for their collaboration during the experimental phase of this research.

Conflicts of Interest: The authors declare no conflict of interest. The funders had no role in the design of the study; in the collection, analyses, or interpretation of data; in the writing of the manuscript; or in the decision to publish the results.

References

1. Roitsch, T.; Cabrera-Bosquet, L.; Fournier, A.; Ghamkhar, K.; Jiménez-Berni, J.; Pinto, F.; Ober, E.S. New sensors and data-driven approaches—A path to next generation phenomics. *Plant Sci.* **2019**, *282*, 2–10. [CrossRef] [PubMed]
2. Herrero-Huerta, M.; Bucksch, A.; Puttonen, E.; Rainey, K.M. Canopy roughness: A new phenotypic trait to estimate above-ground biomass from unmanned aerial system. *Plant Phenomics* **2020**, *2020*, 1–10. [CrossRef] [PubMed]
3. Herrero-Huerta, M.; Rodriguez-Gonzalvez, P.; Rainey, K.M. Yield prediction by machine learning from UAS-based multi-sensor data fusion in soybean. *Plant Methods* **2020**, *16*, 78. [CrossRef] [PubMed]
4. Symonova, O.; Topp, C.N.; Edelsbrunner, H. DynamicRoots: A software platform for the reconstruction and analysis of growing plant roots. *PLoS ONE* **2015**, *10*, e0127657. [CrossRef] [PubMed]
5. Moreira, F.F.; Hearst, A.A.; Cherkauer, K.A.; Rainey, K.M. Improving the efficiency of soybean breeding with high-throughput canopy phenotyping. *Plant Methods* **2019**, *15*, 139–148. [CrossRef]

6. Herrero-Huerta, M.; Meline, V.; Iyer-Pascuzzi, A.S.; Souza, A.M.; Tuinstra, M.R.; Yang, Y. 4D Structural root architecture modeling from digital twins by X-Ray Computed Tomography. *Plant Methods* **2021**, *17*, 1–12. [CrossRef]
7. Gerth, S.; Claußen, J.; Eggert, A.; Wörlein, N.; Waininger, M.; Wittenberg, T.; Uhlmann, N. Semiautomated 3D root segmentation and evaluation based on X-Ray CT imagery. *Plant Phenomics* **2021**, *2021*, 8747930. [CrossRef]
8. Jin, S.; Su, Y.; Gao, S.; Wu, F.; Ma, Q.; Xu, K.; Guo, Q. Separating the structural components of maize for field phenotyping using terrestrial lidar data and deep convolutional neural networks. *IEEE Trans. Geosci. Remote Sens.* **2019**, *58*, 2644–2658. [CrossRef]
9. York, L.M. Functional phenomics: An emerging field integrating high-throughput phenotyping, physiology, and bioinformatics. *J. Exp. Bot.* **2019**, *70*, 379–386. [CrossRef]
10. Bucksch, A.; Fleck, S. Automated detection of branch dimensions in woody skeletons of fruit tree canopies. *Photogramm. Eng. Remote Sens.* **2011**, *77*, 229–240. [CrossRef]
11. Bucksch, A. A practical introduction to skeletons for the plant sciences. *Appl. Plant Sci.* **2014**, *2*, 1400005. [CrossRef] [PubMed]
12. Cornea, N.D.; Silver, D.; Min, P. Curve-skeleton properties, applications, and algorithms. *IEEE Trans. Vis. Comput. Graph.* **2007**, *13*, 530. [CrossRef] [PubMed]
13. Au, O.K.C.; Tai, C.L.; Chu, H.K.; Cohen-Or, D.; Lee, T.Y. Skeleton extraction by mesh contraction. *ACM Trans. Graph.* **2008**, *27*, 1–10. [CrossRef]
14. Brandt, J.W.; Algazi, V.R. Continuous skeleton computation by Voronoi diagram. *CVGIP Image Underst.* **1992**, *55*, 329–338. [CrossRef]
15. Marie, R.; Labbani-Igbida, O.; Mouaddib, E.M. The delta medial axis: A fast and robust algorithm for filtered skeleton extraction. *Pattern Recognit.* **2016**, *56*, 26–39. [CrossRef]
16. Mohamed, W.; Hamza, A.B. Reeb graph path dissimilarity for 3D object matching and retrieval. *Vis. Comput.* **2012**, *28*, 305–318. [CrossRef]
17. Herrero-Huerta, M.; Lindenbergh, R.; Gard, W. Leaf Movements of indoor plants monitored by Terrestrial LiDAR. *Front. Plant Sci.* **2018**, *9*, 189. [CrossRef]
18. Moeslund, J.E.; Clausen, K.K.; Dalby, L.; Fløjgaard, C.; Pärtel, M.; Pfeifer, N.; Brunbjerg, A.K. Using airborne lidar to characterize North European terrestrial high-dark-diversity habitats. *Remote Sens. Ecol. Conserv.* **2022**. [CrossRef]
19. Li, N.; Kähler, O.; Pfeifer, N. A comparison of deep learning methods for airborne lidar point clouds classification. *IEEE J. Sel. Top. Appl. Earth Obs. Remote Sens.* **2021**, *14*, 6467–6486. [CrossRef]
20. Campos, M.B.; Litkey, P.; Wang, Y.; Chen, Y.; Hyyti, H.; Hyyppä, J.; Puttonen, E. A long-term terrestrial laser scanning measurement station to continuously monitor structural and phenological dynamics of boreal forest canopy. *Front. Plant Sci.* **2021**, *11*, 606752. [CrossRef]
21. Lei, L.; Li, Z.; Wu, J.; Zhang, C.; Zhu, Y.; Chen, R.; Yang, G. Extraction of maize leaf base and inclination angles using Terrestrial Laser Scanning (TLS) data. *IEEE Trans. Geosci. Remote Sens.* **2022**, *60*, 1–17. [CrossRef]
22. Lin, C.; Hu, F.; Peng, J.; Wang, J.; Zhai, R. Segmentation and stratification methods of field maize terrestrial LiDAR point cloud. *Agriculture* **2022**, *12*, 1450. [CrossRef]
23. Wu, S.; Wen, W.; Xiao, B.; Guo, X.; Du, J.; Wang, C.; Wang, Y. An accurate skeleton extraction approach from 3D point clouds of maize plants. *Front. Plant Sci.* **2019**, *10*, 248. [CrossRef] [PubMed]
24. Miao, T.; Zhu, C.; Xu, T.; Yang, T.; Li, N.; Zhou, Y.; Deng, H. Automatic stem-leaf segmentation of maize shoots using three-dimensional point cloud. *Comput. Electron. Agric.* **2021**, *187*, 106310. [CrossRef]
25. Xin, X.; Iuricich, F.; Calders, K.; Armston, J.; De Floriani, L. Topology-based individual tree segmentation for automated processing of terrestrial laser scanning point clouds. *Int. J. Appl. Earth Obs. Geoinf.* **2023**, *116*, 103145.
26. Li, M.; Shamshiri, R.R.; Schirrmann, M.; Weltzien, C.; Shafian, S.; Laursen, M.S. UAV oblique imagery with an adaptive micro-terrain model for estimation of leaf area index and height of maize canopy from 3D point clouds. *Remote Sens.* **2022**, *14*, 585. [CrossRef]
27. Tirado, S.B.; Hirsch, C.N.; Springer, N.M. UAV-based imaging platform for monitoring maize growth throughout development. *Plant Direct* **2020**, *4*, e00230. [CrossRef]
28. Du, L.; Yang, H.; Song, X.; Wei, N.; Yu, C.; Wang, W.; Zhao, Y. Estimating leaf area index of maize using UAV-based digital imagery and machine learning methods. *Sci. Rep.* **2022**, *12*, 15937. [CrossRef]
29. Qi, C.R.; Su, H.; Mo, K.; Guibas, L.J. Pointnet: Deep learning on point sets for 3d classification and segmentation. In Proceedings of the IEEE Conference on Computer Vision and Pattern Recognition, Honolulu, HI, USA, 21–26 July 2017; pp. 652–660.
30. Krizhevsky, A.; Sutskever, I.; Hinton, G.E. Imagenet classification with deep convolutional neural networks. *Adv. Neural Inf. Process. Syst.* **2012**, *25*, 1097–1105. [CrossRef]
31. Su, Y.-T.; Bethel, J.; Hu, s. Octree-based segmentation for terrestrial LiDAR point cloud data in industrial applications. *ISPRS J. Photogramm. Remote Sens.* **2016**, *113*, 59–74. [CrossRef]
32. Cao, J.; Tagliasacchi, A.; Olson, M.; Zhang, H.; Su, Z. Point Cloud Skeletons via Laplacian Based Contraction. In Proceedings of 2010 Shape Modeling International Conference, Aix-en-Provence, France, 21–23 June 2010; pp. 187–197.
33. Herrero-Huerta, M.; Meline, V.; Iyer-Pascuzzi, A.S.; Souza, A.M.; Tuinstra, M.R.; Yang, Y. Root phenotyping from X-ray Computed Tomography: Skeleton extraction. *Int. Arch. Photogramm. Remote Sens. Spatial Inf. Sc* **2021**, *XLIII-B4–2021*, 417–422. [CrossRef]
34. Jin, S.; Su, Y.; Zhang, Y.; Song, S.; Li, Q.; Liu, Z.; Ma, Q.; Ge, Y.; Liu, L.; Ding, Y.; et al. Exploring Seasonal and Circadian Rhythms in Structural Traits of Field Maize from LiDAR Time Series. *Plant Phenomics* **2021**, *2021*, 9895241. [CrossRef] [PubMed]

35. Russ, T.; Boehnen, C.; Peters, T. 3D Face Recognition Using 3D Alignment for PCA. In Proceedings of IEEE Computer Society Conference on Computer Vision and Pattern Recognition (CVPR'06), New York, NY, USA, 17–22 June 2006; pp. 1391–1398. [CrossRef]
36. Jin, S.; Su, Y.; Wu, F.; Pang, S.; Gao, S.; Hu, T.; Guo, Q. Stem–leaf segmentation and phenotypic trait extraction of individual maize using terrestrial LiDAR data. *IEEE Trans. Geosci. Remote Sens.* **2018**, *57*, 1336–1346. [CrossRef]
37. Zhou, L.; Gu, X.; Cheng, S.; Yang, G.; Shu, M.; Sun, Q. Analysis of plant height changes of lodged maize using UAV-LiDAR data. *Agriculture* **2020**, *10*, 146. [CrossRef]
38. Elarab, M.; Ticlavilca, A.M.; Torres-Rua, A.F.; Maslova, I.; McKee, M. Estimating chlorophyll with thermal and roadband multispectral high-resolution imagery from an unmanned aerial system using relevance vector machines for precision agriculture. *Int. J. Appl Earth Obs* **2015**, *43*, 32–42.
39. Herrero-Huerta, M.; Rodriguez-Gonzalvez, P.; Lindenbergh, R. Automatic Tree Parameter Extraction by Mobile Lidar System in an Urban Context. *PLoS ONE* **2018**, *13*, e0196004. [CrossRef]

drones

Article

Tree Branch Skeleton Extraction from Drone-Based Photogrammetric Point Cloud

Wenli Zhang [1,*], Xinyu Peng [1], Guoqiang Cui [1], Haozhou Wang [2], Daisuke Takata [3] and Wei Guo [2,*]

1 Information Department, Beijing University of Technology, Beijing 100022, China
2 Graduate School of Agricultural and Life Sciences, The University of Tokyo, Tokyo 188-0002, Japan
3 Faculty of Food and Agricultural Sciences, Fukushima University, Fukushima 960-1296, Japan
* Correspondence: zhangwenli@bjut.edu.cn (W.Z.); guowei@g.ecc.u-tokyo.ac.jp (W.G.)

Abstract: Calculating the complex 3D traits of trees such as branch structure using drones/unmanned aerial vehicles (UAVs) with onboard RGB cameras is challenging because extracting branch skeletons from such image-generated sparse point clouds remains difficult. This paper proposes a skeleton extraction algorithm for the sparse point cloud generated by UAV RGB images with photogrammetry. We conducted a comparison experiment by flying a UAV from two altitudes (50 m and 20 m) above a university orchard with several fruit tree species and developed three metrics, namely the F1-score of bifurcation point (FBP), the F1-score of end point (FEP), and the Hausdorff distance (HD) to evaluate the performance of the proposed algorithm. The results show that the average values of FBP, FEP and HD for the point cloud of fruit tree branches collected at 50 m altitude were 64.15%, 69.94%, and 0.0699, respectively, and those at 20 m were 83.24%, 84.66%, and 0.0474, respectively. This paper provides a branch skeleton extraction method for low-cost 3D digital management of orchards, which can effectively extract the main skeleton from the sparse fruit tree branch point cloud, can assist in analyzing the growth state of different types of fruit trees, and has certain practical application value in the management of orchards.

Keywords: unmanned aerial vehicle; tree phenotyping; open software

Citation: Zhang, W.; Peng, X.; Cui, G.; Wang, H.; Takata, D.; Guo, W. Tree Branch Skeleton Extraction from Drone-Based Photogrammetric Point Cloud. *Drones* **2023**, *7*, 65. https://doi.org/10.3390/drones7020065

Academic Editors: Monica Herrero-Huerta, Jose A. Jiménez-Berni, Shangpeng Sun, Ittai Herrmann and Diego González-Aguilera

Received: 18 December 2022
Revised: 6 January 2023
Accepted: 8 January 2023
Published: 17 January 2023

1. Introduction

Point cloud data are commonly used to describe the 3D structure of objects and have important applications in 3D reconstruction, engineering measurement, and morphological analysis, among other applications. With the widespread use of agricultural automation technology, the demands for point cloud analysis for high-throughput orchard phenotyping and growth modeling, such as plant shape, canopy structure, organ morphology, and stress response, are gradually increasing. Skeletonization is one of the essential steps in point cloud analysis in which a three-dimensional skeleton of a tree point cloud is obtained. This step makes it easier to calculate tree height, branch length, and angle, which comprise useful information for precise fruit tree growth management.

In the existing research on tree skeleton extraction, algorithms usually consist of two stages: firstly, skeleton points from the tree point cloud are extracted, and then skeleton topology connections are formed from the skeleton points. The skeleton points extraction is the fundamental step of the whole process. Current research approaches to skeleton point extraction can be broadly divided into three categories:

Graph structure-based [1–4]: a graph structure is a discrete structure consisting of vertices and edges connecting the vertices. The skeleton of a tree-like point cloud can be regarded as a combination of vertices and edges in a graph structure. The point cloud is converted into a graph structure, the shortest distance from the root node to each point in the point cloud is calculated to determine the skeleton points, and adjacent skeleton points are connected to form the branch skeleton. This method designs a skeleton point extraction algorithm for point clouds with a tree-like structure, which conforms to the structure of tree branches and has a good skeleton extraction effect. However, this method needs to calculate the shortest path on the finely constructed graph structure, which causes high computational costs.

Generic point cloud skeleton extraction algorithms-based [5–14]: generic point cloud skeleton extraction algorithms mainly include generalized rotationally symmetric axis (ROSA), L1-median and Laplace-based contraction (LBC). Many researchers have used this class of algorithms for tree branch skeleton extraction. Since these algorithms have no specific requirements on the structure of point cloud objects, they are adapted to a wider range of point cloud skeleton extraction tasks and can be easily used for skeleton extraction of point clouds of many types of tree branches. However, such methods were not designed for tree point clouds, such algorithms may extract structures such as annular or detached skeletons from point clouds, but these structures do not exist in natural tree branches, which may not guarantee the accuracy of tree skeleton extraction.

Other point cloud skeleton extraction algorithms [15,16]: other tree skeleton extraction methods are based on nearest neighbor shrinkage or clustering to extract the skeleton, and the skeleton points are obtained by the result of nearest neighbor shrinkage or clustering and connected to form the dendritic skeleton, and this kind of methods get better skeleton extraction results on the point cloud model with regular shape. However, these methods require high quality of point clouds, and the algorithms are sensitive to parameters, which need to be finely set to obtain better skeleton extraction results.

In general, point cloud acquisition for branch skeleton extraction uses laser scanning, multi-view reconstruction, and virtual model construction. Tree point clouds acquired using terrestrial laser scanning (TLS) or scanners have high accuracy [1–4,7–14], but this approach is expensive for large scale data acquisition for scenarios such as orchards and is not suitable for actual orchard production. Although the use of unmanned aerial vehicle (UAV) laser scanning (ULS) to acquire point cloud has the advantages of large acquisition range and high speed, the acquisition process is affected by airframe shaking, which makes it difficult to align into higher quality point cloud, and if the same quality of point cloud as TLS is to be acquired, the cost of the acquisition equipment will increase significantly. Although the point cloud acquisition using photogrammetry [15,16] reduces the cost of laser scanning, the spatial information of tree objects needs to be strictly calibrated when acquiring tree point cloud, and the process requires a relatively stable data acquisition environment, which limits its application in complex environmental scenarios such as orchards. The construction of point cloud from virtual 3D models [14,17,18] is not affected by the actual acquisition environment, but the virtual point cloud often does not represent the real 3D spatial structure of a fruit tree. In recent years, with the development of UAV aerial photography technology, the use of UAV equipped with RGB cameras to generate point clouds has been widely used, which combines the advantages of ULS while significantly reducing the equipment cost [19–21]. However, due to the limitation of the resolution of RGB cameras, the point clouds collected by this method are sparse and irregularly distributed in space, which makes fruit tree skeleton extraction challenging. Therefore, an extraction algorithm for sparse RGB point cloud is needed to complete the fruit tree skeleton extraction task.

Meanwhile, reasonable evaluation metrics are important for verifying the performance of the tree point cloud skeleton extraction algorithms. In fruit tree growth management, parameters such as tree height, branch length, and branch angle need to be accurately calculated, and the fruit tree branch skeleton is an important basis for calculating these parameters. Therefore, setting reasonable metrics to evaluate the performance of the skeleton extraction algorithm is of great significance for the extension of the tree skeleton extraction algorithm for use in scenarios such as orchards. In the evaluation work on tree point cloud skeleton extraction, Bucksch et al. [4] graded the branches of tree branches by length, counted the number of extracted tree branch skeletons and real branch skeletons under different length gradations, and evaluated the effectiveness of the skeleton extraction algorithm by calculating the correlation coefficient R^2 between the two. This approach evaluates the performance of skeleton extraction only in terms of quantity, ignoring the spatial information of branches. Li et al. [14] manually measured the real parameters of tree branches and calculated the F1 value (F1-Score), correlation coefficient (R^2), and root mean square error (RMSE) between the extracted tree branch skeleton parameters and the real measured parameters to evaluate the branch skeleton structure parameters for accuracy. Fu et al. [18] proposed a quantitative evaluation of the skeleton extraction effect where the point cloud was generated from a virtual branch skeleton and the evaluation of the skeleton extraction effect relied on the pre-existing virtual tree skeleton. However the tree branch skeleton is a combination of a series of spatial curves, and drawing the curves in 3D space by hand is often inaccurate and subjective, which affects the reliability of the evaluation results. Therefore, this evaluation method is not applicable to the skeleton extraction evaluation task of a real tree point cloud. We observed that there is no unified standard for the evaluation of skeleton extraction algorithms at this stage, and there is a need for evaluation metrics to verify the performance of actual tree point cloud skeleton extraction algorithms.

In summary, in this paper, we propose a branch skeleton extraction algorithm for UAV and photogrammetry-generated sparse RGB fruit tree point cloud and suggest evaluation metrics for tree branch point cloud to verify the performance of the algorithm proposed in this paper. The overall aims of the research were as follows.

(1) To address the problem of low accuracy of existing algorithms in skeleton extraction for sparse point cloud data. In this paper, a spatial density-based regional point cloud aggregation algorithm is designed to aggregate sparse tree point cloud before skeleton point extraction, and the aggregated point cloud can describe the 3D skeleton morphology of branches, which can effectively improve the accuracy of subsequent skeleton point extraction.
(2) To address the problem that the generic point cloud skeleton extraction algorithm is prone to broken branches and self-loops when the skeleton topology is connected which leads to unrealistic tree point cloud skeleton extraction results. In this paper we propose a skeleton topology connection method with spherical shrinkage from the outside to the root node, which can better adapt to the bifurcated tree structure effectively avoiding the appearance of non-tree branch structure, and can effectively extract the initial skeleton of tree branches.
(3) To objectively evaluate the skeleton extraction algorithm and thus verify the performance of the skeleton extraction algorithm on real tree branch point cloud. In this paper, based on the consideration of the characteristics of tree branch structure morphology, the metrics for evaluating the accuracy of skeleton morphology: FBP (F1 score of bifurcation point), FEP (F1-score of end point) and the metric for evaluating the accuracy of skeleton topology: HD (Hausdorff distance) were designed, which can reasonably evaluate the skeleton extraction performance of the algorithm.
(4) To release an easy-to-use software application that helps the community to test the proposed algorithm and use it for other related applications.

2. Materials and Methods

2.1. Data Acquisition

The data set used in this experiment were acquired by an onboard RGB camera (1 inch CMOS, lens FOV 84 degrees, focus length 8.8 mm, Max image size 5472 × 3648 pixels) mounted on a low-cost industrial level UAV (DJI Phantom 4 RTK, DJI, Shenzhen China). The flight altitude was set to 50 m and 20 m, and both the front and side overlap in between photos were set to 80%. Synthesis of RGB point cloud was performed using DJI Terra. All the flights were conducted at an orchard of the Institute for Sustainable Agro-ecosystem Services (ISAS), University of Tokyo Orchard (Tokyo, Japan) (35°44′16.0″ N, 139°32′20.9″ E) on the following dates: 13 January 2022, 14 February 2022, and 4 March 2022 (as shown in Figure 1a). There are multiple types of fruit trees growing in the orchard

In this paper, peach trees, persimmon trees, chestnut trees, and plum trees were selected as experimental objects, and five fruit trees were randomly selected from each type of fruit tree to tag and conduct comparative experiments (as shown in the white dashed box in Figure 1b). More details of the orchard are provided in Table 1.

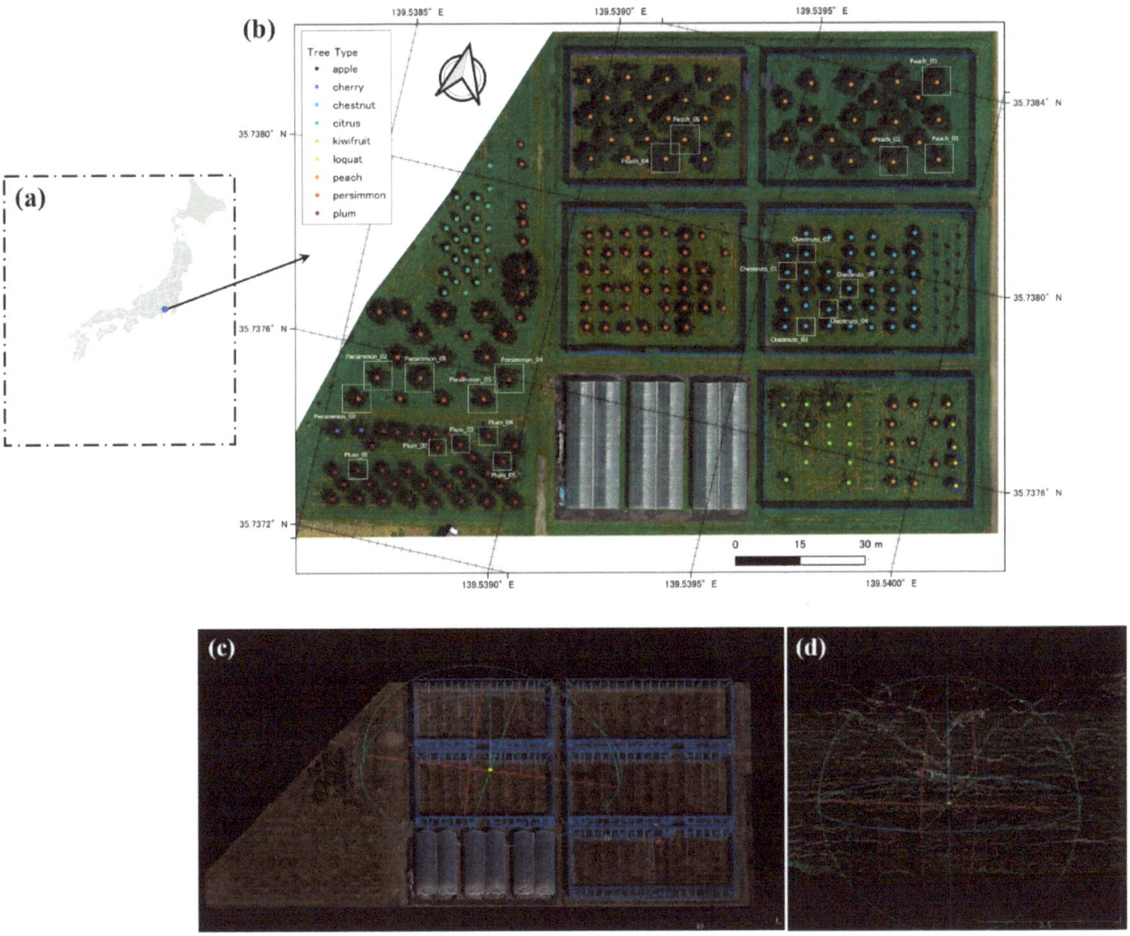

Figure 1. Data acquisition location and orchard point cloud. (**a**) Point cloud collection location (35°44′16.0″ N, 139°32′20.9″ E). (**b**) DOM (Digital Orthophoto Map) generated from UAV images, different color points in the map indicate the planting points of different species of fruit trees, and white dashed boxes and symbols in the map indicate the fruit tree objects selected for the experiment. (**c**) Three-dimensional view of the point cloud of the orchard. (**d**) Close-up view of the point cloud of the branches of the fruit trees.

2.2. General Architecture of the Algorithm

This section describes the general architecture of the algorithm in this paper, and Figure 2 shows the structure of the algorithm. The algorithm consists of the following components.

Table 1. Orchard point cloud containing tree species and number of individuals.

Tree Type	Number of Individuals
Apple	17
Cherry	2
Chestnut	86
Citrus	25
Kiwifruit	15
Loquat	5
Peach	51
Persimmon	95
Plum	46

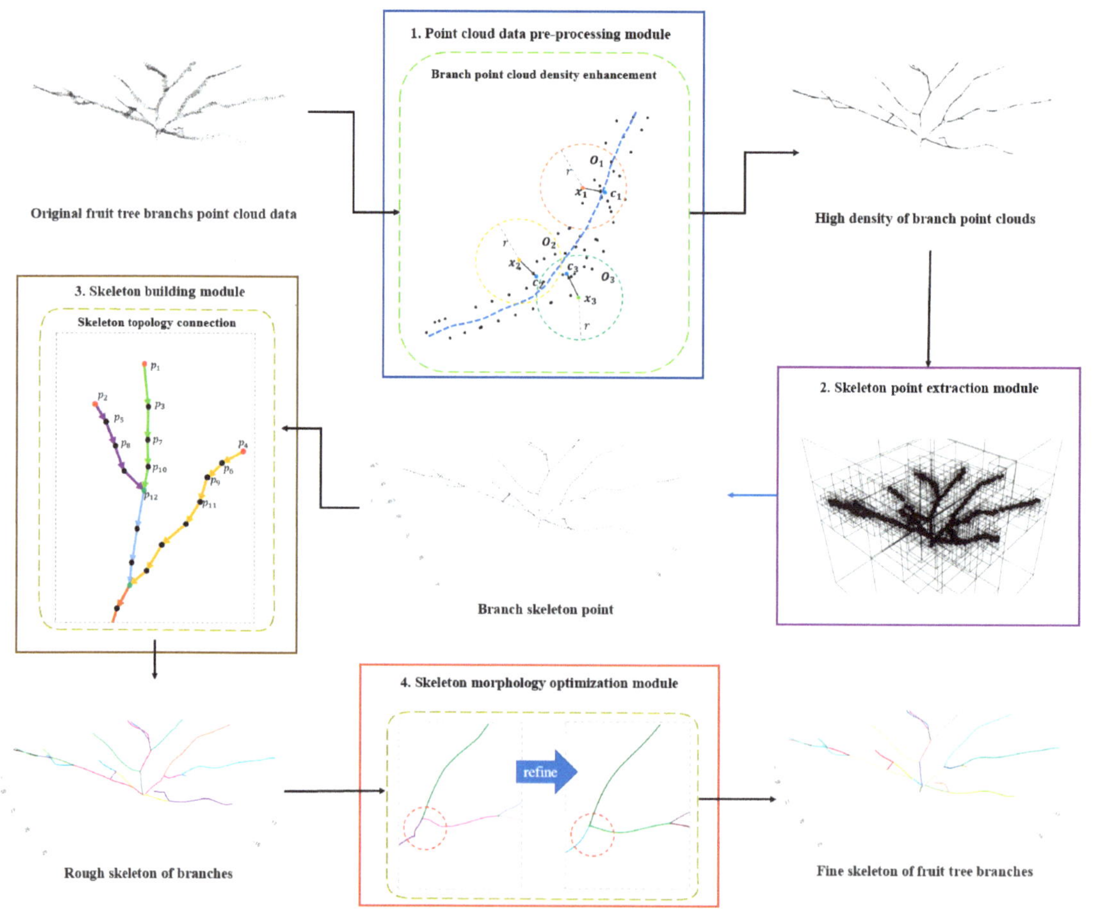

Figure 2. Functional structure of point cloud skeleton extraction algorithm for fruit tree branches.

1. Point cloud pre-processing module: this module first denoises the original branch point cloud data, and then aggregates the sparse branch point clouds using the regional point cloud aggregation method proposed in this study, which can form a dense point cloud that roughly describes the morphology of the branch skeleton and prepares for the subsequent skeleton point extraction.
2. Skeleton point extraction module: this module is based on the octree algorithm to spatially divide the clustered point cloud and extract the branch skeleton points in the divided subspace.
3. Skeleton construction module: this module uses the spherical shrinkage based skeleton topology connection method proposed in this paper to form the thick skeleton of the branch.
4. Branch skeleton morphology optimization module: this module fine-tunes the positioning of key points in the coarse skeleton first, and then smooths the branch coarse skeleton to output the final fine skeleton.

2.3. Point Cloud Pre-Processing Module

To better extract the skeleton points from the original fruit tree branch point cloud, this section first denoises the original branch point cloud and then aggregates the denoised branch point cloud using a regional point cloud aggregation algorithm.

(1) Point cloud denoising processing

In this study, we used the density-based spatial clustering of applications with noise (DBSCAN) algorithm [22] to remove the anomalous points in the original point cloud. The clustering approach is used to remove not only the anomalous outliers generated during the synthesis of point cloud data but also the anomalous point cloud patches (classes of reduced samples within clusters) generated during the reconstruction of point cloud data.

(2) Branch point cloud density enhancement

The branch skeleton is a series of curve combinations describing the centers of fruit tree branches, and the skeleton points are the spatial points constituting these curves [23]. These spatial points are the centers of branch point cloud in a certain segment. Determining the centers of fruit tree branch point cloud in a certain segment of space is the focus of skeleton point extraction. Since the branch point clouds generated by the RGB camera are sparse and not uniform, extracting the skeleton points directly from the denoised point cloud will produce the problem of inaccurate positioning.

If the tree point cloud branches can be aggregated to the centerline region of the branch skeleton (as shown in Figure 3a) in some way to form a higher density point cloud that can describe the 3D curve shape of the skeleton, and then the skeleton points are extracted from it, this can effectively improve the accuracy of skeleton point extraction. Therefore, an algorithm for regional point cloud aggregation is proposed in this paper.

(1) Let the denoised tree branch point cloud be $C_s = \{x_1, x_2 \ldots x_i \ldots x_n \mid x \in R^3\}$, and set the regional aggregation radius r, as shown in Equation (1):

$$r = \frac{\sum_i^n dist(x_i,\ k_i)}{n},\ i \in [1,\ n] \quad (1)$$

where k_i is the k nearest neighbor point of x_i (k is determined according to the empirical value of the preparatory experiment and adjusted according to different tree species).

(2) Traverse each point x_i in the point cloud, set a search region of radius r around x_i, set the set of points contained in this search region as $O_i = \{x_1, x_2 \ldots x_j \ldots x_m \mid x \in R^3\}$, take the shape center c_i of the point set O_i, as shown in Equation (2):

$$c_i = \frac{\sum_j^m x_j}{m},\ j \in [1,\ m] \quad (2)$$

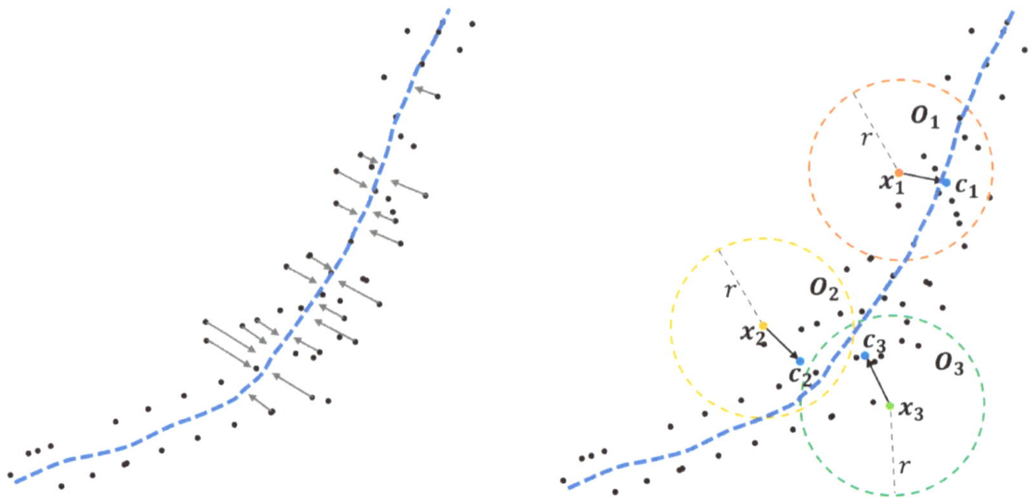

(a) (b)

Figure 3. Schematic diagram of regional point cloud aggregation. (**a**) Branch point cloud aggregation toward the skeleton centerline: the black point is the branch point cloud to be aggregated, the blue dashed line is the assumed skeleton centerline, and the gray arrow is the direction of movement of the point in the aggregation process. (**b**) Regional point cloud aggregation algorithm: red, orange and green dashed lines with different colors indicate the search area of x_i, blue dot c_i is the target location of x_i movement.

As the target location for x_i to move, the new point cloud set $C_a = \{c_1, c_2 \ldots c_j \ldots c_n \mid c \in R^3\}$ obtained after the point cloud traversal is completed, and Figure 3b shows the process of point cloud aggregation in the region.

(3) Keeping the search radius r constant, n iterations of step (2) are carried out to finally obtain the aggregated branch point cloud C_t.

Figure 4 shows the process of branch point cloud aggregation after n iterations ($n = 8$). With increasing n, the distribution of the fruit tree branch point cloud gradually aggregates toward the center line of the skeleton, and since the number of points of the branch point cloud does not change in the process, the density of the branch point cloud gradually increases in the process of aggregation. After n iterations of iterative aggregation, the spatial distribution of branch point cloud is dense and concentrated near the center line of the branch skeleton, which can already show the three-dimensional morphology of the fruit tree branch skeleton. Extracting the skeleton points subsequently on this basis can effectively improve the accuracy of skeleton point positioning.

2.4. Skeleton Point Extraction Module

Skeleton point extraction involves finding spatial points that can describe the branch skeleton from the branch trunk point cloud. Since the spatial distribution of the aggregated fruit tree branch trunk point cloud can already describe the three-dimensional form of the branch skeleton, the skeleton points can be considered as the concentrated expression of the fruit tree branch trunk point cloud after aggregation in a subdivision space. Therefore, this paper proposes a skeleton point extraction method based on subdivision space point cloud shape center calculation, which first divides the aggregated fruit tree branch trunk point cloud spatially, and then calculates the shape center of the point cloud in the subdivision space as the skeleton point (as the previous preparation for the construction of 2.5 section branch trunk skeleton). The specific implementation steps are as follows.

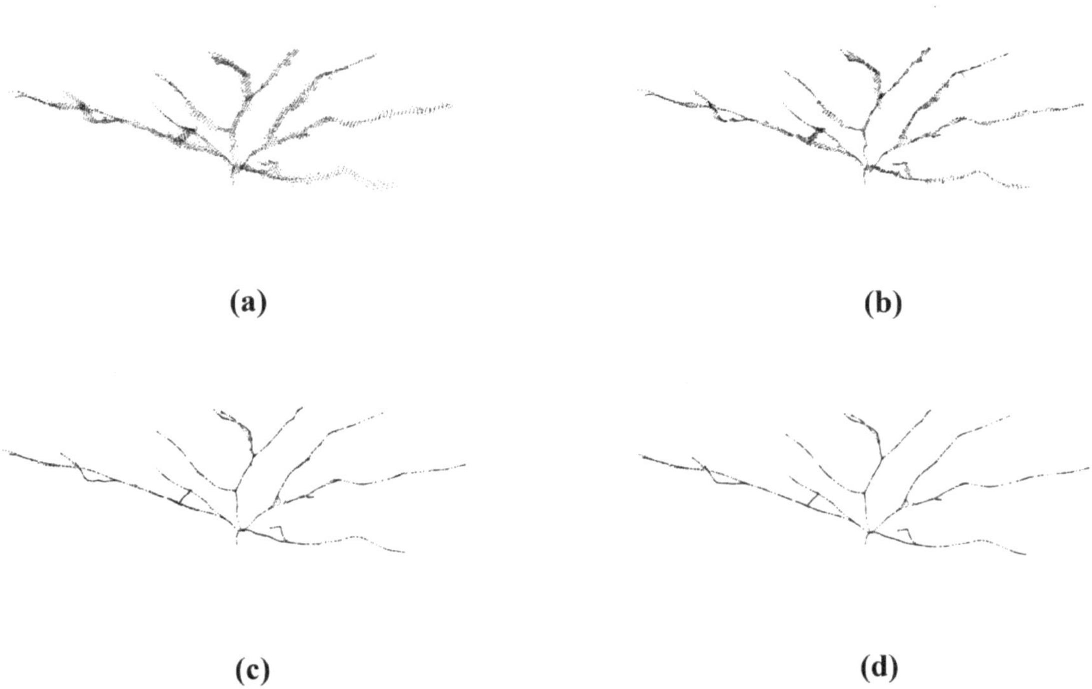

Figure 4. Schematic diagram of point cloud iterative aggregation process: (**a**) Point cloud of fruit tree branches after denoising. (**b**) Point cloud of fruit tree branches after three iterations of regional point cloud aggregation. (**c**) Point cloud of fruit tree branches after five iterations of regional point cloud aggregation. (**d**) Point cloud of fruit tree branches after eight iterations of regional point cloud aggregation.

(1) The octree based spatial partitioning

The octree algorithm [24], as one of the commonly used data structures for 3D spatial partitioning, can accelerate spatial queries and efficiently manage 3D space. In this paper, we first use the spatial octree structure to subspace the aggregated fruit tree branch trunk point cloud, as shown in Figure 5.

S_i: smallest subspace divided using octree, O_i: geometric center of subspace, c_{ij}: branch point cloud location in subspace, k_i: skeleton points in subspace.

(2) Calculation of skeleton points in subspace

The skeleton point of the branch can be regarded as the center of the point cloud geometry of the branch in a certain region. Considering that the center point O_i of the subspace may have a large offset compared with the distribution of the point cloud in the space, it could not represent the distribution of the point cloud in the subspace. Therefore, to better reflect the representativeness of the skeleton point to the point cloud in the subspace, this paper adopts the shape center of the point cloud in the subspace as the skeleton point k_i, which is defined as shown in Equation (3):

$$k_i = \frac{\sum_j^n c_{ij}}{n}, j \in [1, n] \tag{3}$$

where k_i denotes the skeleton points calculated in each subspace, n is the number of point clouds contained in each subspace, and c_{ij} is the location of the branch point cloud in each subspace.

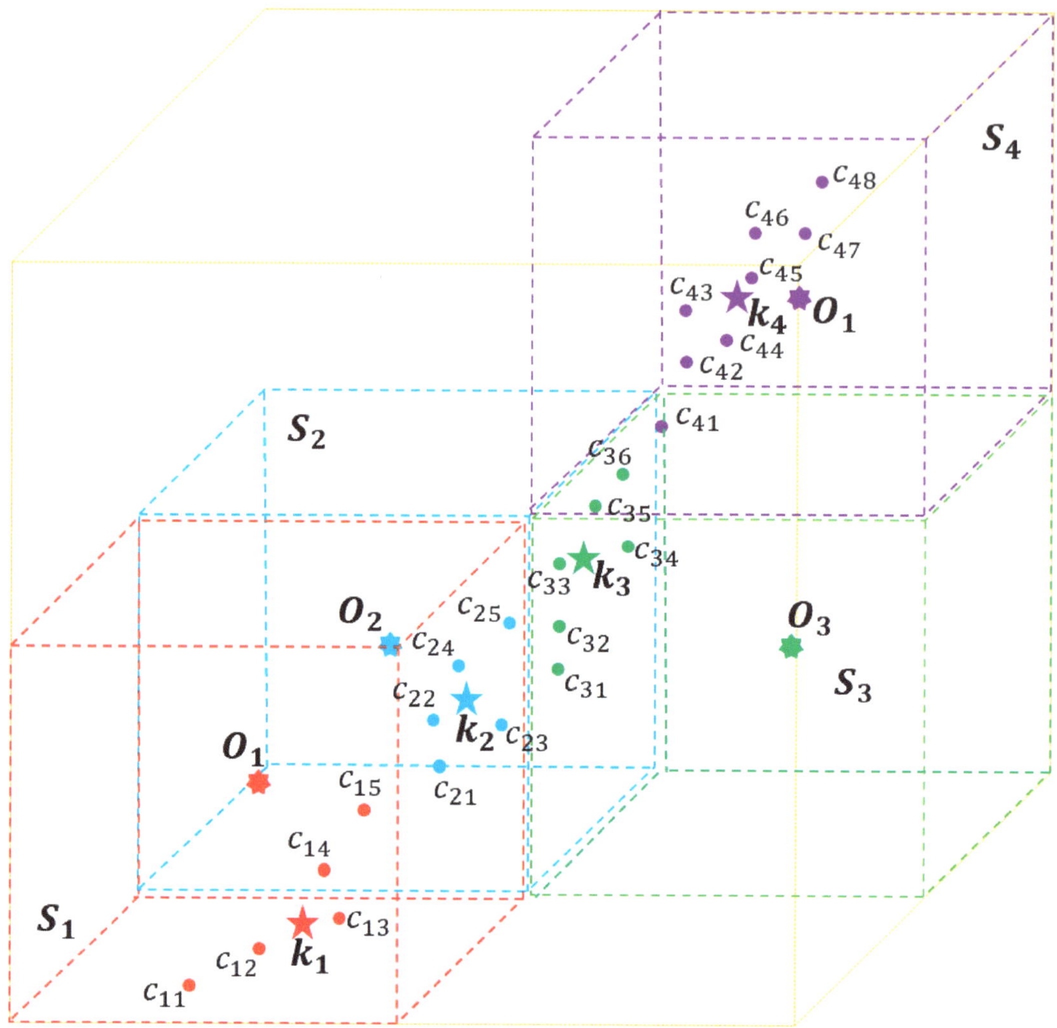

Figure 5. Partitioning of the aggregated branch and stem point cloud using spatial octree.

Calculating the point cloud shape centers as skeleton points in the subspace of the octree subdivision can ensure that the spatial distribution of its skeleton points can better reflect the fruit tree branch trunk morphology, which is ready for the skeleton construction in Section 2.5. Figure 6a shows the effect of spatial partitioning of the aggregated whole branch trunk point cloud using the octree algorithm; Figure 6b shows the extracted fruit tree branch trunk skeleton points, and the extracted skeleton point distribution can still better reflect the fruit tree skeleton morphology.

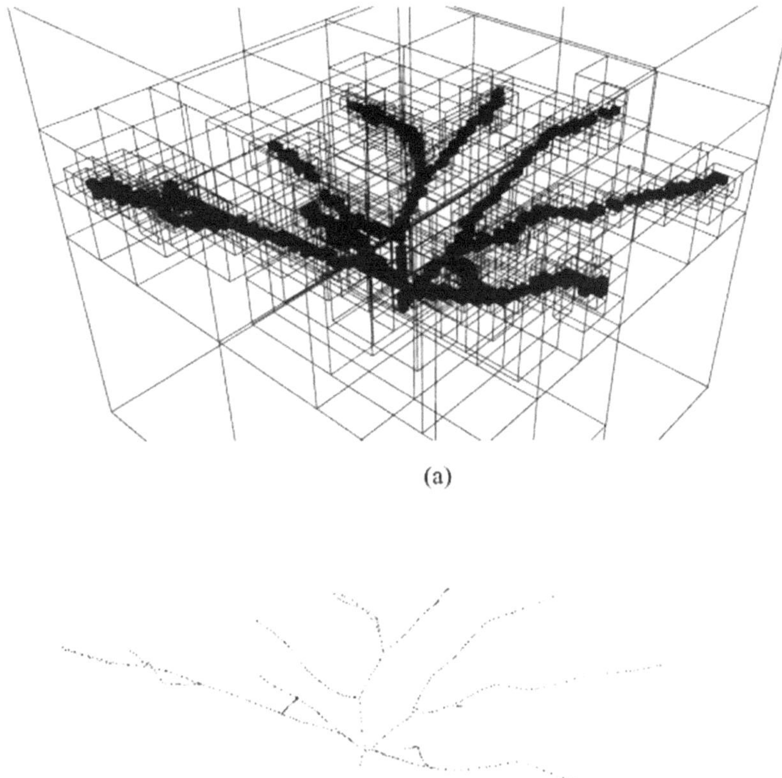

(a)

(b)

Figure 6. Schematic diagram of extraction of skeleton points from a spatial octree. (**a**) Spatial octree partitioning of the point cloud. The black grid in the figure indicates the subspace where the branch point cloud is located. (**b**) Skeleton points extracted using the spatial octree.

2.5. Skeleton Building Module

After extracting the skeleton points of fruit tree branches, the skeleton points need to be topologically connected to form the tree branch skeleton. In previous work, the skeleton of the tree point cloud was usually obtained based on the nearest neighbor or bottom-up topology connection [25], and the number of sub-branches at the bifurcation was not well determined in the case of uneven distribution of skeleton points due to the need to set a reasonable number of nearest neighbor skeleton points, which affects the correctness of the extracted skeleton topology. Since the point cloud data in this paper are obtained from the reconstruction of RGB images collected by an overhead UAV, the point cloud data are sparse and contain less details of branches, which mainly reflect the main branch structure of the fruit trees (e.g., main branches, submain branches, etc.). Therefore, to quickly construct the skeleton topology of the main branches of fruit trees, we designed a spherical shrinkage skeleton topology connection method from the outside to the root node, which can effectively adapt to the bifurcated structure of fruit tree branches, and can extract the key points (end points, bifurcation points) and skeleton segments (end point-bifurcation point, bifurcation point-bifurcation point) of the skeleton while obtaining the main skeleton of fruit tree branches.

Figure 7 shows the spherical contraction topology connection process of the local branch skeleton, and the specific process of the algorithm is as follows.

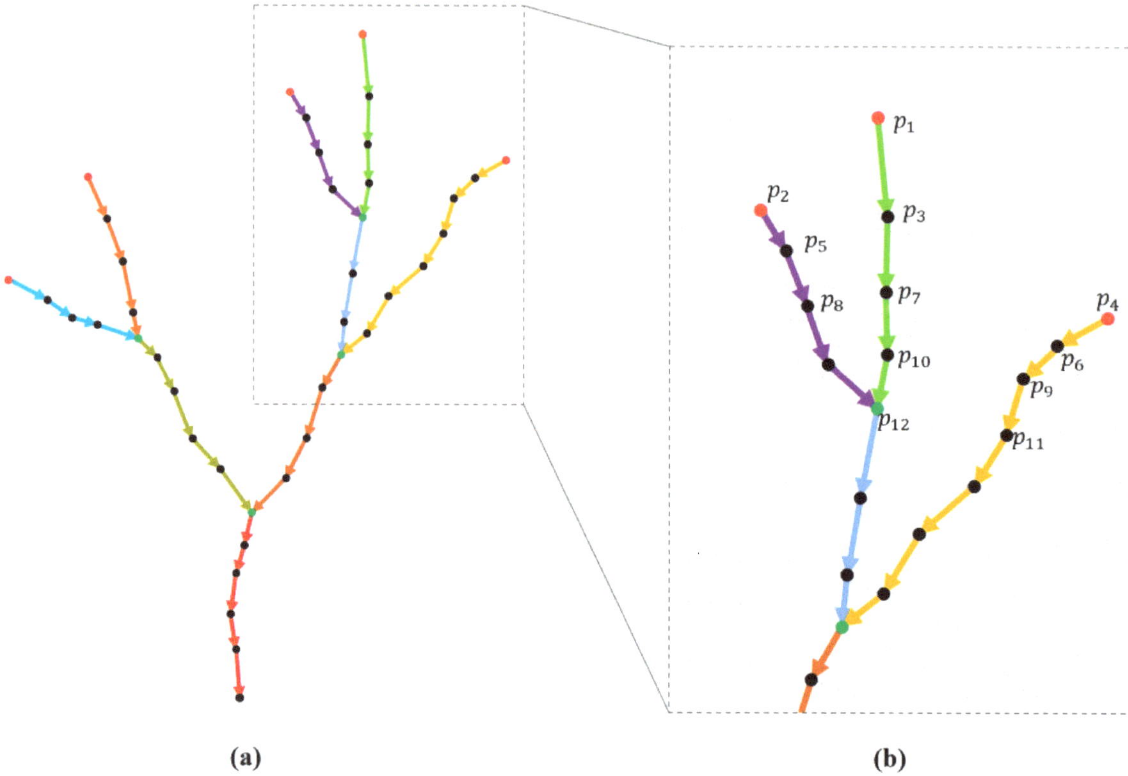

Figure 7. Schematic diagram of spherical contraction for skeleton topology connection. (**a**) Schematic diagram of the local skeleton of fruit tree branches. (**b**) Schematic diagram of spherical contraction for topological connection of local skeleton. Red dots are branch end points, green dots are branch bifurcation points, gray arc-shaped dashed lines indicate the spherical shells where the skeleton points are located, and different colored connecting segments indicate each skeleton segment. p_1–p_{12} are the skeleton points on each skeleton segment.

(1) Firstly, starting from the farthest skeleton point p_1, make a ball with the radius of this point and the root node, connect this point with the nearest skeleton point p_3 which is located inside the sphere shell. Following this, select the next far point p_2 as the starting point, make a ball with the radius of this point and the root node, select the nearest skeleton point p_5 which is located inside the sphere shell to connect, and iterate the above process to traverse all skeleton points.

(2) After all skeleton points are traversed, the *div* of each skeleton point is calculated (*div*: the number of connections between a skeleton point and other skeleton points around it. In this paper, we define $div = 1$ for end points, $div = 2$ for ordinary branch skeleton points, and $div >= 3$ for bifurcation points) to extract the end points and bifurcation points in the skeleton points, and calculate each skeleton segment of the branch according to the connection status between the skeleton points.

(3) Due to the possible redundancy of the extracted skeleton points, there may be shorter burr branches (skeleton segments containing fewer skeleton points) in the topologically connected skeleton. Set the threshold value l. If the skeleton segment contains skeleton points greater than l, the skeleton segment is retained, otherwise the skeleton segment is removed.

After the skeleton topology connection, the algorithm obtains the endpoints, bifurcation points, and the set of skeleton segments of the fruit tree branches along with the whole skeleton. Figure 8 shows the coarse skeleton of a tree formed after the spherical shrinkage topology connection, where red dots indicate the fruit tree branch skeleton end points, green dots indicate the fruit tree branch skeleton bifurcation points, and different color connection segments in the figure represent different segments of the fruit tree branch skeleton.

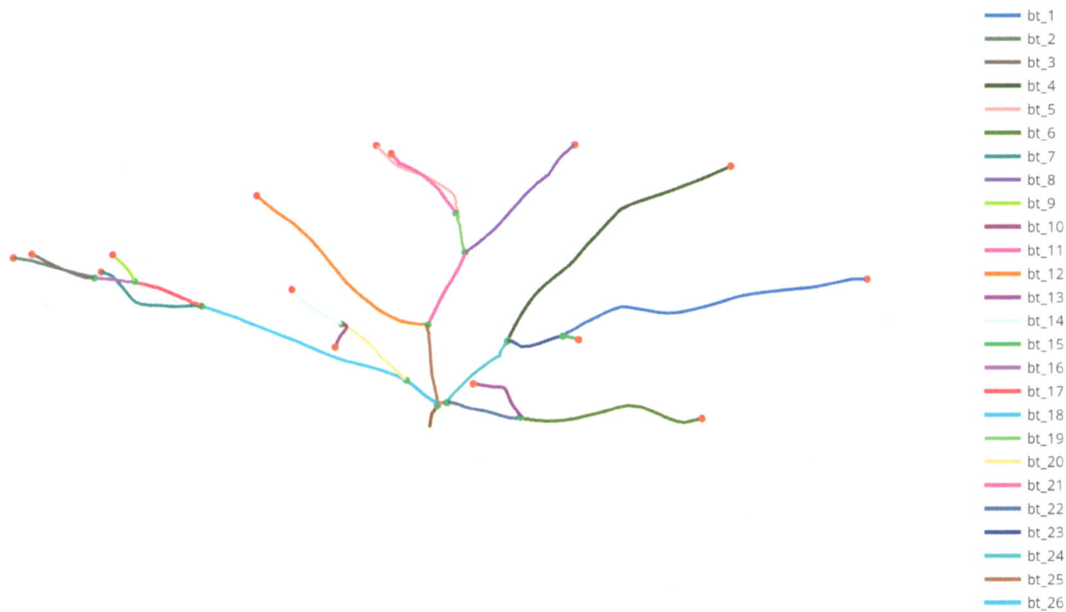

Figure 8. The coarse skeleton formed after topological connection.

Red dots indicate fruit tree branch skeleton endpoints, green dots indicate fruit tree branch skeleton bifurcation points, and the connected segments of different colors in the figure represent different segments of the fruit tree branch skeleton.

2.6. Skeleton Morphology Optimization Module

In 3D space, the real tree branch skeleton can be considered as a combination of multiple smooth spatial curves, and the two endpoints of the curves are always at the endpoints and bifurcation points of the branch skeleton. Therefore, the positions of the skeleton endpoints and bifurcation points need to be kept from moving during the smoothing process of the skeleton. However, the algorithms commonly used for skeleton smoothing (e.g., polynomial smoothing, B spline fitting smoothing, etc.) are not adapted to the prerequisite of fixed skeleton curve endpoints and bifurcation points. Therefore, to solve this problem, the spatial Bezier curve [26] is chosen in this paper to perform spatial curve smoothing on the extracted skeleton. To ensure the correctness of the optimized skeleton morphology, the key points in the thick skeleton of the branches were first fine-tuned in this paper before the pre-processing of the skeleton smoothing using Bezier curves (see Appendix A: branch skeleton key point adjustment strategy).

In the extracted fruit tree branch skeleton, let a skeleton segment be $BH = [P_0, P_1, P_2, \ldots, P_n]$ (n is the number of skeleton points contained in the skeleton segment), where P_0 is the end point (or bifurcation point) of the skeleton segment, P_n is the bifurcation point of the skeleton segment, and P_0 and P_n are used as the starting point and end points, and $P_1, P_2, \ldots, P_{n-1}$ as the control points of the Bezier curve, and the smoothing of this skeleton segment using an $m(m = n - 1)$ order spatial Bezier curve, the smoothed curve $B(t)$ of this skeleton segment can be obtained, and the definition of $B(t)$ is shown in Equation (4):

$$B(t) = \sum_{i=0}^{m} C_m^i (1-t)^{m-i} t^i P_i \,, t \in [0,\ 1] \tag{4}$$

In the equation:

$$C_m^i = \frac{m!}{i! * (m-i)!}$$

As shown in Figure 9b, after the smoothing process of the spatial Bezier curve, the segments of the skeleton are smoother and more natural, which can better reflect the three-dimensional spatial shape of the real fruit tree branches.

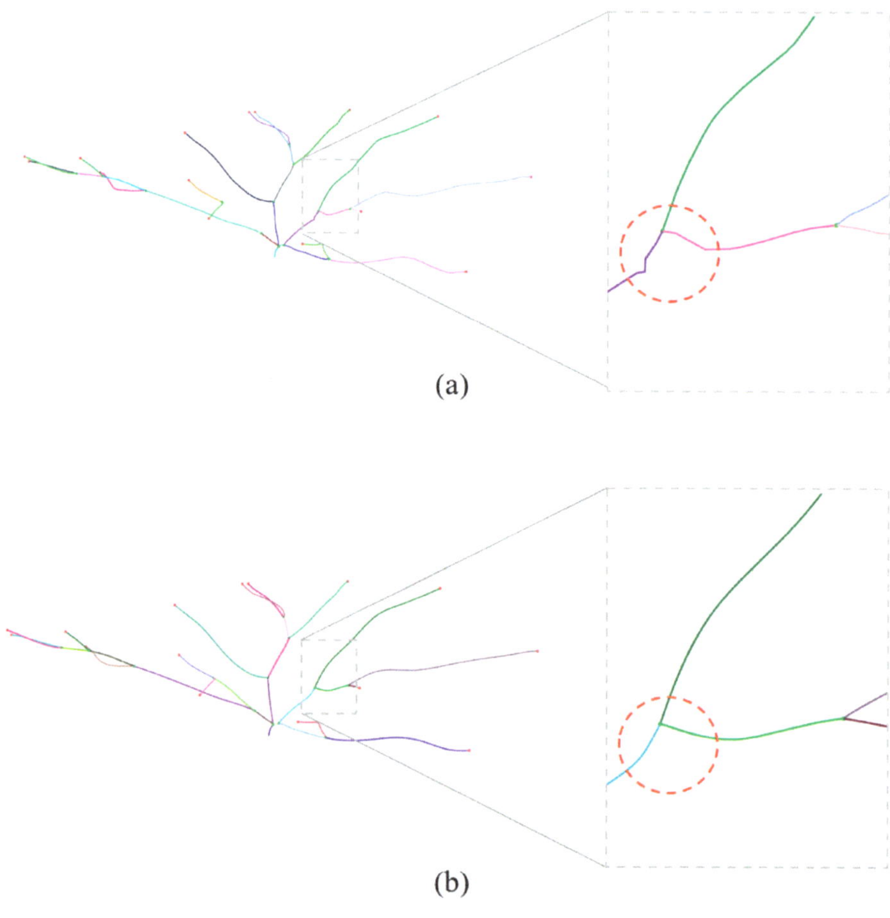

Figure 9. Comparison of branch skeleton before and after smoothing. (**a**) Effect of fruit tree branch skeleton before smoothing. (**b**) Effect of fruit tree branch skeleton after smoothing.

3. Results

In this section, the proposed tree point cloud skeleton extraction algorithm is evaluated. First, evaluation metrics are designed to verify the effectiveness of the tree point cloud skeleton extraction algorithm, and based on the evaluation metrics, the proposed algorithm in this paper is compared with the L_1-medial and the Laplacian based contraction (LBC) algorithm [6,11] for experiments. Section 3.1 introduces the metrics proposed in this paper to evaluate the effectiveness of skeleton extraction, F1-score of bifurcation Point (FBP), F1-score of end Point (FEP), and Hausdorff distance (HD). In Section 3.2, the environment of the comparison experiments and the specific comparison experimental data and results are presented in this paper.

3.1. Evaluation Metrics

We propose FBP and FEP to evaluate the accuracy of skeleton topological connection and HD to evaluate the accuracy of skeleton morphology. Note that the generated branch point cloud and its extracted skeleton of different tree species are normalized in the spatial range prior to evaluation by Equation (5):

$$P_{\text{normalization}} = \left(\frac{x - x_{min}}{x_{max} - x_{min}}, \frac{y - y_{min}}{y_{max} - y_{min}}, \frac{z - z_{min}}{z_{max} - z_{min}} \right) \quad (5)$$

where $[x_{min}, x_{max}, y_{min}, y_{max}, z_{min}, z_{max}]$ indicates the spatial range of the real fruit tree point cloud.

The original fruit tree branch point cloud and the extracted branch skeleton are transformed into the normalized 3D space, and the evaluation of skeleton extraction in this paper is carried out in the normalized 3D space.

3.1.1. Skeleton Topology Accuracy Metrics FBP and FEP

Branching angles are usually obtained by calculating the angles between branch bifurcation points and endpoints, while the accuracy of bifurcation point and endpoint positioning affects the topological accuracy of the fruit tree branch skeleton. Therefore, a reasonable evaluation of the accuracy of bifurcation point and endpoint positioning is an important part of the skeleton evaluation work. Since it is simpler to obtain the bifurcation points and endpoints of branches from real branch point cloud and is not easily influenced by subjective judgment, the spatial coordinates of these points in real branch point cloud obtained by manual labeling are used as real values to evaluate the positioning effect of the points of the skeleton extracted by the algorithm.

In this study, FBP is used to evaluate the extraction effect of bifurcation points of the skeleton, and the definition of FBP is shown in Equation (6):

$$FBP = \frac{2 * PBP * RBP}{PBP + RBP} \quad (6)$$

In the equation:

$$PBP(Precision\ of\ Bifurcation\ Point) = \frac{BP}{BN}$$

$$RBP(Recall\ of\ Bifurcation\ Point) = \frac{BP}{BT}$$

where BP is the number of accurately located skeleton bifurcation points extracted by the algorithm (set the Euclidean distance d between the skeleton bifurcation points extracted by the algorithm and their nearest true skeleton bifurcation points, and if d is less than the set threshold t, the skeleton bifurcation points are considered to be accurately located), BN is the total number of skeleton bifurcation points extracted by the algorithm, and BT is the total number of true branch skeleton bifurcation points.

Similarly, FEP is set in this paper to evaluate the extraction effect of skeleton end points. FEP is defined as shown in Equation (7):

$$FEP = \frac{2 * PEP * REP}{PEP + REP} \quad (7)$$

In the equation:

$$PEP(Precision\ of\ End\ Point) = \frac{EP}{EN}$$

$$REP(Recall\ of\ End\ Point) = \frac{EP}{ET}$$

where EP is the number of accurately located skeleton endpoints extracted by the algorithm (set the Euclidean distance d between the skeleton endpoints extracted by the algorithm and their nearest real skeleton endpoints, and if d is less than the set threshold t, the skeleton endpoints are considered to be accurately located), EN is the total number of skeleton endpoints extracted by the algorithm, and ET is the total number of real branch skeleton endpoints.

3.1.2. Skeleton Morphological Accuracy Metric HD

To evaluate the accuracy of the skeleton morphology extracted by the algorithm, it is necessary to evaluate the degree of fit between the extracted skeleton and the real fruit tree. However, it is difficult and subjective to annotate the 3D skeleton from the real fruit tree branch point cloud, which will affect the accuracy of the skeleton morphology evaluation results. Therefore, in this paper, we generated a point cloud around the extracted skeleton and measured the morphological accuracy of the extracted skeleton by comparing the similarity between the generated fruit tree branch point cloud and the real fruit tree branch point cloud. The HD is a measure to describe the degree of similarity between two point sets [27], and the smaller the HD, the greater is the degree of similarity between the two point sets; therefore, it can be used to measure the degree of similarity between two point clouds.

To evaluate the morphological accuracy of the fruit tree branch skeleton, we set the real branch point cloud after denoising as C_s, generated the point cloud C_g around the fruit tree branch skeleton by random sampling, and calculated the average of HD between C_s and C_g for n times ($n = 10$ in this paper) as the final HD between C_s and C_g, which can reduce the error caused by a single calculation. The HD between C_s and C_g: $H(C_s, C_g)$ is calculated as shown in Equation (8):

$$H(C_s, C_g) = \frac{\sum_i^n \max[h(C_s, C_g), h(C_g, C_s)]}{n} \quad (8)$$

In the equation:

$$h(C_s, C_g) = \max_{P_i \in C_s} \min_{P_j \in C_g} \|P_i - P_j\|$$

$$h(C_g, C_s) = \max_{P_j \in C_g} \min_{P_i \in C_s} \|P_j - P_i\|$$

Figure 10 shows the generated point cloud C_g around the extracted skeleton and the overlap effect with the real point cloud C_s.

3.2. Comparison Experiments

In this section, based on the evaluation metrics proposed in Section 3.1, the algorithm proposed in this paper is compared with the L_1-medial and the LBC. Section 3.2.1 introduces the comparison experiment environment; Section 3.2.2 introduces the comparison experiment data and results, and the analysis and discussion of the results.

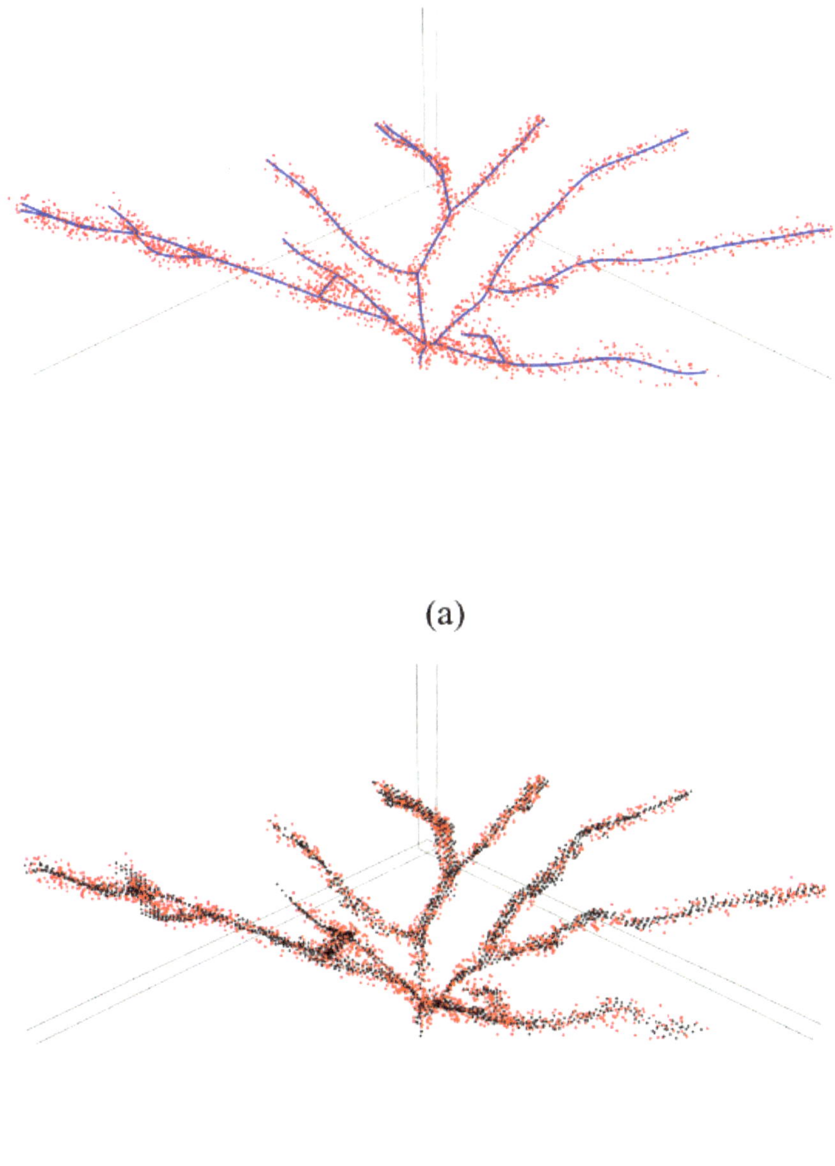

(a)

(b)

Figure 10. Point cloud generation effect diagram (take a peach tree as an example). (**a**) Generated point cloud around the fruit tree branch skeleton. (**b**) Overlap effect of generated point cloud and original point cloud. The blue curve is the fruit tree branch skeleton, the black point is the original branch point cloud C_s, and the red point is the generated branch point cloud C_g.

3.2.1. Experimental Environment

The algorithms used for the experiments in this paper are performed on the same computer, which is configured with an AMD Ryzen 7 4800U processor with 1.80 GHz and 16 GB running memory, and the experimental environment is a Windows 10 operating system. This algorithm was run using python 3.7, and the GUI was developed using C++ and the Qt framework.

3.2.2. Comparison Experiments

Currently, the L_1-medial and the LBC algorithm, as the most commonly used algorithms in point cloud skeleton extraction work, have been largely applied in fruit tree skeleton extraction tasks [7–14]. Further, the existing tree point cloud skeleton extraction work is also usually carried out based on these two algorithms. In this section, on the point cloud of fruit tree branches generated by UAV images at 50 m and 20 m, the point cloud skeleton extraction algorithm of fruit tree branches proposed in this paper is compared with L_1-medial and LBC, and the specific comparison experiments and experimental results are shown in Appendix B Exhibit 2-1.

(1) 50 m altitude comparison experiment

In this study, we first conducted comparative experiments on the point cloud of fruit tree branches collected by UAV at a 50 m altitude to qualitatively evaluate the skeleton extraction effect of fruit tree branches, and quantitatively evaluate the three algorithms in terms of skeleton connection accuracy and skeleton morphology accuracy by combining three metrics, FEP, FBP, and HD.

Figure 11 shows the visualization of the fruit tree branch skeleton extracted by the three algorithms on the point cloud of fruit trees collected at a 50 m altitude (one tree of each fruit tree is selected for display).

From the results of the skeleton extraction in the figure, although the L1-medial and the LBC extracted the skeleton of the fruit tree branches, the skeleton has different degrees of defects (as shown by the orange dashed circles in Figure 11. Using the L1-medial algorithm when processing the sparser RGB point cloud, the skeleton of some branches could not be connected well. Due to the irregular distribution of the point cloud generated by the RGB images, the LBC extracts a ring-like structure in the branch skeleton of fruit trees, which does not exist in real trees, and affects the realism of the algorithm skeleton extraction effect In contrast, the proposed algorithm extracts a more complete branch skeleton and correct topological connections, and the final extracted branch skeleton has a smoother and more natural shape due to the smoothing process of the branch skeleton in the algorithm.

From the average values of the metrics in Table 2, the point cloud skeleton extraction algorithm proposed in this paper has higher FBP and FEP than the other two algorithms reaching 64.15% and 69.94% for the four fruit trees, respectively. Meanwhile, the HD of the extracted skeleton by the algorithm was 0.0699, which is also smaller than those of the other two compared algorithms. Furthermore, from Tables in the Appendix section, the algorithm proposed in this paper performed the best among 47 metrics out of the total 60 metrics for 20 fruit trees. The experimental results show that the algorithm is more accurate in locating the bifurcation points and end points of the skeleton on the point cloud of fruit tree branches collected at a 50 m altitude, and the point cloud generated from the skeleton extracted by the algorithm has higher similarity with the real fruit tree branch point cloud. This suggests that the skeleton of fruit tree branches extracted by the proposed algorithm has better skeleton topology and morphological accuracy, and better reflects the real morphology of fruit tree branches.

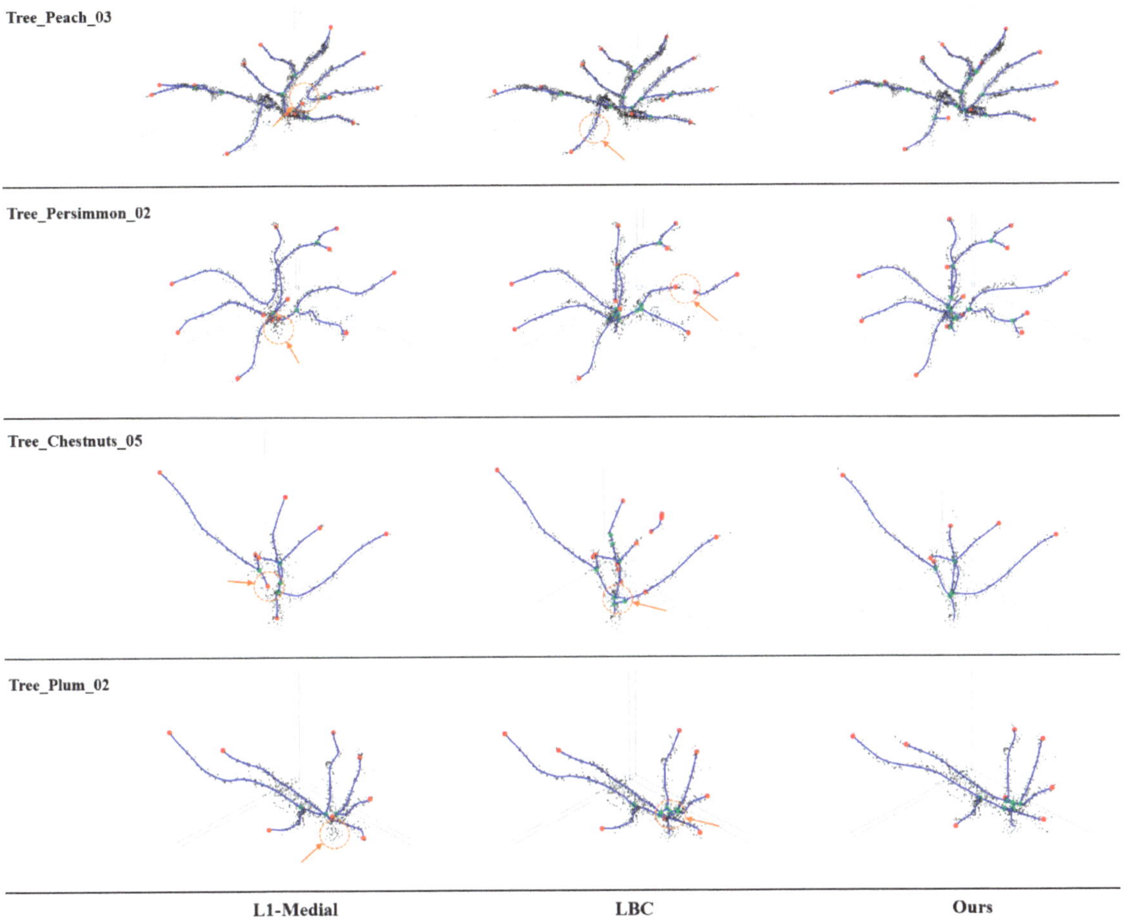

Figure 11. The visualization of the fruit tree branch skeleton extracted by the three algorithms on the point cloud of fruit trees collected at a 50 m altitude. The black points are the original fruit tree branch point cloud, the blue curve is the skeleton extracted by the algorithm, the green points are the bifurcation points of the fruit tree branch skeleton, and the red points are the end points of the fruit tree branch skeleton.

To verify the stability of the skeleton extraction algorithm under different acquisition times, acquisition light conditions, and other environments, we collected point clouds of fruit tree branches at a 50 m altitude on 13 January, 24 February, and 4 March 2022, and conducted comparative experiments, and the specific experimental results are shown in the Appendix section following the paper (Appendix C Exhibit 3-1, Exhibit 3-2, Exhibit 3-3). Appendix C Exhibit 3-4 in the Appendix of the paper show the standard deviations of the different algorithms for each metric of the fruit trees in the three comparative experiments mentioned above. Based on the information presented in Appendix C Exhibit 3-4, we can see that the algorithm of this paper has a smaller standard deviation in the experimental metrics compared with the L_1-medial and the LBC. This indicates that the data fluctuation of the three 50 m altitude comparison experiments of this paper's algorithm is lower. Therefore, the algorithm proposed in this paper can maintain a good skeleton extraction effect under different fruit tree branch point cloud acquisition environments.

Table 2. The average values and variation of metrics for the skeleton extraction results of the three algorithms on the branch point clouds of the four types of fruit trees generated from images collected at a 50 m and 20 m altitude ($t = 0.03$).

		50 m			20 m			Variation		
Tree_Species_Number	**Metric**	**L1-Medial**	**LBC**	**Proposed**	**L1-Medial**	**LBC**	**Proposed**	**L1-Medial**	**LBC**	**Proposed**
Average_Peach	FBP	70.76%	57.60%	83.60%	60.26%	70.41%	86.30%	−10.50%	13.81%	2.71%
	FEP	73.63%	75.02%	83.64%	64.50%	65.41%	83.35%	−9.13%	−9.61%	−0.29%
	HD	0.0647	0.0574	0.0433	0.0762	0.0698	0.0546	0.0115	0.0124	0.0113
Average_Persimmon	FBP	58.04%	58.13%	74.76%	68.80%	70.82%	84.77%	10.76%	13.69%	10.01%
	FEP	61.60%	67.36%	85.45%	70.71%	61.49%	90.04%	9.11%	−5.87%	4.59%
	HD	0.0878	0.063	0.0582	0.0996	0.0688	0.0527	0.0118	0.0058	−0.0055
Average_Chestnuts	FBP	33.74%	30.35%	45.92%	59.05%	56.01%	79.93%	25.31%	26.66%	34.01%
	FEP	42.48%	33.40%	48.54%	55.51%	44.86%	83.64%	13.03%	11.46%	35.10%
	HD	0.1094	0.1032	0.0909	0.0708	0.06	0.0393	−0.0386	−0.0432	−0.0516
Average_Plum	FBP	40.72%	32.43%	52.31%	65.38%	63.26%	81.98%	24.66%	31.83%	29.67%
	FEP	60.20%	36.41%	62.12%	77.85%	52.79%	81.59%	17.65%	16.38%	19.46%
	HD	0.1184	0.1068	0.0871	0.097	0.0736	0.0432	−0.0215	−0.0332	−0.0439
Average_Total	FBP	50.82%	44.63%	64.15%	63.37%	61.13%	83.24%	12.56%	21.50%	19.10%
	FEP	59.48%	53.05%	69.94%	67.14%	56.14%	84.66%	7.67%	3.09%	14.72%
	HD	0.0951	0.0826	0.0699	0.0859	0.0681	0.0474	−0.0092	−0.0146	−0.0224

In summary, combining the effect of fruit tree skeleton extraction and experimental evaluation metrics, it can be seen that in the comparison experiments of RGB point cloud generated from images collected at a 50 m altitude, compared with the other two algorithms the skeleton of fruit tree branches extracted by the algorithm in this paper has a more realistic 3D structure and has a better performance of evaluation metrics on most of the fruit trees. Further, the proposed algorithm can more effectively and stably extract the point cloud of fruit tree branches from the point cloud of fruit tree branches, and ensure realism of the skeleton. However, we also observed that the experimental metrics of all three algorithms on chestnut and plum trees are poor compared to the results on peach and persimmon trees. To verify the effectiveness of our algorithm on the point cloud of chestnut and plum trees with higher quality of point cloud, we also reduced the altitude of image acquisition to 20 m to generate a higher quality RGB point cloud.

(2) 20 m altitude comparison experiment

Figure 12 shows the visualization of the fruit tree branch skeleton extracted by the three algorithms on the point cloud of fruit trees collected at 20 m altitude (one tree of each fruit tree is selected for display).

The black points are the original fruit tree branch point cloud, the blue curve is the skeleton extracted by the algorithm, the green points are the bifurcation points of the fruit tree branch skeleton, and the red points are the end points of the fruit tree branch skeleton.

As shown in the figure, compared with the point cloud of fruit tree branches collected at 50 m, the point cloud collected by the UAV at a flight altitude of 20 m is of higher quality presenting a larger number of fruit tree branches and a more complex branching structure of fruit trees. The effect of skeleton extraction shows that although the skeletons extracted by the L_1-medial and LBC can show the morphology of fruit tree branches, there are still different degrees of broken branches and self-loops (as shown by the orange dashed circles in Figure 12); especially in the peach and persimmon trees with more complex branch structures, the two algorithms show more skeleton topology connection errors, which affect the authenticity of the extracted fruit tree branch skeleton. Compared with the two comparative algorithms, the proposed algorithm extracted more complete branch skeletons and accurate topological connections in terms of the point cloud of fruit tree images captured at an altitude of 20 m with more complex structures, although the proposed algorithm also showed local skeleton linkage errors (as shown in the gray dashed circles in Figure 12).

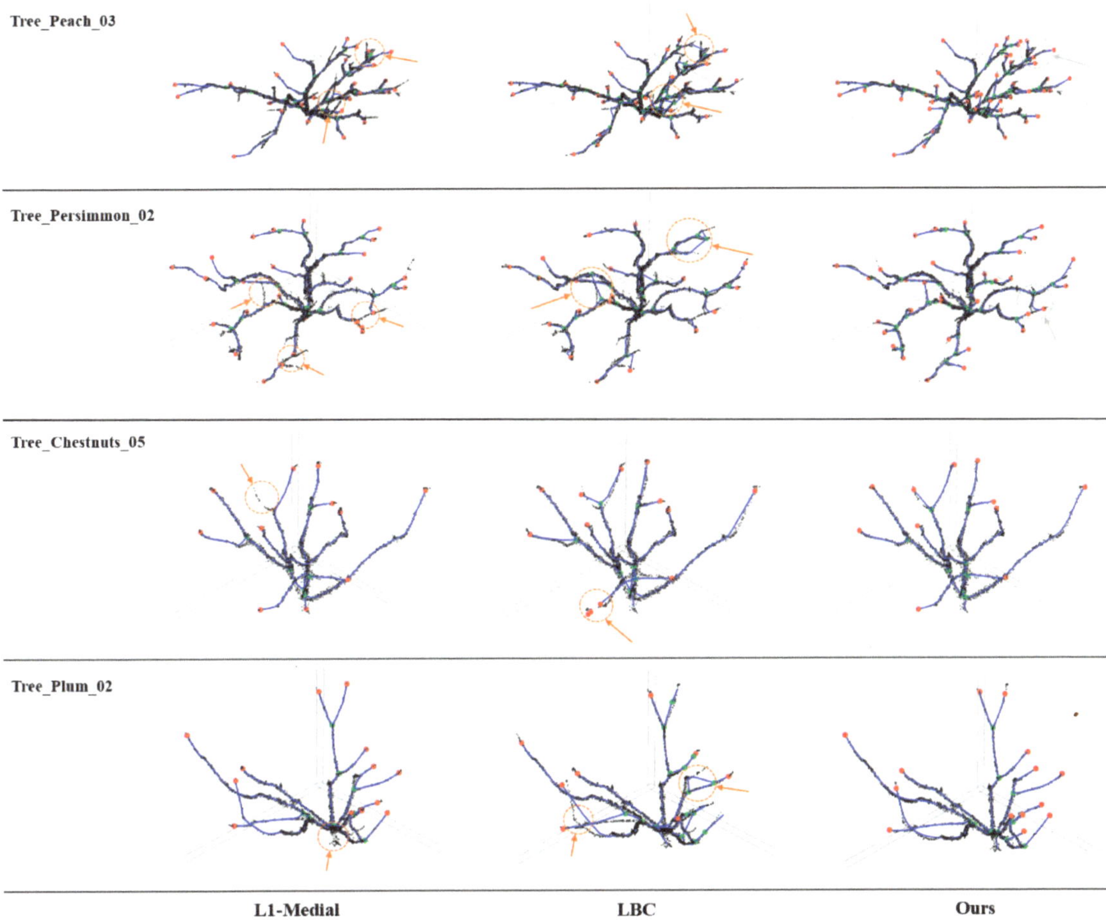

Figure 12. The visualization of the fruit tree branch skeleton extracted by the three algorithms on the point cloud of fruit trees collected at 20 m altitude.

Based on information provided in Table 2, in the comparison experiments on four fruit trees at 20 m altitude, the average value of FBP for the proposed algorithm is 83.24%, which is 19.87% higher than that of L_1-medial and 17.11% higher than that of LBC; the average value of FEP is 84.66%, which is 17.52% higher than that of L_1-medial and 28.52% higher than that of LBC; the average value of HD is 0.0474, which is 0.0385 lower than that of L_1-medial algorithm and 0.0207 lower than that of LBC, which shows that the proposed model achieved the best values in all three metrics. Based on the information presented in Appendix C Exhibit 3-5 in the Appendix text, the algorithm proposed in this paper performs best in 52 metrics among 60 metrics for 20 fruit trees. It can be seen that in the point cloud of fruit tree branches collected at a 20 m altitude, the skeleton of fruit tree branches extracted by the algorithm proposed in this paper can still maintain better topology and morphological accuracy compared with the L_1-medial and LBC. In other words, the skeleton extracted by the proposed algorithm is more realistic in the point cloud of fruit tree branches collected at a 20 m altitude.

(3) Analysis of experimental results at different altitudes

Figure 13 shows the average values of the experiment metrics on all fruit tree branch point clouds generated from images collected at both altitudes, and Table 2 shows the amount of variation in the average values of the metrics for the three algorithms on the four types of fruit trees at both altitudes.

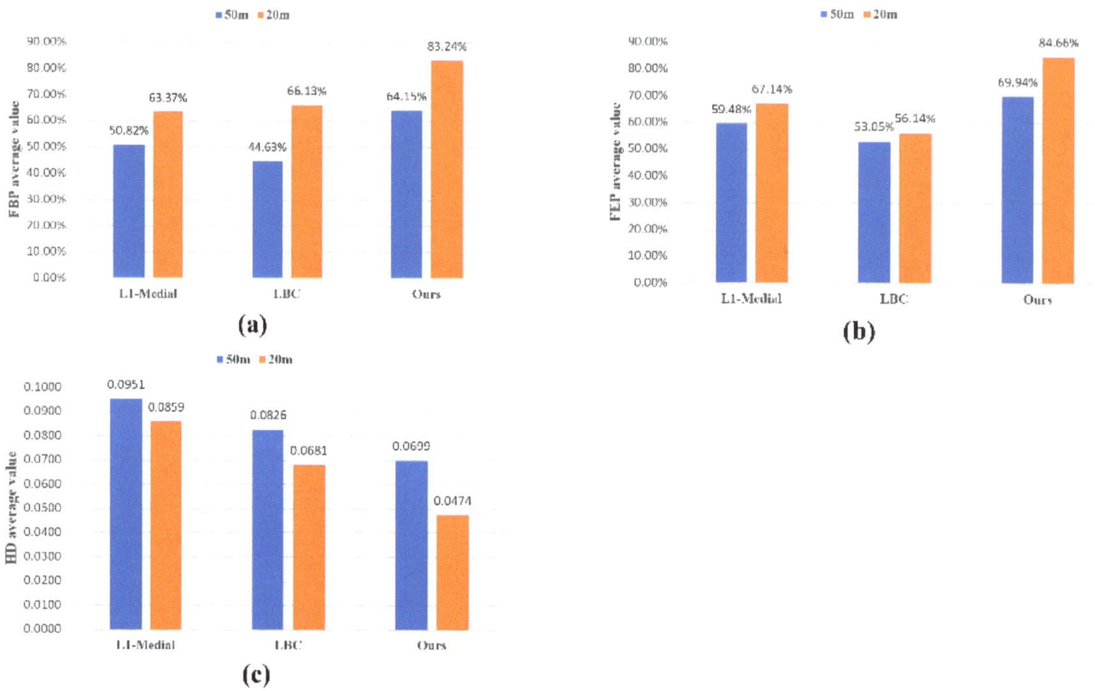

Figure 13. The average values of metrics for the three algorithms on all fruit trees. (**a**) Average FBP (F1-score of bifurcation point) for the different algorithms at 50 m and 20 m. (**b**) Average FEP (F1-score of end point) for the different algorithms at 50 m and 20 m. (**c**) Average HD (Hausdorff distance) for the different algorithms at 50 m and 20 m.

As shown in Figure 13, with a decrease in the fruit tree point cloud collection altitude the average values of FBP and FEP of the three skeleton extraction algorithms on the four fruit trees showed an increasing trend, and the average values of HD showed a decreasing trend, which showed that the quality of the collected fruit tree branch point cloud was improved with a reduction in the data collection altitude.

From the column of the amount of variation in Table 2, chestnut and plum trees whose metrics performed poorly in the comparison experiment at a 50 m altitude, showed significant improvement in the skeleton extraction results in the comparison experiment with images taken at a 20 m altitude. For chestnut and plum trees with sparse branches, the improvement of fruit tree branch point cloud quality can effectively improve the accuracy of the algorithm in extracting skeleton topology and morphology. The proposed algorithm improved the average values of FBP, FEP, and HD metrics by 34.01%, 35.10%, and 0.0516 respectively, for the five chestnut trees, and improved the average value by 29.67%, 19.46% and 0.0439, respectively, for the five plum trees. This indicates that the proposed algorithm can improve the skeleton extraction effect to a greater extent with improvement in the quality of fruit tree branch point cloud.

We also noted that the experimental metrics of peach and persimmon trees in the comparison experiment at a 20 m altitude did not improve significantly with the improvement of point cloud quality compared with those of chestnut and plum trees, but deteriorated in some metrics (e.g., the average values of FEP and HD experimental metrics of peach trees). The reason for this is that although the fruit tree branch point clouds collected by lowering the UAV flight altitude to 20 m are of higher quality, the fruit tree branch point cloud of peach and persimmon trees also present a more complex structure, and the fine branches of the fruit tree are more distinct (as shown in Figure 14). These fine branches can easily be ignored in the process of skeleton extraction, which makes the fruit tree branch skeleton extraction more difficult.

Figure 14. Point cloud of branches of the same fruit tree collected at different altitudes. (Tree_Peach_02 is shown as an example). (**a**) Point cloud of branches collected at 50 m altitude. (**b**) Point cloud of branches collected at 20 m altitude.

From Table 2, although the average value of FEP decreased by 0.29% and the average value of HD increased by 0.0113 for peach trees for the proposed algorithm, the metrics deteriorated less compared to that in the L_1-medial and LBC algorithms. The L_1-medial algorithm reduced the FBP metric of peach tree by 10.50%, the LBC reduced the FEP metric of persimmon tree by 5.87%, and both algorithms increased the HD metric of persimmon tree by 0.0118 and 0.0058, respectively, which shows that the proposed algorithm can still maintain a better performance in the complex fruit tree branch point cloud skeleton extraction task compared with the two other algorithms.

4. Discussion and Conclusions

In this paper, a branch skeleton extraction algorithm is proposed for the sparse RGB fruit tree point cloud collected by UAV, which can effectively extract the skeleton from the sparse fruit tree branch point cloud, approximate the 3D structure of the original fruit tree, and more realistically reflect the branch morphology of the fruit tree. The algorithm proposed is based on point clouds generated from images captured by UAV RGB cameras. Compared with the method of extracting skeleton points using point cloud data collected by devices such as LiDAR [7,8,12–14], the algorithm proposed can effectively reduce the cost of point cloud data collection while maintaining the accuracy of skeleton point extraction, such as evaluate the growth of orchard trees by analyzing the tree morphometry using UAVs [28,29].

Combining the comparison experiments at 50 m and 20 m altitudes, it can be found that the algorithm proposed in this paper has better performance in terms of FBP, FEP and HD metrics compared with L_1-medial and LBC. At the same time, the results of comparison experiments at different altitudes show that for fruit trees with sparse branch structure (e.g., chestnut and plum trees), the reduction in UAV flight altitude can improve the quality of the collected fruit tree branch point cloud and improve the skeleton extraction effect of the algorithm. However, for the more complex branch structure (e.g., peach and persimmon trees), with a reduction in flight altitude, the UAV can collect a more complex fruit tree branch point cloud structure, which brings challenges to the skeleton extraction of the algorithm. Therefore, when a more complex fruit tree branch structure is dealt with, whether the skeleton can be effectively extracted from the branch point cloud is an important basis to test the adaptability of the fruit tree point cloud skeleton extraction algorithm. Compared with the L_1-medial and LBC, the algorithm proposed in this paper can still maintain good experimental results on more complex fruit tree branch point cloud, indicating that the algorithm proposed in this paper can also adapt to the extraction task of the skeleton of more complex fruit tree branch point cloud (Figure 15).

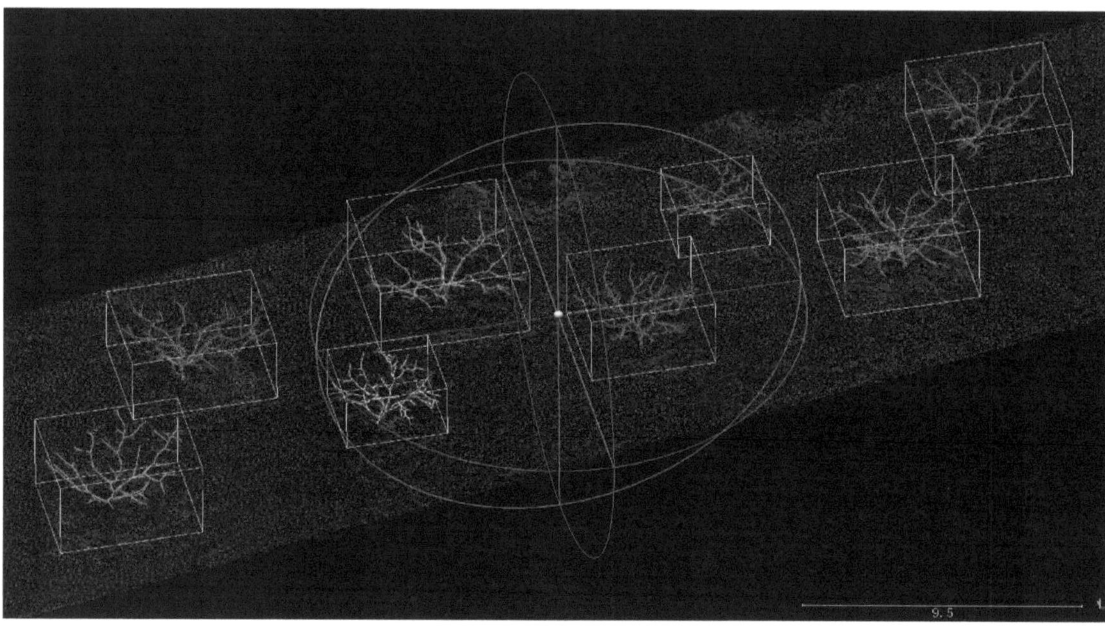

Figure 15. The effect of the algorithm in extracting the branch skeleton of fruit trees in an orchard plot (the different color curves in the figure indicate the branch skeleton of each fruit tree).

As a conclusion, after the experiments and discussions in this paper, the proposed method can effectively extract the skeleton from the point cloud of fruit tree branches measured by UAV photogrammetry. It is undeniable that extracting the skeleton from a complex tree branch point cloud is still a challenging task. Unlike conventional objects with clear 3D structures, tree branches have open surfaces and complex branches, and the collected point cloud often fails to clearly represent the tree branch morphology, causing greater difficulties in the subsequent branch skeleton extraction. Therefore, the algorithm proposed in this paper still has areas for improvement:

First, based on the experimental results, it can be found that the algorithm in this paper shows advantages in fruit tree branch point cloud skeleton extraction. However, there are still some problems, such as poor performance in FEP and FBP metrics of some fruit trees in the comparison experiments. The reason for this is that the algorithm in this paper aggregates the fruit tree branch point cloud, and the skeleton point extraction and subsequent skeleton topology connection are performed on the aggregated point cloud, and the deviation of skeleton key point positioning occurs. In future work, optimizing the positioning of skeleton end points by using the search method with direction and the positioning of skeleton bifurcation points by using the local density judgment can be considered.

Second, at this stage, the parameters need to be set manually during skeleton extraction. For example, the number of points and iterations of regional point cloud aggregation will affect the point cloud aggregation and the recurrence depth of spatial octree, and subspace point threshold will affect the extracted skeleton points, among other such issues. The parameters during skeleton extraction need to be adjusted according to the type and height of fruit trees and other attributes, which may require some experience. It is possible to conduct several experiments on branch point clouds of different tree species to acquire empirical values to apply to the needs of different types of fruit trees, and also to try to project reasonable parameter thresholds based on the projected area and space size of fruit trees in the vertical direction.

In the actual work of fruit tree morphological analysis, the extracted fruit tree skeleton can be used to calculate parameters such as branch length and branch angle (e.g., the angle between main and secondary branches, and other parameters, as shown in Figure 16) to achieve the task of automated fruit tree morphological analysis. On the basis of the extracted fruit tree branch skeleton, phenotypic information of fruit tree canopy can be calculated, which can be used in practical applications for tasks such as orchard canopy cover, biomass density and carbon sequestration estimation [30–32]. In addition, the real 3D tree branch skeleton is the basis for modeling scenes of orchards and woodlands, and the algorithm proposed in this paper can extract a realistic fruit tree branch skeleton, and the reconstructed tree model on this basis can better restore the real 3D morphology of fruit trees. Therefore, the virtual 3D model of fruit trees can be formed based on the extracted skeleton by regularized programming and mapping in future work (as shown in Figure 17) to meet the demand for 3D visualization display of orchards and other scenes [33–36].

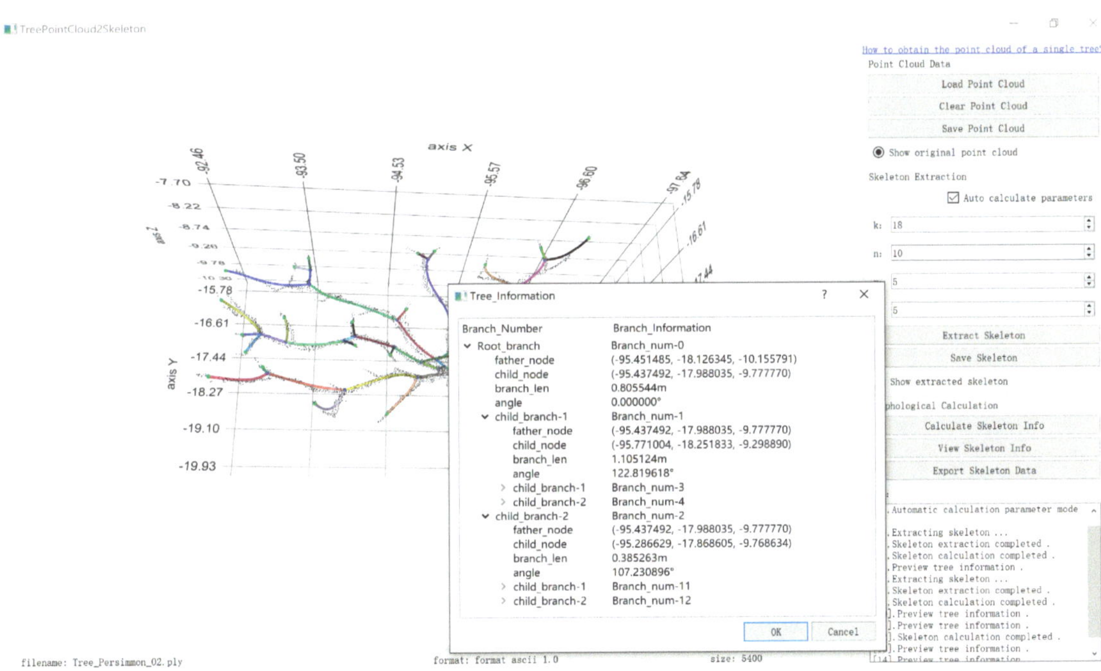

Figure 16. Automated calculation of fruit tree branch angles using the skeleton extraction software in this paper.

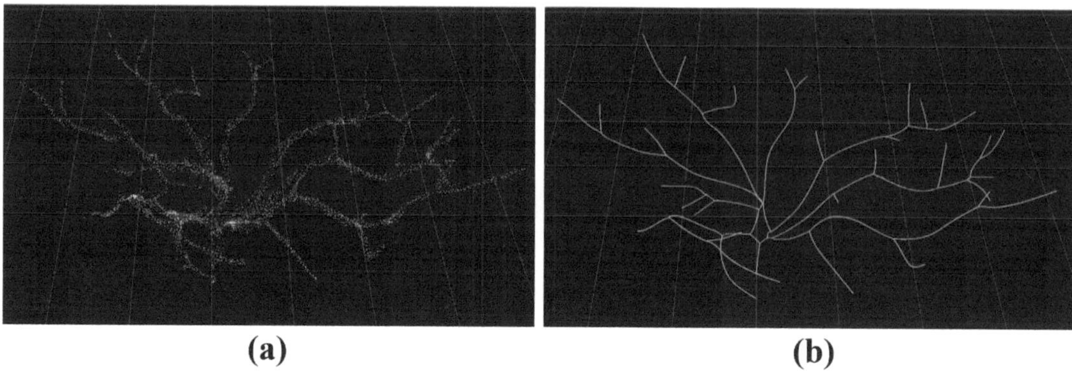

Figure 17. Fruit tree branch reconstruction based on the skeleton extracted by the algorithm in this paper (persimmon tree is shown as an example). (**a**) Original fruit tree branch point cloud. (**b**) Fruit tree branch skeleton extracted by this algorithm.

Author Contributions: W.Z., X.P. and W.G. conceived the ideas and designed the methodology; W.Z., X.P. and W.G. implemented the technical pipeline, conducted the experiments, and analyzed the results; G.C. and H.W. investigated relevant work; D.T. provided the dataset for the experiments; all authors discussed, wrote the manuscript, and gave final approval for publication. All authors have read and agreed to the published version of the manuscript.

Funding: This study was partially supported by the National Natural Science Foundation of China (NSFC) Program 62276009 and Japan Science and Technology Agency (JST) AIP Acceleration Research JPMJCR21U3.

Data Availability Statement: The data and executable files (EasySkeleton) used in this paper will be available upon request here: https://github.com/I3-Laboratory/EasySkeleton, (accessed on 17 December 2022).

Acknowledgments: The authors thank the technical staff of the Institute for Sustainable Agro-ecosystem Services (ISAS), the University of Tokyo for the field management, and Kubota for the fruit tree's location survey.

Conflicts of Interest: The authors declare that we have no conflict of interest.

Appendix A Branch and Trunk Skeleton Key Point Adjustment Strategy

Skeleton bifurcation point adjustment: as the point cloud at the bifurcation point of the fruit tree branches is dense, the extracted skeleton bifurcation points may deviate from the bifurcation center of the branch point cloud. As the subsequent optimization of the skeleton morphology needs to be based on the skeleton to accurately locate the bifurcation points of the skeleton and to better assist this process, the bifurcation points of the skeleton need to be adjusted toward the bifurcation center of the branch point cloud by moving the bifurcation points as far as possible toward the parent branch. Since the positioning of the bifurcation points of the skeleton in 3D space cannot be adjusted by simply fitting straight lines of the skeleton segments to find the intersection points (the intersection points of multiple non-coplanar lines cannot be determined in 3D space). Therefore, the following skeleton bifurcation point adjustment strategy is designed in this paper.

As shown in Figure A1, let the initial extracted coarse skeleton bifurcation point be bp, the sub-branch skeleton points neighboring bp be sp_1, sp_2, and the parent branch skeleton point be sp_f, let the line between sp_1 and sp_f be L_1, the line between sp_2 and sp_f be L_2, and the line between bp and sp_f be L_f. Calculate the projection points f_1 and f_2 of bp on L_1 and L_2, and then calculate the midpoints of the projection points f_1 and f_2 on L_f as the new skeleton bifurcation points bp', and update the skeleton segments where they are located, and the adjusted skeleton bifurcation points are more reasonably located

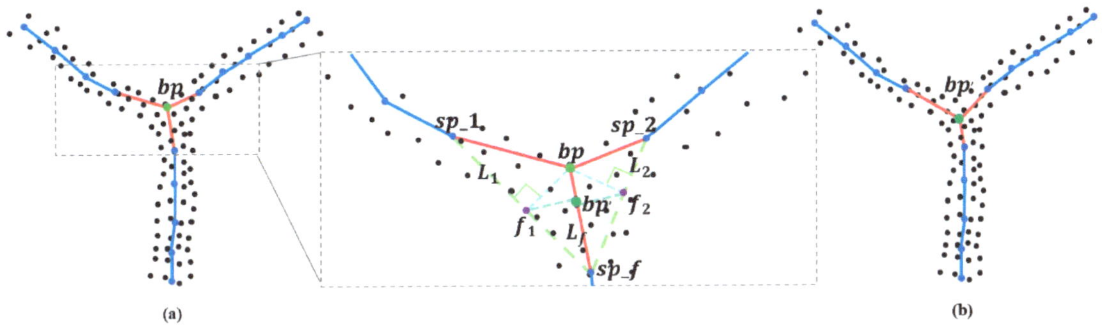

Figure A1. Schematic diagram of skeleton bifurcation point adjustment. (**a**) Positioning of skeleton bifurcation points before adjustment. (**b**) Positioning of skeleton bifurcation points after adjustment.

Skeleton end point adjustment: since the skeleton point extraction operation is built on top of the aggregated fruit tree branch point cloud, the extracted skeleton endpoints will naturally move towards the inside of the branch and cannot be positioned to the end of the fruit tree branch. To locate the skeleton endpoints accurately, it is necessary to extend the skeleton as far as possible to the endpoints of the fruit tree branch point cloud in the vicinity of the extracted skeleton endpoints. For this purpose, the following skeleton endpoint adjustment strategy is designed in this paper.

As shown in Figure A2, a spherical region with radius r (empirical value) is set at the end point ep as the center of the sphere O_E, and the point set C_E in which the point cloud C is in this region after the denoising is searched. The line where the skeleton point sp connected with the end point ep is taken as the normal l, the plane α perpendicular to l and containing the end point ep is calculated, the plane α is used to partition C_E, the set A_E of points in C_E that are on the opposite side from sp are separated and the point in A_E which is farthest from ep as the adjusted skeleton endpoint ep' is selected, and ep is connected with ep', as the extension of this skeleton segment.

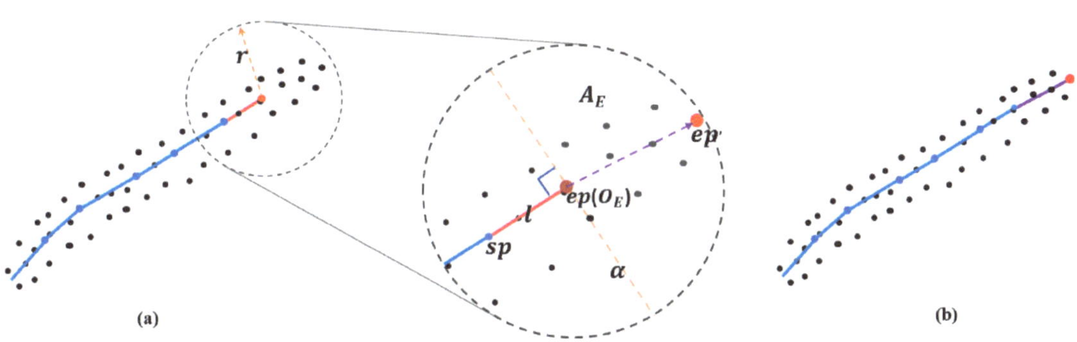

Figure A2. Schematic diagram of skeleton endpoint adjustment. (**a**) Skeleton endpoint positioning before adjustment. (**b**) Skeleton endpoint positioning after adjustment.

Appendix B Overview of Experimental Results

Exhibit 2-1. Comparison experimental contents of fruit tree branch point cloud skeleton extraction.

Comparison Experiments	Experimental Content	Figures and Tables
(1) 50 m altitude comparison experiment	(1-1) Visual analysis of skeleton extraction effect at 50 m altitude	Figure 11
	(1-2) Metric analysis of skeleton extraction effect at 50 m altitude	Table 2 Exhibit 1-1
	(1-3) Algorithm stability analysis (analysis of three 50 m altitude experimental metrics)	Appendix B Exhibit 3-1, 3-2, 3-3 Appendix B Exhibit 3-4
(2) 20 m altitude comparison experiment	(2-1) Visual analysis of skeleton extraction effect at 20 m altitude	Figure 12
	(2-2) Metric analysis of skeleton extraction effect at 20 m altitude	Table 2 Exhibit 3-1
(3) Analysis of experimental results at different altitudes	Analysis of the change of metrics from 50 m to 20 m experimental results	Table 2 Figure 13 Figure 14

Appendix C The Detailed Comparative Experimental Results of This Article

The arrows in the metrics column in the attached table represent the experimental metrics for which the algorithms in this paper performed poorly, and the bold font in each metric represents the value of the metric corresponding to the best algorithm under that metric.

Exhibit 3-1: comparison experimental results of branch point cloud skeleton extraction for fruit trees collected at 50 m altitude (1).

(Comparison experiment time: 13 January 2022.)

Tree_Species_Number	Metric	L_1-Medial	LBC	Proposed	
Tree_Peach_01	FBP	66.67%	58.82%	**85.71%**	
	FEP	42.11%	72.73%	**82.35%**	
	HD	0.0790	0.0756	**0.0631**	
Tree_Peach_02	FBP	81.82%	76.00%	**85.71%**	
	FEP	87.50%	81.25%	**89.66%**	
	HD	0.0897	0.0399	**0.0353**	
Tree_Peach_03	FBP	66.67%	47.62%	**76.19%**	
	FEP	**86.96%**	51.67%	69.57%	↓
	HD	0.0495	0.0584	**0.0432**	
Tree_Peach_04	FBP	72.00%	50.00%	**81.48%**	
	FEP	82.35%	78.57%	**89.66%**	
	HD	0.0502	0.0734	**0.0394**	
Tree_Peach_05	FBP	66.67%	55.56%	**88.89%**	
	FEP	69.23%	**90.91%**	86.96%	↓
	HD	0.0410	0.0398	**0.0354**	
Average_Peach	FBP	70.76%	57.60%	**83.60%**	
	FEP	73.63%	75.02%	**83.64%**	
	HD	0.0647	0.0574	**0.0433**	
Tree_Persimmon_01	FBP	42.86%	62.50%	**80.00%**	
	FEP	81.82%	**92.31%**	86.49%	↓
	HD	0.1043	0.0479	**0.0412**	
Tree_Persimmon_02	FBP	58.52%	55.17%	**68.97%**	
	FEP	62.50%	66.67%	**87.50%**	
	HD	0.0851	0.0772	**0.0750**	

Tree_Species_Number	Metric	L_1-Medial	LBC	Proposed	
Tree_Persimmon_03	**FBP**	58.82%	66.67%	**85.71%**	
	FEP	53.33%	86.96%	**86.96%**	
	HD	0.0963	0.0535	**0.0461**	
Tree_Persimmon_04	**FBP**	**80.00%**	70.59%	78.26%	↓
	FEP	34.48%	55.17%	**77.42%**	
	HD	0.0809	0.0709	**0.0739**	↑
Tree_Persimmon_05	**FBP**	50.00%	35.71%	**60.87%**	
	FEP	75.86%	35.71%	**88.89%**	
	HD	0.0725	0.0658	**0.0547**	
Average_Persimmon	**FBP**	58.04%	58.13%	**74.76%**	
	FEP	61.60%	67.36%	**85.45%**	
	HD	0.0878	0.0630	**0.0582**	
Tree_Chestnuts_01	**FBP**	23.16%	31.58%	**46.15%**	
	FEP	40.00%	22.22%	**47.06%**	
	HD	0.0909	0.1132	**0.0647**	
Tree_Chestnuts_02	**FBP**	22.22%	12.50%	**33.33%**	
	FEP	**47.06%**	38.10%	40.67%	↓
	HD	0.1582	0.1743	**0.1406**	
Tree_Chestnuts_03	**FBP**	33.33%	55.17%	**55.17%**	
	FEP	34.48%	47.62%	**58.82%**	
	HD	**0.0591**	0.0605	0.0807	↑
Tree_Chestnuts_04	**FBP**	**40.00%**	12.50%	22.22%	↓
	FEP	53.33%	19.05%	**46.15%**	
	HD	0.1325	**0.0953**	0.1023	↑
Tree_Chestnuts_05	**FBP**	50.00%	40.00%	**72.73%**	
	FEP	37.50%	40.00%	**50.00%**	
	HD	0.1250	0.0728	**0.0662**	
Average_Chestnuts	**FBP**	33.74%	30.35%	**45.92%**	
	FEP	42.48%	33.40%	**48.54%**	
	HD	0.1094	0.1032	**0.0909**	
Tree_Plum_01	**FBP**	36.36%	28.57%	**42.86%**	
	FEP	**70.59%**	52.17%	58.82%	↓
	HD	0.0845	0.1023	**0.0818**	
Tree_Plum_02	**FBP**	22.22%	54.55%	**61.54%**	
	FEP	66.67%	26.67%	**66.67%**	
	HD	0.1365	0.1036	**0.0770**	
Tree_Plum_03	**FBP**	**60.00%**	26.67%	57.14%	↓
	FEP	30.77%	11.11%	**56.92%**	
	HD	0.1611	**0.0773**	0.0993	↑
Tree_Plum_04	**FBP**	25.00%	9.52%	**40.00%**	
	FEP	**71.43%**	50.00%	66.67%	↓
	HD	0.1154	0.1147	**0.0683**	
Tree_Plum_05	**FBP**	60.00%	42.86%	**60.00%**	
	FEP	61.54%	42.11%	**61.54%**	
	HD	0.129	0.136	**0.109**	

Tree_Species_Number	Metric	L_1-Medial	LBC	Proposed	
Average_Plum	FBP	40.72%	32.43%	**52.31%**	
	FEP	60.20%	36.41%	**62.12%**	
	HD	0.1184	0.1068	**0.0871**	
Average_Total	FBP	50.82%	44.63%	**64.15%**	
	FEP	59.48%	53.05%	**69.94%**	
	HD	0.0951	0.0826	**0.0699**	

Exhibit 3-2: comparison experimental results of branch point cloud skeleton extraction for fruit trees collected at 50 m altitude (2).
(Comparison experiment time: 14 February 2022).

Tree_Species_Number	Metric	L_1-Medial	LBC	Proposed	
Tree_Peach_01	FBP	61.54%	58.82%	**85.71%**	
	FEP	66.67%	**72.73%**	82.35%	
	HD	0.0781	0.0716	**0.0628**	
Tree_Peach_02	FBP	66.67%	76.00%	**85.71%**	
	FEP	85.71%	81.25%	**89.66%**	
	HD	0.0927	0.0417	**0.0367**	
Tree_Peach_03	FBP	63.16%	42.11%	**72.73%**	
	FEP	**83.33%**	51.67%	66.67%	↓
	HD	0.0483	0.0563	**0.0453**	
Tree_Peach_04	FBP	72.00%	59.26%	**88.89%**	
	FEP	80.00%	81.48%	**89.66%**	
	HD	0.0579	0.0397	**0.0359**	
Tree_Peach_05	FBP	66.67%	52.63%	**88.89%**	
	FEP	76.92%	**90.91%**	86.96%	↓
	HD	0.0635	0.0372	**0.0332**	
Average_Peach	FBP	66.01%	57.76%	**84.39%**	
	FEP	78.53%	75.61%	**83.06%**	
	HD	0.0698	0.0493	**0.0428**	
Tree_Persimmon_01	FBP	60.00%	52.94%	**80.00%**	
	FEP	**95.00%**	91.89%	89.47%	↓
	HD	0.1129	0.0547	**0.0450**	
Tree_Persimmon_02	FBP	60.00%	46.67%	**69.57%**	
	FEP	68.57%	73.33%	**76.47%**	
	HD	0.0805	0.0798	**0.0703**	
Tree_Persimmon_03	FBP	53.33%	72.73%	**66.67%**	
	FEP	54.55%	86.96%	**86.96%**	
	HD	0.1277	0.0706	**0.0504**	
Tree_Persimmon_04	FBP	**78.79%**	75.86%	75.86%	↓
	FEP	66.67%	66.67%	**90.91%**	
	HD	0.0784	**0.0689**	0.0694	↑
Tree_Persimmon_05	FBP	70.00%	34.48%	**60.87%**	
	FEP	81.48%	35.71%	**88.89%**	
	HD	0.0710	0.0646	**0.0503**	

Tree_Species_Number	Metric	L_1-Medial	LBC	Proposed	
Average_Persimmon	FBP	64.42%	56.54%	**70.59%**	
	FEP	73.25%	70.91%	**86.54%**	
	HD	0.0972	0.0677	**0.0571**	
Tree_Chestnuts_01	FBP	23.16%	31.58%	**46.15%**	
	FEP	40.00%	22.22%	**47.06%**	
	HD	0.1017	0.1078	**0.0650**	
Tree_Chestnuts_02	FBP	22.22%	11.77%	**33.33%**	
	FEP	**47.06%**	42.11%	40.67%	↓
	HD	0.1563	0.1708	**0.1418**	
Tree_Chestnuts_03	FBP	33.33%	55.17%	**55.17%**	
	FEP	34.48%	47.62%	**58.82%**	
	HD	**0.0573**	0.0619	0.0791	↑
Tree_Chestnuts_04	FBP	**40.00%**	10.53%	22.22%	↓
	FEP	53.33%	17.39%	**46.15%**	
	HD	0.1265	**0.0922**	0.1086	↑
Tree_Chestnuts_05	FBP	50.00%	41.67%	**50.00%**	
	FEP	37.50%	40.00%	**50.00%**	
	HD	0.1293	0.0747	**0.0658**	
Average_Chestnuts	FBP	33.74%	30.14%	**41.38%**	
	FEP	42.48%	33.87%	**48.54%**	
	HD	0.1121	0.1015	**0.0921**	
Tree_Plum_01	FBP	36.36%	28.57%	**42.86%**	
	FEP	**70.59%**	52.17%	58.82%	↓
	HD	0.0864	0.0865	**0.0793**	
Tree_Plum_02	FBP	22.22%	50.00%	**76.92%**	
	FEP	70.59%	37.50%	**66.67%**	
	HD	0.1018	0.0987	**0.0750**	
Tree_Plum_03	FBP	**60.00%**	37.50%	57.14%	↓
	FEP	30.77%	11.11%	**56.92%**	
	HD	0.1736	**0.0801**	0.0984	↑
Tree_Plum_04	FBP	25.00%	9.52%	**36.36%**	
	FEP	**71.43%**	50.00%	66.67%	↓
	HD	0.1113	0.1320	**0.0656**	
Tree_Plum_05	FBP	60.00%	42.86%	**60.00%**	
	FEP	66.67%	42.11%	**72.34%**	
	HD	0.1701	0.1382	**0.1186**	
Average_Plum	FBP	40.72%	33.69%	**54.66%**	
	FEP	62.01%	38.58%	**64.28%**	
	HD	0.1216	0.1071	**0.0874**	
Average_Total	FBP	51.22%	44.53%	**62.75%**	
	FEP	64.07%	54.74%	**70.61%**	
	HD	0.1002	0.0814	**0.0698**	

Exhibit 3-3: comparison experimental results of branch point cloud skeleton extraction for fruit trees collected at 50 m altitude (3).
(Comparison experiment time: 4 March 2022).

Tree_Species_Number	Metric	L_1-Medial	LBC	Proposed	
Tree_Peach_01	**FBP**	76.92%	58.82%	**85.71%**	
	FEP	66.67%	63.64%	**82.35%**	
	HD	0.0786	0.0692	**0.0651**	
Tree_Peach_02	**FBP**	83.33%	76.00%	**88.00%**	
	FEP	85.71%	81.25%	**89.66%**	
	HD	0.0880	0.0407	**0.0390**	
Tree_Peach_03	**FBP**	63.16%	38.10%	**76.19%**	
	FEP	**76.92%**	51.67%	69.57%	↓
	HD	0.0490	0.0521	**0.0456**	
Tree_Peach_04	**FBP**	69.57%	46.15%	**81.48%**	
	FEP	78.57%	85.71%	**89.66%**	
	HD	0.0557	**0.0650**	0.0343	
Tree_Peach_05	**FBP**	80.00%	44.44%	**88.89%**	
	FEP	80.00%	**90.91%**	86.96%	↓
	HD	0.0512	0.0392	**0.0333**	
Average_Peach	**FBP**	74.60%	52.70%	**84.05%**	
	FEP	77.58%	74.64%	**83.64%**	
	HD	0.0669	0.0532	**0.0435**	
Tree_Persimmon_01	**FBP**	53.85%	56.25%	**80.00%**	
	FEP	81.82%	**91.89%**	89.47%	↓
	HD	0.0505	0.0533	**0.0489**	
Tree_Persimmon_02	**FBP**	60.00%	57.14%	**69.57%**	
	FEP	62.07%	75.00%	**76.47%**	
	HD	0.0817	0.0797	**0.0723**	
Tree_Persimmon_03	**FBP**	57.14%	66.67%	**66.67%**	
	FEP	60.87%	86.96%	**86.96%**	
	HD	0.1213	0.0739	**0.0557**	
Tree_Persimmon_04	**FBP**	**78.79%**	68.97%	75.86%	↓
	FEP	90.91%	60.00%	**90.91%**	
	HD	0.0977	**0.0681**	0.0685	↑
Tree_Persimmon_05	**FBP**	55.56%	31.25%	**60.87%**	
	FEP	62.86%	35.71%	**88.89%**	
	HD	0.0800	0.0688	**0.0422**	
Average_Persimmon	**FBP**	61.07%	56.06%	**70.59%**	
	FEP	0.7170	0.6991	**0.8654**	
	HD	0.0862	0.0687	**0.0575**	
Tree_Chestnuts_01	**FBP**	16.67%	40.00%	**42.86%**	
	FEP	33.33%	21.05%	**66.67%**	
	HD	0.0921	0.0904	**0.0625**	
Tree_Chestnuts_02	**FBP**	22.22%	11.77%	**33.33%**	
	FEP	**47.06%**	40.00%	40.67%	↓
	HD	0.1513	0.1542	**0.1458**	
Tree_Chestnuts_03	**FBP**	33.33%	55.17%	**55.17%**	
	FEP	34.48%	47.62%	**58.82%**	
	HD	**0.0656**	0.0685	0.0776	↑

Tree_Species_Number	Metric	L_1-Medial	LBC	Proposed	
Tree_Chestnuts_04	**FBP**	40.00%	9.52%	**22.22%**	↓
	FEP	53.33%	17.39%	**46.15%**	
	HD	0.1189	**0.0790**	0.0977	↑
Tree_Chestnuts_05	**FBP**	66.67%	41.67%	**66.67%**	
	FEP	50.00%	42.86%	**50.00%**	
	HD	0.1036	0.0870	**0.0669**	
Average_Chestnuts	**FBP**	35.78%	31.63%	**44.05%**	
	FEP	43.64%	33.78%	**52.46%**	
	HD	0.1039	0.0958	**0.0901**	
Tree_Plum_01	**FBP**	36.36%	26.09%	**42.86%**	
	FEP	**68.52%**	50.00%	58.82%	↓
	HD	0.0867	0.0973	**0.0810**	
Tree_Plum_02	**FBP**	22.22%	50.00%	**76.92%**	
	FEP	66.67%	25.00%	**72.34%**	
	HD	0.1060	0.1252	**0.0609**	
Tree_Plum_03	**FBP**	60.00%	31.53%	**57.14%**	↓
	FEP	30.77%	11.11%	**47.09%**	
	HD	0.1412	**0.0793**	0.0870	↑
Tree_Plum_04	**FBP**	25.00%	9.52%	**36.36%**	
	FEP	**71.43%**	50.00%	66.67%	↓
	HD	0.1341	0.1242	**0.0976**	
Tree_Plum_05	**FBP**	44.44%	28.57%	**57.14%**	
	FEP	66.67%	42.11%	**66.67%**	
	HD	0.1794	0.1518	**0.1018**	
Average_Plum	**FBP**	37.61%	29.14%	**54.09%**	
	FEP	60.81%	35.64%	**62.32%**	
	HD	0.1295	0.1155	**0.0857**	
Average_Total	**FBP**	52.26%	42.38%	**63.20%**	
	FEP	63.43%	53.49%	**71.24%**	
	HD	0.0966	0.0833	**0.0692**	

Exhibit 3-4: standard deviation of the comparison experimental results of branch point cloud skeleton extraction for fruit trees collected at 50 m altitude.
(Comparison experiment time: 4 March 2022).

Tree_Species_Number	Metric	L_1-Medial	LBC	Proposed
Tree_Peach_01	**FBP**	0.0783	0.0000	**0.0000**
	FEP	0.1418	0.0525	**0.0000**
	HD	0.0004	0.0032	**0.0013**
Tree_Peach_02	**FBP**	0.0922	**0.0000**	0.0132
	FEP	0.0103	0.0000	**0.0000**
	HD	0.0024	0.0009	**0.0019**
Tree_Peach_03	**FBP**	0.0203	**0.0478**	0.0200
	FEP	0.0508	0.0000	**0.0167**
	HD	0.0006	0.0032	**0.0013**

Tree_Species_Number	Metric	L_1-Medial	LBC	Proposed
Tree_Peach_04	FBP	**0.0141**	0.0674	0.0428
	FEP	0.0191	0.0359	**0.0000**
	HD	0.0040	0.0175	**0.0026**
Tree_Peach_05	FBP	0.0770	**0.0576**	0.0000
	FEP	0.0555	0.0000	**0.0000**
	HD	**0.0112**	0.0014	0.0012
Average_Peach	FBP	0.0430	0.0288	**0.0040**
	FEP	0.0260	0.0049	**0.0033**
	HD	0.0025	0.0041	**0.0004**
Tree_Persimmon_01	FBP	0.0868	0.0485	**0.0000**
	FEP	0.0761	**0.0024**	0.0173
	HD	0.0338	0.0036	**0.0039**
Tree_Persimmon_02	FBP	0.0085	0.0557	**0.0035**
	FEP	0.0364	0.0441	**0.0637**
	HD	0.0024	0.0015	**0.0024**
Tree_Persimmon_03	FBP	**0.0281**	0.0350	0.1100
	FEP	0.0405	0.0000	**0.0000**
	HD	0.0166	0.0110	**0.0048**
Tree_Persimmon_04	FBP	**0.0070**	0.0361	0.0139
	FEP	0.2831	**0.0577**	0.0779
	HD	0.0105	**0.0015**	0.0029
Tree_Persimmon_05	FBP	0.1032	0.0231	**0.0000**
	FEP	0.0955	0.0000	**0.0000**
	HD	0.0048	0.0022	**0.0063**
Average_Persimmon	FBP	0.0319	0.0109	**0.0241**
	FEP	0.0633	0.0183	**0.0063**
	HD	0.0059	0.0030	**0.0006**
Tree_Chestnuts_01	FBP	0.0375	0.0486	**0.0190**
	FEP	0.0385	**0.0067**	0.1132
	HD	0.0059	0.0119	**0.0014**
Tree_Chestnuts_02	FBP	0.0000	0.0042	**0.0000**
	FEP	0.0000	0.0201	**0.0000**
	HD	0.0035	0.0108	**0.0027**
Tree_Chestnuts_03	FBP	0.0000	0.0000	**0.0000**
	FEP	0.0000	0.0000	**0.0000**
	HD	0.0044	0.0043	**0.0015**
Tree_Chestnuts_04	FBP	0.0000	0.0151	**0.0000**
	FEP	0.0000	0.0096	**0.0000**
	HD	0.0068	0.0087	**0.0055**
Tree_Chestnuts_05	FBP	0.0962	**0.0096**	0.1177
	FEP	0.0722	0.0165	**0.0000**
	HD	0.0138	0.0077	**0.0005**
Average_Chestnuts	FBP	0.0118	**0.0080**	0.0228
	FEP	0.0067	**0.0025**	0.0226
	HD	0.0042	0.0039	**0.0010**

Tree_Species_Number	Metric	L_1-Medial	LBC	Proposed
Tree_Plum_01	FBP	0.0000	0.0143	**0.0000**
	FEP	0.0119	0.0126	**0.0000**
	HD	**0.0012**	0.0081	0.0013
Tree_Plum_02	FBP	**0.0000**	0.0262	0.0888
	FEP	**0.0226**	0.0679	0.0328
	HD	0.0189	0.0141	**0.0088**
Tree_Plum_03	FBP	0.0000	0.0543	**0.0000**
	FEP	0.0000	**0.0000**	0.0568
	HD	0.0163	0.0015	**0.0068**
Tree_Plum_04	FBP	0.0000	**0.0000**	0.0210
	FEP	0.0000	0.0000	**0.0000**
	HD	0.0122	**0.0086**	0.0178
Tree_Plum_05	FBP	0.0898	0.0825	**0.0165**
	FEP	0.0296	**0.0000**	0.0540
	HD	0.0270	0.0086	**0.0084**
Average_Plum	FBP	0.0180	0.0235	**0.0123**
	FEP	**0.0092**	0.0152	0.0120
	HD	0.0057	0.0050	**0.0009**
Average_Total	FBP	0.0075	0.0127	**0.0071**
	FEP	0.0249	0.0088	**0.0065**
	HD	0.0026	0.0010	**0.0004**

Exhibit 3-5: comparison experimental results of branch point cloud skeleton extraction for fruit trees collected at 20 m altitude.
(Comparison experiment time: 4 March 2022).

Tree_Species_Number	Metric	L_1-Medial	LBC	Proposed	
Tree_Peach_01	FBP	55.56%	59.38%	**82.35%**	
	FEP	62.30%	62.30%	**81.69%**	
	HD	0.0741	0.0751	**0.0651**	
Tree_Peach_02	FBP	65.71%	73.56%	**81.25%**	
	FEP	63.93%	65.00%	**79.25%**	
	HD	0.1025	**0.0862**	0.1002	↑
Tree_Peach_03	FBP	60.61%	72.73%	**83.72%**	
	FEP	71.11%	**71.80%**	71.11%	↓
	HD	0.0503	0.0557	**0.0387**	
Tree_Peach_04	FBP	51.52%	70.73%	**92.31%**	
	FEP	61.54%	62.34%	**89.58%**	
	HD	0.0909	0.0547	**0.0339**	
Tree_Peach_05	FBP	67.93%	80.65%	**91.89%**	
	FEP	63.64%	65.63%	**95.12%**	
	HD	0.0656	0.0775	**0.0350**	

Tree_Species_Number	Metric	L_1-Medial	LBC	Proposed	
Average_Peach	FBP	60.26%	71.41%	**86.30%**	
	FEP	64.50%	65.41%	**83.35%**	
	HD	0.0762	0.0698	**0.0546**	
Tree_Persimmon_01	FBP	75.00%	**83.08%**	82.35%	
	FEP	75.00%	68.85%	**88.57%**	
	HD	0.0855	**0.0612**	0.0675	↑
Tree_Persimmon_02	FBP	61.22%	70.18%	**87.50%**	
	FEP	55.74%	59.26%	**89.55%**	
	HD	0.1263	0.0713	**0.0589**	
Tree_Persimmon_03	FBP	**80.00%**	65.39%	79.25%	↓
	FEP	81.36%	46.81%	**93.55%**	
	HD	0.0860	0.0713	**0.0424**	
Tree_Persimmon_04	FBP	63.16%	60.47%	**80.77%**	
	FEP	83.64%	47.83%	**89.66%**	
	HD	0.0937	0.0836	**0.0394**	
Tree_Persimmon_05	FBP	64.62%	80.00%	**93.98%**	
	FEP	57.83%	84.71%	**88.89%**	
	HD	0.1205	0.0567	**0.0554**	
Average_Persimmon	FBP	68.80%	71.82%	**84.77%**	
	FEP	70.71%	61.49%	**90.04%**	
	HD	0.0996	0.0688	**0.0527**	
Tree_Chestnuts_01	FBP	28.57%	**66.67%**	60.00%	↓
	FEP	38.46%	31.58%	**69.57%**	
	HD	0.0659	0.0686	**0.0532**	
Tree_Chestnuts_02	FBP	66.67%	62.50%	**66.67%**	
	FEP	34.78%	57.14%	**81.67%**	
	HD	0.1004	0.0701	**0.0321**	
Tree_Chestnuts_03	FBP	53.33%	70.59%	**84.12%**	
	FEP	66.67%	42.11%	**91.74%**	
	HD	0.0554	0.0532	**0.0336**	
Tree_Chestnuts_04	FBP	66.67%	50.00%	**94.74%**	
	FEP	47.62%	43.48%	**85.24%**	
	HD	0.0967	0.0560	**0.0291**	
Tree_Chestnuts_05	FBP	80.00%	35.29%	**94.12%**	
	FEP	90.00%	50.00%	**90.00%**	
	HD	**0.0405**	0.0521	0.0483	↑
Average_Chestnuts	FBP	59.05%	57.01%	**79.93%**	
	FEP	55.51%	44.86%	**83.64%**	
	HD	0.0708	0.0600	**0.0393**	
Tree_Plum_01	FBP	66.67%	32.00%	**69.57%**	
	FEP	**92.31%**	57.14%	88.89%	↓
	HD	0.1284	0.0604	**0.0575**	
Tree_Plum_02	FBP	57.14%	71.43%	**82.35%**	
	FEP	75.00%	53.33%	**77.78%**	
	HD	0.0415	0.0544	**0.0285**	

Tree_Species_Number	Metric	L_1-Medial	LBC	Proposed
Tree_Plum_03	FBP	58.82%	**84.21%**	72.73% ↓
	FEP	75.00%	60.00%	**75.00%**
	HD	0.1100	0.1007	**0.0441**
Tree_Plum_04	FBP	70.59%	73.68%	**95.24%**
	FEP	70.00%	50.00%	**78.26%**
	HD	0.1113	0.0630	**0.0407**
Tree_Plum_05	FBP	73.68%	60.00%	**90.00%**
	FEP	76.92%	43.48%	**88.00%**
	HD	0.0621	0.0893	**0.0454**
Average_Plum	FBP	65.38%	64.26%	**81.98%**
	FEP	77.85%	52.79%	**81.59%**
	HD	0.0970	0.0736	**0.0432**
Average_Total	FBP	63.37%	66.13%	**83.24%**
	FEP	67.14%	56.14%	**84.66%**
	HD	0.0859	0.0681	**0.0474**

References

1. Xu, H.; Gossett, N.; Chen, B. Knowledge and heuristic-based modeling of laser-scanned trees. *ACM Trans. Graph. (TOG)* **2007**, *26*, 19-es. [CrossRef]
2. Wang, Z.; Zhang, L.; Fang, T.; Mathiopoulos, P.T.; Qu, H.; Chen, D.; Wang, Y. A structure-aware global optimization method for reconstructing 3-D tree models from terrestrial laser scanning data. *IEEE Trans. Geosci. Remote Sens.* **2014**, *52*, 5653–5669. [CrossRef]
3. Livny, Y.; Yan, F.; Olson, M.; Chen, B.; Zhang, H.; El-Sana, J. Automatic reconstruction of tree skeletal structures from point clouds. In *ACM SIGGRAPH Asia 2010 Papers*; Association for Computing Machinery: New York, NY, USA, 2010; pp. 1–8.
4. Alexander, B.; Lindenbergh, R.; Menenti, M. SkelTre. *Vis. Comput.* **2010**, *26*, 1283–1300.
5. Andrea, T.; Zhang, H.; Cohen-Or, D. Curve skeleton extraction from incomplete point cloud. In *ACM SIGGRAPH 2009 Papers*; Association for Computing Machinery: New York, NY, USA, 2009; pp. 1–9.
6. Huang, H.; Wu, S.; Cohen-Or, D.; Gong, M.; Zhang, H.; Li, G.; Chen, B. L1-medial skeleton of point cloud. *ACM Trans. Graph.* **2013**, *32*, 1–8.
7. Mei, J.; Zhang, L.; Wu, S.; Wang, Z.; Zhang, L. 3D tree modeling from incomplete point clouds via optimization and L1-MST. *Int. J. Geogr. Inf. Sci.* **2017**, *31*, 999–1021. [CrossRef]
8. Song, C.; Pang, Z.; Jing, X.; Xiao, C. Distance field guided $$ L_1 $$ L1-median skeleton extraction. *Vis. Comput.* **2018**, *34*, 243–255. [CrossRef]
9. Bin, L.; Wang, Q.; Fan, X. An Optimized L1-Medial Skeleton Extraction Algorithm. In Proceedings of the 2021 IEEE International Conference on Industrial Application of Artificial Intelligence (IAAI), Harbin, China, 24–26 December 2021; IEEE: Piscataway, NJ, USA, 2021.
10. Cao, J.; Tagliasacchi, A.; Olson, M.; Zhang, H.; Su, Z. Point cloud skeletons via laplacian based contraction. In Proceedings of the 2010 Shape Modeling International Conference, mAix-en-Provence, France, 21–23 June 2010; IEEE: Piscataway, NJ, USA, 2010.
11. Su, Z.; Zhao, Y.; Zhao, C.; Guo, X.; Li, Z. Skeleton extraction for tree models. *Math. Comp. Model.* **2011**, *54*, 1115–1120. [CrossRef]
12. He, G.Z. The Trees Skeleton Extraction Based on point cloud Contraction. In *Applied Mechanics and Materials*; Trans Tech Publications Ltd.: Wollerau, Switzerland, 2014; Volume 475.
13. Wu, S.; Wen, W.; Xiao, B.; Guo, X.; Du, J.; Wang, C.; Wang, Y. An accurate skeleton extraction approach from 3D point clouds of maize plants. *Front. Plant. Sci.* **2019**, *10*, 248. [CrossRef]
14. Li, Y.; Su, Y.; Zhao, X.; Yang, M.; Hu, T.; Zhang, J.; Liu, J.; Liu, M.; Guo, Q. Retrieval of tree branch architecture attributes from terrestrial laser scan data using a Laplacian algorithm. *Agri. Forest Meteor.* **2020**, *284*, 107874. [CrossRef]
15. Jianling, Z.; Liu, J.; Zhang, M. Curve skeleton extraction via k-nearest-neighbors based contraction. *Int. J. Appl. Math. Comp. Sci.* **2020**, *30*, 123–132.
16. Lou, L.; Liu, Y.; Shen, M.; Han, J.; Corke, F.; Doonan, J.H. Estimation of branch angle from 3D point cloud of plants. In Proceedings of the 2015 International Conference on 3D Vision, Lyon, France, 19–22 October 2015; IEEE: Piscataway, NJ, USA, 2015.
17. Yinxi, G.; Yang, Y.; Yang, X. Three-dimensional reconstruction of the virtual plant branching structure based on terrestrial lidar technologies and L-system. *Int. Arch. Photogram. Remote Sens. Spatial Inf. Sci.* **2018**, *42*, 3.
18. Fu, L.; Liu, J.; Zhou, J.; Zhang, M.; Lin, Y. Tree skeletonization for raw point cloud exploiting cylindrical shape prior. *IEEE Access* **2020**, *8*, 27327–27341. [CrossRef]

19. Dell, M.; Stone, C.; Osborn, J.; Glen, M.; McCoull, C.; Rimbawanto, A.; Tjahyono, B.; Mohammed, C. Detection of necrotic foliage in a young Eucalyptus pellita plantation using unmanned aerial vehicle RGB photography–a demonstration of concept. *Aust. Forest.* **2019**, *82*, 79–88. [CrossRef]
20. Maimaitijiang, M.; Sagan, V.; Sidike, P.; Maimaitiyiming, M.; Hartling, S.; Peterson, K.T.; Maw, M.J.; Shakoor, N.; Mockler, T.; Fritschi, F.B. Vegetation metric weighted canopy volume model (CVMVI) for soybean biomass estimation from unmanned aerial system-based RGB imagery. *ISPRS J. Photogram. Remote Sens.* **2019**, *151*, 27–41. [CrossRef]
21. González-Jaramillo, V.; Fries, A.; Bendix, J. AGB estimation in a tropical mountain forest (TMF) by means of RGB and multispectral images using an unmanned aerial vehicle (UAV). *Remote Sens.* **2019**, *11*, 1413. [CrossRef]
22. Ester, M.; Kriegel, H.P.; Sander, J.; Xu, X. A density-based algorithm for discovering clusters in large spatial databases with noise *Inkdd* **1996**, *96*, 226–231.
23. Cornea, N.D.; Silver, D.; Min, P. Curve-skeleton properties, applications, and algorithms. *IEEE Trans. Nisual. Comp. Graphics* **2007** *13*, 530. [CrossRef]
24. Meagher, D. Geometric modeling using octree encoding. *Comp. Graphics Image Process.* **1982**, *19*, 129–147. [CrossRef]
25. Francisco, Y.; Silwal, A.; Kantor, G. Visual 3d reconstruction and dynamic simulation of fruit trees for robotic manipulation. In Proceedings of the IEEE/CVF Conference on Computer Vision and Pattern Recognition Workshops, Seattle, WA, USA, 13–19 June 2020.
26. Josh, H.M.L.; Yam, Y. A skeletonization technique based on delaunay triangulation and piecewise bezier interpolation. In Proceedings of the 2007 6th International Conference on Information, Communications & Signal Processing, Singapore, 10–13 December 2007; IEEE: Piscataway, NJ, USA, 2007.
27. Facundo, M. Gromov-Hausdorff distances in Euclidean spaces. In Proceedings of the 2008 IEEE Computer Society Conference on Computer Vision and Pattern Recognition Workshops, Anchorage, AK, USA, 23–28 June 2008; IEEE: Piscataway, NJ, USA, 2008
28. Krause, S.; Sanders, T.G.; Mund, J.P.; Greve, K. UAV-based photogrammetric tree height measurement for intensive forest 737 monitoring. *Remote Sens.* **2019**, *11*, 758. [CrossRef]
29. Sun, G.; Wang, X.; Yang, H.; Zhang, X. A canopy information measurement method for modern standardized apple orchards 739 based on UAV multimodal information. *Sensors* **2020**, *20*, 2985. [CrossRef]
30. Lefsky, M.A.; Cohen, W.B.; Harding, D.J.; Parker, G.G.; Acker, S.A.; Gower, S.T. Lidar remote sensing of above-ground biomass in 741 three biomes. *Glob. Ecol. Biogeogr.* **2002**, *11*, 393–399. [CrossRef]
31. Lambert, M.; Ung, C.; Raulier, F. Canadian national tree aboveground biomass equations. *Can. J. For. Res.* **2005**, *35*, 1996–2018 [CrossRef]
32. Hilker, T.; van Leeuwen, M.; Coops, N.C.; Wulder, M.A.; Newnham, G.J.; Jupp, D.L.; Culvenor, D.S. Comparing canopy metrics 745 derived from terrestrial and airborne laser scanning in a Douglas-fir dominated forest stand. *Trees* **2010**, *24*, 819–832. [CrossRef]
33. Liu, J.; Zhang, X.; Li, H.; Dai, M. Creation of tree models from freehand sketches by building 3D skeleton point cloud. In *International Conference on Technologies for E-Learning and Digital Entertainment*; Springer: Berlin/Heidelberg, Germany, 2010.
34. Preuksakarn, C.; Boudon, F.; Ferraro, P.; Durand, J.B.; Nikinmaa, E.; Godin, C. Reconstructing plant architecture from 3D laser scanner data. In Proceedings of the 6th International Workshop on Functional-Structural Plant Models, Davis, CA, USA, 12–17 September 2010.
35. Guo, J.; Jiang, H.; Benes, B.; Deussen, O.; Zhang, X.; Lischinski, D.; Huang, H. Inverse procedural modeling of branching structures by inferring L-systems. *ACM Trans. Graph.* **2020**, *39*, 1–13. [CrossRef]
36. Dobson, D.; Dong, H.; van der Horst, N.; Langhorst, L.; van der Vaart, J.; Wu, Z. Tree Reconstruction from a Point Cloud Using an L-System. Master's Thesis, Technische Universiteit Delft, Delft, The Netherlands, 2021.

drones

MDPI

Review

UAV Formation Trajectory Planning Algorithms: A Review

Yunhong Yang [1], Xingzhong Xiong [1,2,*] and Yuehao Yan [2]

1 School of Automation and Information Engineering, Sichuan University of Science and Engineering, Yibin 644000, China
2 Artificial Intelligence Key Laboratory of Sichuan Province, Sichuan University of Science and Engineering, Yibin 644000, China
* Correspondence: xzxiong@suse.edu.cn

Abstract: With the continuous development of UAV technology and swarm intelligence technology, the UAV formation cooperative mission has attracted wide attention because of its remarkable function and flexibility to complete complex and changeable tasks, such as search and rescue, resource exploration, reconnaissance and surveillance. The collaborative trajectory planning of UAV formation is a key part of the task execution. This paper attempts to provide a comprehensive review of UAV formation trajectory planning algorithms. Firstly, from the perspective of global planning and local planning, a simple framework of the UAV formation trajectory planning algorithm is proposed, which is the basis of comprehensive classification of different types of algorithms. According to the proposed framework, a classification method of existing UAV formation trajectory planning algorithms is proposed, and then, different types of algorithms are described and analyzed statistically. Finally, the challenges and future research directions of the UAV formation trajectory planning algorithm are summarized and prospected according to the actual requirements. It provides reference information for researchers and workers engaged in the formation flight of UAVs.

Keywords: heuristic algorithm; machine learning; multi-UAV formation; trajectory planning

Citation: Yang, Y.; Xiong, X.; Yan, Y. UAV Formation Trajectory Planning Algorithms: A Review. *Drones* **2023**, 7, 62. https://doi.org/10.3390/drones7010062

Academic Editors: Yosoon Choi and Diego González-Aguilera

Received: 16 December 2022
Revised: 29 December 2022
Accepted: 9 January 2023
Published: 16 January 2023

1. Introduction

Since its outstanding performance in the Gulf War in 1991, drones have made good achievements in the Afghanistan War, the Iraq War, the fight against the Islamic State (ISIS) terrorist group, the "Neptune Spear" decapitation operation in 2011, and the Russia–Ukraine conflict in 2022. Their success has caused countries around the world to invest a large amount of manpower and financial resources in the research of UAV [1], as shown in Figure 1. After decades of development, UAVs have not only been applied in the military fields of reconnaissance, surveillance, communication relay, electronic countermeasures, combat assessment, harassment, decoy, anti-submarine, target attack, etc. At the same time, they have been widely used in agriculture [2], energy [3], civil [4] and other very important fields. However, there are some problems with a single drone performing its mission. For example, when a single UAV performs a reconnaissance mission, it may be limited by the observation angle and cannot observe the target area from multiple different orientations [5]; when faced with a large-scale search task, a single UAV cannot effectively cover the entire reconnaissance area [6]; during the attack, the combat range, killing radius, destruction capability and attack accuracy are limited, thus affecting the success rate of the entire combat mission [7]; if a single drone fails in the middle of a mission, it must immediately interrupt the mission and return, but in a war, it may delay the aircraft and destroy the entire operation plan. In order to improve combat effectiveness and make up for the deficiency of a single UAV, a multi-UAV cooperative formation (cluster) combat task is proposed. It refers to the formation, maintenance or reconstruction of a certain geometric formation during the execution of a task by multiple UAVs to adapt to the battlefield situation and task requirements.

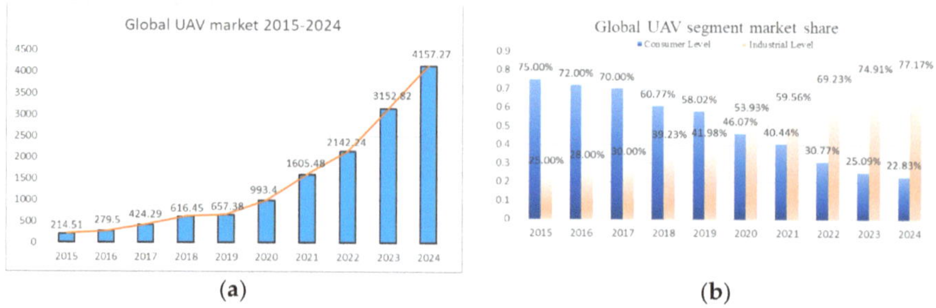

Figure 1. Statistics and forecast of global UAV market size from 2015 to 2024 (data source: Drone II): (**a**) Global UAV market 2015–2024 (blue column: global UAV investment; yellow line: upward trend); (**b**) Global UAV segment market share (blue column: proportion of consumer drones; yellow column: proportion of industrial drones).

Unmanned aerial vehicle formation has incomparable advantages over a single un manned aerial vehicle. When UAV formation is not possible in extreme weather, main taining different formations can improve the aerodynamic efficiency of UAVs to varying degrees, thus reducing the overall flight resistance and saving fuel [8]; it can realize the all-round reconnaissance or observation of the target, such as the enemy target monitoring and reconnaissance, resource exploration and so on, and it can greatly improve the scope of target monitoring; 0069t can also realize simultaneous strikes on multiple mission targets disrupt the enemy's combat command system, improve the lethality and hit rate of targets and improve combat effectiveness, as shown in Figure 2. In the process of use, UAVs are equipped with intelligent devices, which can simulate the transport environment in real time, determine their own position, control their flight status, select effective trajectory points, and calculate safe trajectory. These are important guarantees for UAVs formation to reach the target point from the take-off point as well as important prerequisites for a UAV formation to complete tasks. Therefore, it is important to select a suitable algorithm for UAV formation trajectory planning.

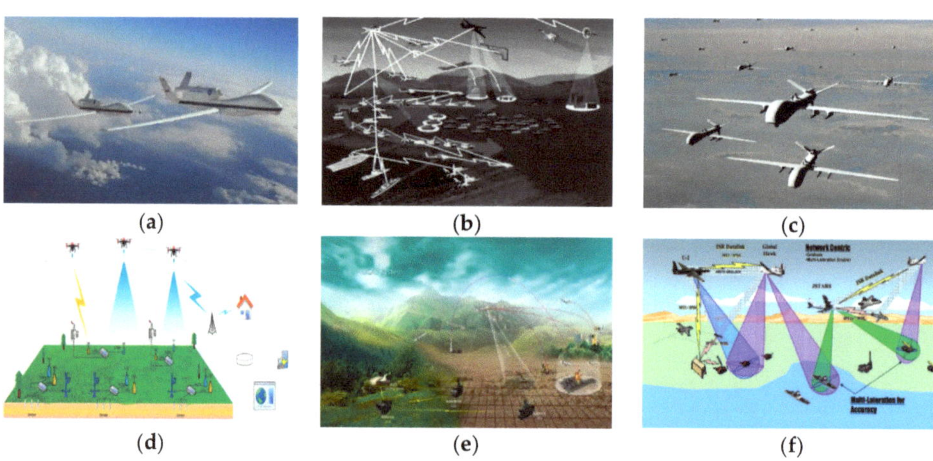

Figure 2. Unmanned aerial vehicles fly in formation to perform tasks. (**a**) UAV formation flight (**b**) UAV formation electronic warfare; (**c**) Unmanned aerial vehicle (UAV) formation performs mission; (**d**) UAV formation communication relay; (**e**) UAV formation strikes target; (**f**) UAV formation reconnaissance.

The purpose of trajectory planning of UAV formation is based on the specific tasks terrain, weather and other environmental factors of each UAV as well as its own flight performance. Under the premise of satisfying multiple constraints, the specified performance index can be optimized or better so that all UAVs in the formation can safely reach the target from the starting point. The trajectory planning of UAV formation is a complex multi-objective optimization and decision problem under multiple constraints. With the increasing number of UAVs, the analytic space of the problem will increase exponentially. In the study of UAV trajectory planning, the algorithm is the soul of UAV trajectory planning, which is directly related to the efficiency and results of trajectory planning. Compared with single UAV trajectory planning, the complexity of UAV formation trajectory planning is mainly reflected as follows:

(1) In many cases, the scope of planning space is large and complex: for example, there are various spatial obstacles and dynamic threats in the modern battlefield environment;
(2) There are many constraints. Not only should the planned flyable trajectory conform to the actual dynamics and kinematic characteristics of the UAV, but also the coordination between time and space and the concealment of the trajectory should be considered;
(3) Multi-UAV trajectory planning can adapt to battlefield dynamic changes and adjust trajectories online in real time.

For the trajectory planning of UAV formation, many papers have proposed solutions from different perspectives, but there are still many unsolved problems and many limitations, resulting in numerous and complex papers without a comprehensive and systematic classification, which is not conducive to research and reading.

The reference [9] classifies and statistically analyzes the cooperative flight path planning of various UAV formations from the three elements of a UAV system (mission, UAV crew and environment) and the three elements of UAV formation cooperative flight path planning (UAV flight path, target and constraint), but it does not discuss the flight path planning algorithm of UAV formation.

Stochastic Heuristic Algorithms (SHA) are reviewed in reference [10], and the characteristics, improvements, applications, advantages and disadvantages of some of them are discussed, but non-SHA algorithms in UAV formation flight paths are not discussed.

The reference [11] divides the flight path planning algorithms of UAV formation into five categories, including optimal algorithm, graph theoretics-based planning method, heuristic information-based planning algorithm, swarm intelligence algorithm and neural network algorithm. Then, a simple description is given to these categories, but no specific algorithms are discussed.

Reference [12] reviews swarm intelligence algorithms from four aspects, such as collision avoidance processing, task allocation, track planning and formation recombination, and it discusses classical algorithms among them. However, it does not discuss non-swarm intelligence algorithms, which has certain limitations.

Compared with many studies in the literature on UAV formation trajectory planning [9–12], the contributions of this paper are as follows.

In this review, the UAV formation trajectory planning algorithms used in recent decades are classified in detail, and the basic principles of these algorithms are described and compared so as to find out the shortcomings of UAV formation trajectory planning algorithms. The challenges and future research directions of the algorithm are summarized and prospected, which provides reference information for researchers and workers engaged in the formation flight of UAVs.

This paper can be divided into the following parts: Firstly, a simple classification framework of the UAV formation trajectory planning algorithm is introduced in Section 2. Then, the global trajectory planning algorithms are summarized in time order in Sections 3 and 4. Among them, Section 3 summarizes the traditional algorithm and Section 4 summarizes the intelligent algorithm. Section 5 summarizes the local trajectory planning algorithm. Section 6 summarizes the challenges the algorithm faces. Section 7 summarizes the focus and direction of future research. Section 8 summarizes the full text.

2. Classification Framework of UAV Formation Trajectory Planning Algorithm

This paper provides a classification framework of the UAV formation trajectory planning algorithm, which includes two elements: global planning and local planning.

The global trajectory planning algorithm belongs to the static programming algorithm which carries out trajectory planning based on existing map information and seeks an optimal trajectory from the starting point to the target point. In this paper, global trajectory planning algorithms are divided into traditional algorithms and intelligent algorithms according to whether they are inspired by natural organisms; the intelligent algorithms are divided into machine learning algorithms and heuristic algorithms according to whether they imitate human behavior or other animal behavior. The global trajectory planning algorithm framework is shown in Figure 3.

Figure 3. Framework diagram of global trajectory planning algorithm.

The local trajectory planning algorithm belongs to the dynamic trajectory planning algorithm, which means that the pilot aircraft collects the current position information and local obstacle information in real time according to the UAV sensor in the formation and then obtains the optimal trajectory between the starting point and the ending point. The local trajectory planning algorithm framework is shown in Figure 4.

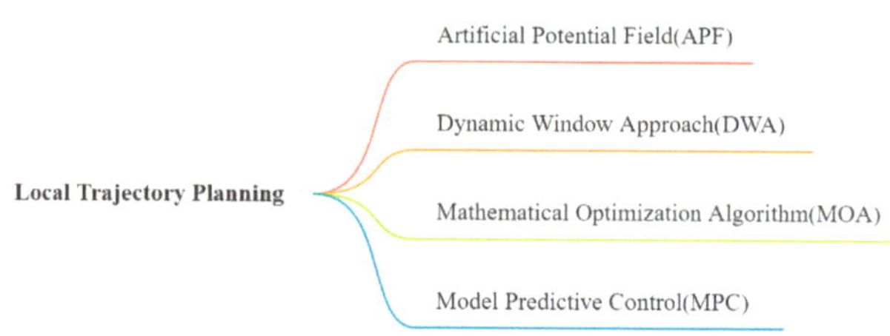

Figure 4. Frame diagram of local trajectory planning algorithm.

3. Traditional Algorithm

Traditional methods must build the map environment for the target before performing trajectory planning. Firstly, the map environment was discretized into graphs, and feasible trajectories were generated by the search algorithm to complete the global trajectory planning of UAV formation. The existing algorithms are the Dijkstra algorithm

Dubins Curve, Floyd algorithm, Voronoi graph method, Probabilistic Roadmaps (PRM), and Rapidly-Exploring Random Tree (RRT).

3.1. Dijkstra Algorithm

The Dijkstra algorithm is the classical shortest trajectory method in the geometric graph method, in which the vertex represents trajectory points, the edge represents a feasible trajectory, the line between nodes is called an edge, and each edge has a corresponding weight, which is the distance or cost of the journey; it is suitable for two-dimensional static obstacle avoidance scenes with a non-negative side weight. The key to using this algorithm is to select effective trajectory points, shorten the planning time, expand from the starting point, find the shortest trajectory for a node in each step, select the node with the smallest distance from the node that has never been visited to register, then traverse the adjacent nodes of the node after the node is included, and then update the distance. The cost diagram of a Dijkstra algorithm is shown in Figure 5.

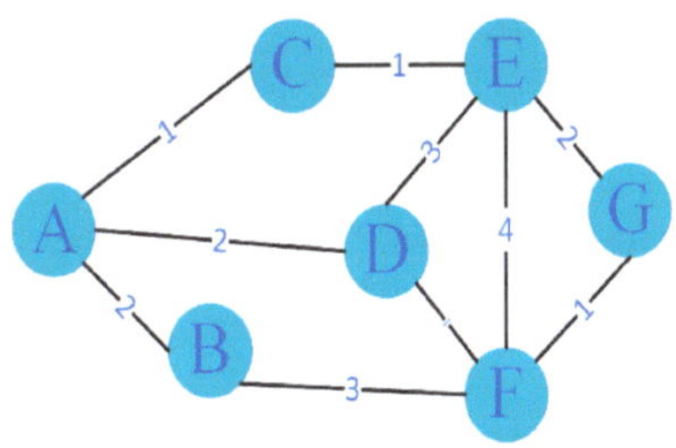

Figure 5. A cost diagram of a Dijkstra algorithm (A–G: nodes; lines: trajectories; numbers: the distance between vertices).

Aiming at the uncertain region search problem, Sujit and Ghose [13] proposed a search algorithm based on the K-shortest trajectory algorithm for UAV to search targets in an unknown environment. It satisfies the requirements of endurance time of each UAV and the location of the base station of UAV operation, and it enables each UAV to search in the area of maximum uncertainty so as to maximize the search benefit.

In order to meet the needs of searching an unknown environment and tracking moving targets in a balanced way, Tin [14] improved on Dijkstra's algorithm and proposed a robust shortest algorithm (ARSP) to deal with arc uncertainty. The influence of information uncertainty and environmental change on the trajectory planning algorithm is overcome, and the trajectory is quickly replanned at the same time.

Ueno and Kwon [15] applied the Dijkstra algorithm to the minimum time reconstruction of UAV formation in order to meet the requirements of optimality and short computing time, and the trajectory generated within the shortest time is close to the optimal trajectory.

Aggarwal et al. [16] proposed an approximate trajectory generation method to generate an approximate trajectory length under the condition of meeting the safety constraints of a UAV. This method is based on the total cost of the Lagrange relaxation (LARAC) algorithm, and it iteratively uses the Dijkstra algorithm (iDijkstra) to modify the edge cost, which solves the safety constraints and flight energy consumption of UAV caused by extreme high temperature.

3.2. Dubins Curve

The Dubins curve is the shortest locus connecting two two-dimensional planes (i.e., the X-Y plane) under the condition that the curvature constraint is satisfied and the tangent directions of the specified beginning and end are satisfied. In 1957, Lester Eli Dubins proved that any locus can consist of a maximum curvature arc or straight segment (the locus between two points must exist). In other words, the shortest path connecting two points will be constituted by the circular arc of the maximum curvature and the straight line segment. The Dubins of any starting point to the end point are composed of not more than three original motions, and the sequence constituted by the three original motions is called a kind of trajectory. As two continuous and identical primordial motions can be combined into one primordial motion, Dubins proved that the optimal trajectory can only be one of the following six combinations: namely, RSR, LSL, RSL, LSR, RLR, LRL. The first four are collectively called a CSC trajectory, and the last two are collectively called a CCC trajectory, where the primordial motion R represents right turn, S stands for straight and L stands for left. Figure 6 is the trajectory diagram of one Dubins curve LRL.

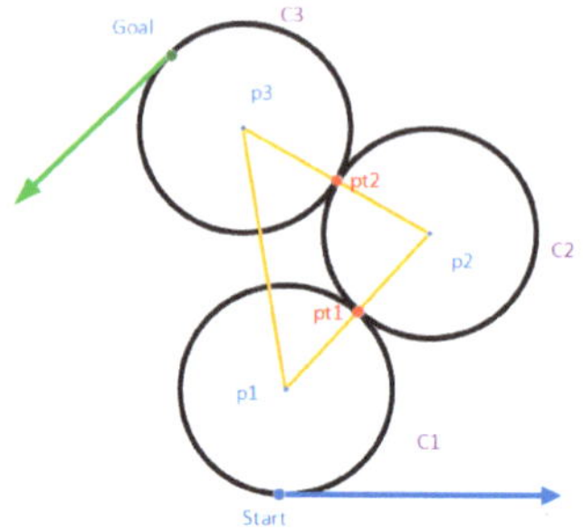

Figure 6. An LRL trajectory diagram of Dubins curve (black circle: circle curvature; yellow lines: the connecting line between the centers of trajectories; blue line: initial flight direction; green line: final flight direction; pt1–pt2: intersection point between curvatures; C1–C3: curvature name).

D'Amato, Mattei, and Notaro [17] modeled the UAV as a Dubins vehicle, using a method based on the Reduced Visibility Graph (RVG), connecting selected nodes by arcs and segments, and adding the Rendez-Vous Waypoints (RVWs). It was based on the leader-follower Stackelberg model's two-layer game theory method to optimize the location of the trajectory point and the trajectory of the UAV as much as possible in order to find the optimal trajectory while maintaining the shape of the formation in many places.

3.3. Floyd Algorithm

The Floyd algorithm, also known as the interpolation method, is a relatively classic algorithm for solving graph theory problems. It is an algorithm to solve the shortest trajectory between vertices in a given weighted graph, and it can correctly handle the shortest trajectory problem of directed graphs. At the same time, it is a dynamic programming algorithm, and the connection weight between nodes can be positive or negative; similar to Dijkstra's algorithm, but different from it is that Floyd's algorithm is used to find the distance between any two points, which is the shortest path of multiple sources, and it can be calculated with negative weights, while Dijkstra's algorithm is used to find the shortest route from one vertex to all other vertices, is the single-source shortest path, and negative weight circuits cannot be calculated.

Faced with the problem of multi-UAV cooperative patrol trajectory planning under constraints such as time windows, mandatory patrol nodes, UAV flight time and imaging sensors, Yang et al. [18] proposed a new cooperative patrol trajectory planning method using the Floyd algorithm to generate the initial trajectory, and then used the improved forward insertion heuristic algorithm (PFIH) to obtain the optimal trajectory.

Zhou and Nie [19] proposed a graph-based trajectory planning method for multi-UAV systems, using the Floyd algorithm to update the adjacent cost matrix and trajectory matrix and solved the problem of UAV formation trajectory planning.

3.4. Fast Marching Method

The fast marching method (FM) is an efficient numerical algorithm for solving the optical path function equation (Eikonal equation), and the optical path function equation is as follows:

$$|\nabla T(x,y)|V(x,y) = 1 \quad (1)$$

where (x, y) is the coordinate of the calculation point in the pose space, $T(x, y)$ is the time when the interface function arrives at the calculation point, $V(x, y)$ is the propagation velocity set by the interface function, and it is a fixed value in trajectory planning. The optical path function solution model is shown in Figure 7.

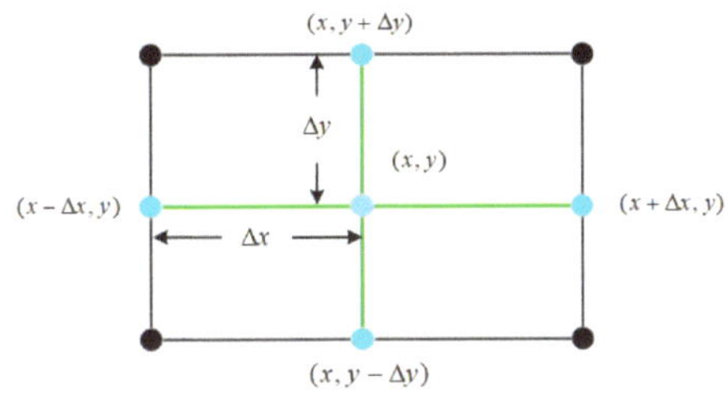

Figure 7. Optical path function solution model diagram (Δx and Δy: spacing in x and y directions on discrete space).

The fast marching method first establishes a rasterized space for storing time values, and then, the time cost will be converted into the distance cost during planning. Then, we set reachable points and unreachable points and complete the minimum value search operation by continuously updating the distance cost to obtain the distance matrix. We use it to construct the potential field and then use the gradient descent method to iterate continuously from the starting point along the direction of the fastest gradient descent in the generated potential field, obtaining a smooth trajectory without collision.

Aiming at the problem that the trajectory obtained by the traditional FM algorithm in the 3D environment will be too close to obstacles and the trajectory is not smooth enough, López et al. [20,21] proposed a fast marching square algorithm (FM2), which improves the FM algorithm by changing the propagation speed in space so that the wave will tend to follow the track travel.

3.5. Voronoi Diagram Method

The Voronoi graph method (also known as Dirichlet tessellation) is a space segmentation algorithm proposed by Russian mathematician Georgy Voronoy. It divides the space into many sub-regions through a series of seed nodes (Seed Points), each sub-region is called a Cell, and the distance between all points in each Cell and the Seed Points in the current Cell is less than Distance to all other Seed Points. According to the distribution of obstacles, the Voronoi diagram squares the free space between the edges of the obstacles, and at the same time, it draws the vertical line of adjacent obstacles to form a polygon around the obstacles so that each side is equidistant from the surrounding obstacles. Then, the origin and destination nodes can be connected into the graph by constructing trajectories from the nodes to the edges closest to each node. Figure 8 is a Voronoi trajectory diagram.

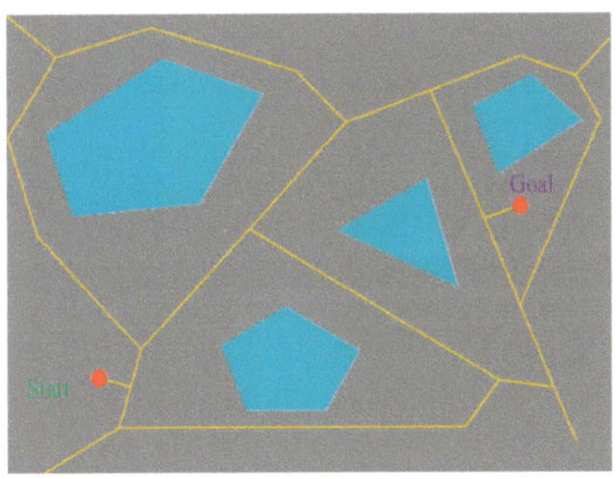

Figure 8. A Voronoi diagram method of trajectory diagram (blue area: obstacles; yellow lines: feasible trajectories).

Unavoidable accidents or environmental interference problems will inevitably occur when UAV formations perform multi-mission planning and collaborative trajectory planning. In order to cope with this situation, Meng et al. [22] proposed an algorithm to deal with multi-UAV multi-task trajectory re-planning in an unexpected event environment. Each UAV uses a Voronoi diagram to plan its own initial, optimal or sub-optimal trajectory; then, it replans its trajectory according to the new multi-task requirements corresponding to some unexpected events.

To solve the coverage problem with average Voronoi partitions, Chen et al. [23] proposed a distributed coverage algorithm to cover the convex area of the average Voronoi partition of the UAV formation. By exchanging local information with neighbors, the Voronoi partition is continuously iteratively updated, and the UAV direction of movement is calculated. The algorithm can theoretically make the area difference infinitely small so as to achieve the actual average area coverage.

Chen et al. [24] proposed a method based on consistency theory, using the Voronoi diagram method to create a threat domain, and designing a cost function for trajectory planning of multiple UAVs, so that multiple UAVs can take off at the same time and reach the specified target Point, solving the problem of UAV formation attacking multiple targets in a static threat environment.

Hu et al. [25] proposed a distributed formation control and collision avoidance method based on the Voronoi partition and traditional artificial potential field, using the Voronoi partition theory to divide the entire space into non-overlapping regions, and further proposed the target switching scheme; this method solves the problem of local optimum when an artificial potential field is used as motion control law.

3.6. Probabilistic Roadmap Algorithm

The Probabilistic Roadmap Algorithm (PRM) is a method based on graph search which converts continuous space into discrete space. Trajectory planning is mainly divided into two stages. In the offline learning stage, a large number of robot pose points are randomly sampled; then, neighbor nodes are searched and connections are established to construct a landmark map. In the online query phase, a feasible trajectory is searched from the landmark map using a heuristic search algorithm based on the starting point target point and landmark map information. Figure 9 is a trajectory diagram of a roadmap algorithm.

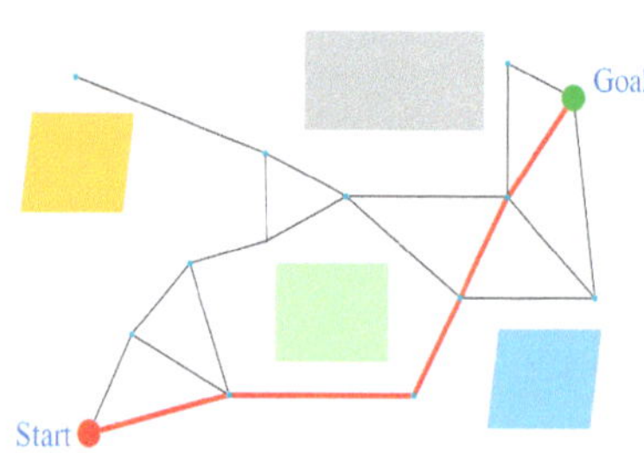

Figure 9. A PRM trajectory map (color areas: obstacles; black lines: feasible trajectories; red line optimal trajectory).

Madridano et al. [26] proposed a multi-trajectory PRM-based planning method by establishing a parameter to define three different modes, so that different UAVs in the UAV formation can achieve different mission goals.

3.7. Rapidly Exploring Random Trees

Rapidly exploring Random Trees (RRT) is a single query random search algorithm based on sampling. Its basic idea is to randomly sample in the state space, use the graph structure or tree structure extension to build a feasible trajectory set, and then find a complete feasible trajectory from the trajectory set. The RRT algorithm takes the starting point in the state space as the root node and then generates a random extended tree by gradually increasing the leaf nodes at random. If the newly generated node conflicts with the obstacle area during the generation process, the node is discarded and reselected. When the target point is included in the leaf node of the random tree, the expansion of the random tree stops, and an obstacle avoidance route from the starting point to the target point can be obtained. A fast random search tree locus is shown in Figure 10.

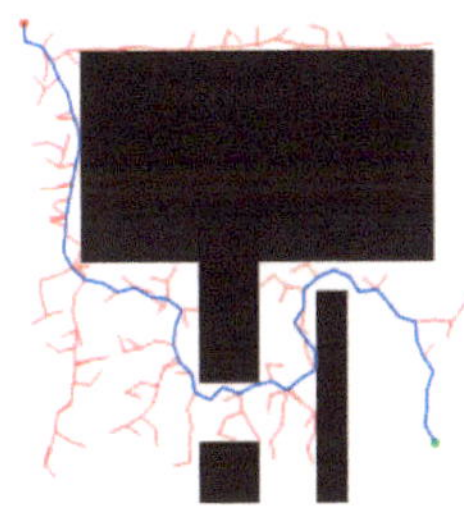

Figure 10. A RRT trajectory diagram (black areas: obstacles; pink lines: feasible trajectories; blue line: optimal trajectory; green: starting point; red: end point).

Aiming at the trajectory planning problem of UAV formation with static, ejection and dynamic obstacles, Kothar et al. [27] proposed a trajectory planning algorithm based on fast search random tree (RRTs) and introduced an anytime algorithm and guidance law based on tracking and line of sight into the algorithm to generate low-cost UAV formation trajectories under kinematic constraints in real time.

Zu et al. [28] proposed an improved Rapid Exploration Random Trees (RRTs) UAV formation collaborative trajectory planning algorithm, using a trajectory pruning method to delete redundant nodes on the trajectory. The UAV uses a trajectory planner, which enables the UAV to share information within the communication range.

When UAV formation faces sudden threat trajectory replanning, the classical RRT algorithm has some problems such as low efficiency, large storage space and slow convergence. Huang and Sun [29] proposed a bidirectional fast search random tree algorithm based on greedy strategy, improved the expansion mode of algorithm nodes, and adopted an adaptive step size rolling detection method to improve the sensitivity of UAV formation to sudden threats.

In addition, the traditional RRT algorithm also has the problem that it cannot adapt to the possible changes in the high-order dynamic characteristics of the autonomous movement of the UAV and the mission process during trajectory planning. In response to this problem, Shi et al. [30] proposed a trajectory generation algorithm based on the integration of the RRT algorithm and the minimum capture algorithm, using the RRT algorithm to generate the initial trajectory, and then using the minimum capture algorithm to smooth the initial trajectory, and using the concept of flight corridors to limit the flight trajectory of drones.

Table 1 summarizes the contents of our survey on traditional trajectory planning algorithms.

Table 1. Content of traditional trajectory planning algorithm in our survey.

Reference	Challenge	Optimization Criteria	Method	Dimension
P. Sujit and D. Ghose [13]	Environment	Trajectory of deviation	KSP	2D
C. Tin [14]	Information, Environment	ARSP	Dijkstra	2D
S. Ueno and S. J. Kwon [15]	Time, Optimality	Optimal control	Dijkstra	2D
R. Aggarwal et al. [16]	Security	LARAC	Dijkstra	2D
E. D'Amato, M. Mattei, and I. Notaro [17]	Environment	RVG, Bi-level optimization	Dubins	3D
J. Yang et al. [18]	Resources	PFIH	Floyd	2D
F. Zhou and H. Nie [19]	Environment	Shortest path	Floyd	2D
B. López et al. [20,21]	Trajectory	Lead–Follow, Multiple applications	FM	3D
B.-b. Meng et al. [22]	Environment	Task allocation	Voronoi + Dijkstra	2D

Table 1. *Cont.*

Reference	Challenge	Optimization Criteria	Method	Dimension
S. Chen et al. [23]	coverage problem	Distributed coverage	Voronoi	2D
X. Chen et al. [24]	Environment, Multiple objectives	Consistency theory	Voronoi	2D
J. Hu et al. [25]	UAV clustering	Target switching	Voronoi + APF	3D
Á. Madridano et al. [26]	Trajectory	Multiple trajectories	PRM	2D/3D
M. Kothari et al. [27]	Environment	Anytime, Guide rate	RRT	2D
W. Zu et al. [28]	Environment	pruning	RRT	2D
J. Huang and W. Sun [29]	Environment	Greedy strategy, Adaptive step size	RRT	3D
B. Shi et al. [30]	Environment	Minimum snap, Flight corridor	RRT	2D

4. Intelligent Algorithm

The intelligent algorithm is based on the principle of bionics computing, simulating the process of group biological behaviors to collaboratively search for the optimal solution in the space; for high-latitude, nonlinear, multi-constrained optimization problems, it can often converge to the optimal value in UAV formation trajectory planning at the same time, and it also solves the problem of UAV formation obstacle avoidance. In this paper intelligent algorithms are divided into two types: heuristic algorithms and machine learning algorithms.

4.1. Heuristic Algorithm

Most heuristic algorithms are optimization algorithms that search approximate optimal solutions based on empirical rules under acceptable computational costs to find solutions to problems. It is not a systematic search for answers, but the use of previous experience to select effective methods, and it cannot guarantee the speed of solutions and optimization degree of feasible solutions [31]. At present, the heuristic algorithms are mainly natural body-like algorithms. The heuristic algorithms used for UAV formation trajectory planning include the Simulated Annealing Algorithm (SA), A* Algorithm, Evolutionary Algorithm (EA), Particle Swarm Optimization (PSO), Pigeon-Inspired Optimization (PIO), Fruit Fly Optimization Algorithm (FOA), Artificial Bee Colony (ABC), Salp Swarm Algorithm (SSA), Ant Colony Optimization algorithm (ACO), Gray Wolf Optimization algorithm (GWO), Harmony Search algorithm (HS), etc.

4.1.1. Simulated Annealing Algorithm

The Simulated Annealing Algorithm (SA) is derived from the annealing of solid matter in physics. Usually, when a solid material is annealed, it is heated to allow its particles to move freely, and then, the particle system descends slowly enough to slow down sufficiently. The system is approximately at a thermodynamic equilibrium point and finally, the particle system will reach its lowest energy state, the ground state, which corresponds to the global minimum of the energy function. The objective function of the optimization problem is equivalent to the energy, and the optimal solution is equivalent to the lowest energy state. The simulated annealing algorithm changes randomly from one state to another state at a given temperature and uses the random acceptance criterion to judge. When the temperature slowly drops to a very low value, it remains at the optimal solution with a probability of 1. When the UAV formation is performing trajectory planning we first define a solution space, arrange the fixed starting point to the end point by unit use the Monte Carlo method as the initial solution, and iterate to create a new solution for the next trajectory point program. The exchange order of the two trajectory points in the obtained solution will generate a new solution. Then, we set the target function of the trajectory length of the UAV and use the simulated annealing criterion to test the cost function according to the data of the distance matrix. We use the difference between the cost functions to determine whether to accept the new trajectory planning and set the cooling process control parameters, initial temperature, cooling coefficient, end temperature, and current temperature iteration number. When the temperature drops to the end temperature the algorithm stops, reaches the minimum temperature, and outputs the formation. The optimal trajectory of the UAV using a simulated annealing algorithm trajectory is shown in Figure 11.

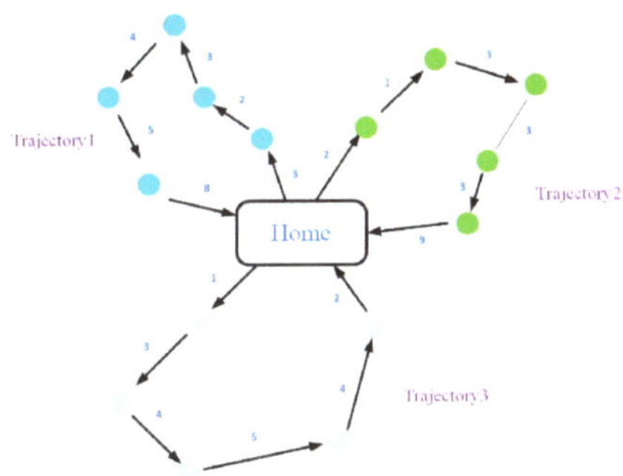

Figure 11. Trajectory diagram of a simulated annealing algorithm (dots: nodes; lines: trajectories; numbers: the distance between vertices).

Turker et al. [32] proposed an alternative method to effectively calculate the cost-fair flight path of a single-station multi-UAV system, using a data parallel computing mechanism to improve the simulated annealing algorithm, and solve the problem of the UAV formation trajectory planning calculation time index problem of growth.

Yue and Zhang [33] proposed a method of UAV formation trajectory planning based on the K-means algorithm and Simulated Annealing (SA) algorithm, using decomposition technology to reasonably decompose the effective area into multiple sub-target points. They use the K-means algorithm to cluster the UAV cruise target points and then use the Simulated Annealing (SA) algorithm for similar sub-target trajectory planning, which solves the problem of UAV cruise distance and scheduling under complex constraints and leads to improved coverage of drones in the sub-target area of the cruise effective area.

4.1.2. A* Algorithm

The A* algorithm is a graph search algorithm that introduces heuristic information factors into the target information of the problem to be solved, making the search direction more accurate and reducing the convergence time. The basic idea of this algorithm for UAV formation trajectory planning is as follows: firstly, the flight space is rasterized and decomposed into some units with regular shapes, and it is judged whether these units are covered by obstacles or intersected with obstacles. Then, find the unit containing the starting point and the target point and use the A* algorithm to find a series of connected units to connect the starting unit and the target unit. The search process of the A* algorithm is based on the value of the heuristic function in the direction of the lower cost; that is, for the node n, the algorithm uses the cost function to evaluate its surrounding nodes and selects the point with the smallest estimated value as the next node. The expression of the cost function is:

$$f(n) = g(n) + h(n) \tag{2}$$

where *h(n)* is the heuristic function; *g(n)* represents the prediction cost function from the current node position to the target point and represents the trajectory cost from the starting point to the current node n; and *f(n)* is the estimated value, which is obtained by adding *h(n)* and *g(n)*. In the grid graph, the heuristic function is usually expressed by the distance between two points. The calculation process of algorithm A* is a step-by-step search process, continuously extending to the direction of the minimum estimated value trajectory, calculating the optimal solution and outputting the optimal trajectory. Figure 12 is a trajectory diagram of an A* algorithm.

The traditional A* algorithm convergence speed is slow, and the trajectory may not be optimal. Hu et al. [34] proposed a distributed velocity perception and trajectory planning algorithm, which introduced a velocity perception strategy and collision prediction into the A* algorithm and carried out trajectory planning of UAV formation.

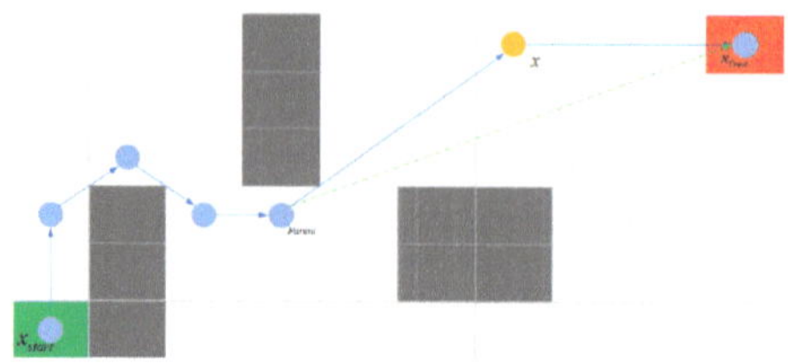

Figure 12. Trajectory diagram of A* algorithm (dots: nodes; lines: trajectories; black areas: obstacles; green dotted line: optimized trajectory).

Su et al. [35] proposed a cooperative search A* algorithm, which introduced cooperation strategies, cooperation constraints and cooperation costs into the constraint model, and they solved the problem of multi-aircraft formation trajectory planning with complex space–time constraints.

Zhang et al. [36] proposed a collaborative tactical planning method of UAV formation based on hierarchical structure, which introduced a hierarchical structure into UAV formation collaborative combat and solved the autonomous control problem of UAV formation in modern air combat.

Haghighi et al. [37] proposed a method based on the cell revisit time value and other effective cost functions such as height, minimum distance, collision avoidance and turning cost to realize multi-objective collaborative trajectory planning of multiple UAVs. A modification of the A* algorithm (MA*) was made to define a new criterion for individual revisit time unit values and extend it to the entire 3D mountain environment area, introducing revisit time and application-specific settings to reduce the computational complexity degree which solves the problems of the traditional A* algorithm, such as high computational complexity, small number of extension units and low ratio of coverage.

Nagasawa et al. [38] proposed a multi-UAV trajectory planning method in the case of three-dimensional building damage investigation or disaster, which combined the fuzzy c-means method of assigning positioning points to UAVs and the A* algorithm to calculate the access sequence of each UAV camera positioning point so as to obtain the feasible trajectory of multiple UAVs, which solves the problem of multi-UAV coverage trajectory planning for the 3D reconstruction of damaged buildings after disasters.

Luo et al. [39] proposed a convergent method to ensure autonomous non-collision trajectory planning of UAVs in the presence of static obstacles and dynamic threats. They extended the jump point search algorithm (JPS), parent node transfer law, seventh-order polynomial interpolation method of minimum capture, virtual gravity field and improved artificial potential field (APF) algorithm to a three-dimensional UAV. Based on a static environment, a collision-free trajectory is generated, which solves the trajectory planning problem of UAV formation flying at low altitude in urban and mountainous areas.

Table 2 summarizes the contents of our review about the simulated annealing algorithm and the A* algorithm.

4.1.3. Evolutionary Algorithm

The Evolutionary Algorithm (EA) is a stochastic optimization search algorithm summed up on the basis of biological evolution in nature. The most widely used algorithm is the Genetic Algorithm (GA). Its main idea is to rasterize the flight space first, find the area covered by obstacles or conflict with obstacles, and then randomly generate starting points in the map. To ensure the collision-free trajectory to the target point, in the trajectory planning process, each collision-free trajectory from the starting point to the goal point is represented as an individual, and each individual has a chromosome, so each collision-free trajectory can also become a chromosome. Each segment in the trajectory is represented as a gene. The collection of all individuals, that is, all generated collision-free trajectories from the starting point to the target point, are called the population. We design the corresponding fitness function to screen out the required individuals from the population. Individuals with high fitness are elite individuals; through the cross-mutation operation between elite individuals, better elite individuals are continuously screened until the termination conditions are met, and finally, what remains is the required obstacle avoidance route. Figure 13 is a diagram of a cross-mutation operation and a trajectory diagram of an Evolutionary Algorithm (EA).

Table 2. Summary of simulated annealing algorithm and A* algorithm in our review.

Reference	Challenge	Optimization Criteria	Method	Dimension
T. Turker et al. [32]	Trajectory	Parallel computing	SA	2D
X. Yue and W. Zhang [33]	coverage problem	K-means	SA	2D
Y. Hu et al. [34]	Trajectory	Speed perception, collision prediction	A*	3D
H. Su et al. [35]	UAV clustering	Constraint model	A*	2D
Z. Zhang et al. [36]	UAV clustering	Hierarchy	A*	3D
H. Haghighi et al. [37]	Trajectory and Multiple objectives	Revisit Time	A*	3D
R. Nagasawa et al. [38]	Environment	Fuzzy c-means method	A*	3D
Y. Luo et al. [39]	Environment	parent node, seventh-order polynomial interpolation	JPS + APF	3D

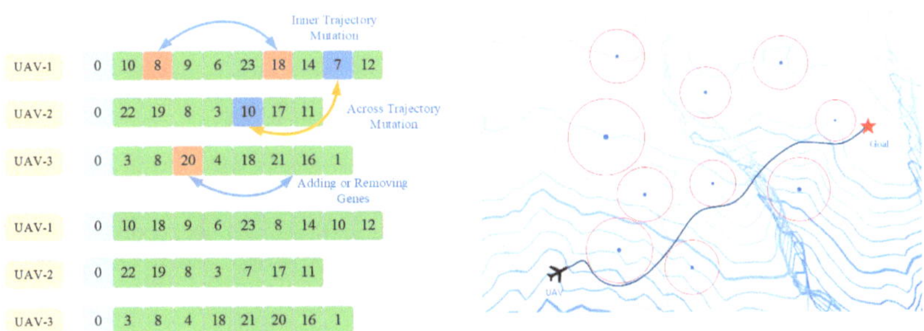

(a) (b)

Figure 13. A cross-variation operation diagram and EA trajectory diagram. (**a**) A cross-mutation operation diagram; (**b**) An Evolutionary Algorithm (EA) trajectory diagram (red circles: obstacles; blue line: optimal trajectory).

Tian et al. [40] proposed an algorithm based on model predictive control (MPC) and the Genetic Algorithm (GA) for multiple UAVs to search for unknown areas cooperatively, combining the flexibility of the Genetic Algorithm and the predictive ability of MPC. The combination avoids the problem where the search process enters into local optimality.

Shen et al. [41] proposed a method based on Genetic Algorithm (GA) to solve the multi-UAV cooperative reconnaissance mission planning problem, introduced integer string chromosome representation and designed a new subsequence crossover algorithm to meet the requirements of reconnaissance resolution. They also inserted mutation operators forward to increase the population diversity, which solves the problems of reconnaissance resolution and the time window when UAV formations perform reconnaissance missions.

Nikolos et al. [42] proposed a trajectory planner suitable for a group of cooperative UAVs to avoid collisions with environmental obstacles, combining b-spline curves, potential fields, and differential evolution (DE) to generate smooth the trajectory curve of the UAV formation, which solves the trajectory planning problem of the UAV formation in a known or unknown static environment.

Lamont et al. [43] proposed a multi-objective evolutionary algorithm (MOEA) for trajectory planning that introduces a Genetic Vector Router (GVR) while combining trajectory tracking capabilities with existing swarm behavior to measure the impact of these capabilities on the impact of swarm characteristics. By using "immigrant" population members to increase the search space and generate trajectories that meet mission requirements, they solved the problem of UAV formations exploring the terrain of larger areas and threatening regional trajectory planning.

Eun and Bang [44] developed an efficient strategy for the assignment and trajectory planning of homogeneous UAVs, combining Voronoi diagrams and Genetic Algorithms (GAs) to generate efficient flyable trajectories in network shapes, solving the problem of task assignment, and trajectory planning in the presence of time constraints is addressed.

Pehlivanoglu and Volkan [45] proposed a new multi-frequency Vibration Genetic Algorithm (mVGA), which constructed a Voronoi diagram using height filtering and fuzzy c-means clustering methods. They generated some initial individuals based on Voronoi vertices to improve the initial population, thereby generating efficient and fast flyable trajectories and solving the local optimization problem in a relatively short optimization period.

Sahingoz [46] proposed a flight-able trajectory planning method for multi-UAV systems, which combines the Genetic Algorithm (GA) and Bezier curves to generate an efficient and feasible trajectory of the UAV formation, solving the problem in which the curve is not smooth when using the traditional Genetic Algorithm (GA) for trajectory planning.

Zhang and Duan [47] proposed an improved constrained Differential Evolution (DE) algorithm, which combines the global search capability of the Differential Evolution (DE) algorithm and the constraint processing technology of level comparison, and they designed a level update strategy that solves the trajectory planning problem of formations under multiple constraints in real scenes.

Cekmez et al. [48] used a parallel Genetic Algorithm on the CUDA architecture to plan feasible trajectories for multiple UAVs; the algorithm first used a clustering method to find a subset of control points and then parallelized it on the programming computing platform. The Genetic Algorithm is used to solve each cluster and generate the feasible trajectory of the UAV formation, which solves the problem of long calculation time of the serial algorithm.

Sørli et al. [49] proposed a co-evolutionary multi-UAV cooperative trajectory planning method, which applied the co-evolutionary Genetic Algorithm to trajectory planning, and they considered the sensors carried by each UAV in the formation quantity and location effects, real-time or near-real-time trajectory planning for each UAV, solving the problem of trajectory planning for UAV formations in dynamic environments.

Chen et al. [50] proposed a parallel optimization method, which uses real coding methods and effective selection operations, crossover operations, and mutation operations to improve the Genetic Algorithm (GA), and at the same time, the Particle Swarm Optimization algorithm (the combination of PSO) and Ant Colony Optimization algorithm (ACO) makes the ants in the PSO-ACO system have particle characteristics. Then, it uses the two algorithms to generate formation trajectories simultaneously, which solves the weak global search ability of the Genetic Algorithm (GA) and the Ant Colony Optimization algorithm (ACO) premature maturation problem.

Binol et al. [51] proposed an improved evolution method of Genetic Algorithm (GA) and Harmony Search (HS); the improved search method utilizes various evolution operators with the same properties at the starting position to determine the overall shortest trajectories, which solves the problem of trajectory planning for drone formations when collecting data from multiple roadside units (RSUs).

Harounabadi et al. [52] proposed a Genetic Algorithm for the trajectory planning of multiple UAVs in message ferry networks. The Genetic Algorithm is used to create node clusters, and then, node scheduling in each cluster is defined according to the traffic between nodes and the message load in nodes. The problem of the average message passing delay of traditional mTSP schemes is solved.

Cao et al. [53] established a global optimization model that takes into account UAVs with various sensors located in different bases and multiple constraints, converts the time into an easily measurable way, and then uses the Genetic Algorithm analysis to solve the optimal detection track problem in the case of a multi-base.

Ma et al. [54] proposed a coordination optimization algorithm combining the Genetic Algorithm and clustering algorithm, using the task time constraint method to determine the number of UAVs required. They find the optimal trajectory for each UAV, solving the problem of multi-task assignment and trajectory planning of multiple UAVs.

Li et al. [55] proposed an improved trajectory planning algorithm based on GA. On the basis of a Genetic Algorithm, the optimal trajectory is obtained by the K-means target clustering algorithm and multi-chromosomal Genetic Algorithm, which solves the trajectory planning problem of multi-UAV maritime target search.

Xiong et al. [56] proposed a trajectory planning algorithm based on Genetic Algorithm with adaptive interference operators. The algorithm can realize the multi-directional attack target by setting intermediate points around the target point. A reasonable fitness function is designed by using the regionalization method, and the adaptive disturbance operator is added to plan the trajectory of each UAV, which solves the trajectory planning problem of multiple UAVs attacking targets in a complex combat environment.

Li et al. [57] proposed an optimized Genetic Algorithm method, which applied the augmented stochastic framework to evaluate the task completion probability (PoC) of the strategy in a three-dimensional grid environment, and then, they used the Genetic Algorithm optimization method to find feasible trajectories that maximize PoC, addressing the Reliability-Aware Multi-Agent Coverage Trajectory Planning (RA-MCTP) problem.

Li et al. [58] proposed a gray Genetic Algorithm, which iteratively uses the Genetic Algorithm to continuously find the agent trajectory that maximizes the PoC and solves the reliability-aware multi-agent coverage trajectory planning in continuous time (RA- MCTP) problem.

Zhang et al. [59] proposed a collaborative trajectory planning model, introduced decision variables into the trajectory cost model, and then improved the Genetic Algorithm to generate a formation flight trajectory, which solved the problems of short effective flight time and low mission success rate when multiple UAVs were threatened.

Asim et al. [60] proposed a variable population size genetic trajectory planning algorithm (GTPA-VP), which improves the Genetic Algorithm through three operators of insertion, replacement and deletion, and updates the stop point adaptively. Using the number and location, on this basis, a multi-color Genetic Algorithm is used to find the association between UAVs and stopping points, the optimal number of UAVs and the optimal order of UAV stopping points. Finally, a Genetic Algorithm is used to construct the flight trajectory of all drones, solving the problem of high energy consumption of drones hovering and flying in IoT services.

Yan et al. [61] proposed an improved Particle Swarm Optimization and Genetic Algorithm (GA-PSO), which introduced partial matching crossover and secondary transposition mutation to the traditional Particle Swarm Optimization (PSO) algorithm and solved the intelligent marine task assignment problem and trajectory planning problem for multiple UAVs.

Wang et al. [62] proposed a trajectory planning method based on the Genetic Algorithm (GA). Through the task analysis of the decision-making part and trajectory planning part, a Genetic Algorithm is used to initialize the trajectory; the fitness value calculation, selection, crossover, mutation and other operations are optimized to obtain the optimal search trajectory, which solves the trajectory planning problem of multiple UAV collaborative search tasks.

Table 3 summarizes the content of evolutionary algorithms in our review.

Table 3. Summary of evolutionary algorithms in our review.

Reference	Challenge	Optimization Criteria	Method	Dimension
J. Tian et al. [40]	Environment	Region division	MPC + GA	2D
L. Shen et al. [41]	UAV clustering	Subsequence crossover, Forward insertion mutation	GA	2D
I. Nikolos et al. [42]	Environment	B-Spline, DE	DE	3D
G. B. Lamont et al. [43]	Environment	GVR, Parallel computing	MOEAs	3D
Y. Eun and H. Bang [44]	Trajectory	Task allocation	Voronoi + GA	2D
Pehlivanoglu and Y. Volkan [45]	Trajectory	Height filtration, Fuzzy c-mean	Voronoi + GA	3D
O. K. Sahingoz [46]	Trajectory	Curves	GA	2D
X. Zhang and H. Duan [47]	Environment	Level update	DE	3D
U. Cekmez et al. [48]	Time, Trajectory	Cluster and parallel computing	GA	2D
J.-V. Sørli et al. [49]	Environment, UAV	Coevolution	GA	2D
J. Chen et al. [50]	Trajectory	Parallel optimization	GA + PSO +ACO	2D
H. Binol et al. [51]	Trajectory, Multiple objectives	Evolutionary operator	GA + HS	2D
M. Harounabadi et al. [52]	Time	Node scheduling	GA	2D
Y. Cao et al. [53]	Trajectory, Multiple objectives	Time conversion	GA	2D
Y. Ma et al. [54]	Trajectory, coverage problem	Clustering, Task time	GA	2D
L. Li et al. [55]	Trajectory, coverage problem	K-means, Multiple chromosome	GA	2D
C. Xiong et al. [56]	Trajectory	Adaptive interference operator, Regionalization	GA	3D
M. Li et al. [57]	coverage problem	Augmented random frame	GA	3D
M. Li et al. [58]	Time, coverage problem	Iterative use	GA	3D
J. Zhang et al. [59]	Environment	Decision variable, Adaptive	GA	2D
M. Asim et al. [60]	Environment, UAV	Variable population	GTPA-VP	2D
M. Yan et al. [61]	Trajectory, Multiple objectives	Partially matched crossover, Secondary transposition mutations	GA + PSO	2D
S. Wang et al. [62]	Multiple objectives, UAV clustering	Task analysis	GA	2D

4.1.4. Particle Swarm Optimization

Particle Swarm Optimization (PSO) is an adaptive stochastic optimization algorithm with a population search strategy developed by simulating the foraging behavior of birds which is used to solve various problems in engineering and science. Particle Swarm Optimization initializes the trajectory planning problem into a group of random particles and then iterates to find the optimal solution. In each iteration, particles update their position and velocity by tracking individual and global extreme values, and then, they use the search space to complete the optimal trajectory planning. Figure 14 shows a particle motion diagram.

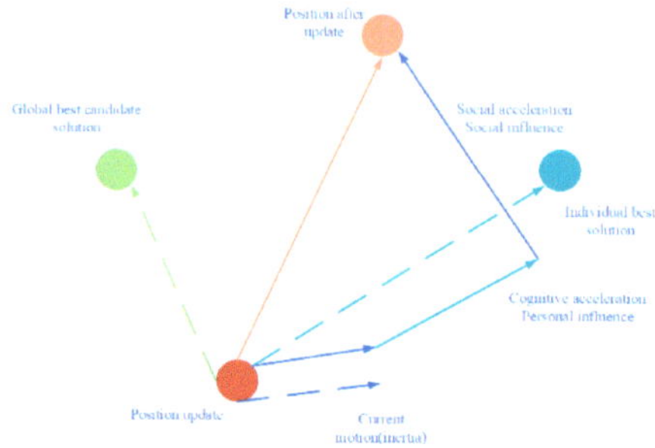

Figure 14. A particle motion diagram.

Sujit et al. [63] proposed a random time algorithm based on the Particle Swarm Optimization algorithm. Tracking guidance law and line of sight guidance law are used to track the trajectory generated by the Particle Swarm Optimization algorithm at any time, which solves the problem of using a pop-up window when multiple UAV tracks may collide and trajectory planning when there are moving obstacles.

Wang et al. [64] proposed a collaborative trajectory planning method for multiple UAVs based on the Particle Swarm Optimization (PSO) algorithm, analyzed the main influencing factors of the cost function after modeling, and carried out collaborative dynamic analysis of multiple UAVs, including static three-dimensional trajectory planning, which solves the problems of unsatisfactory trajectory and poor real-time performance in multi-UAV collaborative trajectory planning.

Alejo et al. [65] proposed a system for automatically planning collision-free four-dimensional trajectories; the system is based on the Particle Swarm Optimization (PSO) algorithm of axis-aligned minimum bounding boxes and stochastic global optimization techniques, and it uses a strategy to quickly calculate the initial point, solving the problems of high computational overhead and slow convergence in evolutionary algorithms.

Liu et al. [66] proposed a cooperative competitive Particle Swarm Optimization (PSO) algorithm, which uses two-stage optimization to reduce the dimensionality of the problem and generates the optimal collaborative trajectory for multiple UAVs in three-dimensional space, solving the trajectory planning problem for UAV formation in cooperative and competitive situations.

Zhang et al. [67] designed an improved PSO algorithm (IPSO), which used a binary value coding matrix and adaptive inertial weight adjustment strategy to generate a feasible trajectory for multiple UAVs, solving the decision-making problem of a multi-UAVs cooperative reconnaissance mission.

In this paper, Li et al. [68] proposed a new trajectory planning method for multiple UAVs by introducing a variable neighborhood drop (VND)-enhanced genetic Particle Swarm Optimization algorithm to optimize flight trajectory with minimum span and solve the problem of limited flight endurance of UAVs in agricultural applications.

Hoang et al. [69] proposed an angle-encoded Particle Swarm Optimization method, which realized the communication between UAVs through the Internet of Things board, minimized the cost function of multiple constraints including the shortest trajectory and the safe operation of UAVs, and found a feasible and frictionless trajectory for the whole formation. The trajectory planning problem of UAV formation in building infrastructure inspection is solved.

Chen et al. [70] proposed a trajectory planning method based on Dubins trajectory and the Particle Swarm Optimization (PSO) algorithm, using Dubins trajectory to reduce the dimensionality of the aircraft kinematics model. Then, using the Particle Swarm Optimization algorithm to optimize the trajectory after the formation reconstruction, it solves the trajectory planning problem when the task adjustment or the environment changes in the UAV formation.

Patley et al. [71] proposed an improved Particle Swarm Optimization method (ODPSO) based on orthogonal design and formulated a point sequence strategy to redefine the objective function. They searched for each the three-dimensional trajectory points within the time step, used the relative particle directivity to improve the search accuracy, and solved the trajectory planning problem of the UAV formation under the conditions of threats and terrain constraints.

Shao et al. [72] proposed a 3D trajectory planning algorithm for UAV formation based on Comprehensive Improved Particle Swarm Optimization (CIPSO). This method uses chaos-based logical mapping to improve the initial distribution of particles, designs commonly used constant acceleration coefficients and maximum speeds to adapt to linear change coefficients, and uses a mutation strategy in which the desired particles replace undesired particles, solving the terrain and threat constraints problems of UAV formation trajectory planning under the condition.

Yang et al. [73] proposed a 4D coordinated trajectory planning algorithm for multiple UAVs, which constructed the solution boundary of the search space and the distance to the destination based on the properties of all threats, and then designed a spatial refinement voting mechanism that solves the problems of local optimum and slow convergence of the standard Particle Swarm Optimization algorithm.

Shao et al. [74] proposed a Distributed Cooperative Particle Swarm Optimization (DCPSO) algorithm with an elite-preserving strategy, which parameterizes the trajectory using a Pythagorean Heatmap (PH) curve. Then, evolutionary theory is used to improve the Particle Swarm Optimization algorithm to generate a flyable and safe trajectory for each UAV, which solves the kinematic constraint problem of multi-UAV trajectory planning.

Liu and Lu [75] proposed an algorithm based on Dubins trajectory and Coevolutionary Particle Swarm Optimization (CCPSO). This algorithm determines the initial reference trajectory by the Dubins trajectory and then converts the time co-constraint into an equal trajectory length, and the trajectory parameters are optimized by CCPSO, which solves the problem of multi-UAV collaborative trajectory planning.

He et al. [76] proposed a new hybrid Particle Swarm Optimization and improved symbiotic search algorithm (HIPSO-MSOS), which introduces a time-stamp segmentation (TSS) model and a multi-objective optimization function to simplify the cost. Using HIPSO-MSOS to generate feasible trajectories and then smoothing trajectories by cubic b-spline curves, the problem of coordinated trajectory planning for multiple UAVs in complex 3D environments is solved.

Ahmed et al. [77] proposed a trajectory planner based on the Particle Swarm Optimization (PSO) algorithm, which uses distributed full coverage and dynamic fitness function to generate the optimal trajectory and solves the problem of trajectory planning for multiple UAVs.

Mobarez et al. [78] proposed an improved Particle Swarm Optimization method, improved the optimization problem by using evolutionary computing technology, added parallel recombination into trajectory planning, and solved the problems of long processing time and non-optimal trajectory in the dynamic trajectory planning of UAV formation.

Xiao et al. [79] proposed a Heterogeneous Adaptive Comprehensive Learning Dynamic Multi-population Particle Swarm Optimization algorithm (HACLDMS-PSO), which incorporated a population dynamic adjustment strategy, disturbance mechanism and adaptive learning probability mechanism into the Particle Swarm Optimization algorithm, which better solved the NP-hard problem in multi-UAV trajectory planning.

Meng-yun et al. [80] proposed a tracking planning method based on multi-strategy improved symbiosis search (MSISOS); this method uses an adaptive strategy and interference strategy to assist the search trajectory and coordinates space–time through UAV information interaction layer constraints. Then, a distributed method is designed for formation trajectory planning, which solves the problems of poor accuracy and slow convergence in multi-UAV trajectory planning in complex battlefield environments.

Chung et al. [81] proposed a trajectory planning algorithm that combines gradient descent-based trajectory planning (GBPP) and Particle Swarm Optimization. The initial trajectory of the algorithm is defined as the input of GBPP, and the hierarchical concept is added to the Particle Swarm Optimization algorithm (HPSO) to generate a feasible trajectory, which solves the problem of long calculation time of the Genetic Algorithm and Particle Swarm Optimization algorithm.

Lu et al. [82] proposed a distributed hybrid Particle Swarm Optimization and differential evolution (DE) technique; this technique adds the nonlinear time-varying method to the Particle Swarm Optimization algorithm (NTVPSO) and adds the adaptive mechanism to the differential evolution (DE) evolution (ADE). Finally, it adopts the distributed method, uses NTVPSO-ADE to realize the collaborative trajectory planning of multiple UAVs, and solves the problem of difficult model establishment and large amount of calculation in formation trajectory planning.

Table 4 summarizes the content of PSO algorithms in our review.

4.1.5. Pigeon-Inspired Optimization

The Pigeon-Inspired Optimization algorithm (PIO) is a swarm intelligent optimization algorithm designed to simulate pigeon homing behavior. First of all, three kinds of pigeon swarm optimization models were proposed: the map model based on the geomagnetic field, the pointer operator model based on the sun and the landmark operator model based on the landmark operator model. Secondly, a general direction was identified through the map and pointer operator, and then, the landmark operator was used to correct the current direction until the best track was found. Figure 15 shows a PIO map and compass operator model and a pigeon flock optimization (PIO) track chart.

Table 4. A summary of Particle Swarm Optimization in our review.

Reference	Challenge	Optimization Criteria	Method	Dimension
P. Sujit et al. [63]	Trajectory, Environment	Tracking, Line of sight guidance law	PSO	3D
G. Wang et al. [64]	Trajectory, Environment	Cost analysis	PSO	3D
D. Alejo et al. [65]	Time	Minimum boundary, Random optimization, One-time strategy	PSO	4D
J. Liu et al. [66]	UAV clustering	Two-stage optimization	PSO	3D
Y.-Z. Zhang et al. [67]	UAV clustering, Multiple objectives	Binary value coding matrices, Adaptive inertia weights	IPSO	2D
X. Li et al. [68]	Trajectory, UAV	VND	GPSO	2D
V. Hoang et al. [69]	Environment, coverage problem	Minimizing cost function	θ-PSO	3D
Q.-y. Chen et al. [70]	Environment	Reduction in dimension	Dubins + PSO	2D
A. Patley et al. [71]	Environment	Point sequence strategy, Inclined plane, Relative particle directivity	ODPSO	2D/3D
S. Shao et al. [72]	Environment	Logical mapping, Adaptive linear change	CIPSO	3D
L. Yang et al. [73]	Trajectory	Spatial refinement voting mechanism	PSO	4D
Z. Shao et al. [74]	Environment, UAV	Pythagorean heat map, Coevolution	DCPSO	3D
Y. Liu and H. Lu [75]	UAV clustering	Constraint conversion	Dubins + CCPSO	2D
W. He et al. [76]	Environment, UAV clustering	TSS, Multiple objective optimization	HIPSO-MSOS	3D
N. Ahmed et al. [77]	UAV clustering	Distributed full coverage, dynamic fitness	PSO	3D
E. Mobarez et al. [78]	Time, Trajectory	Evolutionary computation, Parallel recombination	PSO	3D
J. Xiao et al. [79]	Trajectory	Adaptive, interference, Interactive coordination	MSISOS	3D
W. Chung et al. [81]	Trajectory	GBPP, layered	HPSO	3D
L. Lu et al. [82]	Trajectory	Nonlinear time variation, adaptive, Distributed	NTVPSO-ADE	3D

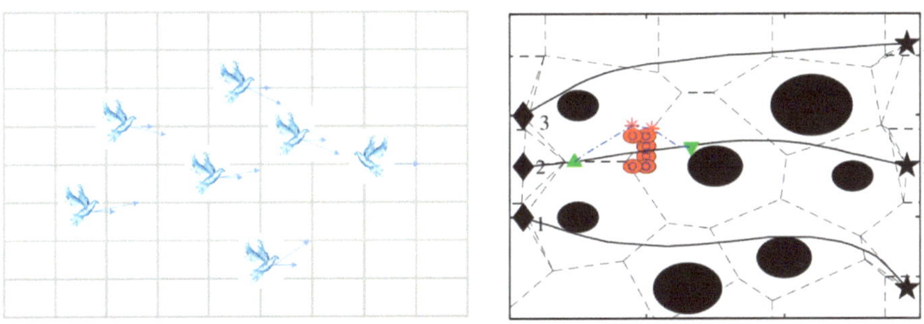

(a) (b)

Figure 15. A PIO map and compass operator model and PIO trajectory map. (**a**) A PIO map compass operator model (arrows: the direction of attraction); (**b**) A Pigeon-Inspired Optimization (PIO) trajectory diagram (diamonds: starting points; Pentagrams: the end points; black areas: obstacles).

Luo et al. [83] proposed a co-evolutionary Pigeon-Inspired Optimization (CPIO) algorithm based on a cooperation–competition mechanism. The search and track (ST) method is introduced to obtain the lowest-cost trajectory, and the dynamic two-stage closed search (DTCSCS) problem of UAV formation under range constraints (RC) and orientation constraints (OC) is solved.

Ruan and Duan [84] proposed a multi-objective social learning pigeon-inspired optimization algorithm (MSLPIO), which uses iterative learning to update waypoint positions, adding social learning factors and dimension-related parameter setting methods, which solves the problem of weak convergence of a traditional Genetic Algorithm.

Duan et al. [85] proposed a dynamic discrete Pigeon-Inspired Optimization algorithm based on hybrid architecture (D^2PIO), constructed and updated the probability mapping by using Bayesian formula, adopted the response threshold S-type function model (RTSM) for target allocation during attack execution, and finally used B-spline curve to generate feasible trajectory. The problem of search–attack task planning for multiple UAVs is solved.

Wang et al. [86] proposed a multi-UAV collaborative trajectory planning method based on the Cauchy mutant pigeon intelligent optimization algorithm (ECM-PIO); the algorithm uses the Cauchy mutation operator for optimization, expanding the search range and reducing the risk of falling into local optimization, which solves the shortcomings of the traditional pigeon swarm algorithm optimization process that has optimization bias and is easy to fall into local optimization.

Yu et al. [87] proposed a mutational pigeon swarm optimization algorithm (MGLPIO) based on swarm learning strategy, which introduces the swarm learning strategy, triple mutation strategy, timestamp segmentation mechanism and coordination cost function into the swarm optimization algorithm (PIO). They used it to solve the optimal trajectory, which solves the problems of low population diversity, weak global search ability and weak convergence of traditional PIO.

Lu et al. [88] proposed an improved Pigeon-Inspired Optimization algorithm (IPIO) based on natural selection and Gauss–Cauchy mutation, established an environment-aware map, and designed an integer encoding method. A discrete compass operator, discrete landmark operators, Gaussian mutation and Cauchy mutation operators are introduced to break away from local optimum. Finally, natural selection is used to accelerate convergence, which solves the problem of collaborative dynamic target search and area coverage of UAV formations in uncertain environments.

Zheng et al. [89] proposed a collaborative search decision-making method based on improved Pigeon-Inspired Optimization, which established a target probability information graph model with a normal distribution, an information graph of the search environment determinism, and a digital information graph. By adding the speed update and correction mechanism and the elite generation mechanism, they improve the traditional Pigeon-Inspired Optimization algorithm. Finally, the improved classification optimization method is used to determine the optimal search flight trajectory of the UAV, which solves the problem of multi-UAV cooperative moving target search.

Luo et al. [90] proposed a closed-loop trajectory planning method based on cooperative Pigeon-Inspired Optimization (CPIO) and artificial potential field (APF). Firstly, a probabilistic graphical model was established, and then, a rolling prediction strategy and CPIO were applied to generate multiple man–machine collaborative target search trajectories, while using Bayesian theorem to update the search probability map, and finally using the APF method to generate return trajectories for each UAV, which solves the multi-UAV cooperative target search problem.

4.1.6. Fruit Fly Optimization Algorithm

The Fruit Fly Optimization Algorithm (FOA) is a new method for deriving global optimization based on the foraging behavior of Drosophila, which uses Drosophila to be superior to other species in sensory perception, especially in the sense of smell and vision. First, fruit flies use their sense of smell to collect the smell in the air. Then, they fly to the vicinity of the food location, where they use vision to find the location where the food and companions gather and fly in that direction, so as to realize the group iterative search of the solution space and complete the multi-UAV trajectory planning. Figure 16 is an FOA iterative evolution search diagram and a Fruit Fly Optimization Algorithm (FOA) trajectory diagram.

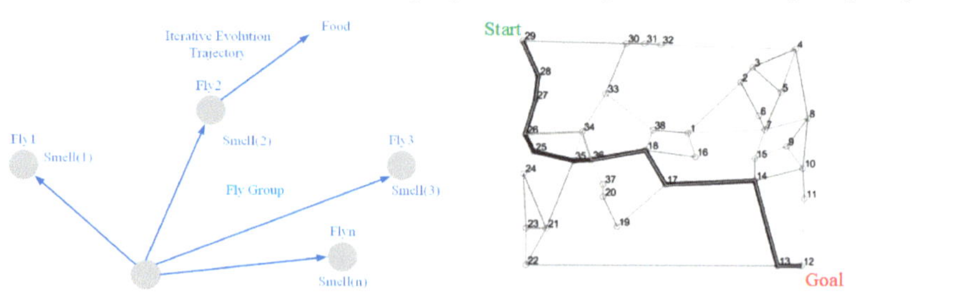

Figure 16. An iterative evolution search diagram and FOA trajectory diagram. (**a**) An iterative evolution search diagram; (**b**) An FOA trajectory diagram (circles: nodes; lines: trajectory; numbers: the distance between vertices).

Shi et al. [91] proposed the multi-swarm Fruit Fly Optimization Algorithm (MSFOA), which divides the entire fruit fly group into multiple multi-task sub-swarms and introduces offspring competition strategies. They propose a collision detection method to solve the problem of slow global convergence, and local optimum of the traditional Fruit Fly Optimization Algorithm is solved.

Li et al. [92] proposed an optimized Fruit Fly Optimization Algorithm (ORPFOA) to determine the optimal number and priority of UAVs while using a change task assignment algorithm combined with reference points and distance–cost matrices. Trajectory planning solves the problem of multi-UAV trajectory planning in a three-dimensional complex environment with online changing tasks.

Mao et al. [93] proposed an improved Fruit Fly Optimization Algorithm (NIFOA) based on Time Stamp Segmentation (TSS). The TSS model was introduced to solve the spatio-temporal coupling problem between multiple UAVs, and the multi-objective problem is transformed into a multi-constraint problem. Finally, the greedy strategy, the restart strategy and the evolutionary strategy of the optimal population are added to complete the multi-UAV trajectory planning, which solves the space–time coupling problem between multi-UAVs and the convergence speed of the traditional Fruit Fly Optimization Algorithm problems with slowness and local optima.

4.1.7. Artificial Bee Colony

The Artificial Bee Colony algorithm (ABC) is an optimization method to imitate the intelligent foraging behavior of bees. The process of the algorithm follows: First, assign a hired bee to the initial honey source and search according to certain rules to generate a new honey source. Then, use the greedy selection method to retain the honey source with high fitness and calculate the probability that the honey source found by the hired bee will be followed. Last, follow the peak using the same method as the hired bee. If the nectar source satisfies the condition of being abandoned, the corresponding hired bee becomes a scout bee and randomly searches in the search space to generate a new nectar source, obtaining the global optimal trajectory through the local optimization behavior of each individual artificial bee. Figure 17 is a trajectory diagram of the Artificial Bee Colony algorithm (ABC).

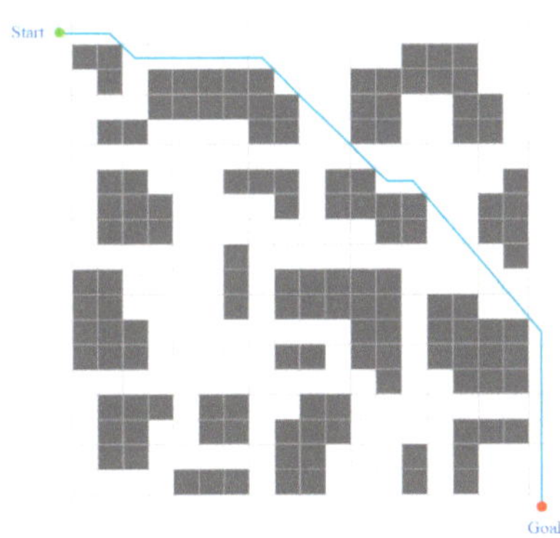

Figure 17. A kind of ABC trajectory diagram (Black areas: obstacles; blue line: optimal trajectory).

Tian et al. [94] proposed an improved Artificial Bee Colony (IABC) algorithm, which optimizes the trajectory points only according to the cost value of the trajectory and solves the problem of long convergence time of the traditional Artificial Bee Colony algorithm.

Bai et al. [95] proposed a hybrid algorithm based on Artificial Bee Colony algorithm (ABC) and A*. The algorithm uses the ABC algorithm to complete the preliminary planning, then uses the A* algorithm to plan the specific trajectory points, and finally combines the adaptive time coordination method to obtain the optimal trajectory, which solves the problem of three-dimensional multi-UAV trajectory planning.

Liu et al. [96] proposed a multi-UAV task assignment and trajectory planning method for disaster medical rescue. The algorithm uses the fitness function considering the current number of iterations and the maximum number of iterations and an Adaptive Genetic Algorithm (AGA) for task allocation; then, a balanced search strategy is added to improve the Artificial Bee Colony algorithm (IABC), and trajectory planning solves the problem of poor convergence efficiency and calculation effect of the traditional Artificial Bee Colony algorithm.

Table 5 summarizes the content of the Pigeon-Inspired Optimization algorithm, Fruit Fly Optimization algorithm and Artificial Bee Colony algorithm in our review.

Table 5. Summarizes the content of our review on PIO, FOA, and ABC.

Reference	Challenge	Optimization Criteria	Method	Dimension
D. Luo et al. [83]	DTSCS	ST	CPIO	2D
W.-y. Ruan and H.-b. Duan [84]	Trajectory	Iterative learning, social learning factor	MSLPIO	2D
H. Duan et al. [85]	UAV clustering	Bayes' formula, RTSM, B-spline curve	D^2PIO	3D
B. Wang et al. [86]	Trajectory	Cauchy mutation operator	ECM-PIO	3D
Y. Yu et al. [87]	Trajectory	triple mutation, timestamp segmentation, coordination costs	MGLPIO	3D
J. Lu et al. [88]	UAV clustering, coverage problem	Environment awareness, integer coding, discrete operators, mutation operators, natural selection	IPIO	2D
W. Zheng et al. [89]	UAV clustering	Probability graph model, pheromone graph, speed update, correction, elite generation	PIO	2D
D. Luo et al. [90]	UAV clustering	Probability graph model, rolling prediction, Bayes' theorem	CPIO + APF	2D
K. Shi et al. [91]	Trajectory	Offspring competition, collision detection	MSFOA	2D/3D
K. Li et al. [92]	Environment	Mission change and distance cost	ORPFOA	3D
Y. Mao et al. [93]	Trajectory	TSS, greedy strategy, restart strategy and evolution strategy	NIFOA	3D
G. Tian et al. [94]	Time	Trajectory cost	IABC	2D
X. Bai et al. [95]	UAV clustering	Adaptive time coordination	ABC + A*	3D
H. Liu et al. [96]	Trajectory	Fitness function, balanced search	AGA + IABC	2D

4.1.8. Salp Swarm Algorithm

The Salp Swarm Algorithm (SSA) is a new method to deduce and seek global optimization based on the swarming behavior of salps when navigating and foraging in the ocean. The SSA algorithm acts as an approximate global optimum by initializing a number of salps at random locations. Then, it calculates the fitness of each salp, finds the salp with the best fitness, assigns the position of the best salp to a variable as the source food to be chased by the salp chain, and uses the formula to update the fitness coefficient. For each dimension, the positions of the leading jumping body and the following jumping body are updated iteratively, and the search space determines the global optimal trajectory. Figure 18 is a Salp Swarm Algorithm (SSA) trajectory diagram.

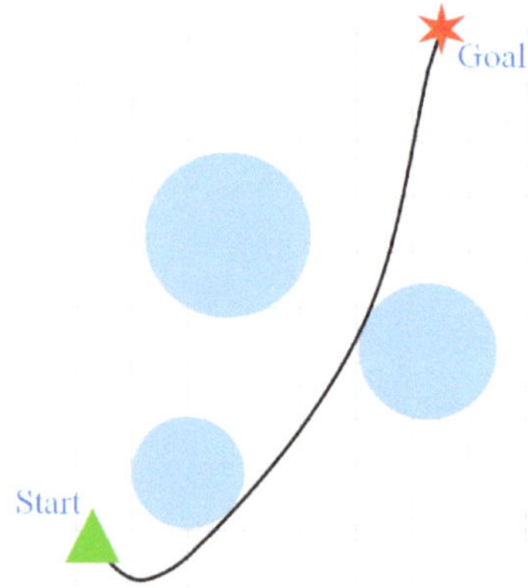

Figure 18. A SSA trajectory diagram (circles: obstacles).

Dewangan and Saxena [97] proposed a new Salp group algorithm (SSA), which uses multiple random operators to solve the problems of slow convergence and poor real-time performance of other heuristic algorithms in multi-UAV trajectory planning.

4.1.9. Ant Colony Optimization Algorithm

Ant Colony Optimization (ACO) is a heuristic global optimization algorithm derived from the trajectory behavior of ants in the process of searching for food. The ant colony algorithm uses the trajectories of ants to represent the feasible solution of the problem to be optimized. All trajectories of the entire ant colony constitute the solution space of the problem to be optimized, and ants with shorter trajectories release more pheromones. The concentration of pheromone accumulated on the shorter trajectory gradually increases, and the number of ants choosing this trajectory increases; eventually, all the ants will concentrate on the best trajectory under the action of positive feedback, and the corresponding trajectory is the optimal solution to the problem. Figure 19 is a trajectory diagram of an Ant Colony Optimization algorithm (ACO).

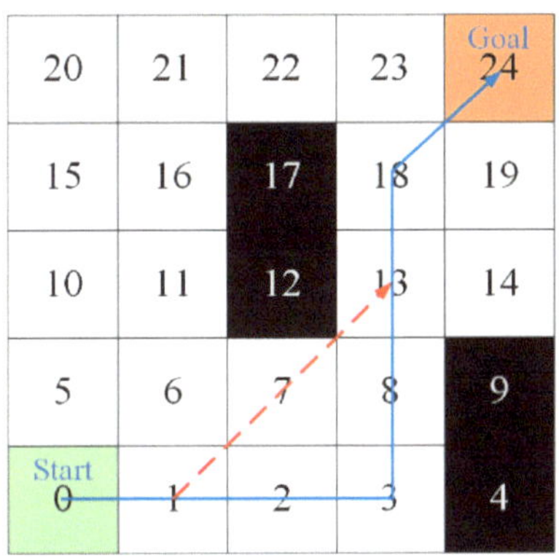

Figure 19. An ACO trajectory diagram ("cell" is the map block after rasterizing the map, "number" is the number of the map block, and the black part represents obstacles.).

Cekmez et al. [98] proposed a parallel Ant Colony Optimization algorithm (ACO) to calculate the trajectory of a UAV. This algorithm implements ACO on CUDA architecture, which gives full play to the parallel characteristics of ACO on GPU and solves the problem of slow convergence of a traditional ant colony algorithm.

Qiannan et al. [99] proposed an intelligent method based on the improved Ant Colony Optimization (ACO) algorithm, which cuts the trajectory generated by ACO and solves the problem that the trajectory of the traditional ACO algorithm may not be optimal.

Huang et al. [100] proposed a coordinated trajectory planning method for multiple UAVs based on K-degree smoothing. In this method, a Voronoi diagram is used to redefine the edge cost, and the redefined heuristic information function and pheromone updating method are used to change the Ant Colony Optimization algorithm. Finally, the K-degree smoothing method is used to smooth the trajectory, which solves the problem of strong coordination and weak coordination in the collaborative trajectory planning of multi-UAVs.

Li [101] proposed a multi-UAV multi-communication target and message priority UAV cooperative communication trajectory method. This method combines a delay tolerant network (DTN), light grid and ant colony algorithm. The trajectory planning is carried out to solve the contradiction between the intermittent link of the underlying communication of multi-UAVs and the continuous demand of the upper-level communication in a highly hostile battlefield.

Perez-Carabaza et al. [102] proposed a new method based on the improvement of Ant Colony Optimization (ACO), which added a new minimum time search (MTS) heuristic function to ACO to solve the traditional ACO problem of slow algorithm convergence and low initial trajectory quality.

Zhen et al. [103] proposed a multi-UAV cooperative search attack method (ISOA) based on an intelligent self-organizing algorithm, using a new state transition rule and a distributed method to improve the Ant Colony Optimization algorithm. Then, using the Dubins curve to smoothly connect the trajectory points generated by ACO, the trajectory planning problem of multiple UAVs under the constraints of maneuverability, collision avoidance and maximum range is solved.

Cekmez et al. [104] proposed an enhanced Ant Colony Optimization (ACO) algorithm, which performs multi-core computing on the parallel computing platform CUDA to solve the trajectory planning problem of multiple UAVs in complex environments.

Lin et al. [105] proposed a multi-objective optimization model of coverage and task time and introduced the similarity measure in an immune optimization algorithm into the Ant Colony Optimization algorithm to solve the problem that the traditional Ant Colony Optimization algorithm is insufficient in track repeatability.

Liu et al. [106] proposed an improved Ant Colony Optimization algorithm, which introduced the location allocation method and the new node selection strategy into the Ant Colony Optimization algorithm and solved the problem of slow trajectory planning optimization speed of the Ant Colony Optimization algorithm in formation transformation

Ali et al. [107] proposed a hybrid meta-heuristic algorithm, which combined maximum-minimum Ant Colony Optimization and differential evolution to solve the problem of slow global convergence of traditional Ant Colony Optimization algorithms and maximum-minimum Ant Colony Optimization algorithms.

Xia et al. [108] proposed a system framework for multi-UAV collaborative task assignment and tracking planning. The framework uses a Particle Swarm Optimization algorithm based on a guidance mechanism to solve the combinatorial optimization model Then, adaptive parameter adjustment, encounter point prediction, bidirectional search and online replanning are introduced into the Ant Colony Optimization algorithm for trajectory planning (SAP-ACO), which solves the cooperative task assignment and tracking planning problems of UAV formations facing moving targets.

Ali et al. [109] proposed a bionic optimization algorithm, which combines the maximum-minimum Ant Colony Optimization (MMACO) and the Cauchy mutant (CM) operator and they use the CM operator to enhance the MMACO algorithm to solve the problems of slow convergence and possible local optimum in traditional ACO and MMACO.

Wei and Xu [110] proposed a distributed trajectory planning algorithm based on dual decomposition of UAV communication chains. This algorithm improves the traditional Ant Colony Optimization algorithm (ACO) from the aspects of trajectory selection, pheromone update, rollback strategy, etc., and solves the problems of poor efficiency, adaptability and robustness of the ACO algorithm.

Li et al. [111] proposed an asynchronous Ant Colony Optimization (AACO) algorithm The visibility matrix and test track coverage matrix are added into the ACO algorithm. The search order of the population track primitive is changed from the current fitness value and the previous fitness value to the current fitness value. Finally, the incentive value is introduced to avoid track repetition, which solves the problem of optimal trajectory planning for multiple UAVs in three-dimensional space.

Majeed and Hwang [112] proposed a multi-objective coverage flight trajectory planning algorithm, which added the fitting sensor footprint scanning (SFS) and sparse trajectory point graph (SWG) to the Ant Colony Optimization (ACO) algorithm. Traversing the area of interest (AOI) solves the problem of high cost of multi-UAV coverage trajectory planning in urban environments.

4.1.10. Gray Wolf Optimization Algorithm

The Gray Wolf Optimization algorithm (GWO) is a new swarm intelligence optimization algorithm inspired by the predation behavior of gray wolves. The Gray Wolf

Optimization algorithm divides gray wolf individual fitness into four different levels: optimal solution, suboptimal solution, third solution and candidate solution according to the calculated fitness; the individuals with the top three fitness guide other wolves towards the goal Iterative search, while continuously updating the solution level and position, until the best trajectory to the target point is found. Figure 20 contains a Gray Wolf Optimization algorithm position update model and a Gray Wolf Optimization algorithm (GWO) trajectory diagram.

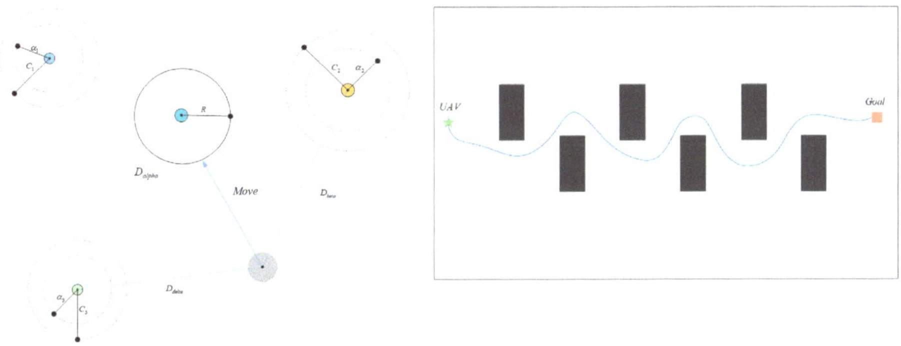

(a) (b)

Figure 20. A position update model and GWO trajectory map. (**a**) A location updating model of the Gray Wolf Optimization algorithm; (**b**) Trajectory diagram of a Gray Wolf Optimization algorithm (black areas: obstacles; blue line: optimal trajectory).

Radmanesh et al. [113] proposed a Bayesian algorithm based on Gray Wolf Optimization, which added dynamic Bayes and range-based value function (DBVF) into GWO to solve the trajectory planning and collision avoidance problems of multiple UAVs with fixed and moving obstacles in uncertain environments.

Dewangan et al. [114] proposed a multi-UAV trajectory planning method based on the Gray Wolf Optimization algorithm (GWO) to solve the problems of slow convergence, high trajectory calculation cost and local optimization of other meta-heuristic and deterministic algorithms in multi-UAV trajectory planning.

Xu et al. [115] proposed an improved Gray Wolf Optimization algorithm (GWO), which improved the population initialization, attenuation factor updating and single position updating of the Gray Wolf Optimization algorithm and solved the NP-hard problem of multi-UAV collaborative trajectory planning.

Yang et al. [116] proposed a trajectory planning method based on multi-population chaotic Gray Wolf Optimization (MP-CGWO). The multi-population concept and chaotic search strategy are added into the Gray Wolf Optimization algorithm (GWO), which solves the problem that the traditional GWO algorithm is easy to fall into local optimization.

Huang et al. [117] proposed a hybrid discrete intelligence algorithm (HDGWO) based on gray wolf optimizer. The algorithm uses the discrete gray wolf update operator and uses integer coding and a greedy algorithm to transform between the gray wolf space and the discrete problem space. Then, it adds the center position operation and the stagnation compensation gray wolf update operation, and finally, it adds an azimuth to improve the gray wolf algorithm, which solves the GWO problems of poor global convergence ability and local search ability.

Jiaqi et al. [118] proposed an adaptive multi-UAV trajectory planning method to improve the Gray Wolf Optimization algorithm (AP-GWO). This method adds the spiral update position and self-adaptive adjustment mechanism to the Gray Wolf Optimization algorithm, which solves the problems of relatively long convergence time, relatively unsmooth trajectory and possibly not optimal trajectory of the traditional GWO algorithm.

4.1.11. Harmony Search Algorithm

Harmony Search (HS) is a music-based heuristic optimization algorithm. The Harmony Search algorithm mimics the process of musical improvisation, in which musicians continually adjust the pitch of their instruments to achieve better harmony. The search process of the global trajectory planning problem is similar to the music improvisation process; that is, each decision variable constantly updates its own value during the search process so as to converge to the global optimum and obtain the optimal trajectory. Figure 21 is a Harmony Search algorithm (HS) trajectory diagram quoted from reference [119].

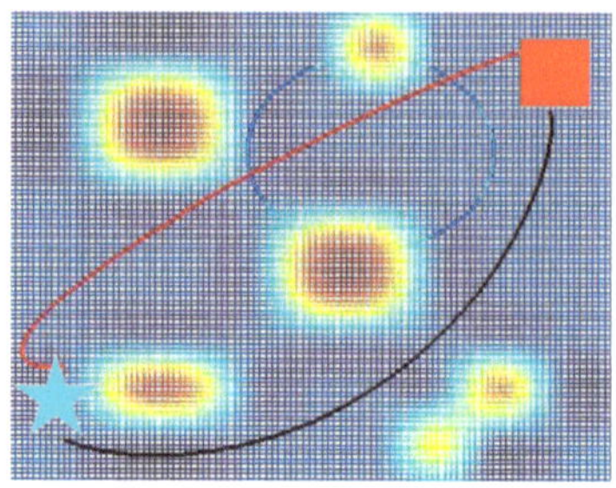

Figure 21. A kind of HS trajectory diagram (colored areas: obstacles).

Wu et al. [119] proposed an improved Harmony Search algorithm (MHS), which introduced an intersection mutation operator and Pythagorean heat map curve (PH) to improve the HS algorithm and solved the traditional HS problem of slow algorithm convergence.

Table 6 summarizes the contents of the Salp Swarm Algorithm, Ant Colony Optimization algorithm, Gray Wolf Optimization algorithm and Harmony Search algorithm in our review.

Table 6. Summary of the contents of SSA, ACO, GWO and HS in the review.

Reference	Challenge	Optimization Criteria	Method	Dimension
R. K. Dewangan and P. Saxena [97]	Time, Trajectory	Random operator	SSA	3D
U. Cekmez et al. [98]	Time	Parallel computing	ACO	2D
Z. Qiannan et al. [99]	Trajectory	Trajectory cutting	IACO	2D
L. Huang et al. [100]	UAV clustering	Redefine, k degree smoothing	Voronoi + ACO	2D
Z. Li [101]	Communication, UAV clustering	DTN, Light lattice diagram	ACO	2D
S. Perez-Carabaza et al. [102]	Time, Trajectory	MTS	ACO	3D
Z. Zhen et al. [103]	UAV, Trajectory	State transitions, Distributed	ISOA	2D
U. Cekmez et al. [104]	Environment	Parallel computing	ACO	3D
W. Lin et al. [105]	Trajectory	Similarity measure	ACO	2D
G. Liu et al. [106]	Time, UAV clustering	Location allocation, Node selection	ACO	2D
Z. A. Ali et al. [107]	Time, Environment	Mixed inspiration	MMACO + DE	3D
C. Xia et al. [108]	UAV clustering, Multiple objectives	Guidance mechanism, Adaptive, Bidirectional search	BSAPACO	2D
Z. A. Ali et al. [109]	Time, Trajectory	CM	MMACO	3D
X. Wei and J. Xu [110]	Time, Trajectory	Pheromone update and rollback policies	ACO	2D
H. Li et al. [111]	Trajectory	Visibility matrix, Coverage matrix, Fitness, Reward	AACO	3D
A. Majeed and S. O. Hwang [112]	Environment, Cost	SFS, SWG	ACO	3D
R. K. Dewangan et al. [113]	Environment	DBVF	GWO	2D
R. K. Dewangan et al. [114]	Trajectory	Mapping	GWO	3D
C. Xu et al. [115]	Trajectory	Initialization, Attenuation factor, Position update	IGWO	3D
L. Yang et al. [116]	Trajectory	Multi-population, Chaotic search	MP-CGWO	3D
G. Huang et al. [117]	Trajectory	Update operators, Greedy algorithms, Stagnation compensation	HDGWO	2D
S. Jiaqi et al. [118]	Trajectory	Spiral update position, Adaptive adjustment	AP-GWO	3D
J. Wu et al. [119]	Time	Intersecting mutation operator, PH	MHS	3D

4.2. Machine Learning Algorithm

The machine learning algorithm mainly simulates or realizes human learning behavior, transforms the UAV formation trajectory planning problem into a decision-making problem and formulates optimal or near-optimal search strategies through continuous learning and interaction in complex environments. With the rapid development of multi-agent algorithms, machine learning algorithms have gradually begun to be applied in UAV formation trajectory planning. The machine learning algorithms currently used for UAV formation trajectory planning include the neural network (NN) algorithm, reinforcement learning (RL) algorithm and deep reinforcement learning (DRL) algorithm.

4.2.1. Neural Network

The neural network (NN) algorithm is based on the information obtained by each UAV sensor, and it quickly obtains the actions that the UAV should take. The neural network has a nonlinear complex network structure composed of a large number of nonlinear unit connections. By simulating the control and feedback functions of the human brain function, a nonlinear mapping system is formed to obtain the mapping relationship between the state space and the action space; then, it completes the UAV formation trajectory planning through its own powerful learning ability and rapid planning ability. Figure 22 is a neural network model.

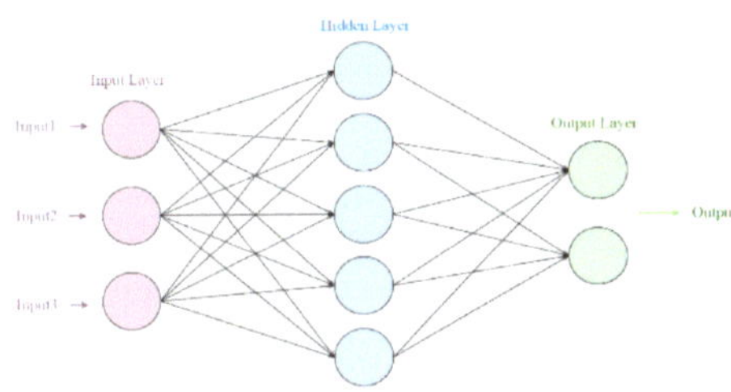

Figure 22. A neural network model (connection: different combinations).

Xia and Yudi [120] designed a fast trajectory planning method using an improved neural network algorithm. This method combined a dynamic adjustable step size with a neural network and added adaptive learning factors for trajectory planning, which solved the problem that the trajectory of a traditional neural network algorithm may not be optimal in the presence of threats.

Sanna et al. [121] proposed a method to cover the trajectory planning problem of UAVs driven by artificial intelligence, which combined a distributed artificial neural network (ANN) and A* algorithm to solve the problems of inadequate grid resolution and low trajectory efficiency of traditional methods.

4.2.2. Reinforce Learning

The reinforcement learning (RL) algorithm is a new learning method which combines dynamic programming with supervised learning. The reinforcement learning algorithm keeps learning in the interaction with the environment, implements the "reward–punishment" mechanism, maximizes the reward index through the feedback evaluation, and realizes the optimal decision output in the UAV formation trajectory planning. Figure 23 shows a reinforcement learning (RL) model.

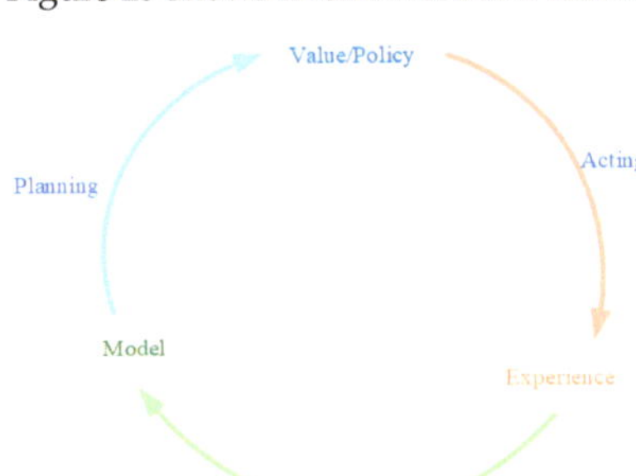

Figure 23. A reinforcement learning model.

Luo et al. [122] proposed a strategy-based Deep-Sarsa algorithm, which combined traditional Sarsa and neural network to find the optimal trajectory of UAV formation and improved the poor trajectory planning ability of heuristic algorithm in dynamic environment.

Qie et al. [123] proposed a multi-agent reinforcement learning algorithm. The algorithm combines the Multi-Agent Deep Deterministic Policy Gradient Algorithm (MADDPG) and the Simultaneous Target Assignment and Trajectory Planning (STATP) method to solve the Multi-Agent Deep Deterministic Policy Gradient (MUTAPP) problem in dynamic environments.

Zhao et al. [124] proposed a q-learning based decentralized multi-UAV cooperative reinforcement learning algorithm (DMUCRL). The algorithm enables UAVs to independently choose their cruising strategy and charging scheduling and at the same time share the learning results in the communication network according to the specified time, which solves the problem of efficient content coverage for multi-UAV trajectory planning.

Wang et al. [125] proposed a collaborative trajectory planning method for multiple UAVs based on attentional reinforcement learning. This method uses a neural network with an attention mechanism to generate a UAV cooperative reconnaissance strategy (AM) and uses a reinforcement algorithm to test a large amount of simulation data and optimize the attention network. It solves the problem in which it is difficult for traditional heuristic algorithms to extract empirical models from large sample terrain data in time.

Liu et al. [126] proposed a trajectory planning method based on the fusion of the Sparse Search Algorithm (SSA) and Biologically Inspired Neural Network (BINN). The algorithm uses SSA to find the node with the lowest comprehensive cost and then uses the b-spline curve to fit it; then, it uses the improved BINN method to replan the local trajectory, which solves the problem where the trajectory planning stability of the heuristic algorithm is poor in a dynamic environment, or the trajectory is probably not the optimal question.

4.2.3. Deep Reinforcement Learning

Deep Reinforcement Learning (DRL) algorithms combine reinforcement learning with deep learning. The optimization goal is obtained through reinforcement learning and environment exploration, the system operation mechanism is obtained by using deep learning, and the specific state characteristics and problem solving are obtained at the same time. Relying on the perception ability of deep learning, this method uses a certain strategy to map the current state into corresponding actions; even for high-dimensional raw data input, through continuous iterative learning, the optimal strategy for UAV formation trajectory planning can finally be obtained. Figure 24 is a deep reinforcement learning (DRL) model.

Wang et al. [127] proposed a trajectory control algorithm based on multi-agent Deep Reinforcement Learning, using multi-agent deep deterministic policy gradient (MADDPG) and a low-complexity method to optimize the UAV trajectory, which solves the problem that traditional dynamic algorithms include both integer variables and continuous variables in Mobile Edge Computing (MEC).

Zhang et al. [128] proposed a constrained deep Q-network (cDQN) algorithm. The algorithm formulates the three-dimensional dynamic motion problem of the UAV under the coverage constraint as a constrained Markov decision process (CMDP). Then, it uses prior information to eliminate invalid actions in the deep Q network (DQN) to maximize the unmanned and real-time downlink connection capability between drones, solving the problem of low capacity of the drone formation communication system under coverage constraints.

Bayerlein et al. [129] proposed a dual-deep Q-network (DDQN) based on a multi agent reinforcement learning (MARL) approach, which transformed the trajectory planning problem into a decentralized partially observable Markov decision process (Dec-POMDP and then solved it by the Deep Reinforcement Learning method optimizing Dec-POMDP to obtain the optimal trajectory, which solves the problem where the UAV formation is difficult to collect data in distributed IoT devices.

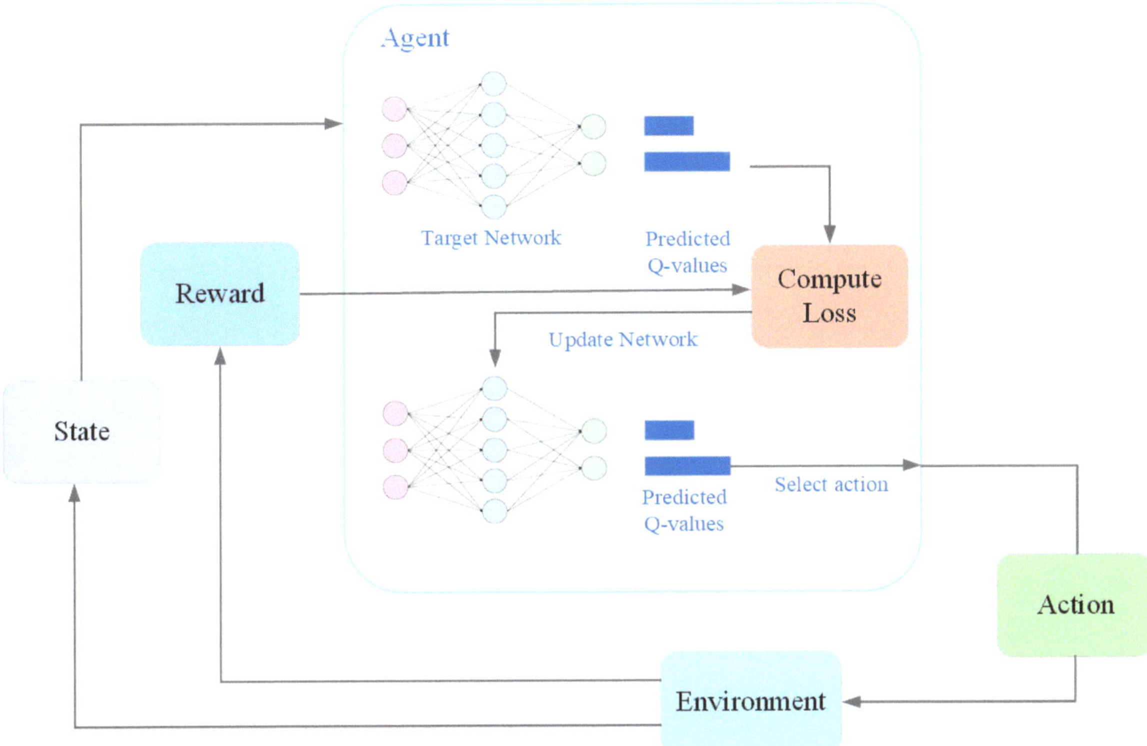

Figure 24. A Deep Reinforcement Learning Model.

Tianle et al. [130] proposed a multi-UAV trajectory planning method based on Deep Reinforcement Learning. This method uses the improved attention dynamic clustering algorithm to optimize the trajectory planning network model and then combines the Particle Swarm Optimization algorithm (PSO) and the Deep Reinforcement Learning (IA-DRL) algorithm to perform trajectory planning, which solves the slow convergence speed of traditional neural network algorithms.

Table 7 summarizes what we surveyed about machine learning algorithms.

Table 7. A summary of the content of the survey about machine learning algorithms.

Reference	Challenge	Optimization Criteria	Method	Dimension
C. Xia and A. Yudi [120]	Trajectory	Dynamic step size, Adaptive learning	NN	3D
G. Sanna et al. [121]	Trajectory	Supervised learning	ANN + A*	2D
W. Luo et al. [122]	Environment	Multi-agent	Deep-Sarsa	3D
H. Qie et al. [123]	Environment	STATP	MADDPG	2D
C. Zhao et al. [124]	coverage problem	Adaptive, Information sharing	DMUCRL	2D
T. Wang et al. [125]	Environment	Attention network	AM	2D
Q. Liu et al. [126]	Environment, Trajectory	SSA, B-spline curve	BINN	3D
L. Wang et al. [127]	MEC	Low complexity	MADDPG	2D
W. Zhang et al. [128]	Communication	CMDP	cDQN	3D
H. Bayerlein et al. [129]	Collect Data	Dec-POMDP	DDQN	2D
S. Tianle et al. [130]	Time	Note dynamic clustering	PSO + IA-DRL	2D

5. Local Trajectory Planning Algorithm

The local trajectory planning algorithm belongs to the dynamic programming algorithm. According to the UAV sensors in the UAV formation, the current location information and local obstacle information are collected in real time so as to dynamically plan the optimal trajectory from the starting point to the target point. The algorithms for UAV formation local trajectory planning usually include the artificial potential field method (APF), dynamic window approach (DWA), mathematical optimization algorithm (MOA), and Model Predictive Control (MPC).

5.1. Artificial Potential Field

The artificial potential field method (APF) was first proposed by Khatib as a virtual force method. The artificial potential field method assumes that each UAV is moving in an artificial potential field. For UAVs, the target point generates an attractive field and obstacles generate a repulsive force field; under the action of the gravitational field and the repulsive field, the UAV generates a feasible trajectory along the direction of the potential field. In general, in order to simplify the calculation, by calculating the negative gradient of the gravitational potential function and the repulsive potential function, the gravitational and repulsive forces on the UAV in the potential field can be obtained, and then the resultant force on the UAV can be obtained. Then, calculating according to the resultant force, each UAV makes the control amount required for attitude adjustment so as to guide the UAV formation to avoid obstacles and complete trajectory planning. Figure 25 is a schematic diagram of artificial potential field forces.

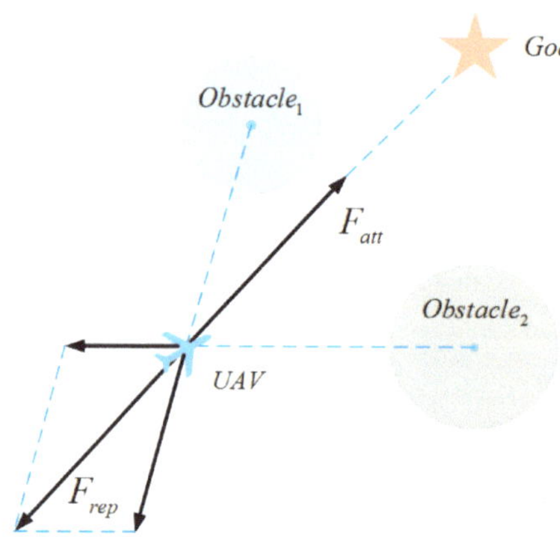

Figure 25. Schematic diagram of artificial potential field.

Rasche et al. [131] proposed a trajectory planning method in a 3D environment. This method adds a multi-UAV distributed work and inter-machine communication to the artificial potential field method (APF) and solves the problem of multi-UAV coordination and task assignment when exploring disaster areas.

Li et al. [132] proposed a trajectory planning method combining the artificial potential field method (APF) and Dubins curve. This method introduces the virtual leader UAV into the UAV formation and uses the Dubins curve to plan its trajectory. Then, it uses the APF to plan the trajectory of the wingman and finally completes the trajectory planning of the UAV formation by constraining the flight trajectory of the virtual leader. It solves the problem where the lead aircraft may have out-of-control failure and the UAV is restricted by the turning radius.

Tang et al. [133] proposed an optimized artificial potential field algorithm. This method simulates other UAVs as dynamic obstacles and at the same time introduces the climbing strategy and dynamic step adjustment method into APF, which solves the problems of inter-machine collision and excessive flight step length in traditional APF under the complex space conditions of multiple UAVs.

Chen et al. [134] proposed an improved artificial potential field method (IAPF). This method introduces the judging mechanism of local minimum points and the jump-out mechanism of 90° movement along the target direction into APF; it solves the problems of unreachable targets near obstacles, local minimum points and UAV track oscillations in traditional APF.

Sun et al. [135] proposed a trajectory planning algorithm for dense UAV formations based on an artificial potential field (APF). The algorithm improves the APF by improving the repulsive force model, adding the target exchange algorithm and adding constraints, and it solves the problems of traditional APF trajectory oscillation, unreachable targets and local minimum points.

Dongcheng and Jiyang [136] proposed a multi-UAV trajectory planning method based on the improved artificial potential field method (IAPF). This method introduces an improved distance factor potential field function and dynamic step size adjustment method into APF, and at the same time, it considers the influence of the force between UAVs; it solves the problem where the traditional APF target cannot be reached: it is easy to fall into a local minimum, and the trajectory shakes the problem.

Wang et al. [137] proposed a collaborative formation distributed trajectory planning method based on the improved artificial potential field (IAPF) and consensus theory. This method introduces the dynamic model and communication network topology, coordination gain factor, repulsion force and planning angle influence factor into APF. Then, the position and velocity variables in the consensus protocol are improved to solve the problem of UAV formation in 3D obstacle environment trajectory planning and position–velocity consistency problems.

Dai et al. [138] proposed a consensus algorithm for distributed cooperative formation trajectory planning. The algorithm introduces the potential field function including distance items and communication effects into APF (IAPF) and then combines the second-order system dynamic model, consistency theory and IAPF for UAV formation collaborative trajectory planning. This solves the problems that traditional APF encounters in UAV including poor convergence problems related to consistency, relative distance and velocity in formation cooperative trajectory planning.

Li et al. [139] proposed a new trajectory planning method using the improved artificial potential field algorithm (IAPF). This method increases the repulsive force between UAVs and defines the front center of mass of the cluster as another source of gravity, which solves the problem where the traditional APF target is unreachable and easily falls into a local minimum.

Wei et al. [140] proposed a UAV time-varying formation trajectory planning method with an interactive topology. This method introduces the improved potential field into APF (IAPF) and then combines distributed time-varying formation control, IAPF and model predictive control (MPC) for UAV formation trajectory planning. It solved the problem where the UAV formation has poor ability to deal with complex environments during flight.

Wang et al. [141] introduced a multi-UAV trajectory planning method based on an adaptive extended potential field. In this method, the gravitational influence factor and the repulsive force influence factor are introduced into the layered potential field function, and the auxiliary force is added to improve the APF. It solves the problems related to the slow convergence speed of a layered potential field algorithm and unreachable target of a traditional APF, easily falling into local minimum, inability to avoid obstacles and lack of a trajectory optimization strategy.

Pan et al. [142] proposed a trajectory planning method based on artificial potential functions (IAPF) for multi-UAV systems. This method introduces the improved artificial potential function (IAPF) of the rotating potential field and at the same time adds the leader–follower UAV model for UAV formation trajectory planning. It solves the problems of poor stability, local minimum and oscillation in the traditional APF system.

Table 8 summarizes the contents of our survey regarding artificial potential field methods.

Table 8. Summary of the methods of artificial potential field in the survey.

Reference	Challenge	Optimization Criteria	Method	Dimension
C. Rasche et al. [131]	UAV clustering	Distributed, UAV communication	APF	3D
X. Li et al. [132]	UAV	Constrain virtual leads	APF + Dubins	2D
J. Tang et al. [133]	UAV clustering, Trajectory	Climb strategy, Dynamic step size	APF	3D
H. Chen et al. [134]	Trajectory	Judging mechanism, 90° jump mechanism	IAPF	3D
H. Sun et al. [135]	Trajectory	Improve the model, Target exchange, Add constraints	APF	3D
L. Dongcheng and D. Jiyang [136]	Trajectory, UAV clustering	Improved function, Dynamic step size	IAPF	3D
N. Wang et al. [137]	Environment, UAV clustering	Communication topology, Coordination gain, Impact factor	IAPF	3D
J. Dai et al. [138]	Trajectory, UAV clustering	Second-order model, Consistency theory	IAPF	3D
R. Li et al. [139]	Trajectory	Increase the repulsive force, Gravity source	IAPF	2D
B. Wei et al. [140]	Environment	Distributed time variation	IAPF + MPC	2D
N. Wang et al. [142]	Trajectory	Impact factor, Auxiliary force	IAPF	3D
Z. Pan et al. [141]	Trajectory	Rotational potential field, Leader–Follow	IAPF	3D

5.2. Dynamic Window Approach

The dynamic window method (DWA) is a classic UAV local trajectory planning algorithm. It determines a sampling velocity space that satisfies the hardware constraints of the UAV in the velocity space according to the current position state and velocity state of the mobile UAV and transforms the local trajectory planning problem into a motion constraint problem in space. Then, it calculates the UAV trajectory of the drone moving for a certain period of time under these speed conditions and evaluates the trajectory through the evaluation function. It selects the trajectory with the best evaluation and the corresponding speed as the movement speed of the UAV; finally, through the motion constraints, it selects the locally optimal trajectory and so on until the UAV reaches the target point. Figure 26 is a schematic diagram of the DWA velocity vector space.

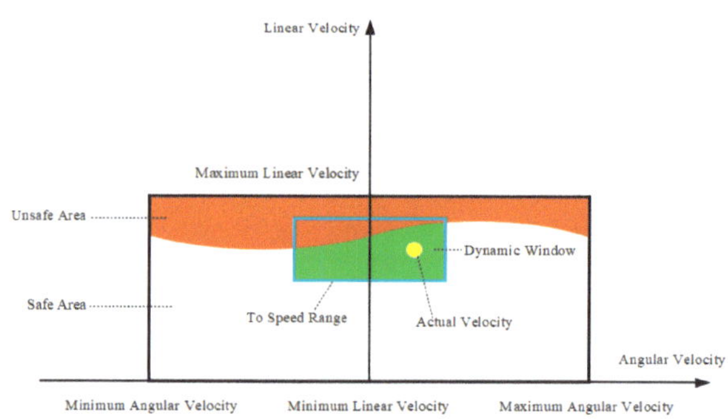

Figure 26. DWA velocity vector space diagram.

Zhang et al. [143] proposed a multi-UAV consistent formation trajectory planning algorithm based on an improved dynamic window method (DWA). The algorithm introduces a new rotation cost evaluation function, A* algorithm and azimuth-related variable weight factors to improve DWA and finally adds a leader–following UAV model for UAV formation trajectory planning. It solves the problems of frequent large-angle rotation and low search efficiency in the traditional DWA algorithm.

5.3. Mathematical Optimization Algorithm

The mathematical optimization algorithm (MOA) is based on the established UAV trajectory planning model, using nonlinear optimization, mixed integer linear programming (MILP), mixed integer nonlinear programming (MINLP) and dynamic programming (DP) to solve the optimal control problem into an easily solvable model to generate feasible trajectories for formation UAVs. Figure 27 is a mathematical optimization algorithm (MOA) model.

Bellingham et al. [144] proposed CPLEX, which is a collaborative trajectory planning method for UAV formation. This method combines the failure probability of each UAV with the selected task and puts forward a new formula to solve the problem of mission failure caused by UAV loss in UAV formation trajectory planning.

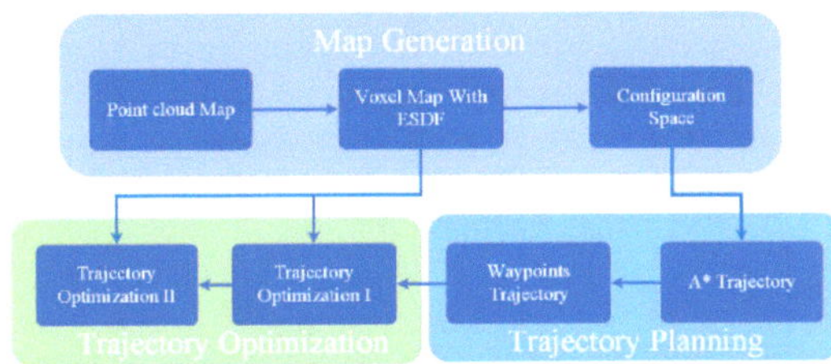

Figure 27. A mathematical optimization algorithm model.

Maza and Ollero [145] proposed a method based on polygonal area decomposition and efficient coverage (PADEC). This method introduces the concepts of regional division and computational scanning into the UAV formation coverage trajectory planning and solves the problem of multi-UAV cooperative search.

Dehghan et al. [146] proposed a trajectory planning method based on multi-UAV for RF source localization. The method combines the differential received signal strength indicator (DRSSI) method, the extended Kalman filter (EKF) and the Cramer–Rao lower bound (CRLB) objective function; finally, using the local value of the CRLB in the current waypoint and the next possible waypoint to determine the optimal trajectory, it solves the problem of slow convergence of the heuristic algorithm.

Wang et al. [147] proposed a decoupled sequence convex programming (SCP) collaborative trajectory planning method for UAV formations. This method represents the UAV formation trajectory planning problem as a non-convex optimal control problem; then, it uses the decoupled sequence convex programming (SCP) method to parameterize the problem into a non-convex programming sub-problem and solves it in parallel to obtain the UAV best trajectory for formation coordination tasks. The problem of insufficient efficiency of the sequence quadratic programming algorithm (SQP) in UAV formation cooperative trajectory planning is solved.

Causa et al. [148] proposed an algorithm for multi-UAV trajectory planning under heterogeneous Global Navigation Satellite System (GNSS) coverage. The algorithm conceives the multi-UAV formation as a reconfigurable distributed system and then introduces methods such as edge definition and cost evaluation, custom target assignment, UAV timing and polynomial trajectory (PT) for formation trajectory planning. It solves the problem of low efficiency of UAV formation trajectory planning task assignment in a three-dimensional heterogeneous environment.

Pengfei et al. [149] proposed an optimal trajectory planning method for multiple UAVs based on the pseudospectral method. This method uses the pseudospectral method to transform the optimal control problem with complex constraints into a nonlinear programming problem; at the same time, it uses the distributed solution and the Nash optimal coordination strategy to solve the multi-UAV trajectory planning problem under complex and multi-constrained conditions.

Li et al. [150] proposed a trajectory planning method for multi-UAV scan coverage with minimum time maximum coverage. This method introduces a Weighted Target Scan Coverage (WTSC) algorithm for greedy target assignment, which solves the problem of insufficient task time and coverage performance of two algorithms, CycleSplit and G-MSCR [151].

Xia et al. [152] proposed a gradient-based sequential minimum optimization (GB-SMO) algorithm, which uses time segmentation instead of traditional waypoint segmentation to establish a trajectory optimization model and introduces virtual line segments to adapt to the trajectory length. Constraints are converted into cost functions and then minimized using GBSMO, which solves the problem of insufficient computational performance of commonly used trajectory planning algorithms considering constraints.

Wang et al. [153] proposed a real-time trajectory planning method for UAV formation transformation based on safe flight corridors, introducing safe flight corridors to avoid UAV collisions, while considering time and space efficiency models. They addressed an issue where drones could collide when performing a formation change.

Cho et al. [154] proposed a multi-UAV search trajectory planning method covering nodes in the shortest time. The method introduces a mixed integer linear programming (MILP) model of hexagonal grid decomposition and at the same time uses the optimization time as a search function to obtain the trajectory of the formation UAVs in iterations. The trajectory planning problem of UAV formation searching for catastrophic marine accidents is solved.

Sun et al. [155] proposed a 4D trajectory planning method with temporal and spatial constraints. This method transforms the arrival time into state adaptation and at the same time transforms the collaborative penetration trajectory planning into a single-objective optimization problem. Then, it uses the multi-leader search distribution estimation algorithm (MLSEDA) to solve the problem, which solves the trajectory planning problem of UAV cooperative penetration.

Cheng et al. [156] proposed a decentralized multi-UAV trajectory planning method for obstacle environments. In this method, the UAV rendezvous trajectory planning problem under constraints is modeled as a non-convex optimal control problem, and then, the consensus protocol and sequential convex programming two-layer collaborative framework are used to solve the UAV formation trajectory. It solves the problems of low calculation efficiency and poor adaptive ability of the traditional UAV formation trajectory planning method.

Yanmaz [157] proposed a hybrid planner that uses joint optimization methods, decoupling optimization methods, and hybrid methods to calculate UAV formations. They generated feasible trajectories under two different requirements of time constraints and connectivity, solving the problem where the connection parameters are difficult to trade off and the resource utilization rate is low in the formation task.

Table 9 summarizes the contents of our survey about the dynamic window method and mathematical optimization algorithm.

Table 9. Summary of dynamic windowing methods and mathematical optimization algorithms in our survey.

Reference	Challenge	Optimization Criteria	Method	Dimension
S. Zhang et al. [143]	Trajectory, Time	New cost function, Variable weight factor, Lead–follow	A* + DWA	2D
J. S. Bellingham et al. [144]	UAV	Failure probability, Task selection	CPLEX	2D
I. Maza and A. Ollero [145]	UAV clustering	Area divided and conquer, Computational scan	PADEC	2D
S. M. M. Dehghan et al. [146]	Trajectory	EKF, CRLB	DRSSI	2D
Z. Wang et al. [147]	Trajectory	Non-convex optimal, Parallel solution	SCP	2D
F. Causa et al. [148]	Environment	GNSS, Custom target allocation	Multiple step path	3D
J. Pengfei et al. [149]	UAV, Environment	Distributed solutions and Nash optimal coordination	Pseudo-spectral Method	3D
J. Li et al. [150]	Trajectory	Greedy goal assignment	WTSC	2D
Q. Xia et al. [152]	Computing performance	Time division, Constraint transformation	GB-SMO	3D
G. Wang et al. [153]	Environment, UAV clustering	Efficiency model	Safe flight corridor	2D
S.-W. Cho et al. [154]	Environment	Hexagonal grid decomposition	MILP	2D
P. Sun et al. [155]	Environment, Time	State adaptation, Single objective optimization	MLSEDA	4D
Z. Cheng et al. [156]	Trajectory	Consensus protocol, Sequential convex programming	Non-convex optimal control	3D
E. Yanmaz [157]	Resource utilization	Joint optimization, Decoupling optimization, Hybrid method	Hybrid planner	2D

5.4. Model Predictive Control

Model Predictive Control (MPC) is a typical online planning method with planning and execution at the same time. In other words, it is a rolling time window approach. In the local planning process, the algorithm first updates the environmental information in the current trajectory search domain and predicts the information change trend in the trajectory search domain on this basis. Then, it searches out the local reference trajectory according to the motion model of the UAV and executes. In the process, the trajectory is corrected according to the motion constraints and control errors of the UAV so as to make full use of real-time feedback information to gradually generate a global trajectory. Figure 28 is a schematic diagram of MPC.

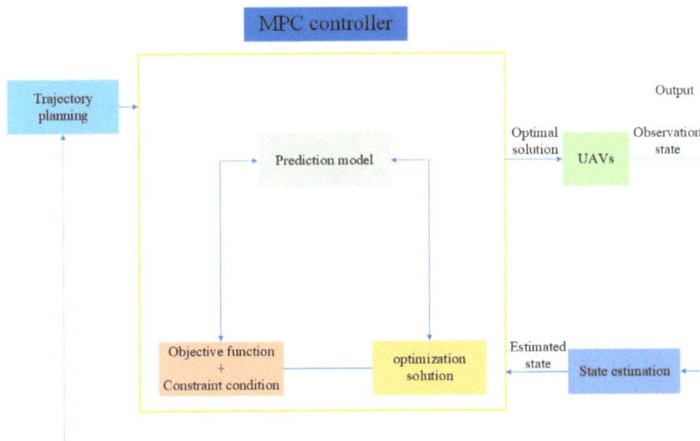

Figure 28. Schematic diagram of Model Predictive Control.

Oh et al. [158] proposed a decentralized nonlinear model predictive control trajectory planning strategy (DNMPC), which introduces filtering technology and decentralized optimization into MPC to realize UAV formation for trajectory planning in relay communication to solve the problem of poor connectivity of the wireless network between the fleet of naval ships.

Cui et al. [159] proposed a multi-object tracking algorithm based on task assignment consensus. It uses the dynamic task allocation model to update the tracking tasks and uses the intermittent asynchronous communication principle to realize the sharing of local observation information. At the same time, it uses the MPC algorithm to complete the tracking trajectory planning and solved the problem of UAV formation tracking multiple moving targets within a limited communication range.

Wu et al. [160] proposed a UAV trajectory planning model (Poc-KF) based on collision probability and Kalman filter. The model uses the collision probability algorithm and the Kalman filter algorithm for UAV collision probability calculation and formation state estimation, and it calculates feasible trajectories for UAV formation in real time. It also addressed possible trajectory conflicts in high-density drone formations.

Wu et al. [161] proposed a behavior tree (BT) model. The specific operation is to combine the model prediction with the decision tree to obtain the behavior tree (BT). Then, they add the virtual target-based tracking (VTB-T) method and use the behavior tree (BT) organization trajectory planning method to construct a feasible trajectory for the UAV. The problem of multi-UAV trajectory planning in the target tracking scenario is solved.

Wang et al. [162] proposed a new trajectory planning algorithm for model predictive control (NMPC). They introduced a virtual target to move along the patrol trajectory at a predetermined speed and designed a decentralized estimator for each UAV to estimate the state of the virtual target. Then, they used a new model prediction algorithm to calculate a feasible trajectory for the formation of UAVs, which addresses the problem of formation reconstruction and trajectory planning in multi-UAV aerial patrol missions.

Chen and Liu [163] proposed a model for predicting flushing force under drones (PMDFF). It regards each UAV as a virtual structure to form a cylindrical UAV model and then uses the cluster and Optimal Interactive Collision Avoidance (ORCA) algorithm to solve the collision-free trajectory. They addressed an issue where downwash effects have an impact on neighboring drones in UAV formation trajectory planning.

Table 10 summarizes the contents of our survey on model predictive control algorithms.

Table 10. Summary of MPC algorithms in our survey.

Reference	Challenge	Optimization Criteria	Method	Dimension
H. Oh et al. [158]	Communication	Filtering technology, Dispersion optimization	DNMPC	2D
Y. Cui et al. [159]	Communication, Multiple objectives	Dynamic update, Intermittent asynchronous communication	MPC	2D
Z. Wu et al. [160]	UAV clustering	Probability calculation, KF	Poc-KF	2D
W. Wu et al. [161]	Environment	VTB-T	BT	3D
Y. Wang et al. [162]	UAV clustering, Trajectory	Virtual target, Decentralized estimates	NMPC	3D
C.-C. Chen and H. H. Liu [163]	UAV clustering	Virtual structure, ORCA	PMDFF	3D

6. Problems of UAV Formation Trajectory Planning Algorithm

With the advancement of low-airspace reforms and the innovation of artificial intelligence and information technology [164], new theories and new achievements related to swarm intelligence continue to emerge, and the improvement of UAV formation trajectory planning algorithms is facing many challenges.

6.1. Physical Constraints of UAV Formation

In the process of trajectory planning, the UAV in the formation is usually simplified into three degrees of freedom particles, ignoring its own constraints such as minimum turning radius, rolling angle and other restrictions on the running state. As a result, the current trajectory planning algorithm is difficult to adapt to the UAV with high maneuverability, there are errors between the release route and the planned route when the actual formation UAV performs the task, and the execution effect will also be affected.

6.2. Performance Problems of UAV Formation Carrying Equipment

In the UAV formation trajectory planning algorithm, the performance of the UAV itself is not considered enough. An insufficient consideration of problems such as fuel consumption, load, and onboard sensor errors in practice makes it difficult to accurately detect complex environments, and the trajectory planning that can be achieved by simulation cannot be realized or has poor robustness in the actual environment.

6.3. Complex Environment Modeling Problem

Most of the current formation trajectory planning algorithms are hypothetical ideal obstacles, but the actual operating environment of UAVs is complex and diverse, especially the detection and description of scenes such as complex concave obstacle environments and dense dynamic obstacles, which need further exploration.

6.4. Algorithm Real-Time Problems

Realistic environmental information is usually time-varying. Regarding UAV formation trajectory planning in an unknown environment, the success rate of trajectory planning strategies used by traditional algorithms and local trajectory planning algorithms in the face of emergencies and dynamic environments is low; in addition, the amount of calculation is large, and the trajectory is not optimal. It is difficult for drones to complete real-time trajectory updates.

6.5. Adaptability of UAV Formation Route Planning Algorithm

When performing formation trajectory planning in a complex dynamic environment there is a lot of information exchange between UAVs, which leads to an increase in the amount of calculations, and intelligent algorithms are prone to fall into problems such as local optima that exist in themselves.

6.6. UAV Formation Communication Problem

With the development of science and technology, the application scenarios of drones in the future will become more and more complex. In certain scenarios, there will be communication interference problems, which will cause the UAV to fail to work normally and even cause irreversible damage.

7. Future Research Focus and Direction

7.1. Improved Model

Constraints such as six degrees of freedom, minimum turning radius, roll angle, and the onboard sensor error of each UAV are added to the modeling to enhance the robustness of actual control. For complex environment modeling, the influence of multiple factors in the complex environment on the effect of trajectory planning must be considered; reliable and accurate data must be obtained through specific measurements or the use of accurate 3D maps, and at the same time, the data must be used to verify the model to make the simulation closer to reality. It can also be better applied to the actual platform in the future.

7.2. Real time Planning

In the face of increasingly complex environments and tasks, in order to meet the requirements of fast optimal solution, computational complexity, convergence speed and rationality, the computational memory is allocated reasonably. Executing trajectory planning algorithms to generate efficient trajectories in the case of limited computing power of UAVs is of great significance for UAV formations to complete tasks in complex environments.

7.3. Fusion Algorithm

It is an important current research trend to integrate different types of trajectory planning algorithms to make up for the defects and deficiencies of existing single methods.

For example, the local trajectory planning method can be combined with artificial intelligence technology represented by machine learning to complement each other. On the one hand, it can solve the problem of easy falling into local optimum in the local trajectory planning method, and on the other hand, it can also make up for the poor real-time performance of the machine learning-based track planning algorithm to a certain extent. It is also possible to combine the characteristics of heuristic algorithms and machine learning algorithms that are easy to integrate with each other to help analyze the performance of the algorithm and expand the application range of the algorithm. At the same time, experiments show that the hybrid algorithm has better adaptability.

7.4. New Algorithm

At present, the existing UAV formation trajectory planning algorithms have more or less defects. Therefore, developing an algorithm that reduces computational requirements, saves time, allows real-time planning, and is more efficient in terms of energy is also a direction worth exploring.

7.5. Fault Tolerance Mechanism

Since the maneuvering area of each UAV is very small, once a collision occurs, it will affect the adjacent UAVs, and a chain effect will be generated between the UAV clusters, which will cause the mission to fail. Therefore, the fault-tolerant redundancy mechanism is an important link to ensure the safe operation of the UAV system. At present, there is no fault-tolerant mechanism design for the core and weak links of the trajectory planning algorithm so as to improve the fault-tolerant ability of unmanned formation flight. In future research, we should focus on the design of the fault-tolerant mechanism when the function of the UAV fails to avoid uncontrollable events.

7.6. Hybrid Frame

It is unrealistic to use a centralized framework to solve problems in the process of carrying out missions in large-scale UAV formations. Therefore, a hybrid framework should be adopted; how to design an appropriate conflict resolution mechanism and how to effectively combine UAV formation trajectory planning with collaborative control to generate feasible flight trajectories are topics worthy of further study.

7.7. Behavior Decision

Recently, UAV swarm-to-swarm dynamic confrontation has become a hot research direction. At the same time, the autonomous decision-making behavior of UAV formations such as autonomous reconnaissance and detection, autonomous target recognition, and autonomous task coordination in complex terrain such as cities and mountainous areas can effectively reduce the loss of manpower and material resources. How to plan and generate the optimal trajectory of UAV formation from the perspective of game theory is undoubtedly a problem worthy of further exploration.

7.8. Allocation of Resources

In the future, drone formations will be widely used in battlefields and anti-terrorism operations. The environment in which UAVs perform these tasks may become very complex, resulting in increased mission difficulty, and the environment may provide extremely strong support for UAV formations, such as satellite links and energy supplies for continuous flight. How to reasonably allocate available resources to each UAV during mission execution is also a challenging problem.

7.9. Communication Networking

The confrontation between UAVs has become more and more information-based. When the UAV formation is performing tasks, it is necessary to ensure that the UAVs can communicate and share information normally and at the same time deal with external communication interference. Although some scholars have noticed related problems, the problem of how to solve communication interference in UAV formation trajectory planning is still a difficult problem.

8. Conclusions

From the perspective of the two key elements of global planning and local planning, this paper proposes a framework for UAV formation trajectory planning algorithms, comprehensively classifies different types of algorithms, and describes different types of algorithms and their variants in a unified way. Then, a review and statistical analysis were carried out on the basis of classification. We found the shortcomings in the UAV formation trajectory planning algorithm methods and put forward the focus and direction of future research. This paper provides reference information for the next step of research work for researchers and workers engaged in UAV formation flight-related work. We believe that with the innovation of various theories and the iterative development of technologies, the UAV formation trajectory planning algorithm will enter a new era.

Author Contributions: Conceptualization, X.X.; methodology; investigation, Y.Y. (Yunhong Yang); writing—original draft preparation, Y.Y. (Yunhong Yang); writing—review and editing, X.X. and Y.Y. (Yuehao Yan); supervision, X.X. and Y.Y. (Yuehao Yan); project administration, X.X. and Y.Y. (Yuehao Yan); funding acquisition, X.X. and Y.Y. (Yuehao Yan). All authors have read and agreed to the published version of the manuscript.

Funding: This work was supported in part by the Innovation Fund of Chinese Universities under grant 2021ZYA09002, and in part by the Postgraduate Innovation Fund Project of Sichuan University of Science and Engineering under grant Y2022137.

Data Availability Statement: No new data were created or analyzed in this study. Data sharing is not applicable to this article.

Acknowledgments: I would like to thank my good friend Junlin Li. His contributions in this paper are as follows: supervising the planning and implementation of research activities and revising the grammar of the article.

Conflicts of Interest: The authors declare that they have no conflict of interest.

References

1. Shirshikova, Z.A. Comparative Analysis of the US-China Artificial Intelligence Architecture and Effects of Autonomous UAVs on the Future of the Battlefield. Master's Thesis, Harvard University, Cambridge, MA, USA, 2022.
2. Norasma, C.; Fadzilah, M.; Roslin, N.; Zanariah, Z.; Tarmidi, Z.; Candra, F. Unmanned aerial vehicle applications in agriculture. *IOP Conf. Ser. Mater. Sci. Eng.* **2019**, *506*, 012063. [CrossRef]
3. Mohsan, S.A.H.; Khan, M.A.; Noor, F.; Ullah, I.; Alsharif, M.H. Towards the unmanned aerial vehicles (UAVs): A comprehensive review. *Drones* **2022**, *6*, 147. [CrossRef]
4. Shakhatreh, H.; Sawalmeh, A.H.; Al-Fuqaha, A.; Dou, Z.; Almaita, E.; Khalil, I.; Othman, N.S.; Khreishah, A.; Guizani, M. Unmanned aerial vehicles (UAVs): A survey on civil applications and key research challenges. *IEEE Access* **2019**, *7*, 48572–48634. [CrossRef]
5. Yun, W.J.; Park, S.; Kim, J.; Shin, M.; Jung, S.; Mohaisen, D.A.; Kim, J.-H. Cooperative multiagent deep reinforcement learning for reliable surveillance via autonomous multi-UAV control. *IEEE Trans. Ind. Inform.* **2022**, *18*, 7086–7096. [CrossRef]
6. Shi, Y.; Liu, Y.; Ju, B.; Wang, Z.; Du, X. Multi-UAV cooperative reconnaissance mission planning novel method under multi-radar detection. *Sci. Prog.* **2022**, *105*, 00368504221103785. [CrossRef]
7. Xu, D.; Chen, G. The research on intelligent cooperative combat of UAV cluster with multi-agent reinforcement learning. *Aerosp. Syst.* **2022**, *5*, 107–121. [CrossRef]
8. Thibbotuwawa, A.; Bocewicz, G.; Radzki, G.; Nielsen, P.; Banaszak, Z. UAV mission planning resistant to weather uncertainty. *Sensors* **2020**, *20*, 515. [CrossRef] [PubMed]
9. Zhang, H.; Xin, B.; Dou, L.-H.; Chen, J.; Hirota, K. A review of cooperative path planning of an unmanned aerial vehicle group. *Front. Inf. Technol. Electron. Eng.* **2020**, *21*, 1671–1694. [CrossRef]
10. Zhao, C.; Liu, Y.; Yu, L.; Li, W. Stochastic Heuristic Algorithms for Multi-UAV Cooperative Path Planning. In Proceedings of the 2021 40th Chinese Control Conference (CCC), Shanghai, China, 26–28 July 2021; pp. 7677–7682.
11. Sun, W.; Hao, M. A Survey of Cooperative Path Planning for Multiple UAVs. In Proceedings of the International Conference on Autonomous Unmanned Systems, Shanghai, China, 26–28 July 2021; pp. 189–196.
12. Tang, J.; Duan, H.; Lao, S. Swarm intelligence algorithms for multiple unmanned aerial vehicles collaboration: A comprehensive review. *Artif. Intell. Rev.* **2022**, 1–33. [CrossRef]
13. Sujit, P.; Ghose, D. Search using multiple UAVs with flight time constraints. *IEEE Trans. Aerosp. Electron. Syst.* **2004**, *40*, 491–509. [CrossRef]
14. Tin, C. Robust multi-UAV Planning in Dynamic and Uncertain Environments. Master's Thesis, Massachusetts Institute of Technology, Cambridge, MA, USA, 2004.
15. Ueno, S.; Kwon, S.J. Optimal reconfiguration of UAVs in formation flight. In Proceedings of the SICE Annual Conference 2007, Takamatsu, Japan, 17–20 September 2007; pp. 2611–2614.
16. Aggarwal, R.; Soderlund, A.A.; Kumar, M. Multi-UAV Path Planning in a Spreading Wildfire. In Proceedings of the AIAA Scitech 2021 Forum, Virtual, 11–22 January 2021; p. 0866.
17. D'Amato, E.; Mattei, M.; Notaro, I. Bi-level flight path planning of UAV formations with collision avoidance. *J. Intell. Robot. Syst.* **2019**, *93*, 193–211. [CrossRef]
18. Yang, J.; Xi, J.; Wang, C.; Xie, X. Multi-base multi-UAV cooperative patrol route planning novel method. In Proceedings of the 2018 33rd Youth Academic Annual Conference of Chinese Association of Automation (YAC), Nanjing, China, 18–20 May 2018; pp. 688–693.
19. Zhou, F.; Nie, H. Quick Path Planning Based on Shortest Path Algorithm for Multi-UAV System in Windy Condition. *Preprints* **2021**, 2021080491.
20. López, B.; Muñoz, J.; Quevedo, F.; Monje, C.A.; Garrido, S.; Moreno, L.E. Path planning and collision risk management strategy for multi-UAV systems in 3D environments. *Sensors* **2021**, *21*, 4414. [CrossRef] [PubMed]
21. Muñoz, J.; López, B.; Quevedo, F.; Monje, C.A.; Garrido, S.; Moreno, L.E. Multi UAV Coverage Path Planning in Urban Environments. *Sensors* **2021**, *21*, 7365. [CrossRef] [PubMed]
22. Meng, B.-B.; Gao, X.; Wang, Y. Multi-mission path re-planning for multiple unmanned aerial vehicles based on unexpected events. In Proceedings of the 2009 International Conference on Intelligent Human-Machine Systems and Cybernetics, Hangzhou, China, 26–27 August 2009; pp. 423–426.
23. Chen, S.; Li, C.; Zhuo, S. A distributed coverage algorithm for multi-UAV with average voronoi partition. In Proceedings of the 2017 17th International Conference on Control, Automation and Systems (ICCAS), Jeju, Republic of Korea, 18–21 October 2017; pp. 1083–1086.

24. Chen, X.; Li, G.-Y.; Chen, X.-M. Path planning and cooperative control for multiple UAVs based on consistency theory and Voronoi diagram. In Proceedings of the 2017 29th Chinese control and decision conference (CCDC), Chongqing, China, 28–30 May 2017; pp. 881–886.
25. Hu, J.; Wang, M.; Zhao, C.; Pan, Q.; Du, C. Formation control and collision avoidance for multi-UAV systems based on Voronoi partition. *Sci. China Technol. Sci.* **2020**, *63*, 65–72. [CrossRef]
26. Madridano, Á.; Al-Kaff, A.; Gómez, D.M.; de la Escalera, A. Multi-path planning method for uavs swarm purposes. In Proceedings of the 2019 IEEE International Conference on Vehicular Electronics and Safety (ICVES), Cairo, Egypt, 4–6 September 2019; pp. 1–6
27. Kothari, M.; Postlethwaite, I.; Gu, D.-W. Multi-UAV path planning in obstacle rich environments using rapidly-exploring random trees. In Proceedings of the the 48h IEEE Conference on Decision and Control (CDC) held jointly with 2009 28th Chinese Control Conference, Shanghai, China, 15–18 December 2009; pp. 3069–3074.
28. Zu, W.; Fan, G.; Gao, Y.; Ma, Y.; Zhang, H.; Zeng, H. Multi-uavs cooperative path planning method based on improved rrt algorithm. In Proceedings of the 2018 IEEE International Conference on Mechatronics and Automation (ICMA), Changchun, China, 5–8 August 2018; pp. 1563–1567.
29. Huang, J.; Sun, W. A method of feasible trajectory planning for UAV formation based on bi-directional fast search tree. *Optik* **2020** *221*, 165213. [CrossRef]
30. Shi, B.; Zhang, Y.; Mu, L.; Huang, J.; Xin, J.; Yi, Y.; Jiao, S.; Xie, G.; Liu, H. UAV Trajectory Generation Based on Integration of RRT and Minimum Snap Algorithms. In Proceedings of the 2020 Chinese Automation Congress (CAC), Shanghai, China, 6–8 November 2020; pp. 4227–4232.
31. Yahia, H.S.; Mohammed, A.S. Path planning optimization in unmanned aerial vehicles using meta-heuristic algorithms: A systematic review. *Environ. Monit. Assess.* **2023**, *195*, 30. [CrossRef]
32. Turker, T.; Yilmaz, G.; Sahingoz, O.K. GPU-accelerated flight route planning for multi-UAV systems using simulated annealing. In Proceedings of the International Conference on Artificial Intelligence: Methodology, Systems, and Applications, Varna, Bulgaria, 7–10 September 2016; pp. 279–288.
33. Yue, X.; Zhang, W. UAV path planning based on k-means algorithm and simulated annealing algorithm. In Proceedings of the 2018 37th Chinese Control Conference (CCC), Wuhan, China, 25–27 July 2018; pp. 2290–2295.
34. Hu, Y.; Yao, Y.; Ren, Q.; Zhou, X. 3D multi-UAV cooperative velocity-aware motion planning. *Future Gener. Comput. Syst.* **2020** *102*, 762–774. [CrossRef]
35. Su, H.; Yun, H.; He, J.; Zhang, F.; Wang, Y. Multi-aircraft path planning method based on cooperative search A-star algorithm In Proceedings of the 2019 IEEE International Conference on Unmanned Systems (ICUS), Beijing, China, 17–19 October 2019 pp. 267–272.
36. Zhang, Z.; Wu, J.; Dai, J.; Ying, J.; He, C. Cooperative Tactical Planning Method for UAV Formation. In Proceedings of the 2020 39th Chinese Control Conference (CCC), Shenyang, China, 27–29 July 2020; pp. 1542–1547.
37. Haghighi, H.; Asadi, D.; Delahaye, D. Multi-objective cooperated path planning of multiple unmanned aerial vehicles based on revisit time. *J. Aerosp. Inf. Syst.* **2021**, *18*, 919–932. [CrossRef]
38. Nagasawa, R.; Mas, E.; Moya, L.; Koshimura, S. Model-based analysis of multi-UAV path planning for surveying postdisaster building damage. *Sci. Rep.* **2021**, *11*, 18588. [CrossRef]
39. Luo, Y.; Lu, J.; Zhang, Y.; Qin, Q.; Liu, Y. 3D JPS Path Optimization Algorithm and Dynamic-Obstacle Avoidance Design Based on Near-Ground Search Drone. *Appl. Sci.* **2022**, *12*, 7333. [CrossRef]
40. Tian, J.; Zheng, Y.; Zhu, H.; Shen, L. A MPC and genetic algorithm based approach for multiple UAVs cooperative search. In Proceedings of the International Conference on Computational and Information Science, Xi'an, China, 15–19 December 2005 pp. 399–404.
41. Tian, J.; Shen, L.; Zheng, Y. Genetic algorithm based approach for multi-UAV cooperative reconnaissance mission planning problem. In Proceedings of the International Symposium on Methodologies for Intelligent Systems, Bari, Italy, 27–29 September 2006; pp. 101–110.
42. Nikolos, I.; Tsourveloudis, N.; Valavanis, K. Evolutionary algorithm based path planning for multiple UAV cooperation. In *Advances in Unmanned Aerial Vehicles*; Springer: Dordrecht, The Netherlands, 2007; pp. 309–340.
43. Lamont, G.B.; Slear, J.N.; Melendez, K. UAV swarm mission planning and routing using multi-objective evolutionary algorithms In Proceedings of the 2007 IEEE Symposium on Computational Intelligence in Multi-Criteria Decision-Making, Honolulu, HI USA, 1–5 April 2007; pp. 10–20.
44. Eun, Y.; Bang, H. Cooperative task assignment and path planning of multiple UAVs using genetic algorithm. In Proceedings of the AIAA Infotech@ Aerospace 2007 Conference and Exhibit, Rohnert Park, CA, USA, 7–10 May 2007; p. 2982.
45. Pehlivanoglu, Y.V. A new vibrational genetic algorithm enhanced with a Voronoi diagram for path planning of autonomous UAV *Aerosp. Sci. Technol.* **2012**, *16*, 47–55. [CrossRef]
46. Sahingoz, O.K. Flyable path planning for a multi-UAV system with Genetic Algorithms and Bezier curves. In Proceedings of the 2013 International Conference on Unmanned Aircraft Systems (ICUAS), Atlanta, GA, USA, 28–31 May 2013; pp. 41–48.
47. Zhang, X.; Duan, H. An improved constrained differential evolution algorithm for unmanned aerial vehicle global route planning *Appl. Soft Comput.* **2015**, *26*, 270–284. [CrossRef]

48. Cekmez, U.; Ozsiginan, M.; Sahingoz, O.K. Multi-UAV path planning with parallel genetic algorithms on CUDA architecture. In Proceedings of the 2016 on Genetic and Evolutionary Computation Conference Companion, Denver, CO, USA, 29–24 July 2016; pp. 1079–1086.
49. Sørli, J.-V.; Graven, O.H.; Bjerknes, J.D. Multi-UAV cooperative path planning for sensor placement using cooperative coevolving genetic strategy. In Proceedings of the International Conference on Swarm Intelligence, Fukuoka, Japan, 27 July–1 August 2017; pp. 433–444.
50. Chen, J.; Ye, F.; Li, Y. Travelling salesman problem for UAV path planning with two parallel optimization algorithms. In Proceedings of the 2017 Progress in Electromagnetics Research Symposium-Fall (PIERS-FALL), Singapore, 19–22 November 2017; pp. 832–837.
51. Binol, H.; Bulut, E.; Akkaya, K.; Guvenc, I. Time optimal multi-UAV path planning for gathering its data from roadside units. In Proceedings of the 2018 IEEE 88th Vehicular Technology Conference (VTC-Fall), Chicago, IL, USA, 27–30 August 2018; pp. 1–5.
52. Harounabadi, M.; Bocksberger, M.; Mitschele-Thiel, A. Evolutionary path planning for multiple UAVs in message ferry networks applying genetic algorithm. In Proceedings of the 2018 IEEE 29th Annual International Symposium on Personal, Indoor and Mobile Radio Communications (PIMRC), Bologna, Italy, 9–12 September 2018; pp. 1–7.
53. Cao, Y.; Wei, W.; Bai, Y.; Qiao, H. Multi-base multi-UAV cooperative reconnaissance path planning with genetic algorithm. *Clust. Comput.* **2019**, *22*, 5175–5184. [CrossRef]
54. Ma, Y.; Zhang, H.; Zhang, Y.; Gao, R.; Xu, Z.; Yang, J. Coordinated optimization algorithm combining GA with cluster for multi-UAVs to multi-tasks task assignment and path planning. In Proceedings of the 2019 IEEE 15th International Conference on Control and Automation (ICCA), Edinburgh, UK, 16–19 July 2019; pp. 1026–1031.
55. Li, L.; Gu, Q.; Liu, L. Research on path planning algorithm for multi-UAV maritime targets search based on genetic algorithm. In Proceedings of the 2020 IEEE International Conference on Information Technology, Big Data and Artificial Intelligence (ICIBA), Chongqing, China, 6–8 November 2020; pp. 840–843.
56. Xiong, C.; Xin, B.; Guo, M.; Ding, Y.; Zhang, H. Multi-UAV 3D Path Planning in Simultaneous Attack. In Proceedings of the 2020 IEEE 16th International Conference on Control & Automation (ICCA), Singapore, 9–11 October 2020; pp. 500–505.
57. Li, M.; Richards, A.; Sooriyabandara, M. Reliability-Aware Multi-UAV Coverage Path Planning using a Genetic Algorithm. In Proceedings of the 20th International Conference on Autonomous Agents and MultiAgent Systems, Virtual, 3–7 May 2021; pp. 1584–1586.
58. Li, M.; Richards, A.; Sooriyabandara, M. Asynchronous Reliability-Aware Multi-UAV Coverage Path Planning. In Proceedings of the 2021 IEEE International Conference on Robotics and Automation (ICRA), Xi'an, China, 30 May–5 June 2021; pp. 10023–10029.
59. Zhang, J.; Li, N.; Zhang, D.; Wei, X.; Zhang, X. Multi-UAV cooperative Route planning based on decision variables and improved genetic algorithm. *J. Phys. Conf. Ser.* **2021**, *1941*, 012012. [CrossRef]
60. Asim, M.; Mashwani, W.K.; Belhaouari, S.B.; Hassan, S. A novel genetic trajectory planning algorithm with variable population size for multi-UAV-assisted mobile edge computing system. *IEEE Access* **2021**, *9*, 125569–125579. [CrossRef]
61. Yan, M.; Yuan, H.; Xu, J.; Yu, Y.; Jin, L. Task allocation and route planning of multiple UAVs in a marine environment based on an improved particle swarm optimization algorithm. *EURASIP J. Adv. Signal Process.* **2021**, *2021*, 94. [CrossRef]
62. Wang, S.; Jiang, Z.; Bao, X. Autonomous Trajectory Planning Method for Multi-UAV Collaborative Search. In Proceedings of the 2021 5th International Conference on Automation, Control and Robots (ICACR), Nanning, China, 25–27 September 2021; pp. 84–88.
63. Sujit, P.; Beard, R. Multiple UAV path planning using anytime algorithms. In Proceedings of the 2009 American Control Conference, St. Louis, MO, USA, 10–12 June 2009; pp. 2978–2983.
64. Wang, G.; Li, Q.; Guo, L. Multiple UAVs routes planning based on particle swarm optimization algorithm. In Proceedings of the 2010 2nd International Symposium on Information Engineering and Electronic Commerce, Ternopil, Ukraine, 25–25 July 2010; pp. 1–5.
65. Alejo, D.; Cobano, J.; Heredia, G.; Ollero, A. Particle swarm optimization for collision-free 4d trajectory planning in unmanned aerial vehicles. In Proceedings of the 2013 International Conference on Unmanned Aircraft Systems (ICUAS), Atlanta, GA, USA, 28–31 May 2013; pp. 298–307.
66. Liu, J.; Shi, T.; Li, P.; Ren, X.; Ma, H. Trajectories planning for multiple UAVs by the cooperative and competitive PSO algorithm. In Proceedings of the 2015 IEEE Intelligent Vehicles Symposium (IV), Seoul, Republic of Korea, 28 June–1 July 2015; pp. 107–114.
67. Zhang, Y.-Z.; Li, J.-W.; Hu, B.; Zhang, J.-D. An improved PSO algorithm for solving multi-UAV cooperative reconnaissance task decision-making problem. In Proceedings of the 2016 IEEE International Conference on Aircraft Utility Systems (AUS), Beijing, China, 10–12 October 2016; pp. 434–437.
68. Li, X.; Zhao, Y.; Zhang, J.; Dong, Y. A hybrid PSO algorithm based flight path optimization for multiple agricultural UAVs. In Proceedings of the 2016 IEEE 28th International Conference on Tools with Artificial Intelligence (ICTAI), San Jose, CA, USA, 6–8 November 2016; pp. 691–697.
69. Hoang, V.; Phung, M.D.; Dinh, T.H.; Ha, Q.P. Angle-encoded swarm optimization for uav formation path planning. In Proceedings of the 2018 IEEE/RSJ International Conference on Intelligent Robots and Systems (IROS), Madrid, Spain, 1–5 October 2018; pp. 5239–5244.
70. Chen, Q.-Y.; Lu, Y.-F.; Jia, G.-W.; Li, Y.; Zhu, B.-J.; Lin, J.-C. Path planning for UAVs formation reconfiguration based on Dubins trajectory. *J. Cent. South Univ.* **2018**, *25*, 2664–2676. [CrossRef]

71. Patley, A.; Bhatt, A.; Maity, A.; Das, K.; Ranjan Kumar, S. Modified particle swarm optimization based path planning for multi-UAV formation. In Proceedings of the AIAA Scitech 2019 Forum, San Diego, CA, USA, 7–11 January 2019; p. 1167.
72. Shao, S.; Peng, Y.; He, C.; Du, Y. Efficient path planning for UAV formation via comprehensively improved particle swarm optimization. *ISA Trans.* **2020**, *97*, 415–430. [CrossRef]
73. Yang, L.; Zhang, X.; Zhang, Y.; Xiangmin, G. Collision free 4D path planning for multiple UAVs based on spatial refined voting mechanism and PSO approach. *Chin. J. Aeronaut.* **2019**, *32*, 1504–1519.
74. Shao, Z.; Yan, F.; Zhou, Z.; Zhu, X. Path planning for multi-UAV formation rendezvous based on distributed cooperative particle swarm optimization. *Appl. Sci.* **2019**, *9*, 2621. [CrossRef]
75. Liu, Y.; Lu, H. A strategy of multi-UAV cooperative path planning based on CCPSO. In Proceedings of the 2019 IEEE International Conference on Unmanned Systems (ICUS), Beijing, China, 17–19 October 2019; pp. 328–333.
76. He, W.; Qi, X.; Liu, L. A novel hybrid particle swarm optimization for multi-UAV cooperate path planning. *Appl. Intell.* **2021**, *51* 7350–7364. [CrossRef]
77. Ahmed, N.; Pawase, C.J.; Chang, K. Distributed 3-D Path Planning for Multi-UAVs with Full Area Surveillance Based on Particle Swarm Optimization. *Appl. Sci.* **2021**, *11*, 3417. [CrossRef]
78. Mobarez, E.; Sarhan, A.; Ashry, M. Obstacle avoidance for multi-UAV path planning based on particle swarm optimization. *IOP Conf. Ser. Mater. Sci. Eng.* **2021**, *1172*, 012039. [CrossRef]
79. Xiao, J.; Sun, H.; Chai, X.; Qu, B.; Wen, P.; Zhou, Y.; Wang, H.; Wang, D. Multi-UAV 3D Path Planning Based on Improved Particle Swarm Optimizer. In Proceedings of the 2021 International Conference on Computer, Internet of Things and Control Engineering (CITCE), Guangzhou, China, 12–14 November 2021; pp. 144–149.
80. Meng-yun, S.L.; Dai, J.-Y.; Ying, J.; Lu, L.-L.; Tian, G.-J.; Tang, Q. Multi-UAV coordinated track planning method based on MSISOS algorithm. In Proceedings of the International Conference on Intelligent and Human-Computer Interaction Technology (IHCIT 2022), Tashkent, Uzbekistan, 30 August–10 September 2022; pp. 313–323.
81. Chung, W.; Kim, M.; Lee, S.; Lee, S.-P.; Park, C.-S.; Son, H. Hierarchical Particle Swarm Optimization for Multi UAV Waypoints Planning under Various Threats. *J. Korean Soc. Aeronaut. Space Sci.* **2022**, *50*, 385–391. [CrossRef]
82. Lu, L.; Dai, J.; Ying, J. Distributed multi-UAV cooperation for path planning by an NTVPSO-ADE algorithm. In Proceedings of the 2022 41st Chinese Control Conference (CCC), Hafei, China, 25–27 July 2022; pp. 5973–5978.
83. Luo, D.; Shao, J.; Xu, Y.; You, Y.; Duan, H. Coevolution pigeon-inspired optimization with cooperation-competition mechanism for multi-UAV cooperative region search. *Appl. Sci.* **2019**, *9*, 827. [CrossRef]
84. Ruan, W.-Y.; Duan, H.-B. Multi-UAV obstacle avoidance control via multi-objective social learning pigeon-inspired optimization *Front. Inf. Technol. Electron. Eng.* **2020**, *21*, 740–748. [CrossRef]
85. Duan, H.; Zhao, J.; Deng, Y.; Shi, Y.; Ding, X. Dynamic discrete pigeon-inspired optimization for multi-UAV cooperative search-attack mission planning. *IEEE Trans. Aerosp. Electron. Syst.* **2020**, *57*, 706–720. [CrossRef]
86. Wang, B.; Zhao, Y.; Li, R.; Jing, L.; Liu, Z. Cooperative Path Planning Algorithm of UAV in Urban Environment Based on Improved Pigeon Swarm Algorithm. In Proceedings of the 2021 5th Chinese Conference on Swarm Intelligence and Cooperative Control Springer: Singapore, 2023; pp. 639–650.
87. Yu, Y.; Deng, Y.; Duan, H. Multi-UAV Cooperative Path Planning via Mutant Pigeon Inspired Optimization with Group Learning Strategy. In Proceedings of the International Conference on Swarm Intelligence, Qingdao, China, 17–21 July 2021; pp. 195–204.
88. Lu, J.; Jiang, J.; Han, B.; Liu, J.; Lu, X. Dynamic Target Search of UAV Swarm Based on Improved Pigeon-Inspired Optimization In Proceedings of the 2021 5th Chinese Conference on Swarm Intelligence and Cooperative Control; Springer: Singapore, 2023 pp. 361–371.
89. Zheng, W.; Luo, D.; Zhou, Z.; Xu, Y.; Chen, Y. Multi-UAV Cooperative Moving Target Search Based on Improved Pigeon-Inspired Optimization. In Proceedings of the 2021 5th Chinese Conference on Swarm Intelligence and Cooperative Control; Springer Singapore, 2023; pp. 921–930.
90. Luo, D.; Li, S.; Shao, J.; Xu, Y.; Liu, Y. Pigeon-inspired optimisation-based cooperative target searching for multi-UAV in uncertain environment. *Int. J. Bio-Inspired Comput.* **2022**, *19*, 158–168. [CrossRef]
91. Shi, K.; Zhang, X.; Xia, S. Multiple swarm fruit fly optimization algorithm based path planning method for multi-UAVs. *Appl. Sci.* **2020**, *10*, 2822. [CrossRef]
92. Li, K.; Ge, F.; Han, Y.; Xu, W. Path planning of multiple UAVs with online changing tasks by an ORPFOA algorithm. *Eng. Appl. Artif. Intell.* **2020**, *94*, 103807. [CrossRef]
93. Mao, Y.; Huang, D.; Qin, N.; Zhu, L.; Zhao, J. Cooperative 3D Path Planning of Multi-UAV Via Improved Fruit Fly Optimization *Res. Sq.* **2021**. [CrossRef]
94. Tian, G.; Zhang, L.; Bai, X.; Wang, B. Real-time dynamic track planning of multi-UAV formation based on improved artificial bee colony algorithm. In Proceedings of the 2018 37th Chinese control conference (CCC), Wuhan, China, 25–27 July 2018 pp. 10055–10060.
95. Bai, X.; Wang, P.; Wang, Z.; Zhang, L. 3D multi-UAV collaboration based on the hybrid algorithm of artificial bee colony and A. In Proceedings of the 2019 Chinese Control Conference (CCC), Guangzhou, China, 27–30 July 2019; pp. 3982–3987.
96. Liu, H.; Ge, J.; Wang, Y.; Li, J.; Ding, K.; Zhang, Z.; Guo, Z.; Li, W.; Lan, J. Multi-UAV Optimal Mission Assignment and Path Planning for Disaster Rescue Using Adaptive Genetic Algorithm and Improved Artificial Bee Colony Method. *Actuators* **2022**, *11*, 4. [CrossRef]

97. Dewangan, R.K.; Saxena, P. Three-dimensional route planning for multiple unmanned aerial vehicles using Salp Swarm Algorithm. *J. Exp. Theor. Artif. Intell.* **2022**, 1–20. [CrossRef]
98. Cekmez, U.; Ozsiginan, M.; Sahingoz, O.K. A UAV path planning with parallel ACO algorithm on CUDA platform. In Proceedings of the 2014 International Conference on Unmanned Aircraft Systems (ICUAS), Orlando, FL, USA, 27–30 May 2014; pp. 347–354.
99. Qiannan, Z.; Ziyang, Z.; Chen, G.; Ruyi, D. Path planning of UAVs formation based on improved ant colony optimization algorithm. In Proceedings of the 2014 IEEE Chinese Guidance, Navigation and Control Conference, Yantai, China, 8–10 August 2014; pp. 1549–1552.
100. Huang, L.; Qu, H.; Ji, P.; Liu, X.; Fan, Z. A novel coordinated path planning method using k-degree smoothing for multi-UAVs. *Appl. Soft Comput.* **2016**, *48*, 182–192. [CrossRef]
101. Li, Z. Multi-UAV coordinate communication path planning based on grid map and ant-algorithm. In Proceedings of the 2017 IEEE 2nd Advanced Information Technology, Electronic and Automation Control Conference (IAEAC), Chongqing, China, 25–26 March 2017; pp. 1584–1587.
102. Perez-Carabaza, S.; Besada-Portas, E.; Lopez-Orozco, J.A.; Jesus, M. Ant colony optimization for multi-UAV minimum time search in uncertain domains. *Appl. Soft Comput.* **2018**, *62*, 789–806. [CrossRef]
103. Zhen, Z.; Xing, D.; Gao, C. Cooperative search-attack mission planning for multi-UAV based on intelligent self-organized algorithm. *Aerosp. Sci. Technol.* **2018**, *76*, 402–411. [CrossRef]
104. Cekmez, U.; Ozsiginan, M.; Sahingoz, O.K. Multi-UAV path planning with multi colony ant optimization. In Proceedings of the International Conference on Intelligent Systems Design and Applications, Delhi, India, 14–16 December 2017; pp. 407–417.
105. Lin, W.; Zhu, Y.; Zeng, W.; Wang, S. Track planning model for multi-UAV based on new multiple ant colony algorithm. In Proceedings of the 2018 Chinese Automation Congress (CAC), Xi'an, China, 30 November–2 December 2018; pp. 3862–3867.
106. Liu, G.; Wang, X.; Liu, B.; Wei, C.; Li, J. Path planning for multi-rotors UAVs formation based on ant colony algorithm. In Proceedings of the 2019 International Conference on Intelligent Computing, Automation and Systems (ICICAS), Chongqing, China, 6–8 December 2019; pp. 520–525.
107. Ali, Z.A.; Zhangang, H.; Zhengru, D. Path planning of multiple UAVs using MMACO and DE algorithm in dynamic environment. *Meas. Control* **2020**, 0020294020915727. [CrossRef]
108. Xia, C.; Yongtai, L.; Liyuan, Y.; Lijie, Q. Cooperative task assignment and track planning for multi-UAV attack mobile targets. *J. Intell. Robot. Syst.* **2020**, *100*, 1383–1400. [CrossRef]
109. Ali, Z.A.; Zhangang, H.; Hang, W.B. Cooperative path planning of multiple UAVs by using max–min ant colony optimization along with cauchy mutant operator. *Fluct. Noise Lett.* **2021**, *20*, 2150002. [CrossRef]
110. Wei, X.; Xu, J. Distributed path planning of unmanned aerial vehicle communication chain based on dual decomposition. *Wirel. Commun. Mob. Comput.* **2021**, *2021*, 6661926. [CrossRef]
111. Li, H.; Chen, Y.; Chen, Z.; Wu, H. Multi-UAV Cooperative 3D Coverage Path Planning Based on Asynchronous Ant Colony Optimization. In Proceedings of the 2021 40th Chinese Control Conference (CCC), Shanghai, China, 26–28 July 2021; pp. 4255–4260.
112. Majeed, A.; Hwang, S.O. A multi-objective coverage path planning algorithm for UAVs to cover spatially distributed regions in urban environments. *Aerospace* **2021**, *8*, 343. [CrossRef]
113. Radmanesh, M.; Kumar, M.; Sarim, M. Grey wolf optimization based sense and avoid algorithm in a Bayesian framework for multiple UAV path planning in an uncertain environment. *Aerosp. Sci. Technol.* **2018**, *77*, 168–179. [CrossRef]
114. Dewangan, R.K.; Shukla, A.; Godfrey, W.W. Three dimensional path planning using Grey wolf optimizer for UAVs. *Appl. Intell.* **2019**, *49*, 2201–2217. [CrossRef]
115. Xu, C.; Xu, M.; Yin, C. Optimized multi-UAV cooperative path planning under the complex confrontation environment. *Comput. Commun.* **2020**, *162*, 196–203. [CrossRef]
116. Yang, L.; Guo, J.; Liu, Y. Three-dimensional UAV cooperative path planning based on the MP-CGWO algorithm. *Int. Int. J. Innov. Comput. Inf. Control* **2020**, *16*, 991–1006.
117. Huang, G.; Cai, Y.; Liu, J.; Qi, Y.; Liu, X. A Novel Hybrid Discrete Grey Wolf Optimizer Algorithm for Multi-UAV Path Planning. *J. Intell. Syst.* **2021**, *103*, 49. [CrossRef]
118. Jiaqi, S.; Li, T.; Hongtao, Z.; Xiaofeng, L.; Tianying, X. Adaptive multi-UAV path planning method based on improved gray wolf algorithm. *Comput. Electr. Eng.* **2022**, *104*, 108377. [CrossRef]
119. Wu, J.; Yi, J.; Gao, L.; Li, X. Cooperative path planning of multiple UAVs based on PH curves and harmony search algorithm. In Proceedings of the 2017 IEEE 21st international conference on computer supported cooperative work in design (CSCWD), Wellington, New Zealand, 26–28 April 2017; pp. 540–544.
120. Xia, C.; Yudi, A. Multi—UAV path planning based on improved neural network. In Proceedings of the 2018 Chinese Control And Decision Conference (CCDC), Shenyang, China, 9–11 June 2018; pp. 354–359.
121. Sanna, G.; Godio, S.; Guglieri, G. Neural network based algorithm for multi-UAV coverage path planning. In Proceedings of the 2021 International Conference on Unmanned Aircraft Systems (ICUAS), Athens, Greece, 15–18 June 2021; pp. 1210–1217.
122. Luo, W.; Tang, Q.; Fu, C.; Eberhard, P. Deep-sarsa based multi-UAV path planning and obstacle avoidance in a dynamic environment. In Proceedings of the International Conference on Swarm Intelligence, Shanghai, China, 17–22 June 2018; pp. 102–111.
123. Qie, H.; Shi, D.; Shen, T.; Xu, X.; Li, Y.; Wang, L. Joint optimization of multi-UAV target assignment and path planning based on multi-agent reinforcement learning. *IEEE Access* **2019**, *7*, 146264–146272. [CrossRef]

124. Zhao, C.; Liu, J.; Sheng, M.; Teng, W.; Zheng, Y.; Li, J. Multi-UAV trajectory planning for energy-efficient content coverage: A decentralized learning-based approach. *IEEE J. Sel. Areas Commun.* **2021**, *39*, 3193–3207. [CrossRef]
125. Wang, T.; Zhang, B.; Zhang, M.; Zhang, S. Multi-UAV Collaborative Path Planning Method Based on Attention Mechanism. *Math. Probl. Eng.* **2021**, *2021*, 6964875. [CrossRef]
126. Liu, Q.; Zhang, Y.; Li, M.; Zhang, Z.; Cao, N.; Shang, J. Multi-UAV Path Planning Based on Fusion of Sparrow Search Algorithm and Improved Bioinspired Neural Network. *IEEE Access* **2021**, *9*, 124670–124681. [CrossRef]
127. Wang, L.; Wang, K.; Pan, C.; Xu, W.; Aslam, N.; Hanzo, L. Multi-agent deep reinforcement learning-based trajectory planning for multi-UAV assisted mobile edge computing. *IEEE Trans. Cogn. Commun. Netw.* **2020**, *7*, 73–84. [CrossRef]
128. Zhang, W.; Wang, Q.; Liu, X.; Liu, Y.; Chen, Y. Three-dimension trajectory design for multi-UAV wireless network with deep reinforcement learning. *IEEE Trans. Veh. Technol.* **2020**, *70*, 600–612. [CrossRef]
129. Bayerlein, H.; Theile, M.; Caccamo, M.; Gesbert, D. Multi-UAV path planning for wireless data harvesting with deep reinforcement learning. *IEEE Open J. Commun. Soc.* **2021**, *2*, 1171–1187. [CrossRef]
130. Tianle, S.; Wang, Y.; Zhao, C.; Li, Y.; Zhang, G.; Zhu, Q. Multi-Uav Wrsn Charging Path Planning Based on Improved Heed and Ia-Drl. *SSRN* **2022**, 4204756. [CrossRef]
131. Rasche, C.; Stern, C.; Kleinjohann, L.; Kleinjohann, B. A distributed multi-uav path planning approach for 3d environments. In Proceedings of the 5th International Conference on Automation, Robotics and Applications, Wellington, New Zealand, 6–8 December 2011; pp. 7–12.
132. Li, X.; Fang, Y.; Fu, W. Obstacle avoidance algorithm for Multi-UAV flocking based on artificial potential field and Dubins path planning. In Proceedings of the 2019 IEEE International Conference on Unmanned Systems (ICUS), Beijing, China, 17–19 October 2019; pp. 593–598.
133. Tang, J.; Sun, J.; Lu, C.; Lao, S. Optimized artificial potential field algorithm to multi-unmanned aerial vehicle coordinated trajectory planning and collision avoidance in three-dimensional environment. *Proc. Inst. Mech. Eng. Part G J. Aerosp. Eng.* **2019**, *233*, 6032–6043. [CrossRef]
134. Chen, H.; Chen, H.; Liu, Q. Multi-UAV 3D formation path planning based on improved artificial potential field. *J. Syst. Simul.* **2020**, *32*, 414.
135. Sun, H.; Qi, J.; Wu, C.; Wang, M. Path planning for dense drone formation based on modified artificial potential fields. In Proceedings of the 2020 39th Chinese Control Conference (CCC), Shenyang, China, 27–29 July 2020; pp. 4658–4664.
136. Dongcheng, L.; Jiyang, D. Research on Multi-UAV Path Planning and Obstacle Avoidance Based on Improved Artificial Potential Field Method. In Proceedings of the 2020 3rd International Conference on Mechatronics, Robotics and Automation (ICMRA) Shanghai, China, 16–18 October 2020; pp. 84–88.
137. Wang, N.; Dai, J.; Ying, J. UAV formation obstacle avoidance control algorithm based on improved artificial potential field and consensus. *Int. J. Aeronaut. Space Sci.* **2021**, *22*, 1413–1427. [CrossRef]
138. Wang, N.; Dai, J.; Ying, J. Research on Consensus of UAV Formation Trajectory Planning Based on Improved Potential Field. In Proceedings of the 2021 40th Chinese Control Conference (CCC), Shanghai, China, 26–28 July 2021; pp. 99–104.
139. Li, R.; Xiong, Y.; Zhang, T. Intelligent Path Planning Algorithm for UAV Group Based on Machine Learning. *J. Phys. Conf. Ser* **2021**, *1865*, 042118. [CrossRef]
140. Wei, B. UAV formation obstacle avoidance control method based on artificial potential field and consistency. *J. Phys. Conf. Ser* **2021**, *2083*, 042029. [CrossRef]
141. Pan, Z.; Zhang, C.; Xia, Y.; Xiong, H.; Shao, X. An Improved Artificial Potential Field Method for Path Planning and Formation Control of the Multi-UAV Systems. *IEEE Trans. Circuits Syst. II Express Briefs* **2021**, *69*, 1129–1133. [CrossRef]
142. Wang, N.; Dai Jiyang, Y.J.; Li, Y.; Lu, L. Multi-UAV Trajectory Planning Simulation Based on Adaptive Extended Potential Field. *J Syst. Simul.* **2021**, *33*, 2147.
143. Zhang, S.; Xu, M.; Wang, X. Research on Obstacle Avoidance Algorithm of Multi-UAV Consistent Formation Based on Improved Dynamic Window Approach. In Proceedings of the 2022 IEEE Asia-Pacific Conference on Image Processing, Electronics and Computers (IPEC), Dalian, China, 14–16 April 2022; pp. 991–996.
144. Bellingham, J.S.; Tillerson, M.; Alighanbari, M.; How, J.P. Cooperative path planning for multiple UAVs in dynamic and uncertain environments. In Proceedings of the 41st IEEE Conference on Decision and Control, Las Vegas, NV, USA, 10–13 December 2002 pp. 2816–2822.
145. Maza, I.; Ollero, A. Multiple UAV cooperative searching operation using polygon area decomposition and efficient coverage algorithms. In *Distributed Autonomous Robotic Systems 6*; Springer: Tokyo, Japan, 2007; pp. 221–230.
146. Dehghan, S.M.M.; Moradi, H.; Shahidian, S.A.A. Optimal path planning for DRSSI based localization of an RF source by multiple UAVs. In Proceedings of the 2014 Second RSI/ISM International Conference on Robotics and Mechatronics (ICRoM), Tehran Iran, 15–17 October 2014; pp. 558–563.
147. Wang, Z.; Liu, L.; Long, T. Minimum-time trajectory planning for multi-unmanned-aerial-vehicle cooperation using sequential convex programming. *J. Guid. Control Dyn.* **2017**, *40*, 2976–2982. [CrossRef]
148. Causa, F.; Fasano, G.; Grassi, M. Multi-UAV path planning for autonomous missions in mixed GNSS coverage scenarios. *Sensors* **2018**, *18*, 4188. [CrossRef] [PubMed]

149. Pengfei, J.; Yu, C.; Aihua, W.; Zhenqian, H.; Yichong, W. Optimal Path Planning for Multi-UAV Based on Pseudo-spectral Method. In Proceedings of the 2020 International Conference on Big Data, Artificial Intelligence and Internet of Things Engineering (ICBAIE), Fuzhou, China, 12–14 June 2020; pp. 417–422.
150. Li, J.; Xiong, Y.; She, J.; Wu, M. A path planning method for sweep coverage with multiple UAVs. *IEEE Internet Things J.* **2020**, *7*, 8967–8978. [CrossRef]
151. Gao, X.; Fan, J.; Wu, F.; Chen, G. Approximation algorithms for sweep coverage problem with multiple mobile sensors. *IEEE/ACM Trans. Netw.* **2018**, *26*, 990–1003. [CrossRef]
152. Xia, Q.; Liu, S.; Guo, M.; Wang, H.; Zhou, Q.; Zhang, X. Multi-UAV trajectory planning using gradient-based sequence minimal optimization. *Robot. Auton. Syst.* **2021**, *137*, 103728. [CrossRef]
153. Wang, G.; Song, K.; Zhang, X. The Real-Time Trajectory Planning of UAV Formation Transformation Based on Safety Flight Corridor. In Proceedings of the 2021 Chinese Intelligent Systems Conference; Springer: Singapore, 2022; pp. 481–491.
154. Cho, S.-W.; Park, J.-H.; Park, H.-J.; Kim, S. Multi-UAV Coverage Path Planning Based on Hexagonal Grid Decomposition in Maritime Search and Rescue. *Mathematics* **2021**, *10*, 83. [CrossRef]
155. Sun, P.; Wang, X.; Wu, B.; Xuan, Y.; Luo, J. Cooperative 4D Penetration Path Planning of Multi-UAV Using MLS-EDA Algorithm. In Proceedings of the 2021 5th Chinese Conference on Swarm Intelligence and Cooperative Control; Springer: Singapore, 2023; pp. 1409–1415.
156. Cheng, Z.; Zhao, L.; Shi, Z. Decentralized Multi-UAV Path Planning Based on Two-Layer Coordinative Framework for Formation Rendezvous. *IEEE Access* **2022**, *10*, 45695–45708. [CrossRef]
157. Yanmaz, E. Joint or decoupled optimization: Multi-UAV path planning for search and rescue. *Ad Hoc Netw.* **2022**, *138*, 103018. [CrossRef]
158. Oh, H.; Kim, S.; Suk, J.; Tsourdos, A. Coordinated trajectory planning for efficient communication relay using multiple UAVs. *IFAC Proc. Vol.* **2013**, *46*, 119–124. [CrossRef]
159. Cui, Y.; Ren, J.; Du, W.; Dai, J. UAV target tracking algorithm based on task allocation consensus. *J. Syst. Eng. Electron.* **2016**, *27*, 1207–1218. [CrossRef]
160. Wu, Z.; Li, J.; Zuo, J.; Li, S. Path planning of UAVs based on collision probability and Kalman filter. *IEEE Access* **2018**, *6*, 34237–34245. [CrossRef]
161. Wu, W.; Li, J.; Wu, Y.; Ren, X.; Tang, Y. Multi-UAV Adaptive Path Planning in Complex Environment Based on Behavior Tree. In Proceedings of the International Conference on Collaborative Computing: Networking, Applications and Worksharing, Shanghai, China, 16–18 October 2020; pp. 494–505.
162. Wang, Y.; Yue, Y.; Shan, M.; He, L.; Wang, D. Formation reconstruction and trajectory replanning for multi-uav patrol. *IEEE/ASME Trans. Mechatron.* **2021**, *26*, 719–729. [CrossRef]
163. Chen, C.-C.; Liu, H.H. Model of UAV and downwash for multi-UAV path planning. In Proceedings of the AIAA SCITECH 2022 Forum, San Diego, CA, USA, 3–7 January 2022; p. 0757.
164. Xu, C.; Liao, X.; Tan, J.; Ye, H.; Lu, H. Recent research progress of unmanned aerial vehicle regulation policies and technologies in urban low altitude. *IEEE Access* **2020**, *8*, 74175–74194. [CrossRef]

drones

MDPI

Article

Smart Decision-Support System for Pig Farming

Hao Wang [1,2], Boyang Li [2,*], Haoming Zhong [3], Ahong Xu [3], Yingjie Huang [3], Jingfu Zou [3], Yuanyuan Chen [2], Pengcheng Wu [2], Yiqiang Chen [4], Cyril Leung [1,2,5,*] and Chunyan Miao [2,*]

1 China-Singapore International Joint Research Institute, Guangzhou 510700, China
2 School of Computer Science and Engineering, Nanyang Technological University, Singapore 639798, Singapore
3 Webank, Shenzhen 518052, China
4 Institute of Computing Technology, Chinese Academy of Sciences, Beijing 100190, China
5 Faculty of Applied Science, University of British Columbia, Vancouver, BC V6T 1Z3, Canada
* Correspondence: boyang.li@ntu.edu.sg (B.L.); cleung@ece.ubc.ca (C.L.); ascymiao@ntu.edu.sg (C.M.)

Abstract: There are multiple participants, such as farmers, wholesalers, retailers, financial institutions, etc., involved in the modern food production process. All of these participants and stakeholders have a shared goal, which is to gather information on the food production process so that they can make appropriate decisions to increase productivity and reduce risks. However, real-time data collection and analysis continue to be difficult tasks, particularly in developing nations, where agriculture is the primary source of income for the majority of the population. In this paper, we present a smart decision-support system for pig farming. Specifically, we first adopt rail-based unmanned vehicles to capture pigsty images. We then conduct image stitching to avoid double-counting pigs so that we can use image segmentation method to give precise masks for each pig. Based on the segmentation masks, the pig weights can be estimated, and data can be integrated in our developed mobile app. The proposed system enables the above participants and stakeholders to have real-time data and intelligent analysis reports to help their decision-making.

Keywords: smart agriculture; pig farming

Citation: Wang, H.; Li, B.; Zhong, H.; Xu, A.; Huang, Y.; Zou, J.; Chen, Y.; Wu, P.; Chen, Y.; Leung, C.; et al. Smart Decision-Support System for Pig Farming. *Drones* **2022**, *6*, 389. https://doi.org/10.3390/drones6120389

Academic Editors: Monica Herrero-Huerta, Jose A. Jiménez-Berni, Shangpeng Sun and Diego González-Aguilera

Received: 4 November 2022
Accepted: 28 November 2022
Published: 30 November 2022

1. Introduction

Substantial risk and uncertainty exist in the agricultural production process and the associated supply chain [1]. For instance, unpredictable virus spread may seriously harm pig growth conditions, resulting in large-scale pig illness. The financial wellbeing of the farmer, credit provider, and insurer may incur damages from these operational risks [2].

To make informed decisions, a smooth flow of modern agriculture information is necessary [3], which requires the collaboration of multiple participants and stakeholders. This consists of a number of stakeholders, including farmers, who serve as the industry's primary producers; distributors and merchants that store, package, transport, and distribute the harvest to consumers; financial institutions that provide insurance and oversee capital allocation; and regulatory agencies responsible for ensuring food safety, environmental sustainability, and financial stability. Participation in and monitoring of the production process is of interest to each stakeholder. For instance, a farmer may track the pig weights to be aware of their growth conditions; banks may use the data to identify nonperforming loans; and regulatory agencies may adopt the data to evaluate environmental impacts. However, it would be costly if we used traditional methods to collect and distribute such real-time agriculture information.

Making decisions from out-of-date information may have negative effects [4]. The recent epidemic of African swine disease in China is a case in point. The reliance on manual data collection and reporting, which was inefficient and prone to human mistake (whether deliberate or not), caused slow government responses and the virus to spread quickly.

throughout China in just eight months [5]. Dr. Defa Li, Fellow of the Chinese Academy of Engineering, put the estimated economic cost of the pandemic at 1 trillion Yuan [6]. Moreover, insurance companies also incurred significant losses due to the pandemic [7].

Through the use of Artificial Intelligence (AI) [8], Internet-of-Things (IoT) technologies [9,10], and big data analysis [11], the next wave of the agricultural revolution promises to reduce the aforementioned operational risks and promote the flow of information [12,13]. However, the introduction of such technologies in developing nations in Southeast Asia and elsewhere faces a number of particular difficulties, including the predominance of small and medium-sized farms, unusual farm conditions, inadequate capital support, and a lack of skilled management. The current situation of a few farms in Guangdong, China, is shown in Figure 1. We caution the reader that these are small farms and that the farmers are using low-cost designs. While industrialized economies have had the most success with smart farming and IoT in agriculture, the problems in developing nations have received less attention.

(1) Farmlands in the Pearl River Delta (2) A Family-Run Pig Farm in China

Figure 1. Farms in Guangdong, China. (**1**) depicts an aerial view of depicts farms with erratic shapes that rely on natural irrigation and simple plastic mulch (instead of more expensive greenhouses). (**2**) shows a pig farm run by a family with a great number of pigs, which is enclosed by basic fences. One of our co-authors took these pictures while on a field trip.

In this paper, we propose a smart pig farming support system. The platform collects data on agricultural productivity using cutting-edge sensors, analyzes the data automatically and interactively, and helps different stakeholders to make well-informed decisions. To address the specific challenges in developing countries, we propose to use unmanned vehicles with sensors to monitor the growth conditions of pigs in a pig farm. Specifically, the unmanned vehicles can be installed at the top of pigsties with fixed rails, and pig images can be taken. The captured images are further processed and analyzed with machine learning techniques, where we attempt to do image stitching, pig instance segmentation and weight estimation. Since the data can be captured and analyzed in an automatic fashion, we are able to monitor livestock farming in real time.

Our contributions can be summarized as follows:

- We propose to use unmanned vehicles based on fixed rails to capture pigsty images, which are low-cost and easy to maintain.
- We propose to apply state-of-the-art AI techniques to conduct data analysis, including image stitching, pig segmentation and weight estimation.
- We propose to develop an app for data fusion, which integrates the collected and analyzed information for stakeholders' visualization.

In the following sections, we first review the related works (Section 2), including image stitching, pig segmentation, and pig weight estimation methods. Then, in Section 3, we describe each of our proposed smart pig farming system components. We further give experimental results in Section 4.

2. Related Work

2.1. Image Stitching

Image stitching is defined as combining two or more images of the same scene into one full image, which can be called a panoramic image [14]. Due to the advantage of being robust against scene movement and fast processing, many existing works [15,16] adopt feature-based techniques. This method aims to use properly designed features, such as SIFT [17], to get the input image relationships, which are further used to uncover the overlapping areas between input images.

The idea of the feature-based image stitching method is to build the correspondence for points, lines, edges, etc. In order to produce robust stitching results, it is of great significance to adopt scale-invariant and translation-invariant detectors. There are many prevailing feature detectors [18], including SIFT [17], ORB [19], KAZE [20], and AKAZE [21]. To be specific, SIFT [17] is designed based on the Difference-of-Gaussians (DoG) operator, which is invariant to image rotations and scales robustly. Alcantarilla et al. [20] exploit non-linear scale space through non-linear diffusion filtering and propose the KAZE features, which reduce image noise. AKAZE [21] is an improved version of KAZE. Since it uses the modified local difference binary (MLDB) algorithm, AKAZE is a computationally efficient framework. In this paper, we empirically use SIFT as our image stitching approach, since it gives better results than other approach.

2.2. Segmentation Techniques for Pig Images

Semantic image segmentation [22] is a classical computer vision task that aims to classify each of the image pixels into an instance. With the development of deep learning [23], prevailing segmentation models, such as Faster R-CNN [24] and single-shot detection (SSD) [25], perform well on various kinds of datasets. Specifically, DeepLabV3+ [26] improves upon previous work [27] with several improvements, such as adding a simple and effective decoder module to refine the segmentation results. It is proposed to perform multi-class semantic segmentation. Later on, several research works, such as ViTDet [28] and MViTv2 [29], propose to adopt the transformer as segmentation model backbones. MViTv2 [29] is able to perform both image classification and object detection tasks. The aforementioned methods mainly validate their efficacy on the COCO dataset [30].

Pig segmentation is under the umbrella of the general image segmentation task allowing us to produce the fine-grained shapes and locations of all pigs in given images. The segmented results interpret the complex scene information and give intelligent analysis for the given images. Since the produced results can assist both pig body part identification and behavior recognition tasks [31], many research works [32–34] focus on tackling the pig segmentation problem. To be specific, Seo et al. [32] adopt You Only Look Once (YOLO) [35] to separate touching-pigs in real-time. Shao et al. [33] use the segmentation techniques to predict the posture changes of pigs. In this paper, we adopt the pretrained Mask R-CNN [36] model, which is developed on top of Faster R-CNN [24]. We further use our annotated pig images captured in local farms to fine-tune the pretrained model, such that the fine-tuned model is able to give precise masks for pigs.

2.3. Pig Weight Estimation

Weight is an important index in pig rearing [37] and has an effect on managing various stages of the supply chain [38,39]. In terms of the farm scale, one can evaluate a pig's daily gain and nutritional status through pig weights [40]. Specifically, with real-time pig weight data, farmers can raise pigs in good or bad nutrient status separately to meet the uniform marketing weight standard [37], which brings convenience to the pig farming process. Traditional pig weight measurement requires direct contact with pigs, which is limited by its low efficiency [37]. The non-contact measurement of pig weight is challenging and has drawn much attention.

Some research works [41] adopt additional facilities other than cameras for more weight related information capture. To be specific, Shi et al. [41] propose to utilize a binocular stereo vision system to analyze and estimate pig body size and weight. Pezzuolo et al. [42] adopt a Microsoft Kinect v1 depth camera to get the pig body dimensions. However, these methods require high-cost implementations and thus are not applicable to our focus scenarios, which are the rural areas of developing countries. Therefore, we refer to the method proposed by [37], where we estimate pig weights based on pig sizes.

2.4. Summary

In Table 1, we present the comparison between our proposed method and other methods, where we show our strengths and weaknesses. Specifically, our adopted image stitching and image segmentation methods have good performance on the unmanned vehicle captured pig images, where we produce dense detected keypoints to conduct image stitching and achieve high average precision for pig instance segmentation. However, our processing time is relatively longer than other methods. This is acceptable since we do not require instant feedback on the detected results. After we obtain the captured images from our unmanned vehicles, we process these images at our local machines. In terms of the weight estimation component, we utilize a monocular camera, which may yield inferior performance to the depth camera, but it is low-cost and gives low tolerance on the test samples.

Table 1. Comparison between our proposed method and other methods, where we list our strengths and weaknesses.

Component	Strength	Weakness
Image stitching	Our detected keypoints are dense and scattered generally all over the images.	Our processing time is longer than other methods, because of the descriptor size.
Image segmentation	Our average precision is better than other methods.	Our processing time is longer than other methods, because of the backbone complexity.
Weight estimation	Only monocular camera is required, which is low-cost.	Our estimation accuracy is lower than methods using more sensors.

3. Methodology

Specifically, our proposed smart pig farming support platform is intended to standardize agricultural procedures, boost farm output, improve pest and disease management, and lessen operational risks in agriculture. This platform enables us to conduct production optimization based on subject-matter expertise and sensor data. Governments, financial institutions, and regulators can also identify potential operational and financial risks and get ready for them. The proposed system primarily supports livestock pig farming. Pig farms can identify pig positions and pig weights and monitor their weight in real time.

We show the proposed smart pig farming support system architecture in Figure 2, which includes three main components. Firstly, we use unmanned vehicles equipped with on-site sensors to collect data at the farm. Secondly, the data are then processed utilizing inexpensive edge computing nodes that do not have issues with sporadic Internet connections. Thirdly, the data are analyzed, the findings are summarized, and they are made accessible on an open platform that supports an ecosystem. The ecosystem supports a number of business and agricultural activities as well as various operations. We present the details of our system components below.

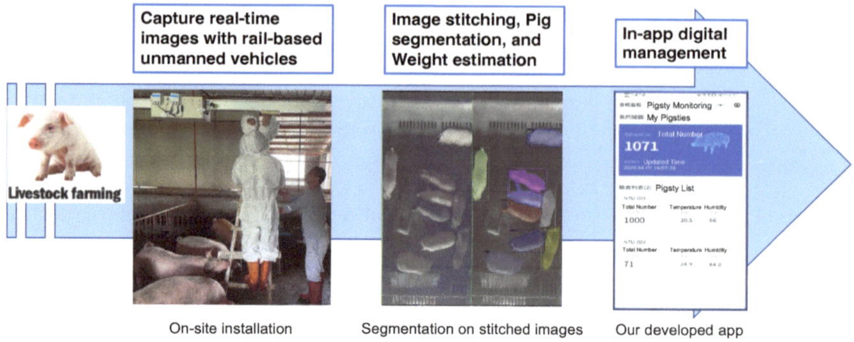

Figure 2. The architecture of our proposed smart pig farming support systems. From left to right we show the images of on-site installation, the segmentation results on stitched images, and the screenshot of our developed app for smart monitoring.

3.1. Data Collection with Unmanned Vehicles

We use rail-based unmanned vehicles that are equipped with surveillance cameras to monitor the livestock pigs. To be specific, the vehicles are installed on the top of pig farms and the rails are designed to be vertical to the pigsty, where the demonstrations are given in Figure 2. The unmanned vehicles move from one end to another end and take consecutive images simultaneously. We show the consecutive raw images taken by our rail-based unmanned vehicles in Figure 3. They are running at regular intervals automatically so that the pig growth conditions can be monitored in real time.

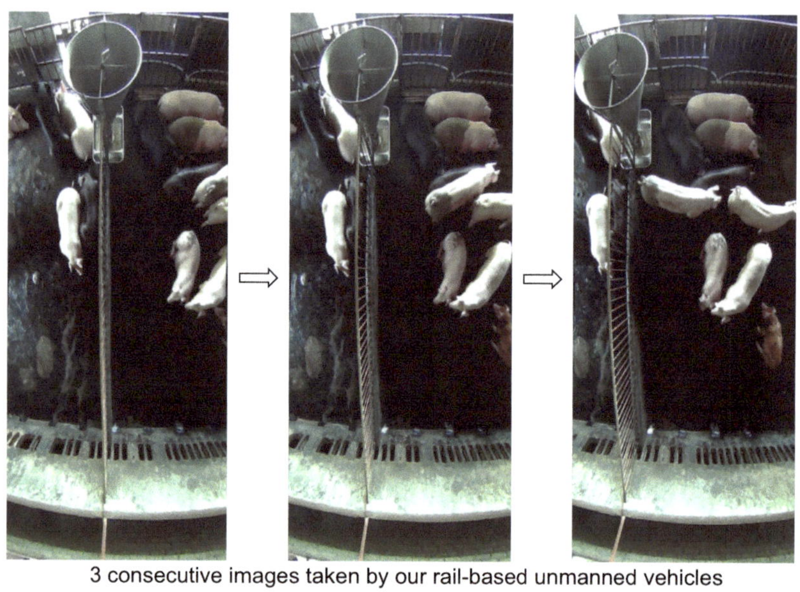

Figure 3. The consecutive raw images taken by our rail-based unmanned vehicles. Based on the collected data, we conduct image stitching, pig segmentation, and weight estimation.

Since we need to train a new pig segmentation model, we hire annotators to perform semantic polygon mask labeling for pig images. In total, we have 636 and 54 images for training and validation, respectively. Moreover, we also collect real pig weight data to estimate the pig weights. Due to our limited on-site manpower and the number of pigs we collected 100 real weight samples; then, we split the collected data into 80 training and 20 test samples.

3.2. Image Stitching

It can be observed that the obtained consecutive scanned images overlap with each other. If we directly use these images for pig segmentation and weight estimation without further processing, the same pigs may be present in different scanned images, which makes pig management difficult. To avoid this, we propose using the image stitching approach to stitch all the images in an attempt to remove overlaps and give an overall perspective on the livestock pigs in the captured pigsties. We follow the methods of [15,16] to combine two or more overlapping images to make one larger image, which is realized in four steps.

In the first step, we aim to detect key points and obtain feature descriptors. Here, we use the SIFT [17] detector since it is rotation-invariant and scale-invariant. In the second step, we match key points with features. Specifically, we define a region around each key point, and then compute local features from normalized regions. Based on the local descriptors, we can perform the matching. In the third step, we firstly sample four random potential matches from the second step, which are used to compute the transform matrix H using direct linear transformation. We use the computed H to obtain the projected points x' from x, which is denoted as

$$x' = Hx. \tag{1}$$

where x and x' are potentially matching pairs. We then count points with a projected distance smaller than a defined threshold, which is set as 0.6. We view these counted points as inliers. We repeat the above process of the third step and return H with the most inliers.

In the fourth step, we aim to stitch two given images, which are depicted as I_1 and I_2 for demonstration. Since we observe that there are the blurry regions in the final panoramic image, we apply linear blending to reduce this effect. Technically, we first define left and right-stitched margins for blending. We then attempt to define weight matrices W_1 and W_2 for I_1 and I_2, respectively. For I_1, we define the weights from the left side of the output panoramic image to the left stitched margin as 1, and the weights from the left stitched margin to the right stitched margin are linearly decremented from 1 to 0. For I_2, we define the weights from the right side of the output panoramic image to the right-stitched margin as 1, and the weights from the left-stitched margin to the right-stitched margin are linearly decremented from 0 to 1. We further apply the weight matrices and combine I_1 and I_2. The stitched images I' are depicted as

$$I' = W_1 I_1 + W_2 I_2. \tag{2}$$

3.3. Pig Segmentation

Here, we aim to locate each instance of the farming pigs, such that we are able to perform individual analysis for them. To this end, we adopt the state-of-the-art semantic segmentation algorithm Mask R-CNN [36]. Technically, there are two stages of Mask R-CNN. First, based on the input image, multiple region proposals are generated where there might be an object. Second, the model predicts the object classes, refines the bounding box positions, and also generates a mask in pixel level on the predicted masks from the first step. Both stages are connected to the backbone structure.

Mask R-CNN uses a ResNet [43] as its backbone. It utilizes standard convolution and FC heads for mask and box prediction, respectively. Moreover, the Feature Pyramid Networks (FPN) [44] is also integrated into the ResNet model, which improves the segmentation accuracy without sacrificing the inference speed. To give the segmentation model a good training start point, we use the ImageNet [45] pretrained models to further improve the segmentation performance. Since there is no publicly available model for pig segmentation, we use our labeled data for model training.

3.4. Pig Weight Estimation

Pig weights can be used to monitor pig growth status. With real-time weight data, farmers are able to know if pigs are in a good or bad nutrient status and change their fodder, which brings convenience to the pig farming process. To estimate pig weights according to our generated segmentation masks, we first fit the obtained masks into ellipses We utilize the major axis and minor axis as pig length h and pig width w, respectively. Then, we follow the equation proposed by [37]. However, considering the particular pig breeds and camera installation conditions in our farms, we need to modify part of the given equation parameters.

Specifically, we denote the weight estimation equations [37] as

$$M = a \times h^b \times w^c, \quad (3)$$

where M denotes the estimated weights and a, b, and c are parameters that should be calculated from our collected data. In order to fit this equation, we first transform Equation (3) into

$$\log(M) = \log(a) + b\log(h) + c\log(w). \quad (4)$$

Since Equation (4) is a linear system, we further adopt the linear regression algorithm to obtain the above a, b, and c parameters. Our final optimized heuristic equation to calculate pig weights is denoted as

$$M = 0.0017 \times h^{1.1908} \times w^{1.0618}. \quad (5)$$

The tolerance of Equation (5) on our test samples is 1.8%.

4. Experiments

4.1. Implementation Details

Unmanned vehicle characteristics. The material of our rail-based unmanned vehicle is iron, the size of which is 80 cm × 40 cm × 30 cm. The rails are installed 3 m above the ground, where the vehicle moves forward at 0.1 m/s. The video camera is attached to the vehicle and points vertically to the ground. The camera takes pictures at the interval of 1 s. The camera resolution is 1080p. To control the unmanned vehicle and camera, we use the Firefly-RK3399 (https://en.t-firefly.com/Product/Rk3399/spec.html (accessed on 3 October 2022)) platform.

Segmentation model. Regarding our adopted pretrained segmentation model, we adopt the Detectron2 (https://github.com/facebookresearch/detectron2 (accessed on 3 October 2022)) implementation. We fine-tune the pretrained segmentation model with a learning rate of 0.00025, and the training iteration number is set as 3000.

Operation cost and efficiency. Our whole on-site system implementation for a one row pigsty costs around 5000 CNY, which is about 702 USD. This is affordable for our deployed local farms. In terms of our operation efficiency, we perform the image stitching and pig weight estimation components in the CPU, and the segmentation component is conducted in the GPU. To be specific, the image stitching and weight estimation components cost about 10 min and 5 s for a pigsty. We use a Nvidia 2080 GPU to obtain the segmented results, and the inference time is 0.3 s per image.

4.2. Image Stitching Results

In Figure 4, we present one resulting image of key points detected by SIFT [17]. We empirically observe that our adopted method gives abundant key point detection results. In Figure 5, we demonstrate a comparison of the matched key points between two images with various features, where we show the key point matching results of ORB [19], KAZE [20], AKAZE [21], and SIFT [17]. Since we do not have ground truth labeling, we only evaluate the performance of these methods qualitatively.

Specifically, it is observed that the SIFT approach produces few intersected matching lines, and the detected key points are diverse. In contrast, ORB and KAZE give many intersected matching lines, and the diversity of AKAZE-generated key points are limited. Key point matching is one of the intermediate steps, and its robust performance is important to our following processing. In Figure 6, we show the qualitative results of the stitched pig images, which cover multiple pigsties under our rail-based unmanned vehicles. The presented qualitative examples demonstrate the efficacy of our adopted method for pig farming image stitching.

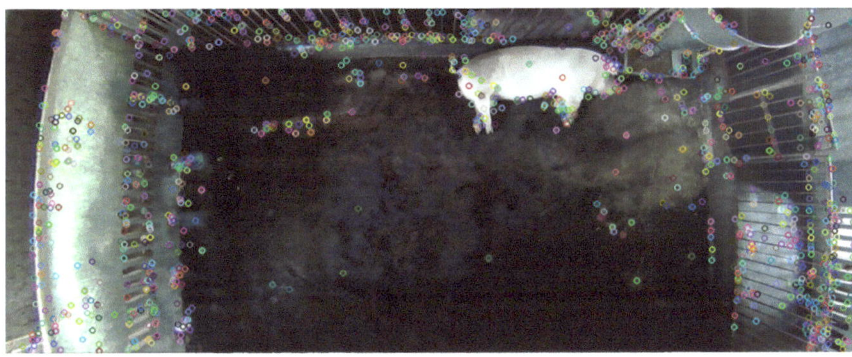

Figure 4. One result of the detected key points by the SIFT detector.

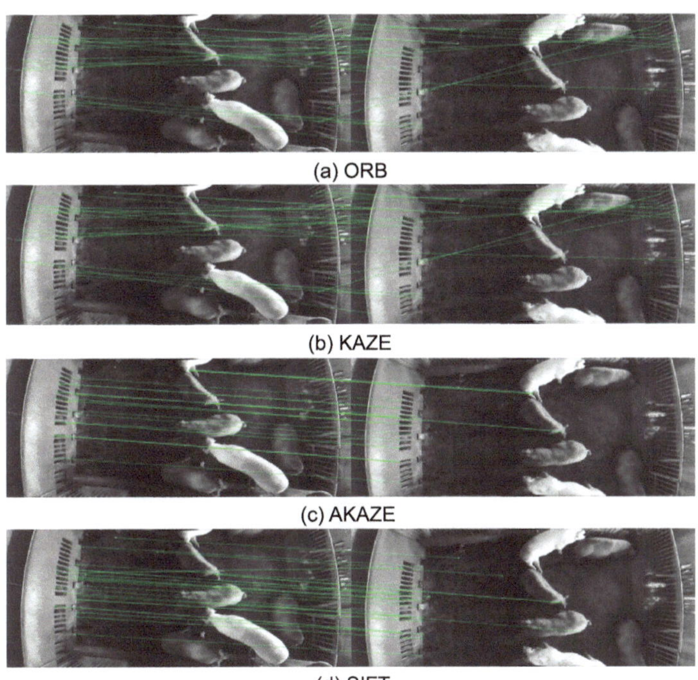

Figure 5. Comparison of different features used for image stitching. We show results of ORB [19], KAZE [20], AKAZE [21], and SIFT [17].

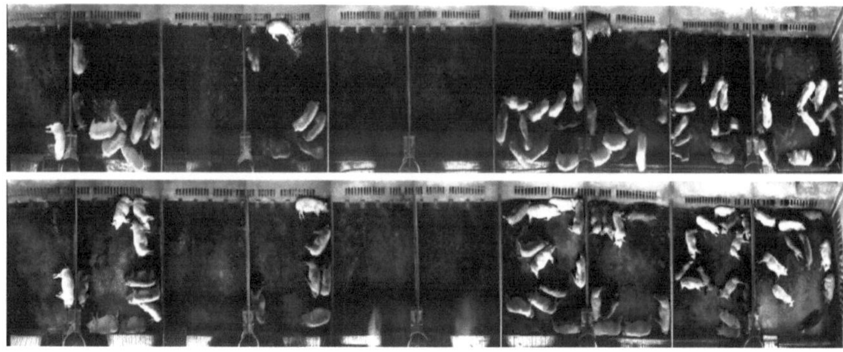

Figure 6. Results of our stitched pigsty images, where each covers one row of the pigsty.

4.3. Segmented Results

In Table 2, we present the quantitative results of our segmentation model, which is evaluated with average precision (AP) under different Intersection over Union (IoU) values. Specifically, we list the quantitative results of various backbone models. DeepLabV3+ [26] improves upon previous work [27] with several improvements, such as adding a simple and effective decoder module to refine the segmentation results. It is proposed to perform multi-class semantic segmentation. However, since we only need to perform segmentation on pig instances, we observe that DeepLabV3+ fails to yield better results than our method. Both ViTDet [28] and MViTv2 [29] adopt the transformer as their model backbones. Based on the experimental results, our adopted ResNet [43] + FPN [44] architecture generalizes to our particular scenarios and gives the best performance among our listed methods. When IoU = 0.50:0.95, our average precision achieves 77.5% with the ResNet-101 model, which indicates that our model performance is consistent under various conditions. ResNet-101 [43] yields better results than ResNet-50 due to its higher complexity.

Table 2. Quantitative results for our pig segmentation validation set. IoU denotes Intersection over Union. FPN denotes Feature Pyramid Network. We show average precision (AP) results of various backbone models.

Method	Average Precision		
	IoU = 0.50	**IoU = 0.75**	**IoU = 0.50:0.95**
DeepLabV3+ [26]	71.1	37.2	36.5
ViTDet [28]	67.0	36.6	37.4
MViTv2 [29]	78.2	40.2	42.3
ResNet50+FPN (Ours)	97.1	87.7	72.4
ResNet101+FPN (Ours)	**98.1**	**90.2**	**77.5**

In Figure 7, we show the pig segmentation results of images taken by our unmanned vehicle, where the results are from various scenarios. The challenge of pig segmentation in our system is that we can only take images from the top view, which suffers from different light conditions. Moreover, the livestock pigs have various poses, and some of them overlap with each other. However, we observe that our adopted Mask R-CNN framework still has achieved over 90% average precision for both IoU = 0.5 and IoU = 0.75, showing the robustness of our system. Specifically, in the second instance of Figure 7, although only parts of the pigs are shown in the image, our model still successfully captures the precise location of these pigs. In the third instance, there are multiple pigs with different poses and our model also gives correct segmentation results.

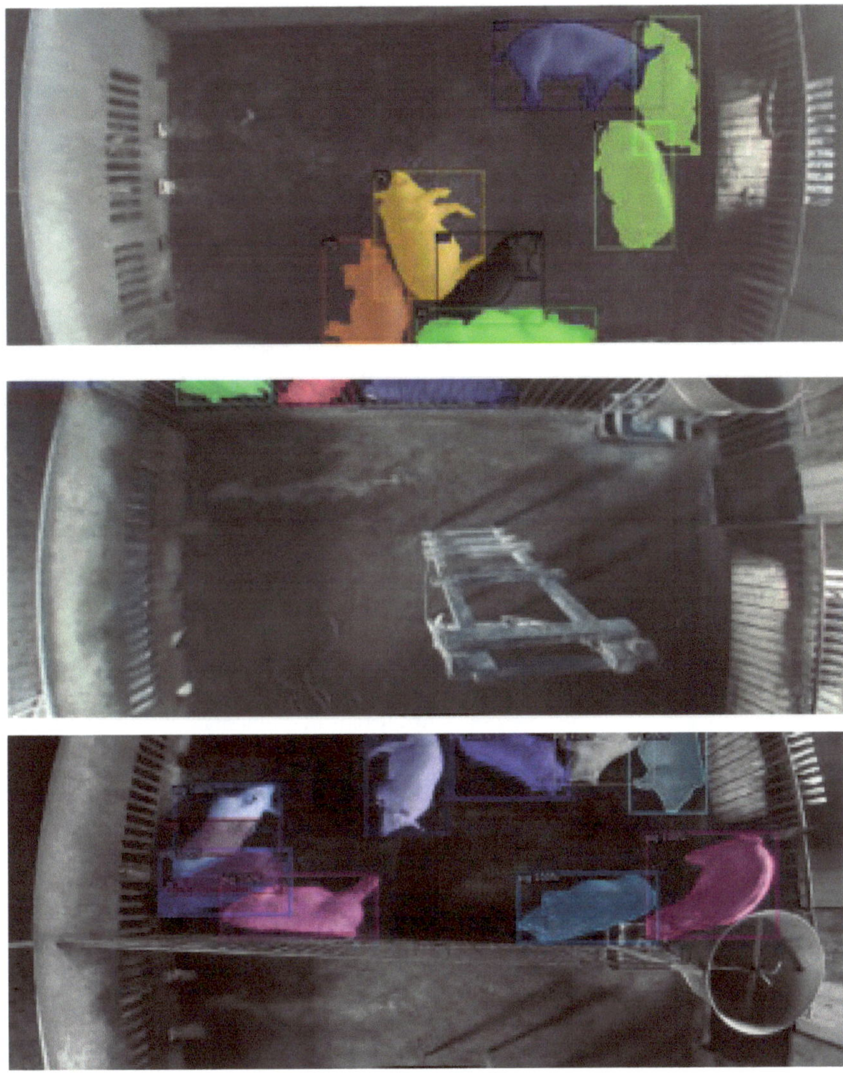

Figure 7. Pig segmentation results in images taken by our rail-based unmanned vehicle.

4.4. Weight Estimation Results

In Figure 8, we present visualizations for the weight estimation of each pig, giving an overview to farmers. The tolerance of our adopted Equation (5) on our test set is 1.8%. The quantitative and qualitative results both validate the usefulness of our adopted pig weight estimation method. The advantage of our system is that it gives real-time feedback on livestock pigs automatically. With the real-time captured weight data of each pig, farmers are able to know the nutrition status and growth conditions of pigs.

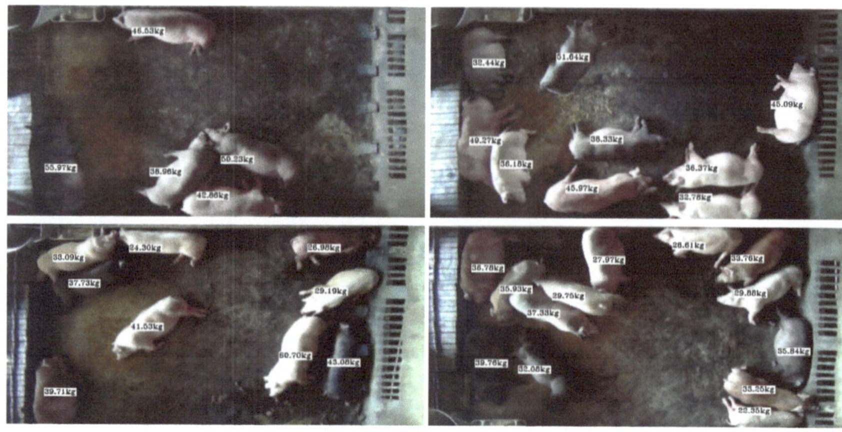

Figure 8. The estimated weight results of each pig for demonstration.

5. Limitations

The proposed system has two main limitations. First, the image stitching results may fail due to pig movement, as shown in Figure 9. This is because our image stitching method assumes that images for stitching contain a static scene and objects only; when a pig moves too much across frames, failure cases may occur. However, our current implementation aims to get the overall estimated pig weights, instead of tracking each pig. Hence, we ran our unmanned vehicle multiple times a day to capture pig images in an attempt to alleviate this issue.

Second, our current implementation only estimates the pig weights based on the segmentation masks, which means we can only obtain an overview of the pig weights. It is useful to uncover some extreme cases. However, we cannot track the weight information and growth status for each individual pig. To improve this, we need to develop pig recognition algorithms so that we are able to track the real-time data and obtain the continuous weight change for each pig. We leave this to our future work.

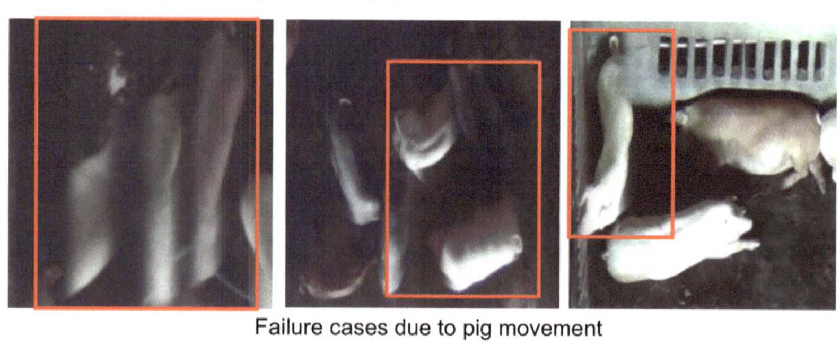

Figure 9. Failure cases due to pig movement.

6. Conclusions

The next phase of the agricultural revolution will increase agricultural output and enhance the sustainability and effectiveness of current farming methods. To support pig farming in developing countries, collecting real-time data and performing relevant analysis is critical. In this paper, we present a smart decision-support system for pig farming, which is low-cost and affordable for developing countries. Specifically, we first adopt on-site rail-based unmanned vehicles to capture all pigsty images at certain intervals. To avoid overlapped pigs in captured images, we conduct image stitching; then, we use an image segmentation model to output pig segmentation masks. Based on the extracted masks, the pig weights can be estimated, and data can be integrated in our developed mobile app. Our proposed system enables pig farming participants and stakeholders to have real-time data reports to help their decision-making.

Author Contributions: Methodology, H.W.; software, Y.C. (Yuanyuan Chen); experiment design, H.Z.; equipment installation, Y.H. and J.Z.; data collection, A.X.; investigation, P.W.; resources, Y.C. (Yiqiang Chen), C.L. and C.M.; writing—original draft preparation, H.W.; writing—review and editing, B.L.; supervision, B.L.; project administration, P.W., Y.C. (Yiqiang Chen) and C.L. All authors have read and agreed to the published version of the manuscript.

Funding: This research is supported by the Joint NTU-WeBank Research Centre on Fintech (Award No: NWJ-2020-007), Nanyang Technological University, by the AI Singapore Programme (Award No: AISG-GC-2019-003) and the NRF Investigatorship Programme (Award No. NRF-NRFI05-2019-0002) of the National Research Foundation, Singapore, and by the the China-Singapore International Joint Research Institute (Award No. 206-A021002).

Data Availability Statement: The data presented in this study are openly available in (https://github.com/hwang1996/Pig-Image-Data (accessed on 3 October 2022)).

Acknowledgments: Any opinions, findings and conclusions or recommendations expressed in this material are those of the authors and do not reflect the views of the funding agencies.

Conflicts of Interest: The research was partially supported by Webank. Webank employees participated in the design of the study, equipment installation and maintenance at the pig farms, and the data collection; they are listed as authors.

References

1. Weis, A.J.; Weis, T. *The Global Food Economy: The Battle for the Future of Farming*; Zed Books: London, UK, 2007.
2. Despommier, D. *The Vertical Farm: Feeding the World in the 21st Century*; Macmillan: New York, NY, USA, 2010.
3. Janssen, S.J.; Porter, C.H.; Moore, A.D.; Athanasiadis, I.N.; Foster, I.; Jones, J.W.; Antle, J.M. Towards a new generation of agricultural system data, models and knowledge products: Information and communication technology. *Agric. Syst.* **2017**, *155*, 200–212. [CrossRef] [PubMed]
4. Waldman, K.B.; Todd, P.M.; Omar, S.; Blekking, J.P.; Giroux, S.A.; Attari, S.Z.; Baylis, K.; Evans, T.P. Agricultural decision making and climate uncertainty in developing countries. *Environ. Res. Lett.* **2020**, *15*, 113004. [CrossRef]
5. Hu, Y. Graphics: The Real Situation of African Swine Fever in China. Available online: https://news.cgtn.com/news/3d3d774e3559444f33457a6333566d54/index.html (accessed on 3 October 2022).
6. Sun, L. Academician Li Defa: The Direct Loss of African Swine Fever in China Is Estimated to be 1 Trillion Yuan. Available online: https://finance.sina.com.cn/money/future/agri/2019-09-26/doc-iicezzrq8551138.shtml (accessed on 3 October 2022).
7. Wu, Y.; Tang, Z. China's Insurers Squeal as Swine Fever Hits Profits. Available online: https://www.caixinglobal.com/2019-12-12/chinas-insurers-squeal-as-swine-fever-hits-profits-101493551.html (accessed on 3 October 2022).
8. Zhang, Y.; Wang, H.; Xu, R.; Yang, X.; Wang, Y.; Liu, Y. High-Precision Seedling Detection Model Based on Multi-Activation Layer and Depth-Separable Convolution Using Images Acquired by Drones. *Drones* **2022**, *6*, 152. [CrossRef]
9. Li, J.; Long, B.; Wu, H.; Hu, X.; Wei, X.; Zhang, Z.; Chai, L.; Xie, J.; Mei, H. Rapid Evaluation Model of Endurance Performance and Its Application for Agricultural UAVs. *Drones* **2022**, *6*, 186. [CrossRef]
10. Bai, A.; Kovách, I.; Czibere, I.; Megyesi, B.; Balogh, P. Examining the Adoption of Drones and Categorisation of Precision Elements among Hungarian Precision Farmers Using a Trans-Theoretical Model. *Drones* **2022**, *6*, 200. [CrossRef]
11. Wolfert, S.; Ge, L.; Verdouw, C.; Bogaardt, M.J. Big data in smart farming—A review. *Agric. Syst.* **2017**, *153*, 69–80. [CrossRef]
12. Zhang, N.; Wang, M.; Wang, N. Precision agriculture—A worldwide overview. *Comput. Electron. Agric.* **2002**, *36*, 113–132. [CrossRef]
13. Vasisht, D.; Kapetanovic, Z.; Won, J.; Jin, X.; Chandra, R.; Sinha, S.; Kapoor, A.; Sudarshan, M.; Stratman, S. FarmBeats: An IoT Platform for Data-Driven Agriculture. In Proceedings of the 14th USENIX Symposium on Networked Systems Design and Implementation (NSDI 17), Boston, MA, USA, 27–29 March 2017; pp. 515–529.
14. Adel, E.; Elmogy, M.; Elbakry, H. Image stitching based on feature extraction techniques: A survey. *Int. J. Comput. Appl.* **2014**, *99*, 1–8. [CrossRef]
15. Brown, M.; Lowe, D.G. Recognising panoramas. In Proceedings of the International Conference on Computer Vision (ICCV), Nice, France, 13–16 October 2003; Volume 3, p. 1218.
16. Brown, M.; Lowe, D.G. Automatic panoramic image stitching using invariant features. *Int. J. Comput. Vis.* **2007**, *74*, 59–73. [CrossRef]
17. Lowe, D.G. Distinctive image features from scale-invariant keypoints. *Int. J. Comput. Vis.* **2004**, *60*, 91–110. [CrossRef]

18. Tareen, S.A.K.; Saleem, Z. A comparative analysis of sift, surf, kaze, akaze, orb, and brisk. In Proceedings of the 2018 International Conference on Computing, Mathematics and Engineering Technologies (iCoMET), Sukkur, Pakistan, 3–4 March 2018; pp. 1–10.
19. Rublee, E.; Rabaud, V.; Konolige, K.; Bradski, G. ORB: An efficient alternative to SIFT or SURF. In Proceedings of the 2011 International Conference on Computer Vision, Tokyo, Japan, 25–27 May 2011; pp. 2564–2571.
20. Alcantarilla, P.F.; Bartoli, A.; Davison, A.J. KAZE features. In Proceedings of the European Conference on Computer Vision, Florence, Italy, 7–13 October 2012; pp. 214–227.
21. Alcantarilla, P.F.; Solutions, T. Fast explicit diffusion for accelerated features in nonlinear scale spaces. *IEEE Trans. Pattern Anal. Mach. Intell.* **2011**, *34*, 1281–1298.
22. Asgari Taghanaki, S.; Abhishek, K.; Cohen, J.P.; Cohen-Adad, J.; Hamarneh, G. Deep semantic segmentation of natural and medical images: A review. *Artif. Intell. Rev.* **2021**, *54*, 137–178. [CrossRef]
23. LeCun, Y.; Bengio, Y.; Hinton, G. Deep learning. *Nature* **2015**, *521*, 436–444. [CrossRef]
24. Ren, S.; He, K.; Girshick, R.; Sun, J. Faster R-CNN: Towards Real-Time Object Detection with Region Proposal Networks. *IEEE Trans. Pattern Anal. Mach. Intell.* **2016**, *39*, 1137–1149. [CrossRef] [PubMed]
25. Liu, W.; Anguelov, D.; Erhan, D.; Szegedy, C.; Reed, S.; Fu, C.Y.; Berg, A.C. Ssd: Single shot multibox detector. In Proceedings of the European Conference on Computer Vision, Amsterdam, The Netherlands, 11–14 October 2016; pp. 21–37.
26. Chen, L.C.; Zhu, Y.; Papandreou, G.; Schroff, F.; Adam, H. Encoder-decoder with atrous separable convolution for semantic image segmentation. In Proceedings of the European Conference on Computer Vision (ECCV), Munich, Germany, 8–14 September 2018, pp. 801–818.
27. Chen, L.C.; Papandreou, G.; Kokkinos, I.; Murphy, K.; Yuille, A.L. Deeplab: Semantic image segmentation with deep convolutional nets, atrous convolution, and fully connected crfs. *IEEE Trans. Pattern Anal. Mach. Intell.* **2017**, *40*, 834–848. [CrossRef] [PubMed]
28. Li, Y.; Mao, H.; Girshick, R.; He, K. Exploring plain vision transformer backbones for object detection. *arXiv* **2022**, arXiv:2203.16527
29. Li, Y.; Wu, C.Y.; Fan, H.; Mangalam, K.; Xiong, B.; Malik, J.; Feichtenhofer, C. MViTv2: Improved Multiscale Vision Transformers for Classification and Detection. In Proceedings of the IEEE/CVF Conference on Computer Vision and Pattern Recognition New Orleans, LA, USA, 19–20 June 2022; pp. 4804–4814.
30. Lin, T.Y.; Maire, M.; Belongie, S.; Hays, J.; Perona, P.; Ramanan, D.; Dollár, P.; Zitnick, C.L. Microsoft coco: Common objects in context. In Proceedings of the European Conference on Computer Vision, Zurich, Switzerland, 6–12 September 2014; pp. 740–755
31. Li, J.; Green-Miller, A.R.; Hu, X.; Lucic, A.; Mohan, M.M.; Dilger, R.N.; Condotta, I.C.; Aldridge, B.; Hart, J.M.; Ahuja, N. Barriers to computer vision applications in pig production facilities. *Comput. Electron. Agric.* **2022**, *200*, 107227. [CrossRef]
32. Seo, J.; Sa, J.; Choi, Y.; Chung, Y.; Park, D.; Kim, H. A yolo-based separation of touching-pigs for smart pig farm applications. In Proceedings of the 2019 21st International Conference on Advanced Communication Technology (ICACT), PyeongChang, Korea 17–20 February 2019; pp. 395–401.
33. Shao, H.; Pu, J.; Mu, J. Pig-posture recognition based on computer vision: Dataset and exploration. *Animals* **2021**, *11*, 1295 [CrossRef]
34. Hu, Z.; Yang, H.; Lou, T. Dual attention-guided feature pyramid network for instance segmentation of group pigs. *Comput Electron. Agric.* **2021**, *186*, 106140. [CrossRef]
35. Redmon, J.; Divvala, S.; Girshick, R.; Farhadi, A. You only look once: Unified, real-time object detection. In Proceedings of the IEEE Conference on Computer Vision and Pattern Recognition, Las Vegas, NV, USA, 27–30 June 2016; pp. 779–788.
36. He, K.; Gkioxari, G.; Dollár, P.; Girshick, R. Mask r-cnn. In Proceedings of the IEEE International Conference on Computer Vision Venice, Italy, 22–29 October 2017; pp. 2961–2969.
37. Li, Z.; Luo, C.; Teng, G.; Liu, T. Estimation of pig weight by machine vision: A review. In Proceedings of the International Conference on Computer and Computing Technologies in Agriculture, Beijing, China, 18–20 September 2013; pp. 42–49.
38. Wongsriworaphon, A.; Arnonkijpanich, B.; Pathumnakul, S. An approach based on digital image analysis to estimate the live weights of pigs in farm environments. *Comput. Electron. Agric.* **2015**, *115*, 26–33. [CrossRef]
39. Pezzuolo, A.; Milani, V.; Zhu, D.; Guo, H.; Guercini, S.; Marinello, F. On-barn pig weight estimation based on body measurements by structure-from-motion (SfM). *Sensors* **2018**, *18*, 3603. [CrossRef] [PubMed]
40. Doeschl-Wilson, A.; Whittemore, C.; Knap, P.; Schofield, C. Using visual image analysis to describe pig growth in terms of size and shape. *Anim. Sci.* **2004**, *79*, 415–427. [CrossRef]
41. Shi, C.; Teng, G.; Li, Z. An approach of pig weight estimation using binocular stereo system based on LabVIEW. *Comput. Electron. Agric.* **2016**, *129*, 37–43. [CrossRef]
42. Pezzuolo, A.; Guarino, M.; Sartori, L.; González, L.A.; Marinello, F. On-barn pig weight estimation based on body measurements by a Kinect v1 depth camera. *Comput. Electron. Agric.* **2018**, *148*, 29–36. [CrossRef]
43. He, K.; Zhang, X.; Ren, S.; Sun, J. Deep residual learning for image recognition. In Proceedings of the IEEE Conference on Computer Vision and Pattern Recognition, Las Vegas, NV, USA, 27–30 June 2016; pp. 770–778.
44. Lin, T.Y.; Dollár, P.; Girshick, R.; He, K.; Hariharan, B.; Belongie, S. Feature pyramid networks for object detection. In Proceedings of the IEEE Conference on Computer Vision and Pattern Recognition, Honolulu, HI, USA, 21–26 July 2017; pp. 2117–2125.
45. Deng, J.; Dong, W.; Socher, R.; Li, L.J.; Li, K.; Fei-Fei, L. Imagenet: A large-scale hierarchical image database. In Proceedings of the 2009 IEEE Conference on Computer Vision and Pattern Recognition, Miami, FL, USA, 20–25 June 2009; pp. 248–255.

drones

MDPI

Article

Auto-Encoder Learning-Based UAV Communications for Livestock Management

Mohammed A. Alanezi [1], Abdullahi Mohammad [2,3], Yusuf A. Sha'aban [2,4], Houssem R. E. H. Bouchekara [4,*] and Mohammad S. Shahriar [4]

1 Department of Computer Science and Engineering Technology, University of Hafr Al Batin, Hafr Al Batin 31991, Saudi Arabia
2 Department of Computer Engineering, Ahmadu Bello University, Zaria 810001, Nigeria
3 Department of Electronic and Electrical Engineering, University College London, London WC1E 7JE, UK
4 Department of Electrical Engineering, University of Hafr Al Batin, Hafr Al Batin 31991, Saudi Arabia
* Correspondence: bouchekara.houssem@gmail.com

Abstract: The advancement in computing and telecommunication has broadened the applications of drones beyond military surveillance to other fields, such as agriculture. Livestock farming using unmanned aerial vehicle (UAV) systems requires surveillance and monitoring of animals on relatively large farmland. A reliable communication system between UAVs and the ground control station (GCS) is necessary to achieve this. This paper describes learning-based communication strategies and techniques that enable interaction and data exchange between UAVs and a GCS. We propose a deep auto-encoder UAV design framework for end-to-end communications. Simulation results show that the auto-encoder learns joint transmitter (UAV) and receiver (GCS) mapping functions for various communication strategies, such as QPSK, 8PSK, 16PSK and 16QAM, without prior knowledge.

Keywords: unmanned aerial vehicle; convolutional auto-encoder; livestock farming; deep neural networks

Citation: Alanezi, M.A.; Mohammad, A.; Sha'aban, Y.A.; Bouchekara, H.R.E.H.; Shahriar, M.S. Auto-Encoder Learning-Based UAV Communications for Livestock Management. *Drones* **2022**, *6*, 276. https://doi.org/10.3390/drones6100276

Academic Editors: Monica Herrero-Huerta, Jose A. Jiménez-Berni, Shangpeng Sun and Diego González-Aguilera

Received: 28 July 2022
Accepted: 7 September 2022
Published: 25 September 2022

1. Introduction

Unmanned Aerial Vehicles (UAVs), known as drones, are self-driven aircraft that work without a human pilot on board [1]. Different types of UAVs are employed for various intents [2]. Initially, the military used the technology for anti-aircraft target techniques, intelligence gathering and surveillance of enemy territories [2–5]. Moreover, UAV technology has evolved beyond its initial purpose. It has, in recent years, gained prominence in diverse spheres of human endeavour. The ease of operating drone technology results in the widespread applications of UAVs in diverse fields, thus making it a prosperous technology [6]. Livestock farming is one of the promising applications of UAVs, where UAVs simplify various operations for efficient animal management [7–10]. Over the years, livestock farming has faced environmental, economic, technical and strategic planning due to varying climatic conditions, population growth and intense competition for land and other natural resources [11]. Nevertheless, the use of advanced technologies such as Artificial Intelligence (AI), the Internet of Things (IoT), Machine Learning (ML), cutting-edge sensors, etc., integrated with UAVs has recently resulted in the widespread adoption of drone technology amongst livestock farmers [9]. As an illustration, Figure 1 shows a typical UAV conceptual design framework of a livestock farming management system (LFMS). The system consists of four development stages; the water examination system, Long-Range Wide-Area Network (LoRaWAN)-based network planning, drone mounted with sensors and cameras and drone path planning optimization. The cattle are fitted with transceivers around their necks. This provides a means for sharing information between the drone and the ground station.

The UAVs are controlled either remotely or manually by a pilot at a ground station, guided using a pre-programmed flight procedure. Wireless communication is one of the critical technologies for UAV wireless communication and is classified into a command and control link and a data link [12]. The command and control link provides essential information about the environment, operating conditions and control instructions for a UAV's safe operation. Therefore, it requires high reliability and low latency. Compared to the command and control link, the data link often maintains the target-related information and thus supports higher data rates

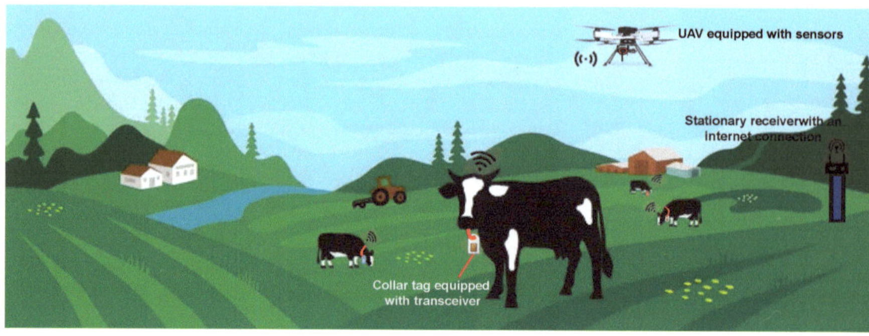

Figure 1. Conceptual framework of UAV-based farm monitoring system [13].

1.1. Motivation

The use of UAVs for livestock location, detection, activity monitoring, anomaly detection and rearing requires onboard sensors operating at radio frequencies (RF) of 2.4 and 5.8 GHz for reliable data transfer between the UAVs and the ground stations [1,4]. However, due to traffic of the target-related UAV data, the tasks become more strenuous and demanding as the resolution of the onboard sensors becomes higher, specifically in the backbone network [14]. Accordingly, this imposes enormous demands on the communication system as well as the challenge of decoding the transmitted data with minimum error probability at the user end. Livestock farming requires extensive farmland, located mainly in rural areas. It is known that many rural areas, even in developed countries, are significantly under-connected with mobile wireless technology. Therefore deploying 5G test beds in rural areas can motivate service providers to improve internet connectivity. Recently, the United Kingdom (UK) launched a project called 5G Rural Integrated Test bed (5GRIT) to create test beds for 5G in rural areas [3]. The project seeks to demonstrate the role 5G networks can play in consolidating farming and tourism sectors using an integrated system of UAVs and AI technologies. Therefore, designing a robust communication system that will provide reliable data transfer between UAVs and the GCS is necessary for UAV-based livestock farming.

1.2. Related Works

Beyond using multirotor UAVs for aerial surveillance, substantial research has been performed on UAVs to ease livestock and agricultural farming [15–17]. Drones have explicitly become helpful in monitoring and enhancing crop and livestock production due to real-time data that help farmers respond more quickly to weed incursion, pest infestation, output projections, livestock health conditions and other issues [18]. UAVs and other related technologies, such as the Internet of things (IoT), have been extensively used for smart agriculture and animal farming [9,19–22].

Different types of UAVs are applied in pest control, crop irrigation, animal health monitoring, animal rearing and other agriculture-related activities [19]. The joint application that both IoT and UAVs can play in smart-driven agriculture was also discussed in [20]. Maddikunta et al. [9] have explored the architecture, adaption and usage of UAVs for smart agriculture. The authors highlighted the UAV's applications and related technologies to efficiently enhance and optimize diverse agricultural processes using smart Bluetooth-enabled sensors. However, reliable data transfer was the major drawback of this approach due to the short range of the Bluetooth UAV-enabled system. In some scenarios, UAVs could be used as tools for mechanized agriculture to ameliorate disorders in various fields through commercial, scientific, agricultural and livestock enhancement [21]. Specifically, the paper focused on providing details of mechanized agriculture using UAV systems for pesticides and fertilizer application in farms that were obstacle rich. Other issues related to the lack of awareness and special education on precision agriculture in animal farming using UAV technology were also highlighted. Furthermore, Alanezi et al. [22] presented a comprehensive review of the state-of-the-art techniques incorporated with UAVs for livestock. The authors highlighted various pressing issues, challenges and opportunities associated with livestock management.

AI and ML have drawn growing research interests and are ubiquitously emerging in many fields due to their capability to model systems through learning from data [23]. Recently, studies and findings have unveiled the potential benefits of deploying AI and machine learning techniques and UAVs for effective livestock farming [2,3,8,24–26]. Studies have shown the feasibility of using UAV video monitoring to predict the food eating behaviour of rangeland-raised Raramuri Criollo non-nursing beef cows [2]. To address the problem associated with animal counting, a computer vision pipeline that uses DNN architecture for automated Holstein Friesian cattle detection and identification was proposed in [24]. The authors introduced a video processing mechanism to efficiently monitor dynamic cattle footage filmed by UAVs. However, the UAV was manually flown and only captured data within small-sized and relatively spaced herds. Rivas et al. [25] presented the use of artificial intelligence techniques for real-time analysis and cattle monitoring using the information captured by drones. The authors used a camera installed in the drone to take images that were later analysed using Convolutional Neural Networks (CNNs) for cattle identification captured in the images. However, the model could not determine the number of animals in a cluster with utmost precision. Furthermore, a test bed implementation that used deep learning algorithms was designed for precision livestock detection and counting from aerial images captured by drones. In the same vein, the use of UAVs to track the postural position of cattle and sheep was studied in [8] to find the optimal number of UAVs that minimizes the UAV-animal distance using a streaming K-means clustering algorithm. All the targeted herds were fitted with global positioning system (GPS) neckbands to monitor their movements. A dual-stream deep learning (DL) architecture that combined exploration strategies learned from previous experiences with instantaneous sensory inputs was proposed to capture the movement of the cattle [26].

Nonetheless, accurate livestock counting in a multi-path crossing by the same animal is still an open problem. While many works of the literature mainly focus on combining ML algorithms with UAVs for efficient smart farming and livestock management, little or no attention is paid to the part that involves data transfer within the UAV communication network. ML techniques, specifically DL, have been used to solve many physical layer communication problems [27–29]. Therefore, this paper proposes a learning framework for an efficient and reliable communications system for UAV-based livestock management. Our contributions are summarized below:

- We built an auto-encoder for end-to-end wireless communications for UAV-assisted livestock management systems. We showed that learning the entire transmitter (UAV) and receiver (GCS or UAV) implementations for a given communication channel link optimized for a chosen loss function (e.g. minimizing BER) is possible. The basic idea is to describe the transmitter, channel, and receiver as a single deep CNN that can be trained as an auto-encoder. Interestingly, this technique can be used as a model approximator to approximate optimal solutions for systems with unknown channel models and loss functions.
- We simulated the communication links with a different set of communication rates to learn various communication schemes, such as QPSK, 8PSK and 16QAM.

- For a (7, 4) communication rate, the proposed auto-encoder performance matched the optimal Hamming code maximum likelihood decoding scheme.

The remainder of the paper is structured as follows: The system model and problem are presented in Section 2. The proposed methodology is described in Section 3. Section 4 presents the simulation results and discussions. Finally, Section 5 summarizes and concludes the paper.

Notations: We use bold uppercase symbols for matrices, bold lowercase symbols for vectors and lowercase symbols for scalars. Finally, notation $\mathscr{L}(\cdot)$ is reserved for the loss function.

2. System Model and Problem Formulation

Reliable communication among UAVs monitoring livestock is critical for efficient and accurate data transmission for managing large herds. Figure 2 portrays a typical high-speed local architecture network constructed over long-distance WiFi access points to establish communication with UAVs and the ground control stations used for cattle and sheep rearing. The UAVs are equipped with onboard cameras and sensors used for taking images of herds and territorial surveillance. Information about the locations of animals is exchanged between the UAVs and the ground control station (GCS), which could be monitored manually or remotely by a human operator.

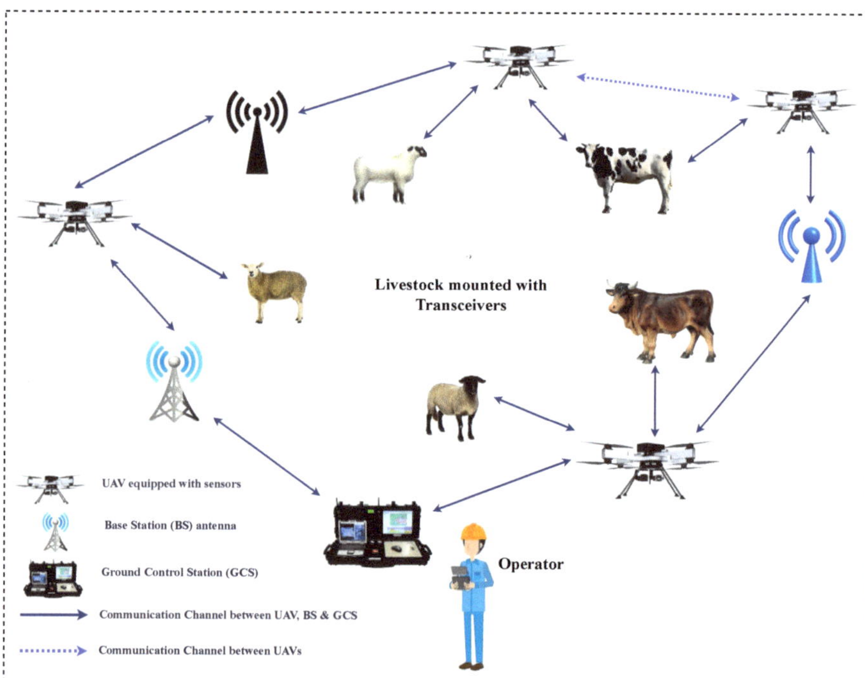

Figure 2. Communications network architecture of UAV-based livestock management system.

Throughout this section, we assume a known perfect channel state information (CSI) between the UAVs and the GCS. The communication links between the UAVs and the GCS can be viewed as a simple communications system consisting of a transmitter, a channel and a receiver. Suppose the UAV wants to send one out of M possible messages $s \in \mathbb{M} = \{1, 2, \cdots, M\}$ to another UAV or GCS through a wireless fading channel. Then the received message is modelled as

$$y = hx + n_0 \quad (1)$$

where x, h and n are the message. The complex baseband message is converted to its equivalent real format using the transformation $f : \mathbb{M} \rightarrow \mathbb{R}^{2n}$ to the message s to generate the transmitted signal $x = f(s) \in \mathbb{R}^{2n}$. It should be noted that the UAV imposes some constraints on the x based on either average energy or average power of the transmitted message as follows [30]

$$||\mathbf{x}||_2^2 \leq 2n, \tag{2}$$

$$\mathbb{E}[|x_i|^2] \leq 1. \tag{3}$$

For simplicity, we use Quadrature phase shift keying (QPSK) and 8-phase shift keying (8PSK) modulation schemes because of their abilities to transmit two bits and three per symbol, respectively. As an illustration, for QPSK, the transmitted symbols $s \in \mathbb{M} = \{\frac{1}{\sqrt{2}} \pm j\frac{1}{\sqrt{2}}, -\frac{1}{\sqrt{2}} \pm j\frac{1}{\sqrt{2}}\}$. Compared to ordinary phase shift keying (PSK), QPSK conveys twice as much information using the same bandwidth [31]. The rate at which the message is sent over a communications channel, known as the communication rate, is given by

$$R = \frac{k}{n} \text{ [bit/channel]}, \tag{4}$$

where $k = \log_2 (M)$ and M is the number of symbols or modulation index. Intuitively, (4) shows that the communication system transmits one out of $M = 2^k$ messages through n active channels or channel uses. This is usually presented by the notation (n, k) [32].

3. Proposed Methodology

Generally, a simple communications system can be viewed as a particular type of auto-encoder from the deep learning viewpoint [30,33]. An auto-encoder is an unsupervised learning model that learns to squeeze and reconstruct the input. Therefore, it can be considered a dimensionality reduction framework that allows the input reconstruction at the output with minimal error. However, in our case, the auto-encoder is used for end-to-end communication to learn the representations of the messages s that are robust to the channel impairments mapping $\mathbf{x}$ to $\mathbf{y}$, such that the transmitted information can be recovered with a minimal probability of error. Contrary to redundancy removal from the input data for compression, our proposed auto-encoder usually adds redundancy, learning an intermediate representation robust to channel variations for reliable data transfer.

Firstly, the UAV flies above the livestock to capture data (usually real-time images) about the livestock and send it to the GCS for analysis. The reliable data transfer requires that the UAV communication system be divided into a sequence of communication blocks, which are traditionally optimized individually. Such an approach depends on complex mathematical models that are usually intractable. However, the communication blocks are jointly optimized as a single learning block to simplify the process while ensuring reliable data transfer from the UAVs to the GCS. The proposed auto-encoder is shown in Figure 3. Here, the transmitter, which could be a UAV, is the encoder consisting of feedforward convolutional neural network (CNN) layers followed by a normalization layer that guarantees that the physical constraints on $\mathbf{x}$ are met based on (2) or (3). Accordingly, the input s to the encoder is encoded as a one-hot vector $\mathbf{1}s \in \mathbb{R}^M$, having an M-dimensional vector, the s-th element of which is equal to one and zero otherwise. The wireless channel layer is represented by a fading channel obtained from a random normal distribution with zero mean and unit variance and an Additive White Gaussian Noise (AWGN).

Similarly, the receiver (decoder), which could be the GCS, is also implemented as a feedforward CNN. The decoder's final layer uses a softmax activation whose output $\mathbf{p} \in (0,1)^M$ is a probability vector over all possible messages. The estimated message corresponds to the index of the element of $\mathbf{p}$ with the highest probability value. Consequently, the conditional probability density function $\mathbf{p}(\mathbf{y}|\mathbf{x})$ defines the channel, where $\mathbf{y} \in \mathbb{R}^{2n}$ designates the received signal. Once $\mathbf{y}$ is received, the decoder applies the transformation $g : \mathbb{R}^{2n} \rightarrow \mathbb{M}$ to yield the estimate of the transmitted message $\hat{s}$. We train the auto-encoder end-to-end using stochastic gradient descent (SGD) over all possible messages $s \in \mathbb{M}$ using the appropriate categorical cross-entropy loss function to generate the predicted output or reconstructed estimate of the transmitted message. Therefore, the loss function is given by

$$\mathscr{L}(\boldsymbol{\theta}) = -\frac{1}{N}\sum_{i=1}^{N} s_i \log(\hat{s}), \tag{5}$$

where N is the number of samples, $\boldsymbol{\theta}$ is the model parameters (weights of the neural network), s is the original transmitted message, and $\hat{s}$ is the estimated message.

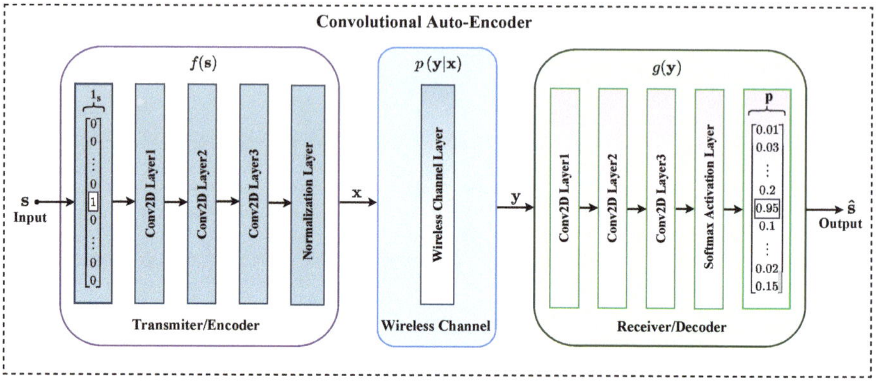

Figure 3. UAV communications system over fading channel depicted as an auto-encoder with an input message s encoded as a one-hot vector.

Accordingly, this end-to-end learning concept is applied to UAV communications where information about herds (usually images) from the UAVs is sent through a wireless channel to the GCS for effective monitoring. The details of the proposed auto-encoder UAV communications system are shown in Figure 4, and its architectural layout is provided in Table 1.

Table 1. Layout of the proposed UAV-based auto-encoder.

Layer	Output
Input	$(M, M, 1)$
2D Convolution + ReLU	$(M_{E1}, M_{E1}, 2n)$, $M_{E1} = \frac{M}{2}$
2D Convolution + ReLU	$(M_{E2}, M_{E2}, 2n)$, $M_{E2} = \frac{M}{2}$
2D Convolution + ReLU	$(M_{E3}, M_{E3}, 2n)$, $M_{E3} = \frac{M}{2}$
Flatten	$(M_{E3} \times M_{E3} \times 2n)$
Normalization	$(M_{E3} \times M_{E3} \times 2n)$
Wireless channel + Noise	$(M_{E3} \times M_{E3} \times 2n)$
Fully Connected + ReLU	$(M_{E3} \times M_{E3} \times 2n)$
2D Convolution + ReLU	$(M_{D3}, M_{D3}, 2n)$, $M_{D3} = \frac{M}{2}$
2D Convolution + ReLU	$(M_{D2}, M_{D2}, 2n)$, $M_{D2} = \frac{M}{2}$
2D Convolution + ReLU	$(M_{D1}, M_{D1}, 2n)$, $M_{D1} = \frac{M}{2}$
Fully Connected + softmax	$(M, M, 1)$

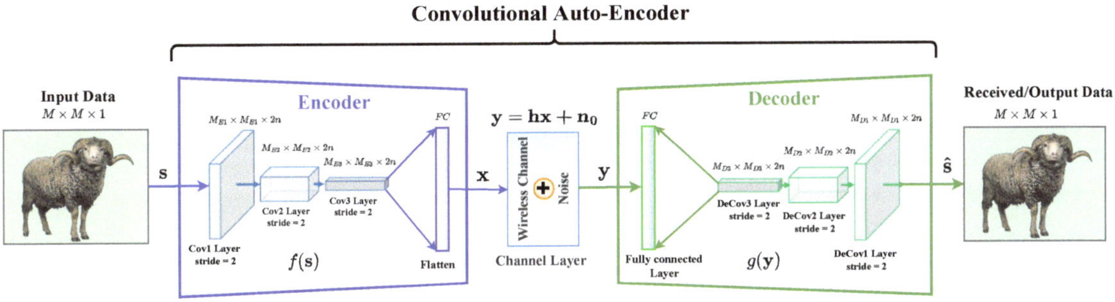

Figure 4. UAV and GCS information communications system over wireless channel presented as an auto-encoder.

Data Generation, Training and Inference

We have generated 50,000 message samples, from which 30,000 are the training samples, and 15,000 samples are used for validation and inference each. The auto-encoder is trained with a fixed signal-to-noise ratio (SNR) or energy per bit to noise power spectral density ($\frac{E_b}{N_0}$) of 7 dB using an Adam optimizer [33] with a learning rate of 0.001. The training was performed with various training batch sizes to determine the appropriate size that gives the best performance. The performance of the trained auto-encoder was tested over different SNR values. We have implemented the auto-encoder model in Keras Tensorflow 2 DL framework and 3.8 Python.

4. Results and Discussions

This section presents the simulation results and discussions based on the performance metric. For performance evaluation, bit-error-rate (BER), $\Pr(\hat{s} \neq s)$ is used as a performance metric for assessing the efficacy of our proposed learning model.

Figure 5 compares the BER performance of a communications system using uncoded QPSK (4, 4) and a Hamming (7,4) code with the optimal maximum likelihood decoding (MLD) against the BER gained by the auto-encoder (7,4) with different training samples and average fixed energy constraints. It can be seen that the performance of the auto-encoder trained with 25,000 training samples matches the Hamming (7, 4) maximum likelihood decoding scheme. We also observe that the auto-encoder's performance falls as the number of training samples decreases. We ensure that the system operates at a 7/4 communication rate for fair performance evaluation. This result reveals that the auto-encoder has learned the UAV and GCS information mapping (i.e., an encoder and decoder function) that achieves the same performance as the Hamming (7,4) code with MLD without prior knowledge.

Figure 6 compares the BER produced when the input data are modulated with binary phase-shift keying (BPSK), QPSK and 8PSK against the BER achieved by the trained auto-encoder with an average fixed energy constraint based on (2). Generally, the number of encoded bits depends on the number of encoded phases. The BPSK uses two distinct phases shifted by 180^0 compared to QPSK, which utilizes four phases to encode the data. Therefore, the QPSK transmits 2-bit data, twice the data transmitted by BPSK per symbol cycle. In contrast, 8PSK uses eight phases, described by a 3-bit transmitting 3-bit symbols per cycle. Accordingly, while the uncoded QPSK (4, 4) produces a BER lower than the BPSK modulation scheme, the Hamming (7, 4) hard decision decoding scheme outperforms all three modulation schemes. Interestingly, the auto-encoder trained with QPSK and 8PSK-modulated symbols performs better than all the baseline communication schemes. We have observed that the performance gap between the auto-encoders trained with QPSK and 8PSK tends to close between 2 and4 dB SNRs. At these SNRs, the communication rate of the auto-encoder trained with QPSK-modulated symbols decreases; it is thus forced to learn a lower modulation scheme. Beyond the 4 dB SNR, a significant decline in BER is observed, suggesting that the auto-encoder trained with the 8PSK modulation scheme learns to transmit more bits per symbol cycle.

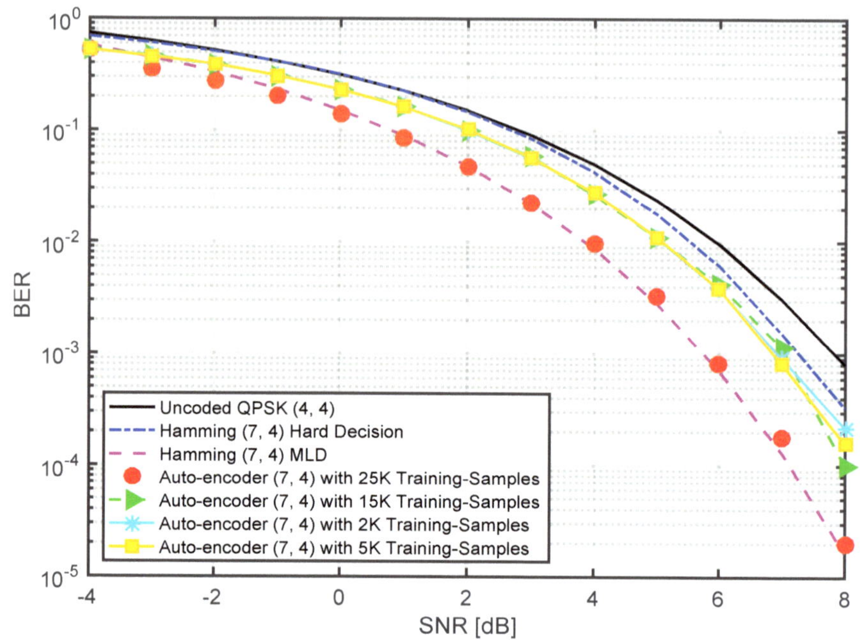

Figure 5. BER vs. SNR for the auto-encoder trained with different amounts of samples against various benchmark communication techniques.

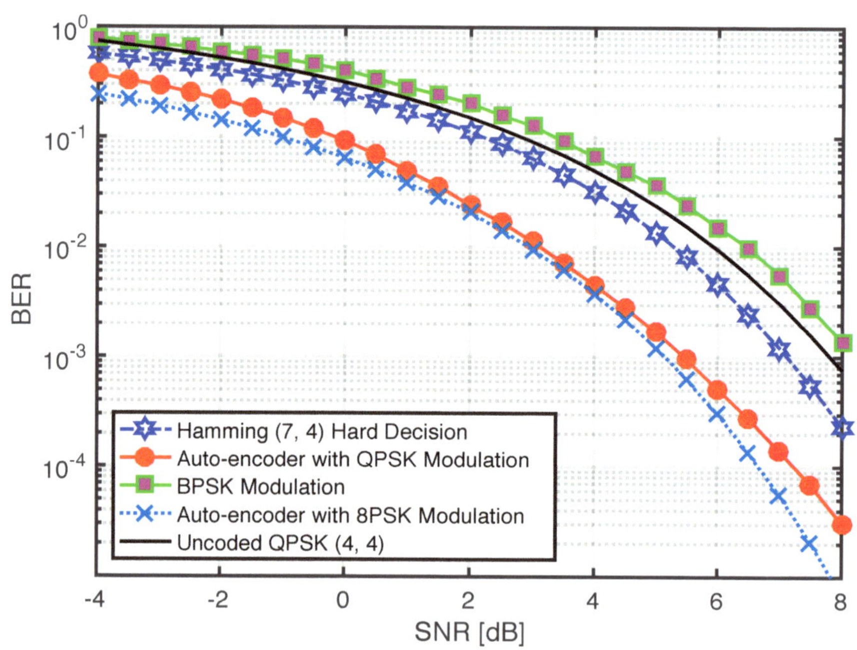

Figure 6. BER vs. SNR for the auto-encoder trained with 25,000 training samples using QPSK and 8PSK modulation schemes against baseline modulation schemes.

Figures 7 and 8 portray the effects of both fixed and varying SNR on the auto-encoder's performance. Figure 7 shows that the BER falls faster when the auto-encoder is trained with a fixed SNR and then saturates at the 25*th* epoch. With this, a relatively small training effort is required for the auto-encoder to learn various communication schemes. However, when the auto-encoder is trained with a varying SNR from −4 to 10 dB, the BER slowly decreases with the training epoch and finally saturates at the 10th epoch, as shown in Figure 8. From these results, we can deduce that an auto-encoder trained with fixed SNR for end-to-end communications converges faster than the one trained with varying SNRs.

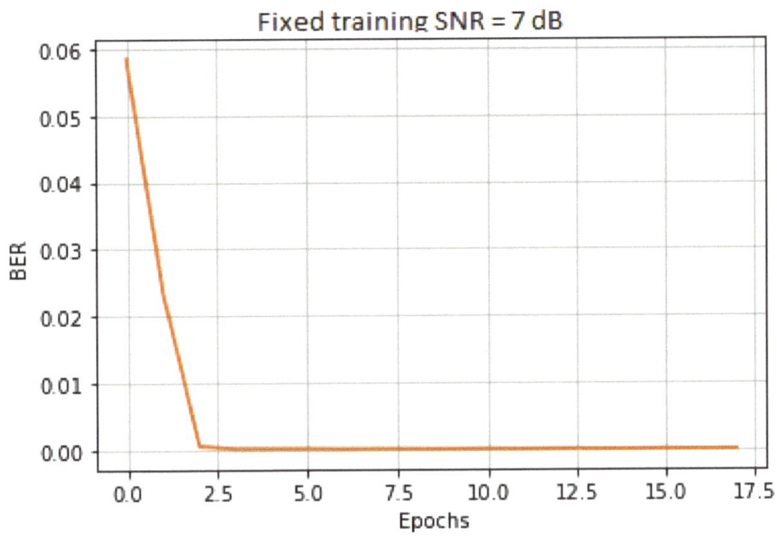

Figure 7. BER vs. the number of epochs of the auto-encoder trained via fixed $SNR =$ dB.

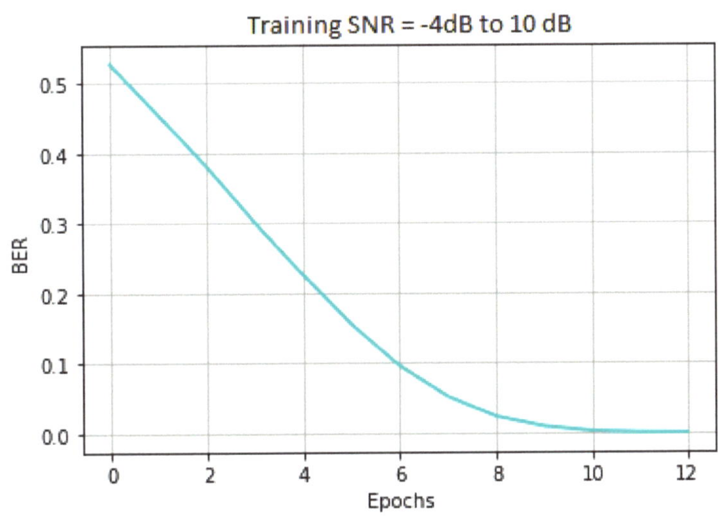

Figure 8. BER vs. the number of epochs of the auto-encoder trained via variable SNR values.

To investigate whether the auto-encoder has learnt some communication schemes without prior knowledge of the channel model, we show the learned constellation representations $\mathbf{x}$ of all messages for different values of (n, k). Figure 9 depicts a typical $(2, 2)$ system, which assembles rapidly to a classical QPSK constellation rotation. The symbol constellations spread over four possible carrier phases within a unit circle, as in the case of a classical QPSK modulation technique.

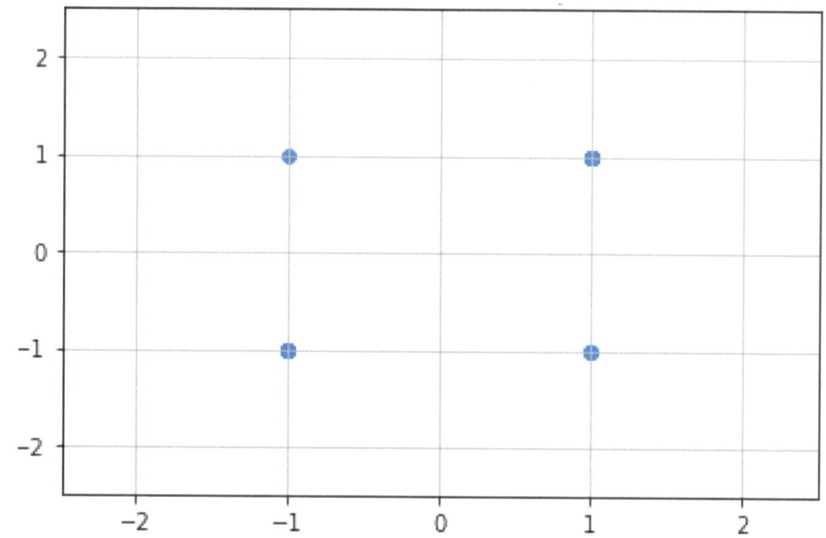

Figure 9. Learned constellation produced by auto-encoders using a (2, 2) communication rate with an average energy constraint.

Correspondingly, Figure 10 illustrates a $(4, 2)$ communication system that produces a rotated 16PSK constellation. Interestingly, with a fixed, average energy constraint, the resulting constellation produced by the learned auto-encoder is similar to the one constructed by the classical 16PSK. The effect of the choice of normalization function for a transmit message under some constraints is noticeable from Figure 11 for the same settings but with an average power normalization rather than an average fixed energy constraint. This produces an intriguing hybrid pentagonal/hexagonal grid structure similar to the performance from a distorted 16QAM constellation with a symbol near the origin surrounded by five equally spaced nearest neighbours. Therefore, this shows that an auto-encoder trained with the average energy constraint produces a much more regular and well-defined communication scheme that matches a particular classical communication technique.

To find whether the proposed auto-encoder is doing well during the learning phase, we compare the training loss against the validation loss in Figure 12. It can be observed that both the training loss and validation loss converge at the eighth iteration. This further demonstrates that the auto-encoder is doing well in learning various communication strategies for efficient data transfer between the UAVs and the GCS.

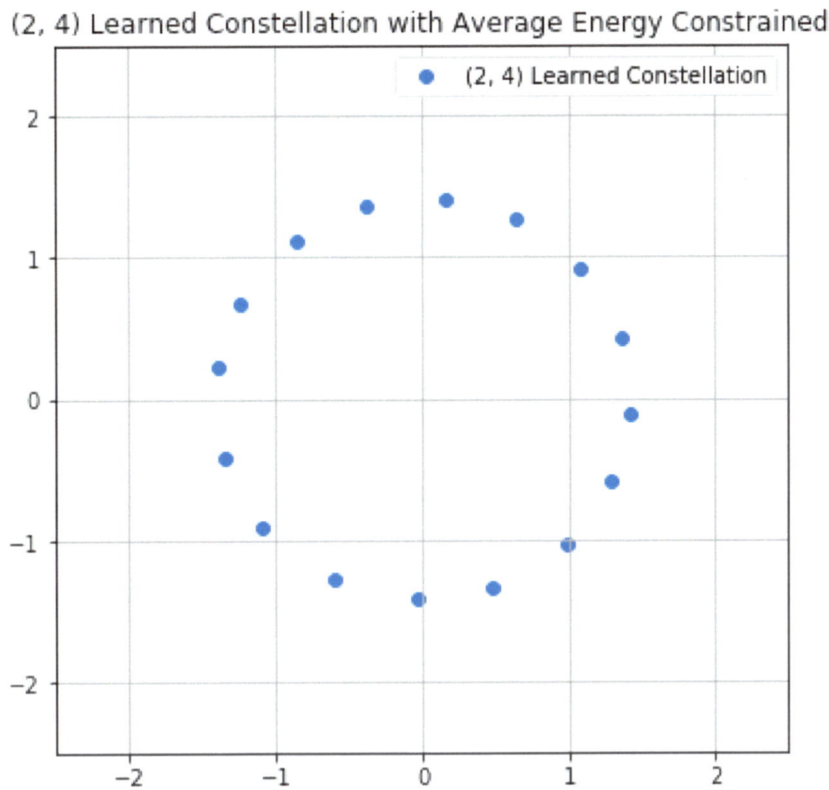

Figure 10. BER vs. SNR for the auto-encoder test with different amounts of test samples against various benchmark communication techniques.

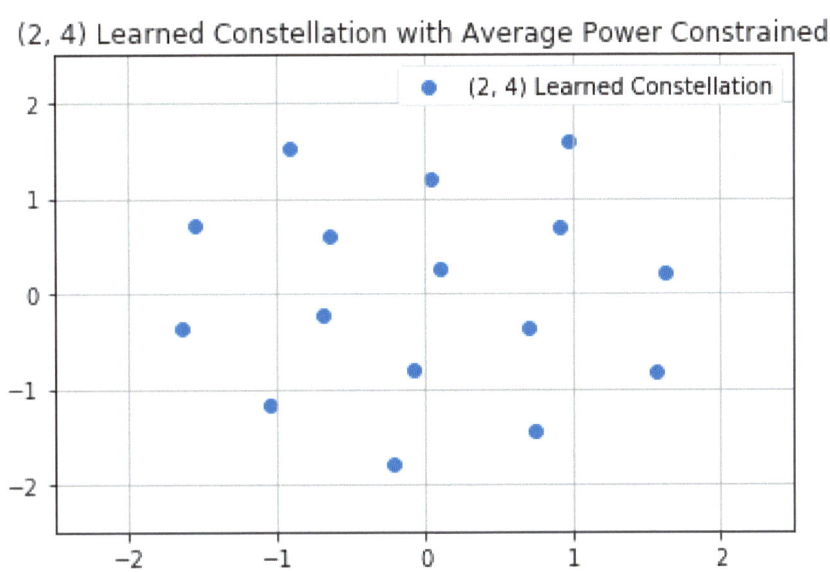

Figure 11. BER vs. SNR for the auto-encoder test with different amounts of test samples against various benchmark communication techniques.

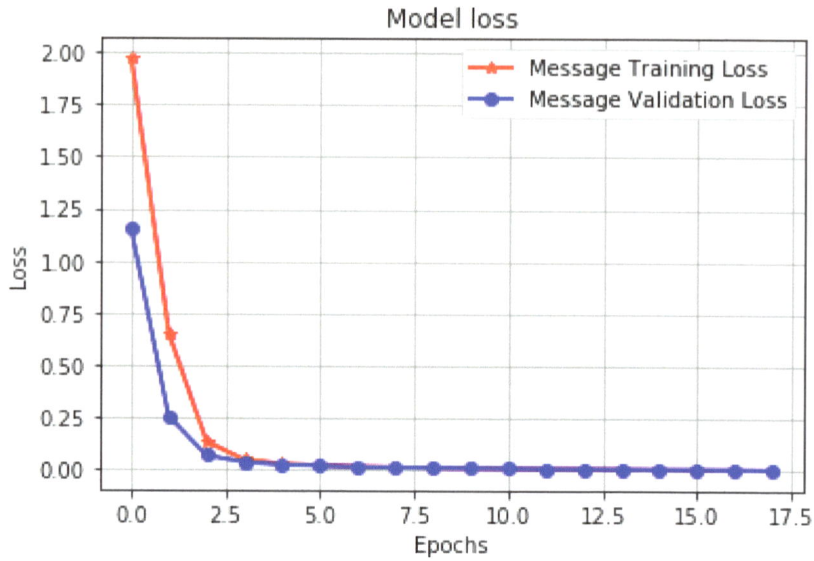

Figure 12. BER vs. SNR for the auto-encoder test with different amounts of test samples against various benchmark communication techniques.

5. Conclusions

This paper presents a communications system as an end-to-end optimization scheme using an auto-encoder to jointly learn UAV (transmitter) and GCS (receiver) signal processing implementations without prior knowledge. The auto-encoder was trained with fixed and varying SNR values and an input message modulated with different modulation schemes, such as BPSK, QPSK and 8PSK, and various communication rates. We have seen from the results that the proposed learning-based communication framework can learn standard and distorted classical modulation techniques when trained with average and average power constraints, respectively. Comparisons with conventional baselines in various scenarios unveil a competitive BER performance against traditional communication techniques. From the results, we find that the proposed learning approach demonstrates its efficacy in terms of reliability in learning optimal communication schemes in a challenging environment or situations where the optimal strategies are not known. This could be used to design UAV communication links for reliable data transfer and efficient smart livestock farming. Future work should consider various channel types and communication rates for a more practical UAV communication system for different baseline schemes. An extension to a multiantenna UAV system with particular attention to beamforming, secrecy, channel interference and energy efficiency is appealing.

Author Contributions: Conceptualization, A.M. and Y.A.S.; methodology, A.M., Y.A.S. and H.R.E.H.B.; software, A.M.; validation, Y.A.S., M.S.S. and H.R.E.H.B.; formal analysis, Y.A.S. and M.S.S.; investigation, A.M.; resources, A.M. and Y.A.S.; writing—original draft preparation, M.A.A., A.M., Y.A.S., H.R.E.H.B. and M.S.S.; writing—review and editing, M.A.A., A.M., Y.A.S., H.R.E.H.B. and M.S.S.; visualization, Y.A.S.; supervision, H.B. and M.A.A; project administration, M.A.A.; funding acquisition, M.A.A. All authors have read and agreed to the published version of the manuscript.

Funding: This research was funded by the Deputyship for Research and Innovation, Ministry of Education in Saudi Arabia, grant number IFP-A-201-2-1.

Institutional Review Board Statement: Not applicable.

Informed Consent Statement: Not applicable.

Data Availability Statement: Not applicable.

Acknowledgments: The authors extend their appreciation to the Deputyship for Research and Innovation, Ministry of Education in Saudi Arabia for funding this research work through the project No IFP-A-201-2-1.

Conflicts of Interest: The authors declare no conflict of interest. The funders had no role in the design of the study; in the collection, analyses, or interpretation of data; in the writing of the manuscript; or in the decision to publish the results.

References

1. Zeng, Y.; Zhang, R.; Lim, T.J. Wireless communications with unmanned aerial vehicles: Opportunities and challenges. *IEEE Commun. Mag.* **2016**, *54*, 36–42. [CrossRef]
2. Nyamuryekung'e, S.; Cibils, A.F.; Estell, R.E.; Gonzalez, A.L. Use of an unmanned aerial vehicle-mounted video camera to assess feeding behavior of Raramuri Criollo cows. *Rangel. Ecol. Manag.* **2016**, *69*, 386–389. [CrossRef]
3. Razaak, M.; Kerdegari, H.; Davies, E.; Abozariba, R.; Broadbent, M.; Mason, K.; Argyriou, V.; Remagnino, P. An integrated precision farming application based on 5G, UAV and deep learning technologies. In Proceedings of the International Conference on Computer Analysis of Images and Patterns, Salerno, Italy, 6 September 2019; pp. 109–119.
4. Xiao, Z.; Zhu, L.; Xia, X.G. UAV communications with millimeter-wave beamforming: Potentials, scenarios, and challenges. *China Commun.* **2020**, *17*, 147–166. [CrossRef]
5. Gura, D.; Rukhlinskiy, V.; Sharov, V.; Bogoyavlenskiy, A. Automated system for dispatching the movement of unmanned aerial vehicles with a distributed survey of flight tasks. *J. Intell. Syst.* **2021**, *30*, 728–738. [CrossRef]
6. Miao, W.; Luo, C.; Min, G.; Wu, L.; Zhao, T.; Mi, Y. Position-based Beamforming Design for UAV communications in LTE networks. In Proceedings of the ICC 2019-2019 IEEE International Conference on Communications (ICC), Shanghai, China, 20–24 May 2019; pp. 1–6.
7. Vroegindeweij, B.A.; van Wijk, S.W.; van Henten, E. Autonomous unmanned aerial vehicles for agricultural applications. In Proceedings of the International Conference of Agricultural Engineering (AgEng), Zurich, Switzerland, 6–10 July 2014.
8. Li, X.; Xing, L. Use of unmanned aerial vehicles for livestock monitoring based on streaming K-means clustering. *Ifac-Papersonline* **2019**, *52*, 324–329. [CrossRef]
9. Maddikunta, P.K.R.; Hakak, S.; Alazab, M.; Bhattacharya, S.; Gadekallu, T.R.; Khan, W.Z.; Pham, Q.V. Unmanned aerial vehicles in smart agriculture: Applications, requirements, and challenges. *IEEE Sens. J.* **2021**, *21*, 17608–17619. [CrossRef]
10. Alanezi, M.A.; Sadiq, B.O.; Sha'aban, Y.A.; Bouchekara, H.R.E.H. Livestock Management on Grazing Field: A FANET Based Approach. *Appl. Sci.* **2022**, *12*, 6654. [CrossRef]
11. Sejian, V.; Silpa, M.; Lees, A.M.; Krishnan, G.; Devaraj, C.; Bagath, M.; Anisha, J.; Reshma Nair, M.; Manimaran, A.; Bhatta, R.; et al. Opportunities, Challenges, and Ecological Footprint of Sustaining Small Ruminant Production in the Changing Climate Scenario. In *Agroecological Footprints Management for Sustainable Food System*; Springer: Singapore, 2021; pp. 365–396.
12. Faraji-Biregani, M.; Fotohi, R. Secure communication between UAVs using a method based on smart agents in unmanned aerial vehicles. *J. Supercomput.* **2021**, *77*, 5076–5103. [CrossRef]
13. Behjati, M.; Mohd Noh, A.B.; Alobaidy, H.A.; Zulkifley, M.A.; Nordin, R.; Abdullah, N.F. LoRa communications as an enabler for internet of drones towards large-scale livestock monitoring in rural farms. *Sensors* **2021**, *21*, 5044. [CrossRef] [PubMed]
14. Zhang, C.; Zhang, W.; Wang, W.; Yang, L.; Zhang, W. Research challenges and opportunities of UAV millimeter-wave communications. *IEEE Wirel. Commun.* **2019**, *26*, 58–62. [CrossRef]
15. Puri, V.; Nayyar, A.; Raja, L. Agriculture drones: A modern breakthrough in precision agriculture. *J. Stat. Manag. Syst.* **2017**, *20*, 507–518. [CrossRef]
16. Jiangpeng, Z.; Haiyan, C.; Liwen, H.; Yong, H. Development and performance evaluation of a multi-rotor unmanned aircraft system for agricultural monitoring. *Smart Agric.* **2019**, *1*, 43.
17. Abdulai, G.; Sama, M.; Jackson, J. A preliminary study of the physiological and behavioral response of beef cattle to unmanned aerial vehicles (UAVs). *Appl. Anim. Behav. Sci.* **2021**, *241*, 105355. [CrossRef]
18. Kaya, S.; Goraj, Z. The Use of Drones in Agricultural Production. *Int. J. Innov. Approaches Agric. Res.* **2020**, *4*, 166–176. [CrossRef]
19. Yinka-Banjo, C.; Ajayi, O. Sky-farmers: Applications of unmanned aerial vehicles (UAV) in agriculture. In *Autonomous Vehicles*; IntechOpen: London, UK, 2019; pp. 107–128.
20. Manlio, B.; Paolo, B.; Alberto, G.; Massimiliano, R. Unmanned Aerial Vehicles for Agriculture: An Overview of IoT-Based Scenarios. In *Autonomous Airborne Wireless Networks*; John Wiley & Sons, Ltd.: Hoboken, NJ, USA, 2021; pp. 217–235.
21. Uche, U.; Audu, S. UAV for Agrochemical Application: A Review. *Niger. J. Technol.* **2021**, *40*, 795–809. [CrossRef]
22. Alanezi, M.A.; Shahriar, M.S.; Hasan, M.B.; Ahmed, S.; Sha'aban, Y.A.; Bouchekara, H.R. Livestock Management with Unmanned Aerial Vehicles: A Review. *IEEE Access* **2022**, *10*, 45001–45028. [CrossRef]
23. Mohammad, A.; Masouros, C.; Andreopoulos, Y. Accelerated learning-based MIMO detection through weighted neural network design. In Proceedings of the ICC 2020-2020 IEEE International Conference on Communications (ICC), Dublin, Ireland, 7–11 June 2020; pp. 1–6.

24. Andrew, W.; Greatwood, C.; Burghardt, T. Visual localisation and individual identification of holstein friesian cattle via deep learning. In Proceedings of the IEEE International Conference on Computer Vision Workshops, Venice, Italy, 22–29 October 2017; pp. 2850–2859.
25. Rivas, A.; Chamoso, P.; González-Briones, A.; Corchado, J.M. Detection of cattle using drones and convolutional neural networks *Sensors* **2018**, *18*, 2048. [CrossRef] [PubMed]
26. Andrew, W.; Greatwood, C.; Burghardt, T. Aerial animal biometrics: Individual friesian cattle recovery and visual identification via an autonomous uav with onboard deep inference. In Proceedings of the 2019 IEEE/RSJ International Conference on Intelligent Robots and Systems (IROS), Macau, China, 3–8 November 2019; pp. 237–243.
27. Mohammad, A.; Masouros, C.; Andreopoulos, Y. Complexity-scalable neural-network-based MIMO detection with learnable weight scaling. *IEEE Trans. Commun.* **2020**, *68*, 6101–6113. [CrossRef]
28. Goutay, M.; Aoudia, F.A.; Hoydis, J. Deep hypernetwork-based MIMO detection. In Proceedings of the 2020 IEEE 21st International Workshop on Signal Processing Advances in Wireless Communications (SPAWC), Atlanta, GA, USA, 26–29 May 2020; pp. 1–5.
29. Mohammad, A.; Masouros, C.; Andreopoulos, Y. An unsupervised learning-based approach for symbol-level-precoding. In Proceedings of the 2021 IEEE Global Communications Conference (GLOBECOM), Madrid, Spain, 7–11 December 2021; pp. 1–6
30. O'shea, T.; Hoydis, J. An introduction to deep learning for the physical layer. *IEEE Trans. Cogn. Commun. Netw.* **2017**, *3*, 563–575 [CrossRef]
31. Feng, Y.; Teng, Z.; Meng, F.; Qian, B. An accurate modulation recognition method of QPSK signal. *Math. Probl. Eng.* **2015**, *2015*, 516081. [CrossRef]
32. Sklar, B. *Digital Communications*; Prentice Hall: Upper Saddle River, NJ, USA, 2001; Volume 2.
33. Goodfellow, I.; Bengio, Y.; Courville, A. *Deep Learning*; MIT Press: Cambridge, MA, USA, 2016.

sensors

MDPI

Article

Automatic Detection of Olive Tree Canopies for Groves with Thick Plant Cover on the Ground

Sergio Illana Rico [1,*], Diego Manuel Martínez Gila [1,2], Pablo Cano Marchal [1,2] and Juan Gómez Ortega [1,2]

1 Robotics, Automation and Computer Vision Group, Electronic and Automation Engineering Department, University of Jaén, 23071 Jaén, Spain
2 Institute for Olive Orchards and Olive Oils, University of Jaén, 23071 Jaén, Spain
* Correspondence: sillana@ujaen.es; Tel.: +34-953-212-448

Abstract: Marking the tree canopies is an unavoidable step in any study working with high-resolution aerial images taken by a UAV in any fruit tree crop, such as olive trees, as the extraction of pixel features from these canopies is the first step to build the models whose predictions are compared with the ground truth obtained by measurements made with other types of sensors. Marking these canopies manually is an arduous and tedious process that is replaced by automatic methods that rarely work well for groves with a thick plant cover on the ground. This paper develops a standard method for the detection of olive tree canopies from high-resolution aerial images taken by a multispectral camera, regardless of the plant cover density between canopies. The method is based on the relative spatial information between canopies.The planting pattern used by the grower is computed and extrapolated using Delaunay triangulation in order to fuse this knowledge with that previously obtained from spectral information. It is shown that the minimisation of a certain function provides an optimal fit of the parameters that define the marking of the trees, yielding promising results of 77.5% recall and 70.9% precision.

Keywords: Delaunay triangulation; high-resolution aerial images; multispectral imagery; olive tree canopy; precision agriculture; remote sensing; thick plant cover; UAV; weeds

Citation: Illana Rico, S.; Martínez Gila, D.M.; Cano Marchal, P.; Gómez Ortega, J. Automatic Detection of Olive Tree Canopies for Groves with Thick Plant Cover on the Ground. *Sensors* **2022**, *22*, 6219. https://doi.org/10.3390/s22166219

Academic Editors: Diego González-Aguilera, Arturo Sanchez-Azofeifa, Jose A. Jiménez-Berni, Shangpeng Sun and Monica Herrero-Huerta

Received: 10 May 2022
Accepted: 2 August 2022
Published: 19 August 2022

Publisher's Note: MDPI stays neutral with regard to jurisdictional claims in published maps and institutional affiliations.

1. Introduction

According to the Food and Agriculture Organization of the United Nations (FAO), in the year 2020 there were 12.8 million hectares dedicated to olive trees in the world [1]. Although far from the figures of wheat, the most cultivated crop in the world with 219 million hectares of dedicated land, olive trees stand in the 22nd position out of 161 in the ranking of primary crops. The European Union (EU) represents 40% of the dedicated land with 5.1 million hectares, with Spain accounting for 2.6 million hectares and 51.4% of the European crop area, followed by Italy with 1.1 million hectares and 22.4%. These figures translated into a total world production of 3.3 million tons of olive oil [2], of which 57.9% (1.9 million tons) were produced by the EU. Within the EU, Spain accounted for 1.1 million tons and 58.6% of the European production, with Italy producing 19.1% (0.37 million tons). In turn, the table olive world production was 2.96 million tons, of which 21.9% (0.65 million tons) were produced by Egypt and 26% (0.77 million tons) by the EU. Within the EU, Spain was the largest producer with 0.46 million tons and 58.6% of the European production.

These figures explain the interest in the olive oil and table olives production from the grove to the final production stages in the olive oil factories. Plenty of research has been devoted to the improvement of the quality of olive oil by means of controlling the key variables of the production process [3–5]. However, since no chemical or biochemical coadjuvants can be employed to improve the olive oil features, as the production process is carried out by mechanical means exclusively, focusing only on the operations on the olive oil factory means that the best that can be achieved is to preserve the quality properties of the olives arriving at the factory. Therefore, current research trends are starting to focus also on the improvement of the quality and productivity in the stages where the olives are still on the tree.

Regarding this, there are works that study different watering strategies [6,7], their effect on the organoleptic properties of the produced olive oil [8] or how the hydric stress affects the olive trees [9,10]. Other works focus on the early detection of plagues, such *Xylella fastidiosa* [11–13] and *Verticillium* [14], or the assessment of the nutritional state of the trees [15,16], while others study the way to evaluate different parameters such as the volume, height and width of the trees [17,18].

Many of these research works share a common factor, namely the acquisition of information on the Earth's surface using different types of sensors, such as visible spectra, multispectral and thermal cameras, *LiDAR* laser sensors or synthetic-aperture radars, mounted on satellites or unmanned aerial vehicles (UAV). Later, this information is used, in conjunction with novel image processing methods [19], in several tasks such as monitoring the temporal evolution of surface deformations [20] or mapping potential landslide runout zones [21]. This is what is known as remote sensing. When these research approaches are focused on the study of fruit tree groves, such as olive trees, a common necessity is to extract features from the pixels of the images belonging to the tree canopy, since, in order to verify the proposed hypothesis, the experimental results need to be compared with some ground truth obtained from the use of other sensors on the trees.

In order to relate these measurements with the features obtained from the pixels of the tree canopies in the images, it is required to delimitate each of these canopies, which is an arduous and tedious process were it to be performed manually [22]. Therefore, most of the time, authors tend to use more or less standard software tools when the simplicity of the images allows it [11,12,23]. Many works employ spectral values of the images to perform this segmentation, which could bias the features extracted afterwards [24,25]. Only a reduced number of articles explicitly address the problem of identifying the tree canopies as their main topic.

Among them, it is worth highlighting [26], in which the authors use the Gram–Schmidt spectral sharpening fusion method to integrate the panchromatic and multispectral images and train a convolutional neural network (CNN) that manages to detect oil palm trees with up to 98.65% precision and 98.88% recall using images obtained from the Quickbird satellite. A different approach is applied in [27], where the authors use the red and infrared spectral bands to compute the normalised difference vegetation index (NDVI) and apply classic computer vision methods such as thresholding and blob detection, obtaining a relative error of between 0.2% and 20.7%.

The next step to increase the ground sample distance (GSD) would be to perform the data acquisition using airborne sensors mounted on aircrafts. In [28], the authors use principal components analysis to process a single band image derived from the 8 bands the sensor captures. They then apply a two-stage approach with edge detection followed by marker-controlled watershed segmentation on an image of a forest of spruce trees, Douglas firs and subalpine firs. According to the number of labelled trees, a recall of 85.3% is achieved, while the recall computed using the number of labelled pixels yields 75.6%. Ref. [29] employs colour-infrared images of forested regions and applies a fuzzy thresholding algorithm to find the seeds that are used by a region growing algorithm. As result, 73% of trees were correctly found, with a variation from 62% to 81%. Some more examples of the use of CNN for individual tree canopy detection can be found in [30], in which the data acquired by an RGB sensor is merged with the information collected by a *LiDAR*. A previous segmentation of the trees of an open woodland of live oak, blue oak and foothill pine is performed using the *LiDAR* information, so later the CNN can be trained with each of these Region Of Interest (ROI). This model has an average tree canopy recall of 0.69 with a precision of 0.61. In [31], citrus and other crop trees are detected from UAV images using a simple CNN algorithm, followed by a classification refinement using superpixels derived from a simple linear iterative clustering algorithm, achieving 94.59% precision and 97.94% recall.

When high-resolution images are captured with UAVs, contrary to the methodology followed in this article, most of them extract structured 3D information using *LiDAR* sensors or applying photogrammetric analysis. These methods can considerably improve the results obtained, especially when it is necessary to differentiate pixels that are spectrally similar but that are at different heights from the ground (tree canopies and weeds). Even so, they have some disadvantages that are discussed in Section 4. For example, [32] uses UAV LiDAR data for individual tree detection in subtropical mixed broadleaf forests in urban scenes. This method improves on the popular local maximum filter by removing those LMs caused by surface irregularities contained in the canopy height model, and obtains an F-score between 73.7% and 93.2%, depending on how irregular the distribution of trees or crown size is. In [33], RGB images of a maize plantation at the seedling stage are combined with *LiDAR* data captured by a UAV. The maize seedlings extracted from the images serve as seeds for the fuzzy C-means clustering algorithm used to segment individual maize plants. The results revealed an accuracy with R^2 greater than 0.95, a mean square error (RMSE) of 3.04–4.55 cm and a mean absolute percentage error of 0.91–3.75%. Other examples of articles using photogrammetry to detect and extract trees from high-resolution UAV RGB images are [34] for citrus trees, [35] for peach trees, [36] for chestnut trees and [37] for papaya trees. The first uses sequential thresholding, Canny edge detection and circular Hough transform algorithms on the Digital Surface Model (DSM), obtaining accuracies that exceeded 80%. The second uses an adaptive threshold and marker-controlled watershed segmentation in the DSM to measure the canopy width and canopy projection area, achieving an RMSE of 0.08–0.47 m and 3.87–4.96 m^2, respectively. The third applies a segmentation stage based on the computation of vegetation indices combined with the canopy height model to extract the candidate tree canopies. The next stage receives these tree canopies to divide those that are greater than a threshold. The last stage extracts the desired features. The segmentation accuracy of this method is above 95%. Finally, in the fourth article, the authors make an improvement with regard to the existing scale-space filtering by applying a Lab colour transformation to reduce overdetection problems associated with the original luminance image. The achieved F-score is larger than 0.94.

As concluded in [38] and evidenced in the previous paragraphs, no algorithm is optimal for all types of images and plant types. For this reason and for clarity, articles related specifically to olive trees and how to detect their canopies have been compared in the Section 3. Although these articles are closely related to the issue at hand, none of them explicitly deals with images from groves with thick plant cover on the ground without using three-dimensional data. Finally, to complement this literature review, mention should also be made of articles dealing with detecting weeds in crops of herbaceous plants such as maize [39], sugar beet [40], sunflower and cotton [41,42] or bean and spinach [43].

The objective of this paper is to develop a method to segment olive tree canopies from high-resolution aerial images that contain information of the visible spectra, specifically, the red, green and blue bands that can cope with high levels of plant cover in the ground between canopies.

The key idea of the approach is to employ not just the spectral information contained in the images but to also consider the relative distance between canopies, computing and extrapolating the planting pattern used by the grower and fusing this information with the spectral data. This paper shows that the minimisation of a certain function provides an optimal fit of the parameters that define the marking of the trees, yielding promising results, without the need to resort to deep learning methods that are difficult to interpret.

The structure of the paper is as follows: Section 2 presents a general diagram divided into blocks that explains the workflow followed to achieve the results of this research. This section is made up of subsections that correspond to each of the blocks in the diagram. Although the methodology followed is explained in detail in all of them, the first block Section 2.1 also focuses on the materials used during data capture and the way in which it was carried out. In Section 3, first, the quality metrics of the results obtained after the application of the developed model are objectively shown, and these numbers are analyzed together with an explanation of the possible causes of the model's failures. Second, the execution times of the building blocks of the model are computed, both for the trainings and for the predictions of the model. Third, a comparison is made with other articles related specifically to olive trees and how to detect their canopies. Finally, in Section 4, the objective of the article is expanded and its usefulness and advantages compared with other methodologies are explained. Possible improvements are also discussed and future work is anticipated.

2. Materials and Methods

The workflow of the method proposed in this paper is presented in Figure 1 and shows the different materials and methods used, as well as the information transference between the different parts that compose it. This workflow can be divided into two blocks: the *Data Preprocessing (DP)* of the images taken in the groves and the *Olive Tree Canopy Detection Model (OTCD)*, which is composed of three submodels, namely, the *Vegetation Classification (VC)*, the *Olive Tree Canopy Estimation (OTCE)* and the *Olive Tree Canopy Marking (OTCM)*.

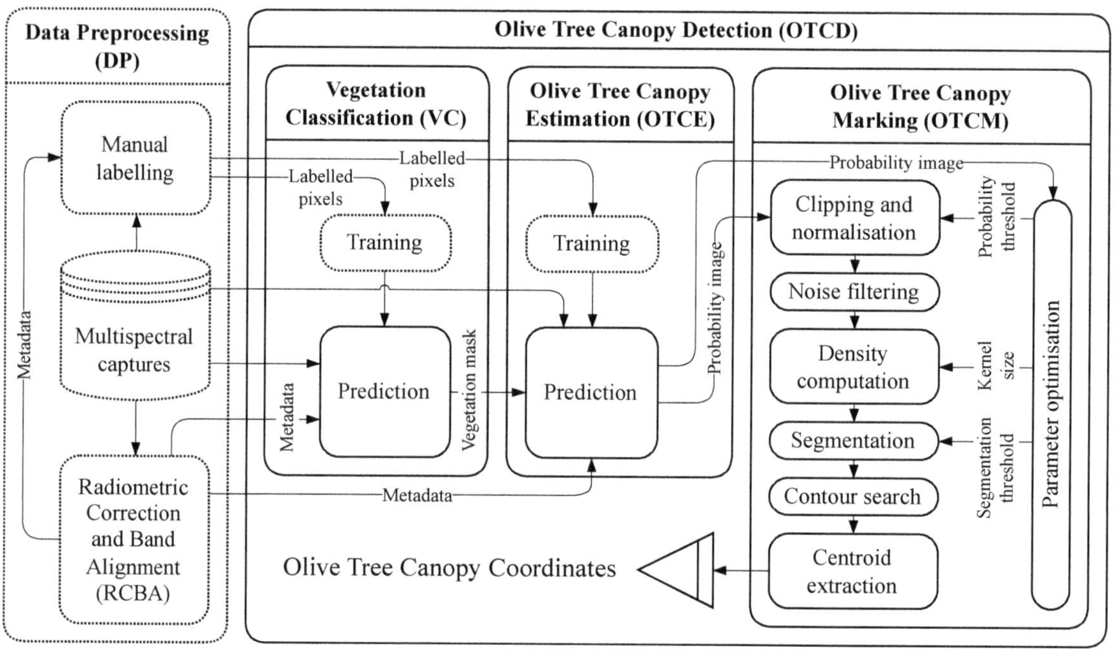

Figure 1. Workflow of the method proposed in this paper. The arrows represent data transfers and the nodes represent functions that are applied to that data. As exceptions, the initial node *Multispectral Captures*, represents the data acquired by the UAV and the final node, *Olive Tree Canopy Coordinates*, represents the most optimal coordinates that the model is capable of predicting. Blocks with dotted outlines represent tasks that are performed beforehand.

The first block *DP* deals with the transformation and separation of the original raw data captured by the sensors to the format required for the input of the model, specifically multispectral images, metadata associated with these multispectral images and pixels that are labelled to their corresponding class.

This preprocessing task, together with the training of the two first submodels of the *OTCD* block, is performed beforehand—in Figure 1 they are shown with a dashed line. This way, the time required to perform these task does not influence the time required to detect the tree canopies of a new image.

The second block *OTCD* deals with the prediction of the coordinates of each of the tree canopies included in the new images provided as inputs.

2.1. Multispectral Captures

The data employed in this work were gathered by the following sensors mounted on a *DJI Matrice 600* UAV:

- A *Micasense RedEdge-M* multispectral camera, capable of capturing 12 bit images with a 1280 by 960 pixel resolution in five bands of the electromagnetic spectrum. These bands are blue (475 nm), green (560 nm), red (668 nm), near infrared (840 nm) and red-edge (717 nm).
- A sunlight sensor *Micasense Downwelling Light Sensor (DLS) 1*, capable of measuring the ambient light for each of the five bands of the camera *Micasense RedEdge-M*.
- A global positioning system (GPS) *Ublox M8N*.

These sensors are completed with a calibrated reflectance panel (CRP) *Micasense RP04-1826404-SC* that takes images of before and after each flight by the UAV in order to be used for the radiometric correction process. The calibrated reflectance values for the specific panel used in this work are 49.1%, 49.3%, 49.4%, 49.3% and 49.4% for the blue, green, red, near-infrared and red-edge bands, respectively.

Each multispectral capture stores the images of each of the five bands in a disk in *.tif* format, together with the metadata provided by the DLS and the GPS.

As commented in the Introduction, the main objective of this paper is to detect olive tree canopies when there is a large amount of plant cover in the ground, so that traditional segmentation methods fail to provide good results. This way, a set of 18 captures (Table 1) with a large amount of plant cover in the ground, were taken from an olive grove in the town of Diezma, province of Granada, in Andalucia, Southern Spain. This is an olive grove of trees of Picual cultivar variety with 29.14 ha of extension, showing a traditional arrangement of the trees. Six flight campaigns were carried out from October 2018 to November 2019. Figure 2 depicts an example of the capture as taken by the multispectral camera, depicting an image for each of the five bands previously mentioned. A thick plant cover can be seen between the olive trees.

Figure 2. Capture 0, as provided by the multispectral camera: one image for each of the blue, green, red, near-infrared and red-edge bands (**left** to **right**), where the large amount of plant cover in the ground is clearly visible.

All these captures, together with the metadata, were used as inputs for the *Radiometric Correction and Band Alignment (RCBA)*, the manual labelling and the first two submodels of the olive tree canopy detection model.

2.2. Radiometric Correction and Band Alignment (RCBA)

Once the multispectral captures are available as inputs to the *RCBA*, the first step is to transform the digital numbers (DN) of the images obtained by the camera sensor to radiance values first and to reflectance values afterwards. This correction, which is carried out for each band, allows the values subsequently employed to be independent of the flight, date, time of the day and climatic conditions, so that different captures can be compared in the same reference frame. Posteriorly, before finishing the *RCBA*, the reflectance images of each band need to be aligned, since the five sensors of the camera have a slight offset from each other.

Table 1. Detailed characteristics of the 18 captures used to train and test the proposed model. *Campaign* is an identifier of the flight number. For each flight made in different months, 3 captures are selected.

ID	Campaign	Date and Time [ISO8601]	Latitude [DD]	Longitude [DD]	Altitude above Sea Level [m]	Altitude above Ground Level [m]	Ground Sample Distance [cm/px]
0		2018-10-04T11:00:00Z	37.316932	−3.354593	1408.694	188.774	13.109
1	2	2018-10-04T11:02:56Z	37.316660	−3.354487	1407.542	187.622	13.029
2		2018-10-04T11:11:48Z	37.314839	−3.357709	1406.342	186.422	12.946
3		2018-11-08T11:33:28Z	37.316566	−3.354403	1403.569	180.834	12.558
4	3	2018-11-08T11:33:32Z	37.316642	−3.354134	1403.604	180.869	12.560
5		2018-11-08T11:59:06Z	37.314442	−3.354405	1402.180	179.757	12.483
6		2018-11-29T09:37:17Z	37.316852	−3.354528	1402.846	182.051	12.642
7	4	2018-11-29T09:40:04Z	37.316620	−3.354438	1402.582	181.787	12.624
8		2018-11-29T10:18:13Z	37.314431	−3.354456	1401.603	181.069	12.574
9		2019-07-18T08:13:38Z	37.316618	−3.354430	1409.757	185.211	12.862
10	5	2019-07-18T08:13:44Z	37.316764	−3.353918	1409.754	185.208	12.862
11		2019-07-18T09:04:03Z	37.314307	−3.354476	1412.117	189.000	13.125
12		2019-10-03T08:16:23Z	37.316612	−3.354427	1408.163	183.848	12.767
13	6	2019-10-03T08:17:19Z	37.316682	−3.353719	1409.065	184.750	12.830
14		2019-10-03T09:16:04Z	37.314328	−3.354328	1408.841	183.622	12.752
15		2019-11-07T08:45:24Z	37.316665	−3.354254	1401.534	177.614	12.334
16	7	2019-11-07T08:46:20Z	37.316640	−3.353892	1400.133	176.213	12.237
17		2019-11-07T11:09:46Z	37.314307	−3.354477	1403.273	178.476	12.394

The computation of the metadata that are obtained from the *RCBA* is computionally expensive, and that is the reason why it is performed beforehand. The results are not applied directly to the multispectral captures but during the training and prediction phase of the models, since the application of the these values is almost instantaneous. This advocates for storing the multispectral captures with these metadata, since it eliminates the need to store all the corrected and aligned images.

For clarity, in the following discussions the dependency of the equations with the wavelength is omitted, but it must be noted that each computation needs to be carried out for each of the five spectral bands captured by the camera.

2.2.1. Raw Images to Radiance Images Conversion

The conversion of the DN into radiance values, L_r, is carried out using Equation (1), as recommended by the camera manufacturer, to each of the pixels x and y, for each of the 5 wavelengths, λ.

$$L_r(x,y) = \frac{a_1}{g \cdot t_e} \cdot V(x,y) \cdot R(y) \cdot \frac{DN(x,y) - DN_{BL}}{DN_{MAX}} \quad (1)$$

In this equation, DN are the raw digital numbers obtained by the camera sensor and DN_{BL} is the average of the raw values of all the covered pixels of the sensor, whose objective is to measure the small amount of signal captured independently of the incident light. The difference between these two values is normalised dividing by DN_{MAX}, which is the maximum digital number achievable, equal to the maximum bit depth of the stored images minus 1. Although the camera captures 12 bit images, they are stored in 16 bits *.tif* format, so the value of DN_{MAX} is $2^{16} - 1$. The terms a_1, g and t_e refer to the first radiometric calibration coefficient, the gain and the exposure time of the camera, respectively.

The correction of the decrease in the light captured by the sensor from its top to the bottom, $R(y)$, is given by Equation (2),

$$R(y) = \frac{1}{1 + (a_2/t_e) \cdot y - a_3 \cdot y}, \quad (2)$$

where a_2 and a_3 are the last two radiometric calibration coefficients.

The vignetting correcting function, $V(x,y)$, is obtained according to Equations (3)–(5),

$$V(x,y) = \frac{1}{k(x,y)}, \quad (3)$$

$$k(x,y) = 1 + k_0 \cdot r(x,y) + k_1 \cdot r(x,y)^2 + k_2 \cdot r(x,y)^3 + k_3 \cdot r(x,y)^4 + k_4 \cdot r(x,y)^5 + k_5 \cdot r(x,y)^6, \quad (4)$$

$$r(x,y) = \sqrt{(x - c_x)^2 + (y - c_y)^2}, \quad (5)$$

where r is the distance of each pixel to the centre of the vignette (c_x, c_y) and k_{0-5} are the polynomial correction factors.

The methods used to obtain Equations (1)–(5) have not been shared by the manufacturer Micasense, but the values can be extracted from the metadata tags embedded in the images. They are shown in Table 2.

2.2.2. Radiance Images to Reflectance Images Conversion

The reflectance is defined as the ratio between the reflected and incoming radiant fluxes. Due to energy conservation considerations, this ratio should always lie between 0 and 1. The most general formulation is given using the bidirectional reflectance distribution function (BRDF), denoted by f_r [44], as:

$$\rho(\omega_i; \omega_r; L_i) = \frac{d\phi_r(\theta_i, \phi_i; \theta_r, \phi_r)}{d\phi_i(\theta_i, \phi_i)} = \frac{\int_{\omega_r} \int_{\omega_i} f_r(\theta_i, \phi_i; \theta_r, \phi_r) \cdot L_i(\theta_i, \phi_i) \cdot d\Omega_i \cdot d\Omega_r}{\int_{\omega_i} L_i(\theta_i, \phi_i) \cdot d\Omega_i}, \quad (6)$$

where the subindex i refers to incoming magnitudes, and r refers to reflected magnitudes. The geometric parameters w, θ, ϕ and Ω are the solid angles, azimuth angles, zenith angles and projected solid angles, respectively.

Table 2. Parameters used to convert raw images to radiance images and their corresponding values and tags extracted from the metadata embedded in the raw images.

Parameter	Metadata Tag	Value
t_e	$ExposureTime$	1/988
g	$ISOSpeed/100$	200/100
DN_{BL}	$(\sum_{i=0}^{4} BlackLevel_i)/4$	$(\sum_{i=0}^{4} 4800)/4$
a_1	$RadiometricCalibration_0$	$1.4511219999999999 \times 10^{-4}$
a_2	$RadiometricCalibration_1$	$1.2972460000000001 \times 10^{-7}$
a_3	$RadiometricCalibration_2$	$-2.9491650000000001 \times 10^{-5}$
c_x	$VignettingCenter_0$	585.34460000000001
c_y	$VignettingCenter_1$	480.0985
k_0	$VignettingPolynomial_0$	2.049875×10^{-7}
k_1	$VignettingPolynomial_1$	$7.0480179999999998 \times 10^{-7}$
k_2	$VignettingPolynomial_2$	$-6.5932030000000001 \times 10^{-9}$
k_3	$VignettingPolynomial_3$	1.907818×10^{-11}
k_4	$VignettingPolynomial_4$	$-2.4725660000000001 \times 10^{-14}$
k_5	$VignettingPolynomial_5$	1.15035×10^{-17}

The BRDF is used to describe the dispersion of a ray of incident light on a surface from an incoming direction towards another outgoing direction, and it is considered an intrinsic property of the surface. Its definition is

$$f_r(\theta_i, \phi_i; \theta_r, \phi_r) = \frac{\mathrm{d}L_r(\theta_i, \phi_i; \theta_r, \phi_r; E_i)}{\mathrm{d}E_i(\theta_i, \phi_i)}, \tag{7}$$

here, $\mathrm{d}L_r$ is the infinitesimal reflected radiance and $\mathrm{d}E_i$ is the infinitesimal incoming irradiance. Given its infinitesimal nature, BRDF is only useful for conceptually understanding other related magnitudes, but it cannot be measured directly, since the raylights do not include any radiant flux. What is measured by the sensors in the multispectral camera and the DLS is the hemispherical–conical reflectance, which is used to obtain the hemispherical-conical reflectance factor (HCRF). The incoming irradiance is hemispheric since, besides the direct solar component, it also considers diffuse components coming from every direction. The reflected radiance is conic due to the fact that each pixel of the camera fills a certain solid angle equal to the instantaneous field of view (IFOV). This IFOV is so small (0.03°) that the measurement of the reflected radiance can be considered directional and not conic, so the HCRF could be approximated using the hemispheric–directional reflectance factor (HDRF).

The HDRF is obtained as the ratio between the radiant flux that is reflected by the surface and the radiant flux that an ideal surface (lossless), and a perfectly diffuse (lambertian) would reflect under the same geometric and luminance conditions, since this would be equal to the incoming flux to this type of surface. This parameter could have values between 0 and $+\infty$, although it usually lies below 1 if the reflections are not specular or close to being so.

$$HDRF = \frac{\mathrm{d}\phi_r(2\pi; \theta_r, \phi_r)}{\mathrm{d}\phi_i(2\pi)} = \frac{\cos\theta_r \cdot \mathrm{d}L_r(2\pi; \theta_r, \phi_r) \cdot \sin\theta_r \cdot \mathrm{d}\theta_r \cdot \mathrm{d}\phi_r \cdot \mathrm{d}A}{\cos\theta_r \cdot \mathrm{d}L_r^{id}(2\pi) \cdot \sin\theta_r \cdot \mathrm{d}\theta_r \cdot \mathrm{d}\phi_r \cdot \mathrm{d}A} = \frac{\mathrm{d}L_r(2\pi; \theta_r, \phi_r) \cdot \mathrm{d}E_i(2\pi)}{\mathrm{d}L_r^{id}(2\pi) \cdot \mathrm{d}E_i(2\pi)}, \tag{8}$$

since the ideal diffuse reflectance is equal to $1/\pi$,

$$HDRF = \frac{\pi \cdot \mathrm{d}L_r(2\pi; \theta_r, \phi_r)}{\mathrm{d}E_i(2\pi)}. \tag{9}$$

The measurements made by the camera are already integrated through the whole solid angle corresponding to each pixel, as are the measurements of the DLS through the whole upper semisphere, so they can be used to compute the HDRF as:

$$HDRF = \frac{\pi \cdot L_r}{E_i} = \frac{\pi \cdot L_r}{E_{dir} + E_{dif}}, \tag{10}$$

here, L_r is the radiance reflected by the ground and vegetation surfaces, which is equal to the radiance measured by each pixel of the multispectral camera after it is corrected using Equation (1), and E_i is the incoming irradiance to these same surfaces. This irradiance is divided into the direct irradiance that is perpendicular to the surface , E_{dir}, and the diffuse irradiance coming from every direction, E_{dif}. The former is computed using Equation (11), where E_s is the same direct irradiance but measured in the direction of the sun, θ_i is the angle between the sun and the horizon, α is the angle between the sun and perpendicular to the irradiance sensor, E_{dls} is the raw measurement of the irradiance sensor, c_f is the coefficient applied to account for the reflected radiance that the sensor cannot measure due to the Fresnel effect (the manufacturer provides a value of 0.9057) and w_{dif} is the percentage of diffuse radiation with respect ot the total radiation (it has a value of 0.167 when the sky is clear and the sun is at its zenith). The computation of the latter is carried out using Equation (12).

$$E_{dir} = E_s \cdot \sin\theta_i = \frac{E_{dls}/c_f}{w_{dif} + \cos\alpha} \cdot \sin\theta_i \tag{11}$$

$$E_{dif} = w_{dif} \cdot E_s \tag{12}$$

The captures taken from the CRP before and after each flight also need to be taken into account. These captures made on the ground are used to compute a correction factor for each aerial capture, f_{CRP}, which is applied to its correspondent incoming irradiance, E_i. This factor is computed by assigning to each capture of each flight an interpolated irradiance and radiance values (using the timestamps) between the irradiance and radiance measured values from their previous and posterior CRP captures and afterwards applying Equation (13).

$$f_{CRP} = \frac{\pi \cdot L_{r,CRP}^{interpolated}}{E_{i,CRP}^{interpolated} \cdot HDRF_{CRP}^{calibrated}} \tag{13}$$

Therefore, in this paper, the terms reflectance and reflectance image refer to the computed values of the corrected HDRF.

$$HDRF_{corr} = \frac{\pi \cdot L_r}{E_i \cdot f_{CRP}} \tag{14}$$

2.2.3. Band Alignment

Since the Micasense camera is composed of five image sensors with their corresponding optics, independently mounted at some distance between each other, it is required to perform an alignment of the images that were captured by each of them. This procedure allows to link the reflectance value of each pixel in the green band with the corresponding values of the rest of the pixels that reside on the same spatial coordinate in the rest of the bands.

To perform this alignment, the camera library provides a method called *align_capture*, based on the algorithm *Enhanced Correlation Coefficient Maximisation* [45]. The results of this alignment are part of the generated metadata.

2.3. Manual Labelling

There is a large variety of software with different types of licenses that can be used for the manual labelling of the images. These are typically web or desktop applications that allow to assign a label to each image or to certain zones within it, usually delimited using rectangles, polygons, or more complex methods that require computer vision or machine learning algorithms. In any case, the labels are always required to be supervised by a human if they are to be used as references for the evaluation of other algorithms.

For the manual labeling of the images in this work the software needs to be able to work with multispectral images, and since there are no applications in the market that can handle this feature [46], a specific application written in *Python* using the library *Matplotlib* was developed (Figure 3).

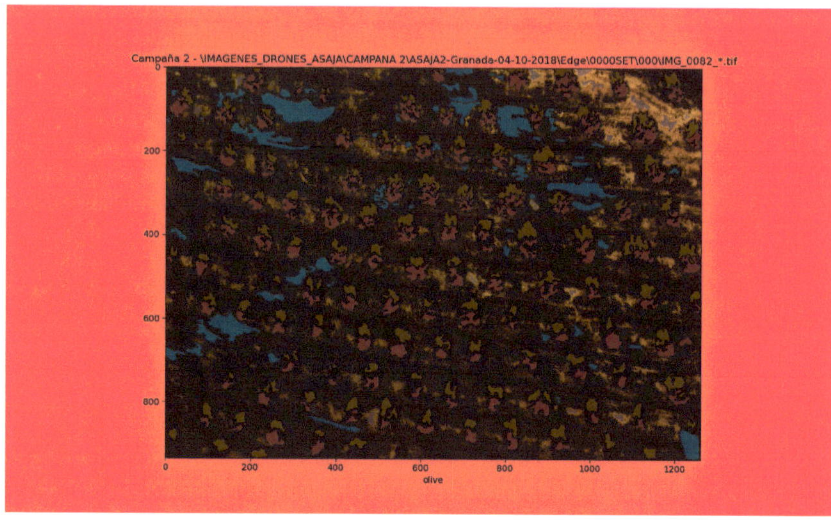

Figure 3. Developed application for the manual labelling of the aerial multispectral images of olive groves. The image shown corresponds to capture 0. The regions are labelled red for the olive tree class, yellow for the shadow class, blue for the weed class and purple for the ground class. The background changes colour to indicate the class selected for labelling.

This application receives as inputs the aerial multispectral captures of the olive groves together with the metadata computed by the *RCBA*, in order to present already corrected and aligned images to the user. Each delimited region of the image can be labelled as one of four possible labels (olive tree, ground, weed or shadow).

Although each labelled pixel is a vector of six components (five band values and the corresponding class), for the rest of the processes only the red, green and blue bands have been taken into account. The reason is that the results obtained using just these three bands are very promising, and this enables the method to be used with simple visible spectra cameras as well. Additionally, these three reflectance values are transformed to the *CIELAB* colour space, so that the colour values are more perceptibly linear, meaning that if the human eye detects two colours as similar, the coordinates in the *CIELAB* colour space are also close together. Conversely, if two colours are perceived by the human eye as different the coordinates are far apart. This conversion was carried out using 32 bits floats to take into account that the reflectance takes values between 0 and $+\infty$.

2.4. Vegetation Classification (VC)

To be able to detect the olive canopies, the first step is to classify each of the pixels of the images to determine which of those correspond to the olive class or, at least, which of those have a high probability of belonging to that class.

Since during the manual labelling process, only positive cases of pixels for each class are labelled, that is, there is no specific label for pixels not belonging to any of the classes and, furthermore, not every pixel in the image is labelled, any classification algorithm could only discern between these four classes. During the prediction stage, if the algorithm is given all the pixels in an image to assign a class to them, it would be forced to assign one of the four possible classes even to pixels that do not belong to those, generating a large amount of false positives.

A possible solution is to include all the nonlabelled pixels into a fifth class called *other* to perform this classification, but this approach conveys two problems that prevented its use. The first one is that, due to the tedious nature of the manual labelling process, a label is assigned only to pixels that are almost 100% sure to belong to a class, according at the criteria of the person that is performing this manual operation. This reduces the fatigue during the process but provokes that the class other is filled with many pixels that actually belong to one of the other classes, potentially confusing the algorithm enough to impair a proper performance. The second problem is that the pixels classified as other are very heterogeneous due to their multiple origins (roads, cars, rocks, buildings, other types of vegetation, etc.), thus having very disperse values in the feature space and complicating the computation of classification boundaries.

Because of the above reasons, the workflow includes the *VC* block, which represents an initial filtering of the pixels to select only those marked as vegetation (olive tree or weed). The task of discerning between these two classes, which is much more difficult, is left for a later stage.

This vegetation classification is implemented using a one-class classification algorithm (OCC) known as *local outlier factor* [47], which is an unsupervised method to detect atypical values computing the local pixel density deviation with respect to its neighbours. In this case, the algorithm is used to predict if each new pixel belongs to the vegetation class or not, since during the training all the pixels received are not atypical.

In addition to the labelled pixels, the algorithm requires the definition of two parameters whose values are not too relevant for the posterior stages. The contamination value has been chosen to have the maximum possible value (0.5) in order to assure that each pixel labelled as vegetation is effectively vegetation, although that means that pixels that could be classified as such are lost. In turn, the number of neighbours evaluated for each pixel is fixed at a high enough value so that the subsequent steps perform adequately without excessively increasing the prediction time (100).

The output of this process is a vegetation mask (Figure 4) for each capture that represents the zone that, with high probability, is selected as olives or weed in the subsequent stages.

2.5. Olive Tree Canopy Estimation (OTCE)

The *OTCE* is based on a decision tree trained with the pixels labelled as olive tree and weed. At prediction time, this tree receives as inputs only the pixels selected by the vegetation mask, in order to classify them as olive tree or weed. However, the results obtained for the olive tree class are not acceptable, nor were they better when different classification algorithms were evaluated. This led to the conclusion that just the spectral information of each pixel independently is not enough to obtain good results, so techniques that take into account the spatial distribution of the pixels in the image were explored.

Nonetheless, this decision provides useful information: the probability image (Figure 5), an image where each pixel of the vegetation mask receives a value between 0 and 1 according to the probability of it belonging to the olive tree class. This probability is computed by dividing the number of pixels classified as olive tree between the total number of pixels inside each leaf of the classification tree.

2.6. Olive Tree Canopy Marking (OTCM)

This block is the most important of the whole canopy detection algorithm and is composed of several connected steps, so that the output of one block is the input to the next.

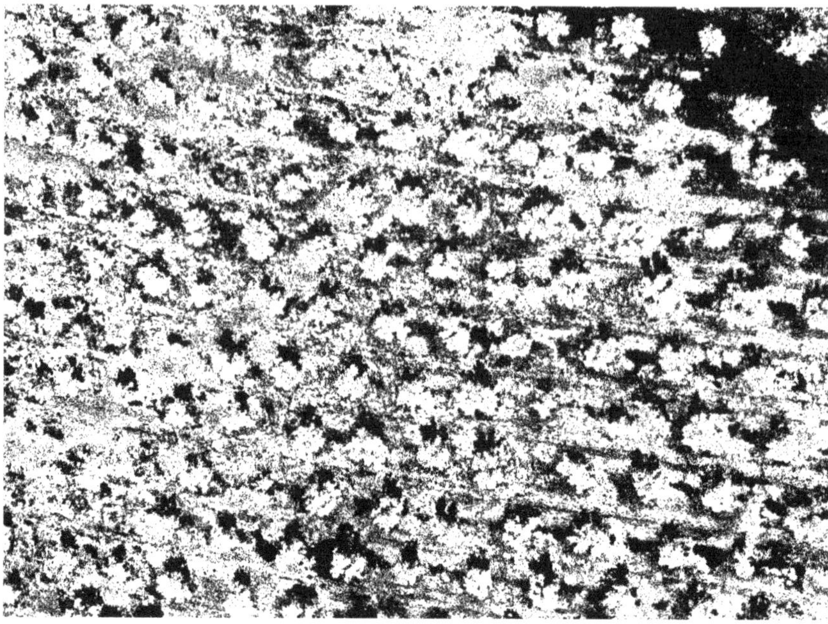

Figure 4. Vegetation mask associated with capture 0. It is a black and white image in which the white pixels were predicted as belonging to the vegetation class by a one-class classification algorithm known as local outlier factor.

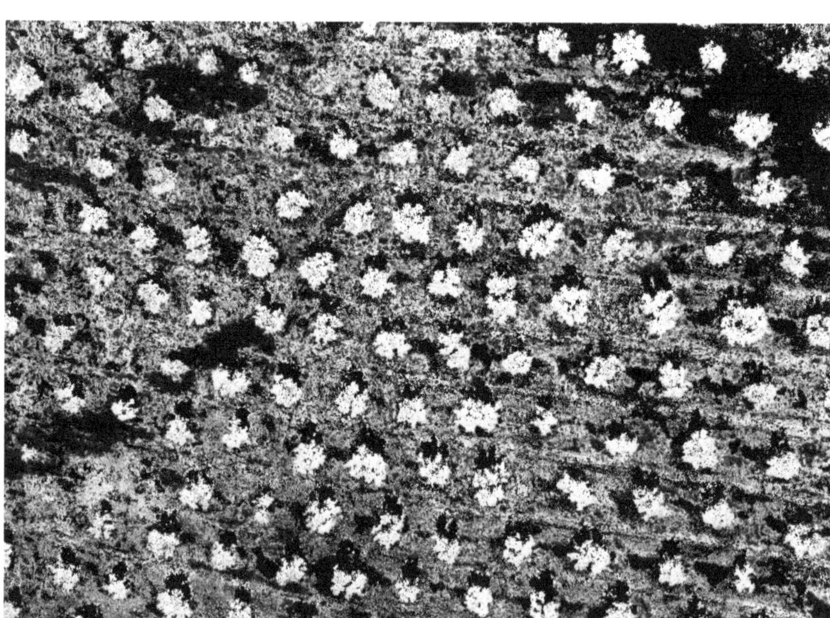

Figure 5. Probability image associated with capture 0. It is a greyscale image in which the values of each pixel indicate the probability of belonging to the olive tree class. This probability is calculated by a decision tree that receives as input only the pixels classified as vegetation during training.

The clipping and normalisation step receives as input the probability image that, basically, is an image with just one channel where the pixel values represent the probability of their belonging to the olive tree class. In this step, the value of all the pixels below a threshold parameter are assigned 0, and the rest of the values are rescaled to work with all the bit depth of the range. This step, therefore, removes those pixels with a probability of belonging to olive tree class below this specified threshold, which is a parameter that needs to be optimized automatically for each image.

The noise-filtering step receives an image where all nonzero pixels have a very high probability of belonging to olive tree class, but they still keep a relative probability information between them (Figure 6a), and applies a simple median filter with the smallest kernel possible (3×3) in order to remove isolated pixels that are most likely noise (Figure 6b).

The next step computes the density for each pixel of the probability values, that is, each pixel is assigned the mean of the probability values of the neighbouring pixels and itself, using a mean filter with circular kernel. This steps allows to homogenise to high values the probabilities of zones with clusters of pixels showing high probability, and to low values areas where there are isolated pixels with low probability. The result is an image where the canopies are homogeneously highlighted and isolated areas have been removed (Figure 6c). The parameter optimisation phase carefully selects the size of the kernel for this step.

The last parameter that has to be optimised in the *OTCM* block is the segmentation threshold that, basically, is the threshold that is used to binarise the image during the segmentation step (Figure 6d). All that is left is to search for the contours of this binary image and obtain a list to extract the centroids (Figure 6e).

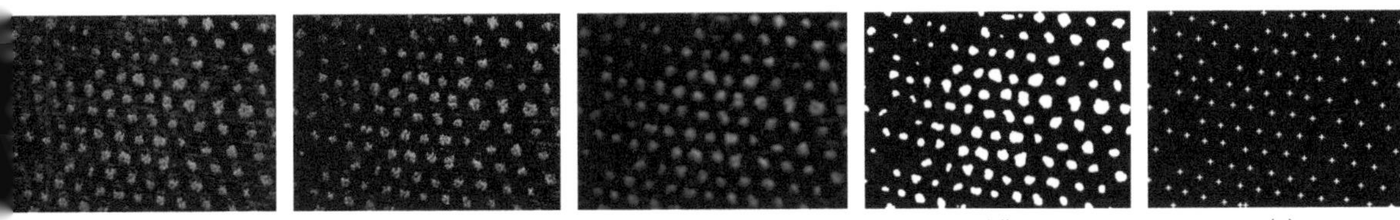

(a) (b) (c) (d) (e)

Figure 6. Images resulting from each stage of the *OTCM* block associated with capture 0. The resulting image of each stage is the one received as input by the next stage. (**a**) Clipping and normalisation. (**b**) Noise filtering. (**c**) Density computation. (**d**) Segmentation. (**e**) Centroid extraction.

This sequence of steps allows to achieve the objective of obtaining the correct coordinates of the olive canopies in each image, but, in order to obtain good results, the three parameters previously mentioned need to be chosen appropriately, as they strongly influence the final result, and the optimum value varies notably between captures.

This is the reason to add the parameter optimisation block, whose task is to search for the most adequate values for the probability threshold p, the kernel size k and the segmentation threshold s for each probability image that it receives as input. In order to do this, a brute force method is applied, where the different steps presented above are performed repeatedly for each parameter combination taken from a regular discretisation of their respective domains. For this work, the set P, considered for the probability threshold, contains 19 values, starting at 0.05 and reaching 0.95 with increments of size 0.05. The set K, related to the kernel size, contains 6 values from 5 to 30 with increments of size 5; finally, the set S contains 50 values for the segmentation threshold from 5 to 250 with increments of 5. These sets provide a total of 5700 iterations. In each of these iterations, the Delaunay triangulation (Figure 7) is computed using the coordinates of the olive canopies found (x, y) and the variation coefficient (C_v) of the set of all of the triangle side length $\{l_1, l_2, \ldots, l_n\}$, removing those that are on the edge (and belong to just one triangle) so that the computation of the ratio between the radius of the inscribed circumference of the adjacent triangle and that of the circumscribed circumference in itself is below 0.1. In order to not leave holes in the triangulation, this procedure is performed recursively until there are no sides that do not fulfill the conditions.

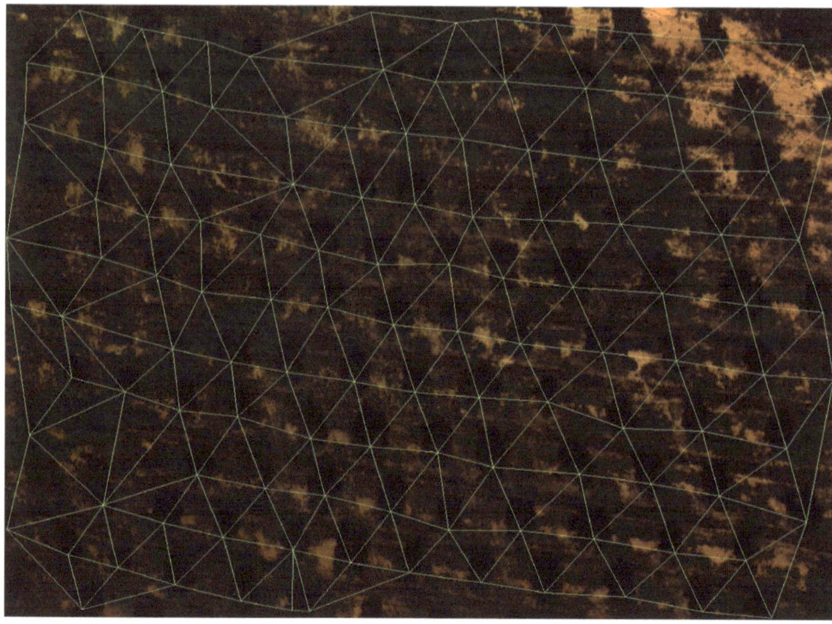

Figure 7. Delaunay triangulation using the coordinates of the olive tree canopies from capture 0 obtained with the optimal combination of parameters calculated by the model whose values are 0.9, 25 and 25 for the probability threshold, the kernel size and the segmentation threshold, respectively.

This procedure can be defined in each capture by means of the discrete function F as:

$$C_v = \{F(p,k,s) \mid (p,k,s) \subset (P \times K \times S)\}, \tag{15}$$

whose value, for each combination of p, k and s, is:

$$C_v = \frac{\sigma}{\bar{l}} = \frac{\sqrt{N \cdot \sum_{n=1}^{N} (l_n - \bar{l})^2}}{\sum_{n=1}^{N} l_n}, \tag{16}$$

with

$$l_n = \sqrt{(x_i - x_j)^2 + (y_i - y_j)^2}, \tag{17}$$

here, i and j refer to the indexes of each end of each side of the Delaunay triangulation and n to the side itself. A graphical representation in $\mathbb{R}^3$ of this discrete function can be visualised if any of the three parameters p, k or s is fixed. For instance, Figure 8a,d shows the function for $p = 0.9$, Figure 8b,e depicts the function for $k = 25$ and Figure 8c,f represents it for $s = 25$. These values are the optimum values computed posteriorly by the model for capture 0.

Figure 8 shows that there are no computed values of the F function for certain combinations of parameters. This is due to the fact that the algorithm filters the results where it cannot compute the Delaunay triangulations, for instance, if the number of detected canopies is fewer than three. It can also be observed that the values that are close to the limits of the domain fluctuate much more between iterations than the central values, without following any clear trend. This happens because, for these parameter combinations, almost no canopies are found if the values of p and s are too large or too small (either all the pixels are removed if the values are too high, or very large contours filling almost the whole image are found if the values are too low), or the number of detected canopies is way too large if the values of k are too low (all the noise is considered as valid regions). This happens in all the images considered in this work and, in order to solve this issue, we consider only the values of C_v that have a number of detected canopies within $\pm 40\%$ of the median number of detected trees for all iterations. That is, a new function G that returns the number of detected canopies is defined in Equation (18), along with the median of the set of values $\{t_1, t_2, \ldots, t_c\}$ computed by G in Equation (19):

$$T = \{G(p,k,s) \mid (p,k,s) \subset (P \times K \times S)\}, \tag{18}$$

$$M = \begin{cases} t_{(c+1)/2} & \text{if } c \text{ is odd} \\ \frac{t_{(c/2)} + t_{(c/2)+1}}{2} & \text{if } c \text{ is even} \end{cases}, \tag{19}$$

and both are used to redefine the function F in Equation (20):

$$C_v = \{F(p,k,s) \mid (p,k,s) \subset (P \times K \times S) \cap G(p,k,s) \in (0.6M, 1.4M)\}, \tag{20}$$

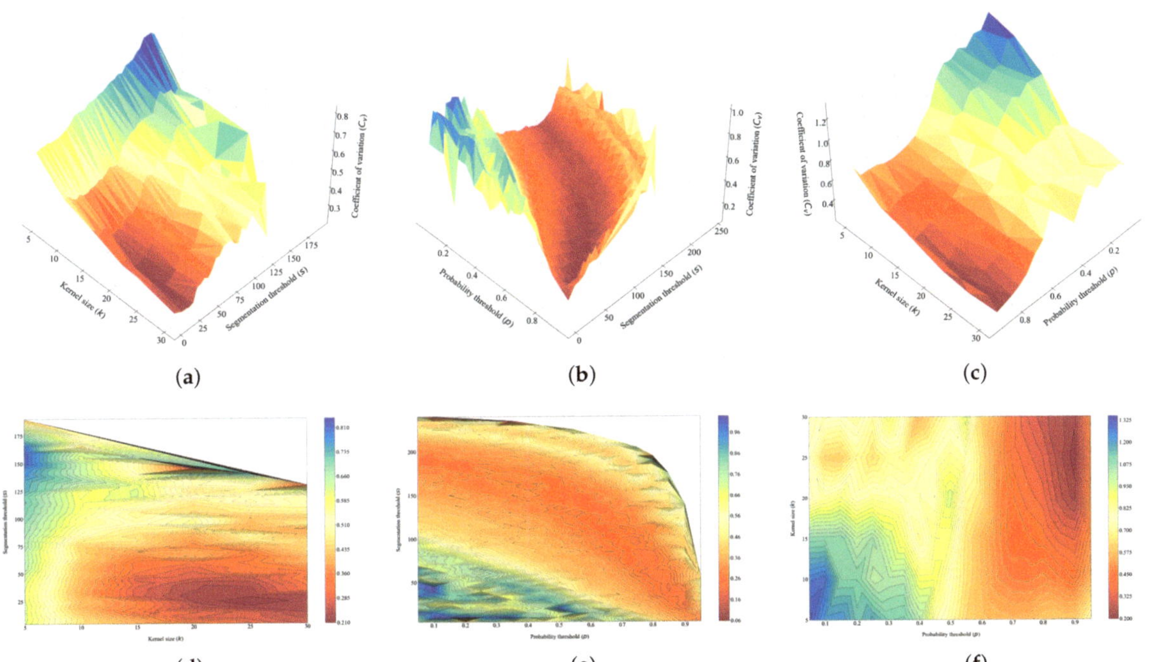

Figure 8. Graphical representation associated with capture 0 in $\mathbb{R}^3$ of the discrete function F, assigning a constant value to one of the parameters in each case. The selected values are the optimum ones computed posteriorly by the model. (**a**) $p = 0.9$. (**b**) $k = 25$. (**c**) $s = 25$. (**d**) $p = 0.9$. (**e**) $k = 25$. (**f**) $s = 25$.

The last step is reduced to an optimisation problem of the redefined function F to find the values of p, k and s and minimise it in Equation (21).

$$(p_{min}, k_{min}, s_{min}) = \arg\min_{p,k,s}(C_v) \tag{21}$$

The graphical representation of the redefined function F is depicted in Figure 9, together with the minimum value of the variation coefficient marked. Many values have been removed from those in Figure 8 due to this redefinition.

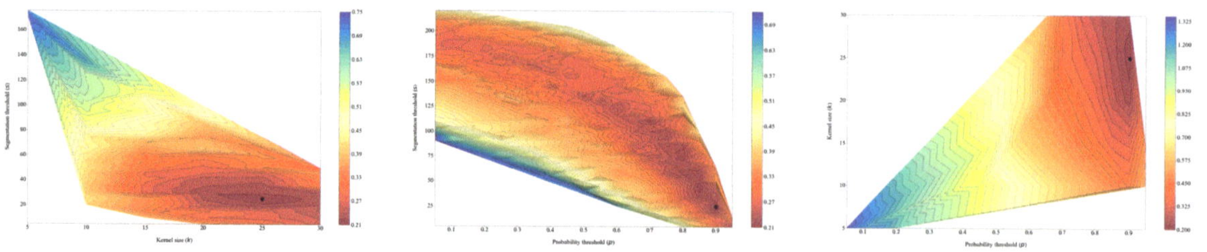

(a) (b) (c)

Figure 9. Graphical representation of the redefined function F associated with capture 0, fixing one of the parameters in each case. It is observed how the domain of the function changes when it is redefined. The black dot represents the minimum value of the function. (**a**) $p = 0.9$. (**b**) $k = 25$ (**c**) $s = 25$.

This methodology, besides using the spectral information of the pixels to detect the tree canopies, includes spatial variables related to pattern detection in images, specifically the optimum iteration is that whose distribution of canopies provides the least variability, that is, the one where the detected canopies are distributed as evenly as possible throughout the image. Additionally, the variation coefficient is used instead of the standard deviation in order to be able to compare captures with different ground sample distances or different plantation distances.

This idea is key for the success of the algorithm, specially for the aerial images of regular plantations with thick plant cover on the ground. The principle employed is similar to the approach used by a human to manually discern whether there is a tree canopy or just plant cover in a certain area of the image: the pattern of the plantation is recognised and extrapolated to contiguous areas. If any extrapolated point is included in the area of interest, then the region is probably classified as canopy. If, on the other hand, in that zone there is no extrapolated point, then most likely the area is just some other type of vegetation. From the grower's point of view, it makes sense ot keep the distances between the olive trees as invariant as possible, since once that the minimum distance is fixed there would be no reason to increase it, as it would imply reducing plantation density and consequently, profit.

3. Results and Discussion

Generally, during the evaluation of the results of a model, the provided prediction for each input data is compared with the ground truth previously labelled by a person. If the model has a training phase, the input data used during this phase cannot be reused later during the evaluation step. For this, the input data are separated into a training dataset and an evaluation dataset. In addition, to guarantee that the results obtained are independent of the way in which both sets are separated, the evaluation is usually carried out through some type of cross-validation.

For the model proposed in this work, it can be seen (Figure 1) that the type of input data is different for the training phase and the prediction phase. In the first case, pixels labelled as olive tree, ground, weed or shadow are used, while the second case only accepts multispectral captures and their metadata as input. This reality means that the evaluation phase is carried out on complete multispectral captures and not on the pixels that make them up, although the first two blocks (*VC* and *OTCE*) extract the pixels from the captures using the metadata. For this reason, it must be taken into account that the labelled pixels used in the training have to be introduced in a grouped way for each multispectral capture in order to guarantee the selection of those groups that are not used later during the evaluation.

The first step is to carry out the manual labelling process to obtain the ground truth for the training and prediction phases of blocks *VC* and *OTCE*. The labelled regions contain pixels with an absolute certainty of belonging to the chosen class, at the cost of leaving possible pixels that still belong to that class unlabelled (Figure 3). Then, the pixels labelled as olive are used as the ground truth in the rest of the model. This ground truth is a binary mask for each multispectral capture, in which the regions belonging to the olive tree canopies are marked (Figure 10). It is obtained by carrying out the morphological operation of dilation with a circular kernel of 11 pixels in diameter applied to the pixels labelled as olive tree in order to cover the entire region of the olive tree canopies. The resulting masks are then inspected to ensure that this value is correct for all captures.

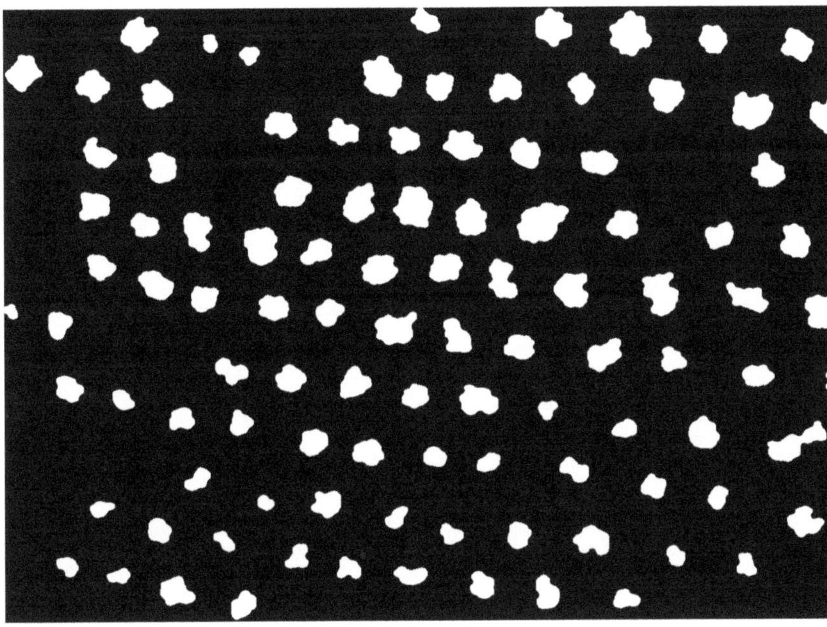

Figure 10. Ground truth associated with capture 0. It is a black and white image in which the white pixels have been manually labeled by a person in order to calculate the quality metrics of the model.

The following concepts are precised in order to discuss the results of this study:

- Positive: Set of connected pixels (contour) marked as olive canopy in the ground truth.
- Negative: Set of connected pixels not marked as olive canopy in the ground truth.
- Predicted positive: Any coordinate in pixels computed by the model corresponding to an olive tree canopy.
- Predicted negative: It is not applicable because the model does not compute coordinates where there are no olive tree canopies.
- True positive: Every predicted positive whose coordinates are within the bounds of some positive.
- True negative: Not applicable since the predicted negatives do not exist.
- False positive: Any predicted positive whose coordinates are not within the bounds of any positive.
- False negative: Any positive for which there is no predicted positive coordinate that is within its bounds.

Table 3 shows the results of performing a 5-fold cross-validation on the model. It shows, from left to right, the ID of the capture (Table 1), the fold in which each capture is part of the test set, the count of positives in the ground truth, the predicted positives, the true positives, the false negatives and the false positives obtained by the model. Finally, three metrics were evaluated: recall, precision and F_1 score. In the last row, the sum of each column is calculated, except for the three metrics, which are calculated in the usual way using the previous sums.

Table 3. 5-fold cross-validation results on the *OTCD* model. In each iteration, the metrics of the model's predictions are evaluated for the captures with the indicated *IDs*, performing the training with the rest of the captures.

ID	Fold	Positives (P)	Predicted Positives (PP)	True Positives (TP)	False Negatives (FN)	False Positives (FP)	Recall	Precision	F_1 Score
2	0	154	166	134	21	32	0.870	0.807	0.838
1		125	141	118	7	23	0.944	0.837	0.887
3		112	115	108	4	7	0.964	0.939	0.952
7		106	98	94	13	4	0.887	0.959	0.922
6	1	100	134	96	4	38	0.960	0.716	0.821
16		100	123	76	26	47	0.760	0.618	0.682
5		238	247	193	45	54	0.811	0.781	0.796
13		103	120	51	52	69	0.495	0.425	0.457
10	2	115	121	110	5	11	0.957	0.909	0.932
12		111	136	90	23	46	0.811	0.662	0.729
15		103	148	67	57	81	0.650	0.453	0.534
0		104	119	101	3	18	0.971	0.849	0.906
8	3	226	238	31	195	207	0.137	0.130	0.134
11		270	285	263	7	22	0.974	0.923	0.948
4		102	119	58	44	61	0.569	0.487	0.525
14	4	263	264	231	32	33	0.878	0.875	0.877
17		225	236	149	80	87	0.662	0.631	0.646
9		123	121	107	16	14	0.870	0.884	0.877
		2680	**2931**	**2077**	**634**	**854**	**0.775**	**0.709**	**0.740**

For a better graphic interpretation of the results, the contours of the olive tree canopies in the ground truth and the marks of the coordinates predicted by the model are combined on single images. Both contours and marks that are true positives are coloured green, while contours and marks that are false negatives and false positives, respectively, are coloured red. Figure 11 shows an example of these images for 5 of the 18 captures.

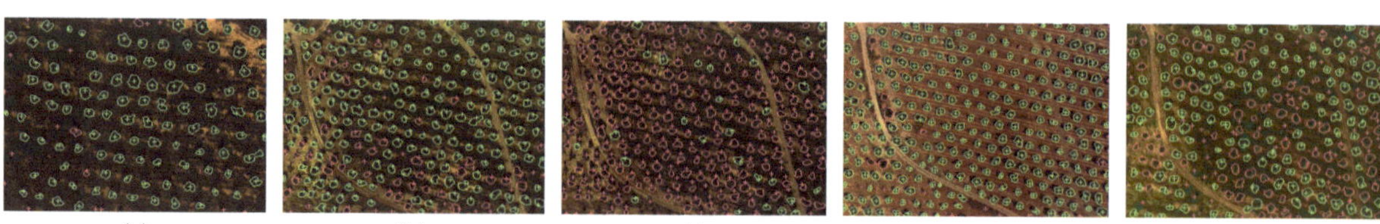

(a) (b) (c) (d) (e)

Figure 11. Resulting images obtained and their comparison with the ground truth. Each green mark and contour set indicates a true positive, each red mark indicates a false positive and each red contour indicates a false negative. (**a**) Capture 0. (**b**) Capture 5. (**c**) Capture 8. (**d**) Capture 11. (**e**) Capture 17.

Analysing the results table (Table 3), it can be seen that 89% of the captures obtain an F_1 score above 0.5, 61% above 0.75, 56% above 0.8 and 28% above 0.9. In 5 of the 18 captures analyzed, practically perfect detection is achieved (Figure 11a,d), in six captures the detection is moderately good (Figure 11b), in five captures the canopies are detected acceptably (Figure 11e) and only in two captures serious detection failures occur (Figure 11c). If the causes of these two failures are analyzed in detail, it is seen that they are not related at all to the main hypothesis proposed in this article—automatic detection of olive tree canopies for groves with thick plant cover is optimised by minimizing the function of coefficients of variation of the lengths of the sides of the Delaunay triangles formed from the coordinates of the canopies predicted by the model. Rather, the causes are generated by the first part of the model, the *VC* in the first instance, classifying shadow pixels as vegetation, and the *OTCE* in the second instance, not being able to exclude them, at least, as weeds. In fact, Figure 11c shows how each and every one of the olive trees is detected, but their shadows are marked instead of the canopies. This is due to the fact that in the probability image the *VC* and the *OTCE* assigned a greater canopy probability to the shadows than to the canopies. The rest of the model from here works well even for images that, due to the time of year they were taken, do not have as much grass on the ground as one would expect.

Table 4 shows the execution times of each model block in the workflow for each capture during the cross-validation, including the *VC* and *OTCE* training time, same at each fold. The last row shows the mean of each column.

Table 4. Execution times of each model block. A training time for each fold and the prediction times for each capture ID are shown. The last row shows the mean of each column.

ID	**Fold**	**VC Training Time (s)**	**OTCE Training Time (s)**	**VC Predict Time (s)**	**OTCE Prediction Time [s]**	**OTCM Prediction Time (s) for 5700 Iterations**
2	0	46.3	6.4	28.7	0.15	56.7
1				40.4	0.24	65.5
3				40.1	0.26	57.7
7				34.8	0.13	55.2
6	1	57.7	6.4	29.7	0.12	60.1
16				46.6	0.16	66.0
5				34.8	0.19	59.6
13				33.8	0.07	51.0
10	2	54.9	7.3	24.5	0.05	43.6
12				28.8	0.11	60.9
15				28.5	0.11	71.7
0				30.8	0.22	61.1
8	3	66.4	6.7	28.2	0.14	56.6
11				26.1	0.04	52.2
4				28.1	0.15	53.1
14	4	63.6	6.8	29.2	0.10	50.9
17				38.7	0.23	54.3
9				29.9	0.09	53.0
		57.0	**6.7**	**32.3**	**0.14**	**57.2**

As expected, one of the two most unfavorable times is the training time with an average of 57 s per fold. This time must not be taken into account to assess the performance of the model since it is a task that is performed only once. However, another of the most critical times occurs during the prediction of the *OTCM* block with a mean of 57.2 s. Although it may seem like a long time, it is necessary to realise that the time is divided for the 5700 iterations of the brute force method applied in this block, that is, each iteration is executed in approximately one hundredth of a second. Furthermore, this time is no longer a concern considering that, for validating and representation purposes, the number of points computed by the *F* function was higher than it was really needed, and it can be drastically reduced. After these, the next highest time is the prediction time for the *VC* block, taking 32.3 s. This time could only be reduced by finding alternative vegetation classification methods.

Finally, although no articles were found that explicitly deal with the detection or segmentation of tree canopies with thick plant cover on the ground without using three-dimensional data, those that are closely related are shown in Tables 5 and 6. They summarise information about the article reference, the method used, the publication date, where the data comes from (dataset), what type of data is used (channels), how much data is used (no. of images) and what quality metrics are achieved (accuracy, precision, recall, omission error, commission error and estimation error) in each of them. They are sorted by publication date, and the first row corresponds to the method presented in this article.

Table 5. Comparison of the results of this study with the results of other closely related articles that although they do not explicitly use images of groves with thick plant cover on the ground, deal with the detection or segmentation of the olive tree canopies. The article reference, the method used the publication date, where the data comes from (dataset) and what type of data is used (channels) are shown.

Reference	Method	Publication Date	Dataset	Channels
-	Proposed	-	MicaSense RedEdge-MX	Red, green and blue
[48]	Deep learning model (SwinTUnet) based on Unet-like networks	15 January 2022	Satellites Pro	Red, green and blue
[49]	Orthophotos + Mask R-CNN	25 February 2021	Parrot Sequoia camera	Red, green, blue and near infrared
[50]	Edge detection + circular Hough transform	1 June 2020	SIGPAC viewer	Red
[25]	Laplacian of Gaussian + improved k-means clustering	26 February 2020	SIGPAC viewer	Red, green and blue
[51]	Colour-based vs. stereo-vision-based segmentation	5 February 2019	DJI Phantom4 camera	Red, green and blue
[52]	Multi-level thresholding + circular Hough transform	4 December 2018	SIGPAC viewer	Red, green and blue
[53]	Radiometrically corrected orthophotos+ Object-based image analysis	4 December 2017	Modified multiSPEC 4C camera	Red, green, red-edge and near-infrared
[54]	Orthophotos + Thresholding + watershed analysis + microbiological cell counting algorithm	27 October 2017	Leica ADS40, ADS80, ADS100 and DMC III cameras	Red, green, blue and near-infrared
[55]	Optical + radar data + object-based classification	19 July 2011	ADS40 Airborne Digital Sensor + RAMSES + TerraSAR-X satellite	Panchromatic, red, green, blue, near-infrared and X
[56]	K-mean clustering	2 July 2010	SIGPAC viewer	Red, green and blue
[57]	Reticular matching	31 August 2007	Quickbird	Panchromatic

Table 6. Comparison of the results of this study with the results of other closely related articles that, although they do not explicitly use images of groves with thick plant cover on the ground, deal with the detection or segmentation of the olive tree canopies. The article reference, how much data is used (no. of images) and what quality metrics are achieved (accuracy, precision, recall, omission error, commission error and estimation error) are shown.

Reference	No. of Images	Accuracy	Precision	Recall	Omission Error Rate	Commission Error Rate	Estimation Error
-	18	N/A	70.9%	77.5%	22.5%	N/A	9.37%
[48]	230	98.3%	N/A	98.8%	1.2%	0.97%	0.94%
[49]	150	N/A	90.07–100%	90.83–100%	0–9.17%	N/A	N/A
[50]	60	N/A	N/A	96%	4%	1.2%	1.27%
[25]	110	97.5%	N/A	99%	1%	4%	0.97%
[51]	10	N/A	99.84%	97.61%	2.39%	N/A	N/A
[52]	N/A	96%	N/A	97%	3%	3%	1.2%
[53]	315	93%	91.6%	95.9%	4.1%	10.57%	4.76%
[54]	4	N/A	N/A	N/A	N/A	N/A	4–27%
[55]	3	76.8–90.5%	N/A	16–66.7%	33.3–84%	0.6–8.4%	N/A
[56]	N/A	N/A	N/A	83.33%	16.67%	0%	
[57]	3	N/A	N/A	93%	7%	5%	1.24%

4. Conclusions

This article presents quite promising results as far as the automatic detection of olive tree canopies is concerned, even more so if we take into account the large amount of plant cover that some of the captures present. This fact makes detection complicated even sometimes for the human eye itself. Considering that olive growing is increasingly oriented towards the search for the quality of the resulting oil, and that organic farming takes advantage of the benefits of maintaining a thick plant cover on the ground, it is concluded that the development of this type of model is an imperative need.

The vast majority of studies carried out with aerial images in the field with olive trees and others fruit trees are based on the search for correlations between the samples collected at the foot of the tree (ground truth) and the features extracted from the images collected by the different types of onboard sensors on the UAV. For this purpose, it is always necessary to extract from the complete images the parts that correspond to the ground truth, usually the canopies. This task receives little attention in scientific works and can be arduous and tedious if performed manually. Even when algorithms are developed to automate this task, they are often used as an aid to the human labeller. In most cases, they work only for groves in which the canopies are clearly delimited by soil with very different spectral features.

The model developed in this article is very useful for performing labelling tasks fully automatically, but due to resulting prediction times of nearly 90 seconds (32.3 + 0.14 + 57.2) per capture on average, it cannot be applied in real time for now, that is, the prediction time is greater than the time that elapses between one capture and the next in the multispectral camera. This time, called period, inverse to frames per second (FPS), was set to 1 second during data collection. Despite this, the proposed method has advantages over alternative methods based on structure from motion and multiview stereo to compute 3D information such as photogrammetry or based on other sensors such as *LiDAR*. The first and most obvious one is that it makes predictions from a single capture, while photogrammetry requires a multitude of them in the same area. This fact allows cheaper data collection by reducing the overlap of the captures from 80% to almost 0% without compromising the accuracy. This consequently also reduces the flight time of the UAV. Another advantage is that more complex and costly sensors with *LiDAR* technology or spectrometers that require advanced knowledge and set-up are not used. In fact, only the red, green and blue bands of the multispectral camera were used.

In this way, as future work, it is intended to explore the viability of the model if the input data comes exclusively from a camera whose sensor works in the visible spectrum without radiometric correction. Other possible future research would be to find out the performance of the proposed method applied to other types of groves such as citrus trees, peach trees, chestnut trees or even plants with other typologies such as vines.

The success rate obtained in this study could be improved by evaluating the influence of tree shadows on the *VC* block, increasing the number of images labelled and used for training, or exploring other one-class classification algorithms that may exclude certain types of data. On the other hand, the number of discrete intervals in which each of the three parameters to be optimised is divided could be reduced with the aim of decreasing the prediction time of the block *OTCM*. Furthermore, a more efficient approach could be applied, such as gradient descent, widely used in the back-propagation algorithm of any neural network.

Finally, the proposed model is developed thinking of performing a single task as best as possible. This task is to detect olive tree canopies, that is, to predict with a very high probability the coordinates of the image in which the pixels belonging to those canopies would be found but without applying segmentation. This last task would have to be optimised in the context of a different model. Furthermore, as future work, the region classification task could be optimised, namely, the identification of the plantation from a structured arrangement of trees against regions that would form part of the environment such as roads, forests, buildings, uninhabited land, etc.

Author Contributions: Conceptualization, S.I.R., D.M.M.G., P.C.M. and J.G.O.; methodology, S.I.R. D.M.M.G., P.C.M. and J.G.O.; writing—original draft preparation, S.I.R., D.M.M.G. and P.C.M. writing—review and editing, S.I.R., D.M.M.G., P.C.M. and J.G.O.; funding acquisition, J.G.O. All authors have read and agreed to the published version of the manuscript.

Funding: This research was partially funded by the Spanish Ministry of Science and Innovation under the project PID2019-110291RB-I00 and developed under the activities of the project Precision Agriculture in Olive Groves using Unmanned Aerial Vehicles (reference GOP3I-JA-16-0015) and partially supported by it.

Institutional Review Board Statement: Not applicable.

Informed Consent Statement: Not applicable.

Data Availability Statement: The data presented in this study are available on request from the corresponding author. The data are not publicly available due to confidentiality reasons derived from the project members.

Acknowledgments: The project was carried out jointly with Fundación Caja Rural de Jaén, Fundación Andaluza de Desarrollo Aeroespacial (FADA), the Centro Provincial de Agricultores de ASAJA Jaén the Instituto de Investigación y Formación Agraria y Pesquera (IFAPA) and the regional delegations of ASAJA in Almería, Córdoba, Granada and Málaga.

Conflicts of Interest: The authors declare no conflict of interest.

Abbreviations

The following abbreviations are used in this manuscript:

BRDF	Bidirectional Reflectance Distribution Function
CNN	Convolutional Neural Network
CRP	Calibrated Reflectance Panel
DSM	Digital Surface Model
DD	Decimal Degrees
DLS	Downwelling Light Sensor
DN	Digital Numbers
DP	Data Preprocessing
EU	European Union
FAO	Food and Agriculture Organization of the United Nations
FN	False Negative
FP	False Positive
FPS	Frames Per Second
GPS	Global Positioning System
GSD	Ground Sample Distance

HCRF	Hemispherical–Conical Reflectance Factor
HDRF	Hemispheric–Directional Reflectance Factor
IFOV	Instantaneous Field of View
LiDAR	Laser Imaging Detection and Ranging
LM	Local Maximum
OCC	One-Class Classification
OTCD	Olive Tree Canopy Detection Model
OTCE	Olive Tree Canopy Estimation
OTCM	Olive Tree Canopy Marking
P	Positive
PN	Predicted Negative
PP	Predicted Positive
RCBA	Radiometric Correction and Band Alignment
RMSE	Root Mean Square Error
ROI	Region Of Interest
TN	True Negative
TP	True Positive
UAV	Unmanned Aerial Vehicles
VC	Vegetation Classification

References

1. FAOSTAT. Available online: https://www.fao.org/faostat/en/#data/QCL (accessed on 8 February 2022).
2. Economic Affairs & Promotion Unit–International Olive Council. Available online: https://www.internationaloliveoil.org/what-we-do/economic-affairs-promotion-unit/#figures (accessed on 8 February 2022).
3. Marchal, P.; Gila, D.; García, J.; Ortega, J. Fuzzy Decision Support System for the Determination of the Set Points of Relevant Variables in the Virgin Olive Oil Elaboration Process. In Proceedings of the 2013 IEEE International Conference on Systems, Man, and Cybernetics, SMC, Manchester, UK, 13–16 October 2013; pp. 3489–3494. [CrossRef]
4. Bordons, C.; Núñez-Reyes, A. Model Based Predictive Control of an Olive Oil Mill. *J. Food Eng.* **2008**, *84*, 1–11. [CrossRef]
5. Furferi, R.; Carfagni, M.; Daou, M. Artificial Neural Network Software for Real-Time Estimation of Olive Oil Qualitative Parameters during Continuous Extraction. *Comput. Electron. Agric.* **2007**, *55*, 115–131. [CrossRef]
6. Fernández, J.; Alcon, F.; Diaz-Espejo, A.; Hernandez-Santana, V.; Cuevas, M. Water Use Indicators and Economic Analysis for On-Farm Irrigation Decision: A Case Study of a Super High Density Olive Tree Orchard. *Agric. Water Manag.* **2020**, *237*, 106074. [CrossRef]
7. Riveros-Burgos, C.; Ortega-Farías, S.; Morales-Salinas, L.; Fuentes-Peñailillo, F.; Tian, F. Assessment of the Clumped Model to Estimate Olive Orchard Evapotranspiration Using Meteorological Data and UAV-based Thermal Infrared Imagery. *Irrig. Sci.* **2021**, *39*, 63–80. [CrossRef]
8. Tovar, M.; Motilva, M.; Romero, M. Changes in the Phenolic Composition of Virgin Olive Oil from Young Trees (Olea Europaea L. Cv. Arbequina) Grown under Linear Irrigation Strategies. *J. Agric. Food Chem.* **2001**, *49*, 5502–5508. [CrossRef]
9. Angelopoulos, K.; Dichio, B.; Xiloyannis, C. Inhibition of Photosynthesis in Olive Trees (*Olea europaea* L.) during Water Stress and Rewatering. *J. Exp. Bot.* **1996**, *47*, 1093–1100. [CrossRef]
10. Brito, C.; Dinis, L.T.; Moutinho-Pereira, J.; Correia, C. Drought Stress Effects and Olive Tree Acclimation under a Changing Climate. *Plants* **2019**, *8*, 232. [CrossRef]
11. Castrignanò, A.; Belmonte, A.; Antelmi, I.; Quarto, R.; Quarto, F.; Shaddad, S.; Sion, V.; Muolo, M.; Ranieri, N.; Gadaleta, G.; et al. A Geostatistical Fusion Approach Using UAV Data for Probabilistic Estimation of Xylella Fastidiosa Subsp. Pauca Infection in Olive Trees. *Sci. Total Environ.* **2021**, *752*, 141814. [CrossRef]
12. Castrignanò, A.; Belmonte, A.; Antelmi, I.; Quarto, R.; Quarto, F.; Shaddad, S.; Sion, V.; Muolo, M.; Ranieri, N.; Gadaleta, G.; et al. Semi-Automatic Method for Early Detection of Xylella Fastidiosa in Olive Trees Using Uav Multispectral Imagery and Geostatistical-Discriminant Analysis. *Remote Sens.* **2021**, *13*, 14. [CrossRef]
13. Adamo, F.; Attivissimo, F.; Di Nisio, A.; Ragolia, M.A.; Scarpetta, M. A New Processing Method to Segment Olive Trees and Detect Xylella Fastidiosa in UAVs Multispectral Images. In Proceedings of the 2021 IEEE International Instrumentation and Measurement Technology Conference (I2MTC), Glasgow, UK, 17–20 May 2021; pp. 1–6. [CrossRef]
14. Blekos, K.; Tsakas, A.; Xouris, C.; Evdokidis, I.; Alexandropoulos, D.; Alexakos, C.; Katakis, S.; Makedonas, A.; Theoharatos, C.; Lalos, A. Analysis, Modeling and Multi-Spectral Sensing for the Predictive Management of Verticillium Wilt in Olive Groves. *J. Sens. Actuator Netw.* **2021**, *10*, 15. [CrossRef]
15. Noguera, M.; Aquino, A.; Ponce, J.; Cordeiro, A.; Silvestre, J.; Calderón, R.; Marcelo, M.; Pedro, J.; Andújar, J. Nutritional Status Assessment of Olive Crops by Means of the Analysis and Modelling of Multispectral Images Taken with UAVs. *Biosyst. Eng.* **2021**, *211*, 1–18. [CrossRef]
16. Cano Marchal, P.; Martínez Gila, D.; Illana Rico, S.; Gómez Ortega, J.; Gámez García, J. Assessment of the Nutritional State for Olive Trees Using Uavs. *Lect. Notes Electr. Eng.* **2021**, *695*, 284–292. [CrossRef]

17. Díaz-Varela, R.; de la Rosa, R.; León, L.; Zarco-Tejada, P. High-Resolution Airborne UAV Imagery to Assess Olive Tree Crown Parameters Using 3D Photo Reconstruction: Application in Breeding Trials. *Remote Sens.* **2015**, *7*, 4213–4232. [CrossRef]
18. Anifantis, A.; Camposeo, S.; Vivaldi, G.; Santoro, F.; Pascuzzi, S. Comparison of UAV Photogrammetry and 3D Modeling Techniques with Other Currently Used Methods for Estimation of the Tree Row Volume of a Super-High-Density Olive Orchard. *Agricultur* **2019**, *9*, 233. [CrossRef]
19. Wang, P.; Wang, L.; Leung, H.; Zhang, G. Super-Resolution Mapping Based on Spatial–Spectral Correlation for Spectral Imagery. *IEEE Trans. Geosci. Remote Sens.* **2021**, *59*, 2256–2268. [CrossRef]
20. Berardino, P.; Fornaro, G.; Lanari, R.; Sansosti, E. A New Algorithm for Surface Deformation Monitoring Based on Small Baseline Differential SAR Interferograms. *IEEE Trans. Geosci. Remote. Sens.* **2002**, *40*, 2375–2383. [CrossRef]
21. Liu, J.; Wu, Y.; Gao, X.; Zhang, X. A Simple Method of Mapping Landslides Runout Zones Considering Kinematic Uncertainties. *Remote Sens.* **2022**, *14*, 668. [CrossRef]
22. Di Nisio, A.; Adamo, F.; Acciani, G.; Attivissimo, F. Fast Detection of Olive Trees Affected by Xylella Fastidiosa from UAVs Using Multispectral Imaging. *Sensors* **2020**, *20*, 4915. [CrossRef]
23. Ortega-Farías, S.; Ortega-Salazar, S.; Poblete, T.; Kilic, A.; Allen, R.; Poblete-Echeverría, C.; Ahumada-Orellana, L.; Zuñiga, M.; Sepúlveda, D. Estimation of Energy Balance Components over a Drip-Irrigated Olive Orchard Using Thermal and Multispectral Cameras Placed on a Helicopter-Based Unmanned Aerial Vehicle (UAV). *Remote Sens.* **2016**, *8*, 638. [CrossRef]
24. Modica, G.; Messina, G.; De Luca, G.; Fiozzo, V.; Praticò, S. Monitoring the Vegetation Vigor in Heterogeneous Citrus and Olive Orchards. A Multiscale Object-Based Approach to Extract Trees' Crowns from UAV Multispectral Imagery. *Comput. Electron. Agric.* **2020**, *175*, 105500. [CrossRef]
25. Waleed, M.; Um, T.W.; Khan, A.; Khan, U. Automatic Detection System of Olive Trees Using Improved K-Means Algorithm. *Remote Sens.* **2020**, *12*, 760. [CrossRef]
26. Li, W.; Fu, H.; Yu, L.; Cracknell, A. Deep Learning Based Oil Palm Tree Detection and Counting for High-Resolution Remote Sensing Images. *Remote Sens.* **2017**, *9*, 22. [CrossRef]
27. Daliakopoulos, I.N.; Grillakis, E.G.; Koutroulis, A.G.; Tsanis, I.K. Tree Crown Detection on Multispectral VHR Satellite Imagery. *Photogramm. Eng. Remote Sens.* **2009**, *75*, 1201–1211. [CrossRef]
28. Wang, L.; Gong, P.; Biging, G.S. Individual Tree-Crown Delineation and Treetop Detection in High-Spatial-Resolution Aerial Imagery. *Photogramm. Eng. Remote Sens.* **2004**, *70*, 351–357. [CrossRef]
29. Erikson, M. Segmentation of Individual Tree Crowns in Colour Aerial Photographs Using Region Growing Supported by Fuzzy Rules. *Can. J. For. Res.* **2003**, *33*, 1557–1563. [CrossRef]
30. Weinstein, B.G.; Marconi, S.; Bohlman, S.; Zare, A.; White, E. Individual Tree-Crown Detection in RGB Imagery Using Semi-Supervised Deep Learning Neural Networks. *Remote Sens.* **2019**, *11*, 1309. [CrossRef]
31. Csillik, O.; Cherbini, J.; Johnson, R.; Lyons, A.; Kelly, M. Identification of Citrus Trees from Unmanned Aerial Vehicle Imagery Using Convolutional Neural Networks. *Drones* **2018**, *2*, 39. [CrossRef]
32. Xu, W.; Deng, S.; Liang, D.; Cheng, X. A Crown Morphology-Based Approach to Individual Tree Detection in Subtropical Mixed Broadleaf Urban Forests Using UAV LiDAR Data. *Remote Sens.* **2021**, *13*, 1278. [CrossRef]
33. Gao, M.; Yang, F.; Wei, H.; Liu, X. Individual Maize Location and Height Estimation in Field from UAV-Borne LiDAR and RGB Images. *Remote Sens.* **2022**, *14*, 2292. [CrossRef]
34. Koc-San, D.; Selim, S.; Aslan, N.; San, B.T. Automatic Citrus Tree Extraction from UAV Images and Digital Surface Models Using Circular Hough Transform. *Comput. Electron. Agric.* **2018**, *150*, 289–301. [CrossRef]
35. Mu, Y.; Fujii, Y.; Takata, D.; Zheng, B.; Noshita, K.; Honda, K.; Ninomiya, S.; Guo, W. Characterization of Peach Tree Crown by Using High-Resolution Images from an Unmanned Aerial Vehicle. *Hortic. Res.* **2018**, *5*, 74. [CrossRef] [PubMed]
36. Marques, P.; Pádua, L.; Adão, T.; Hruška, J.; Peres, E.; Sousa, A.; Sousa, J.J. UAV-Based Automatic Detection and Monitoring of Chestnut Trees. *Remote Sens.* **2019**, *11*, 855. [CrossRef]
37. Jiang, H.; Chen, S.; Li, D.; Wang, C.; Yang, J. Papaya Tree Detection with UAV Images Using a GPU-Accelerated Scale-Space Filtering Method. *Remote Sens.* **2017**, *9*, 721. [CrossRef]
38. Larsen, M.; Eriksson, M.; Descombes, X.; Perrin, G.; Brandtberg, T.; Gougeon, F.A. Comparison of Six Individual Tree Crown Detection Algorithms Evaluated under Varying Forest Conditions. *Int. J. Remote Sens.* **2011**, *32*, 5827–5852. [CrossRef]
39. Peña, J.; Torres-Sánchez, J.; de Castro, A.; Kelly, M.; López-Granados, F. Weed Mapping in Early-Season Maize Fields Using Object-Based Analysis of Unmanned Aerial Vehicle (UAV) Images. *PLoS ONE* **2013**, *8*, e77151. [CrossRef]
40. Sa, I.; Chen, Z.; Popovic, M.; Khanna, R.; Liebisch, F.; Nieto, J.; Siegwart, R. WeedNet: Dense Semantic Weed Classification Using Multispectral Images and MAV for Smart Farming. *IEEE Robot. Autom. Lett.* **2018**, *3*, 588–595. [CrossRef]
41. de Castro, A.; Torres-Sánchez, J.; Peña, J.; Jiménez-Brenes, F.; Csillik, O.; López-Granados, F. An Automatic Random Forest-OBIA Algorithm for Early Weed Mapping between and within Crop Rows Using UAV Imagery. *Remote Sens.* **2018**, *10*, 285. [CrossRef]
42. Pérez-Ortiz, M.; Peña, J.M.; Gutiérrez, P.A.; Torres-Sánchez, J.; Hervás-Martínez, C.; López-Granados, F. A Semi-Supervised System for Weed Mapping in Sunflower Crops Using Unmanned Aerial Vehicles and a Crop Row Detection Method. *Appl. Soft Comput.* **2015**, *37*, 533–544. [CrossRef]
43. Dian Bah, M.; Hafiane, A.; Canals, R. Deep Learning with Unsupervised Data Labeling for Weed Detection in Line Crops in UAV Images. *Remote Sens.* **2018**, *10*, 1690. [CrossRef]

44. Nicodemus, F.E.; Richmond, J.; Hsia, J.; Ginsberg, I.; Limperis, T. *Geometrical Considerations and Nomenclature for Reflectance*; National Bureau of Standards: Gaithersburg, MD, USA, 1977. [CrossRef]
45. Evangelidis, G.D.; Psarakis, E.Z. Parametric Image Alignment Using Enhanced Correlation Coefficient Maximization. *IEEE Trans. Pattern Anal. Mach. Intell.* **2008**, *30*, 1858–1865. [CrossRef]
46. Li, J.; Meng, L.; Yang, B.; Tao, C.; Li, L.; Zhang, W. LabelRS: An Automated Toolbox to Make Deep Learning Samples from Remote Sensing Images. *Remote Sens.* **2021**, *13*, 2064. [CrossRef]
47. Novelty Detection with Local Outlier Factor (LOF). Available online: https://scikit-learn/stable/auto_examples/neighbors/plot_lof_novelty_detection.html (accessed on 8 February 2022).
48. Abozeid, A.; Alanazi, R.; Elhadad, A.; Taloba, A.I.; Abd El-Aziz, R.M. A Large-Scale Dataset and Deep Learning Model for Detecting and Counting Olive Trees in Satellite Imagery. *Comput. Intell. Neurosci.* **2022**, *2022*, e1549842. [CrossRef] [PubMed]
49. Safonova, A.; Guirado, E.; Maglinets, Y.; Alcaraz-Segura, D.; Tabik, S. Olive Tree Biovolume from UAV Multi-Resolution Image Segmentation with Mask R-CNN. *Sensors* **2021**, *21*, 1617. [CrossRef]
50. Waleed, M.; Um, T.W.; Khan, A.; Ahmad, Z. An Automated Method for Detection and Enumeration of Olive Trees Through Remote Sensing. *IEEE Access* **2020**, *8*, 108592–108601. [CrossRef]
51. Salamí, E.; Gallardo, A.; Skorobogatov, G.; Barrado, C. On-the-Fly Olive Tree Counting Using a UAS and Cloud Services. *Remote Sens.* **2019**, *11*, 316. [CrossRef]
52. Khan, A.; Khan, U.; Waleed, M.; Khan, A.; Kamal, T.; Marwat, S.N.K.; Maqsood, M.; Aadil, F. Remote Sensing: An Automated Methodology for Olive Tree Detection and Counting in Satellite Images. *IEEE Access* **2018**, *6*, 77816–77828. [CrossRef]
53. Karydas, C.; Gewehr, S.; Iatrou, M.; Iatrou, G.; Mourelatos, S. Olive Plantation Mapping on a Sub-Tree Scale with Object-Based Image Analysis of Multispectral UAV Data; Operational Potential in Tree Stress Monitoring. *J. Imaging* **2017**, *3*, 57. [CrossRef]
54. Chemin, Y.H.; Beck, P.S.A. A Method to Count Olive Trees in Heterogenous Plantations from Aerial Photographs. *Geoinformatics* **2017**, *Preprint*. [CrossRef]
55. Peters, J.; Van Coillie, F.; Westra, T.; De Wulf, R. Synergy of Very High Resolution Optical and Radar Data for Object-Based Olive Grove Mapping. *Int. J. Geogr. Inf. Sci.* **2011**, *25*, 971–989. [CrossRef]
56. Moreno-Garcia, J.; Linares, L.J.; Rodriguez-Benitez, L.; Solana-Cipres, C. Olive Trees Detection in Very High Resolution Images. In Proceedings of the Information Processing and Management of Uncertainty in Knowledge-Based Systems; Hüllermeier, E., Kruse, R., Hoffmann, F., Eds.; Springer: Berlin/Heidelberg, Germany, 2010; pp. 21–29. [CrossRef]
57. González, J.; Galindo, C.; Arevalo, V.; Ambrosio, G. Applying Image Analysis and Probabilistic Techniques for Counting Olive Trees in High-Resolution Satellite Images. In Proceedings of the Advanced Concepts for Intelligent Vision Systems, Delft, The Netherlands, 28–31 August 2007; Blanc-Talon, J., Philips, W., Popescu, D., Scheunders, P., Eds.; Springer: Berlin/Heidelberg, Germany, 2007; pp. 920–931. [CrossRef]

sensors

Article

UBER: UAV-Based Energy-Efficient Reconfigurable Routing Scheme for Smart Wireless Livestock Sensor Network

Mohammed A. Alanezi [1], Abdulazeez F. Salami [2], Yusuf A. Sha'aban [3], Houssem R. E. H. Bouchekara [3,*], Mohammad S. Shahriar [3], Mohammed Khodja [4,5] and Mostafa K. Smail [6]

1 Department of Computer Science and Engineering Technology, University of Hafr Al Batin, Hafr Al Batin 31991, Saudi Arabia
2 Department of Computer Engineering, University of Ilorin, Ilorin 240103, Nigeria
3 Department of Electrical Engineering, University of Hafr Al Batin, Hafr Al Batin 31991, Saudi Arabia
4 Department of Electronics, College of Engineering, Mustaqbal University, Buraidah 51452, Saudi Arabia
5 Department of Electrical Engineering, Faculty of Technology, M'sila University, M'sila 28000, Algeria
6 Institut Polytechnique des Sciences Avancées, 63 Boulevard de Brandebourg, 94200 Ivry-sur-Seine, France
* Correspondence: bouchekara.houssem@gmail.com

Abstract: This paper addresses coverage loss and rapid energy depletion issues for wireless livestock sensor networks by proposing a UAV-based energy-efficient reconfigurable routing (UBER) scheme for smart wireless livestock sensor networking applications. This routing scheme relies on a dynamic residual energy thresholding strategy, robust cluster-to-UAV link formation, and UAV-assisted network coverage and recovery mechanism. The performance of UBER was evaluated using low, normal and high UAV altitude scenarios. Performance metrics employed for this analysis are network stability (NST), load balancing ratio (LBR), and topology fluctuation effect ratio (TFER). Obtained results demonstrated that operating with a UAV altitude of 230 m yields gains of 31.58%, 61.67%, and 75.57% for NST, LBR, and TFER, respectively. A comparative performance evaluation of UBER was carried out with respect to hybrid heterogeneous routing (HYBRID) and mobile sink using directional virtual coordinate routing (MS-DVCR). The performance indicators employed for this comparative analysis are energy consumption (ENC), network coverage (COV), received packets (RPK), SN failures detected (SNFD), route failures detected (RFD), routing overhead (ROH), and end-to-end delay (ETE). With regard to the best-obtained results, UBER recorded performance gains of 46.48%, 47.33%, 15.68%, 19.78%, 46.44%, 29.38%, and 58.56% over HYBRID and MS-DVCR in terms of ENC, COV, RPK, SNFD, RFD, ROH, and ETE, respectively. The results obtained demonstrated that the UBER scheme is highly efficient with competitive performance against the benchmarked CBR schemes.

Keywords: cattle; herd cluster-based routing; performance analysis; unmanned aerial vehicle; wireless livestock sensor network

Citation: Alanezi, M.A.; Salami, A.F.; Sha'aban, Y.A.; Bouchekara, H.R.E.H.; Shahriar, M.S.; Khodja, M.; Smail, M.K. UBER: UAV-Based Energy-Efficient Reconfigurable Routing Scheme for Smart Wireless Livestock Sensor Network. *Sensors* **2022**, *22*, 6158. https://doi.org/10.3390/s22166158

Academic Editors: Yong He and Somsubhra Chakraborty

Received: 30 April 2022
Accepted: 14 August 2022
Published: 17 August 2022

1. Introduction

Livestock farming (LF) is one of the global economy's backbone industries, but the growing world population has placed tremendous pressure on the demand for food [1–3]. This implies that livestock industries' production capacity and efficiency must scale up to meet this increasing food demand [3–5]. Industrialists have explored arrays of wireless sensor network (WSN) technologies to improve the quality, quantity and efficiency of LF [1,6–9]. Recently, the performance of Internet of things (IoT) devices equipped with embedded sensor nodes (SNs) have been significantly enhanced, which has led to effective control and management of distributed energy supply systems (DESS) [3,10,11]. The DESS acts as an energy source for the WSN while monitoring the livestock's movement activities, and health status [2,3]. It must be mentioned that wearable SNs used for smart livestock monitoring applications are portable battery-constrained devices with information gathering and processing capabilities [3,10,12]. However, with harsh weather conditions the dynamic and random movement of livestock in the field makes livestock monitoring applications based solely on wearable/strap-on SNs face challenges. These challenges include the high cost of frequent SN replacement, erratic loss of coverage, short network lifetime, rapid energy drainage, and the frequent need for human involvement in the system's operation.

These challenges have prompted researchers to explore unmanned aerial vehicle (UAV)-aided solutions to address routing issues associated with monitoring livestock movement and activities [11,13–19]. UAV-aided solutions have created the possibility of designing cost-effective livestock surveillance solutions [2,20–24]. In such surveillance solutions, UAVs serve as: 1.) sinks for collecting time-sensitive data from strap-on SNs and 2.) mobile aerial stations for directly monitoring the livestock [25,26]. Therefore, the massive production of LF products requires innovative integration of WSN and UAV technologies to address the increasing global food demand challenge. Several integrated UAV-WSN solutions have been deployed to increase the yield of LF products, where UAVs act as mobile sinks (MS) collecting livestock data from wearable SNs, and various artificial intelligence (AI) tools are employed for interactive data analytics [3,8,25]. These integrated solutions help increase sales revenue, sustain large-scale production, and ensure the distribution of quality LF products to the end consumers [3,11,26]. Most importantly, extensive unsupervised livestock monitoring applications address the difficulty, monotony, and time-consuming challenges of manual monitoring of large livestock farms. However, challenges of network lifetime issues and frequent disruptive topology changes still exist and demand further research.

Researchers have proffered various generic [27–34] and specific [3,25,35–43] routing algorithms to address these challenges. Among the proposed approaches, cluster-based routing (CBR) techniques have shown more computational simplicity, versatility, robustness and effectiveness in reducing energy consumption and preserving network connectivity/coverage [3,44–46].

Effective management of unattended and large LF fields requires energy-efficient routing techniques and scalable network architectures. This is usually achieved by logically organizing wearable SNs attached to livestock herds into groups called herd clusters (HCs) in this context to achieve energy efficiency and network scalability. From each HC, some set of SNs are elected as HC leads (HCLs) based on their residual energy, relative access and proximity to the MS. HC members (HCMs) transmit livestock data to their HCLs by using single-hop or multi-hop transmission mode. Subsequently, HCLs forward aggregated data to the aerial MS for onward transmission to the base station (BS) [41,43,47,48].

Based on this herd cluster-based routing strategy, this paper presents a UAV-based energy-efficient reconfigurable (UBER) routing algorithm for smart wireless livestock sensor networking applications. UBER relies on a dynamic residual energy thresholding mathematical model, robust cluster-to-UAV link formation strategy, and UAV-assisted network coverage and recovery mechanisms. Simulation experiments were carried out with OMNET++ and MATLAB, while comparative performance analysis was performed with respect to hybrid heterogeneous routing (HYBRID) and mobile sink using directional virtual coordinate routing (MS-DVCR) techniques. The results obtained demonstrated that the proposed UBER scheme is highly energy-efficient with competitive network performance when benchmarked against existing CBR schemes.

The structure of the remainder of this paper is organized in the following manner: Section 2 covers the related research works pertinent to CBR approaches for wireless livestock sensor networking applications. Section 3 provides a technical description of the proposed UBER protocol, while Section 4 presents the obtained simulation results and supporting discussions. Section 5 concludes this paper.

2. Related Works

Traditional CBR techniques are classified as generic based on their mode of network operation and applicability to a wide range of WSN applications. However, generic

CBR techniques often suffer from single HCL failure issues, high energy consumption due to single-hop long-range data transmission, high time complexities, and cluster size challenges [27–34,44,45,49–51].

The grey wolf optimizer (GWO)-based technique was proffered as a nature inspired CBR algorithm for ensuring seamless and robust data transmission in livestock monitoring applications [3]. This algorithm relies on the behavioral pattern of gray wolves and key network parameters, such as transmission range, cluster size, residual energy and data route to minimize energy consumption [3]. The limitation of this algorithm is that the HC configuration process is not flexible and adaptable to varying network requirements.

The Markov decision process (MDP)-based technique was proposed as a path planning CBR technique that utilizes single UAV and static SNs for HC monitoring and data gathering in a large LF field [25]. The CBR technique relies on Markov decision process together with reward and value of information maximization analytical models. The key benefit of this technique is the significant reduction in message delay [25]. The major drawbacks of this technique are the algorithmic computational cost and high energy tax.

The dynamic decentralized/centralized free conflict unmanned aerial vehicle (DDCFC-UAV) technique was proposed as a security-oriented CBR scheme that relies on SNs mounted on UAVs to monitor a defined LF field [35]. The predefined LF field is logically categorized into virtual HCs, and UAVs are assigned to monitor their assigned HC zones HCLs are elected by the UAV at each HC zone using dynamic network and energy requirements criteria stipulated by the CBR scheme [35]. The drawback of this CBR scheme is the challenge of maintaining connectivity for the multi-UAV communication architecture.

The hybrid heterogeneous routing (HYBRID) technique was proffered as a network lifetime improvement CBR technique by employing SNs deployed in harsh LF environments where energy efficiency is achieved by dividing the LF field into HCs and placing the location of the BS at the edge of the LF field [36]. The criteria for HC formation are the residual energy of SNs, which gives SNs having higher residual energy more chance of being elected as HCLs. Inter-HC distance is also considered as a factor for multi-hop data transmission of livestock data in the network [36]. The drawback of this CBR technique is the algorithmic complexity associated with switching between varying energy levels and different modes of transmission.

The mobile sink using directional virtual coordinate routing (MS-DVCR) technique was offered as an energy minimization geographic routing scheme that relies on a directional virtual coordinate strategy in conjunction with the aerial MS operation [37]. The central objective of this scheme is to reduce the frequency of network updates transmitted by the MS to a BS. The main advantage of this scheme is that it provides an alternative solution for dealing with MS localization without carrying out physical distance measurements [37]. The major limitation of this scheme is the high overhead tied to maintaining and exchanging information related to the virtual coordinates within the network.

The lightweight dynamic clustering algorithm (LDCA) was proposed as a real-time low-complexity dynamic CBR scheme for livestock monitoring applications having limited processing resources [38]. This CBR scheme utilizes single-hop transmission, variable cluster size, and multiple parameters (SN-to-MS distance, signal strength, residual energy noise level, environmental factors) for HC configuration to maximize network coverage [38]. The drawbacks of this scheme are SN buffering and regional separation issues associated with rapid MS mobility patterns.

3. Proposed UBER Scheme

This section discusses the integrated UAV-WSN heterogeneous network model, fundamental assumptions, cluster configuration, and data gathering phases.

3.1. Integrated UAV-WSN Network Model

Figure 1 shows the proposed UAV-WSN integrated solution for livestock monitoring applications. In this network model, strap-on SNs acting as HCMs monitor vital parameters (location, perspiration, temperature, heart rate) and transmit sensed parameters to their respective HCLs. The HCLs aggregate all sensed parameters received from their respective HCMs and relay the compressed data to the neighboring UAVs acting as MS. UAVs fly over the LF network perimeter at an altitude within the coverage of the SNs. These MS entities forward aggregated data to the livestock monitoring server (LMS) for processing through a network gateway (NGW)-BS link enabled with ZigBee (ZGB) interface. The LMS analyzes the received livestock data and triggers the required LF controller (LFC) devices, such as temperature regulators, alarms, lighting controllers, and other LFC devices. Authorized mobile end users (MEUs) can check and work on livestock data from the LMS via Internet (INET) connection.

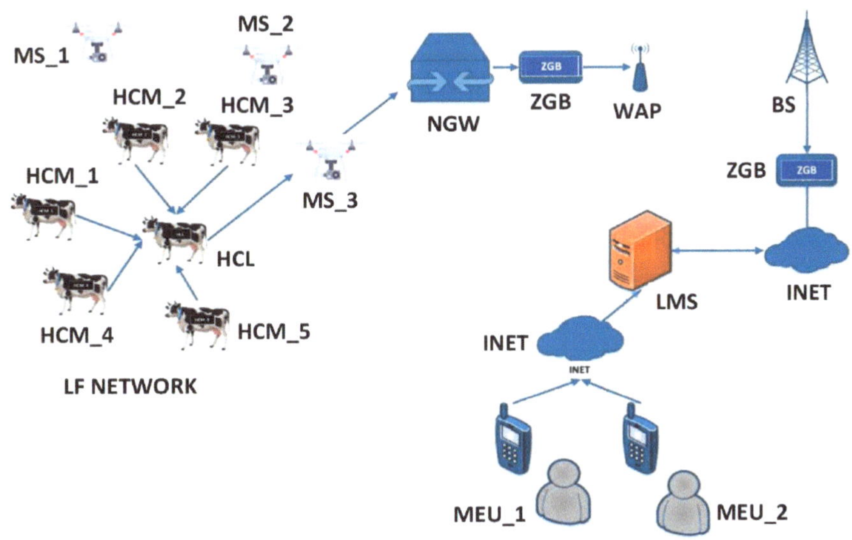

Figure 1. UBER Network Model.

3.2. Assumptions

The fundamental assumptions for the proposed UBER CBR scheme are:

- All SNs are wearable, portable, and identical, with similar energy resources, processing capacities and transceiver characteristics. This assumption describes the physical and electronic properties of the sensor nodes suitable for livestock (cattle) monitoring. It is necessary for the sensors to be wearable and portable to prevent discomfort to the livestock, facilitate sensing through direct body contact, and make sensor replacement easier. It is necessary for the sensors to be identical to avoid tagging/stamping, synchronization and mismatch errors during data aggregation and processing.
- UAVs act as MS with more onboard radios and higher energy resources, processing capacities and transceiver range. This assumption describes the functional and electronic properties of the UAVs suitable for the integrated model presented in Figure 1. The fewer UAVs deployed should have more onboard radios to accommodate the influx of periodic traffic from different clusters.
- UAV's spatial movement is 3-D with a variable velocity profile. This assumption describes the spatial motion capability (in x, y, z direction following straight-line left-to-right up-and-down scanning pattern) of the employed UAV and its limitations (no axial rotation, no angular twists/bends). Velocity profiles for the UAV are stationary, scanning velocity (20 m/s), and data gathering velocity (5 m/s) as stated in Table 1.
- UAVs have embedded intelligence for smart decision-making, and they can be controlled from the LMS. This assumption describes the cognitive capability of the UAV and its limitations (not fully autonomous as its activities can be controlled from the

LMS). The reason for utilizing this semi-autonomous arrangement is to prevent out-of-perimeter straying, which can lead to lost UAV, theft, unrestrained energy loss or device damage.

Table 1. Simulation Parameters.

Symbol	Description	Value
SN-DEP	SNs Deployed	250
LF-NS	LF Network Size	2000 m × 2000 m
PKS	Packet Size	500 bytes
E_{TAX} TL	Energy Tax Threshold Levels	8
E_{tot}	Total Energy of each SN (before depletion)	2 J
E_{idle}	Idle Energy	0.2 μJ
E_{agg}	Aggregation Energy	5 pJ/bit
E_{EC}	Electronic Circuitry Energy	5 nJ/bit
CTR_{max}	Maximum Transmission Range	250 m
A	Path Loss Exponent	2.5
SN-RS	SN Receiver Sensitivity	−95 dBm
MS-ALT	MS Maximum Altitude	230 m
MS-V	MS Velocity	20 m/s
MS-SR	MS Signaling Rate	2 s
MS-TD	MS Tour Duration	960 s
AVG-STAT	Simulation Runs for Statistical Averaging	50

3.3. *Energy Consumption Model*

The traditional energy consumption model is centrally dependent on the transmission distance (l) and the number of packet bits transferred (b), as postulated in [27,28] as the first order radio energy model. The implication of this is that the transceiver energy tax (E_{TAX}) increases exponentially with increasing transmission distance, as expressed in Equation (1).

$$E_{TAX} = b \cdot \left[E_{EC} + \left(E_{PA} \cdot l^2\right)\right] \tag{1}$$

where, E_{EC} and E_{PA} are the circuity energy dissipation and amplifier-dependent energy loss parameter, respectively. Equation (2) defines the transmission distance as [52]:

$$l = \left[\frac{\lambda}{16\pi^2}\right]^{\frac{1}{\alpha}} \tag{2}$$

where, α is the path loss exponent and λ is the wavelength. It must be mentioned that the free space path loss model has limitations for SN-to-UAV communication, especially in the presence of obstacles and weather conditions, which can affect the path loss and increase the path loss exponent. This is one of the technical reasons why the UAV's normal altitude is operated at 230 m, which falls within the SN's transmission range of 250 m in order to preserve network coverage and connectivity. Furthermore, multipath communication has been incorporated into an upcoming sequel paper in order to serve as an improvement. UBER's proposed energy consumption model curbs this exponential increase in E_{TAX} by forming HCs based on the HCM-to-HCL and HCL-to-MS distance parameter estimates which allows SNs nearest to the MS to be elected as HCLs. The effectiveness of this energy conservation strategy is enhanced with the use of dynamic residual energy thresholding (E^{th}_{TAX}) technique as:

$$E^{th}_{TAX} = \begin{cases} b \cdot [E_{EC} + (E_S \cdot l)], \; if \; l \leq l_{th} \\ b \cdot [E_{EC} + (E_L \cdot l)], \; if \; l > l_{th} \end{cases} \tag{3}$$

where, E_S, E_L, and l_{th} are short-range energy transmission, long-range energy transmission and distance threshold. E_S and E_L are approximated based on practical design considerations for IEEE 802.15.4 RF transceivers [53,54]. The distance threshold is defined as:

$$l_{th} = \frac{2 \cdot f_c \cdot h_{tx} \cdot h_{rx}}{k \cdot v} \quad (4)$$

where, k, f_c, v, h_{tx}, and h_{rx} are threshold constant, carrier frequency, signal velocity, transmitting SN antenna height and receiving SN antenna height, respectively. It must be mentioned that Equation (3) has been simplified to a linear form (with the use of practical field approximations [54]) compared to the quadratic form in Equation (1). This adaptive/programmable output power level specification helps to prevent exponential E_{TAX} increase, ensures robust HCL-to-MS data transmission, and reduces distance-dependent interference experienced by SNs in the LF network perimeter.

3.4. Cluster Configuration Phase

The proposed UBER CBR algorithm commences with the network discovery phase. The clustering costs (CC_x) based on Z_{max} (peak value of the received signal strength levels) are shared among neighboring SNs through HELLO packets. Z_{max} is the peak of the received signal power (RSSI) values, which is a reliable indicator of the signal strength. RSSI interpretation is based on proximity as relatively higher RSSI values will be recorded for closer nodes. CC_x is in dBm and it is formulated as:

$$CC_x = max[Z_{x,t}] \quad (5)$$

where, x and $Z_{x,t}$ denote each SN and received signal strength level for SN_x obtained from UAV signals at duration t, respectively. After the network discovery phase, distributed repetitive procedures are used to elect HCLs from candidate SNs. SNs with an established connection with the MS opt to become HCLs by setting their electability probability (HCL_{PR}) as:

$$HCL_{PR} = \begin{cases} max\left[LIM_{UP} \cdot \frac{E_{rsd}}{E_{tot}}, LIM_{LOW}\right], & if\ CC_x > 0 \\ 0, & if\ CC_x = 0 \end{cases} \quad (6)$$

where, LIM_{UP}, LIM_{LOW}, E_{rsd}, and E_{tot} are upper limit probabilistic values for HCL contentions, lower limit probabilistic values for HCL contentions, residual energy, and total energy, respectively. LIM_{UP} is the upper limit set in the network to limit number of HCL competitions/announcements while LIM_{LOW} is the lower limit. Selection of these two parameters is adjusted to allow rapid convergence of HCL_{PR}. E_{rsd} is obtained formally as:

$$E_{rsd} = E_{tot} - E_{TAX}^{th} \quad (7)$$

E_{rsd}/E_{tot} ratio incorporates dynamic residual energy thresholding into the HC configuration process. The formulated electability probability implies that SNs with higher E_{rsd} and established proximal connection to the MS will have a higher chance of being elected as trial HCLs after each round (R), $1 < R < R_{max}$. R means a round of network operation. INFO packets about the trial HCLs are exchanged with neighboring SNs to maintain a set of neighboring trial HCLs. An ordinary node SN_x selects its HCL (MY_HCL) based on the trial HCL with the least CC_x in its neighboring set. The newly elected HCL broadcast POLLING packets, HCL_polling(SN_ID, trial_HCL, CC_x) to its neighboring SNs. After successful completion of the network operation round, the status of the trial HCL is configured to final HCL and POLLING packets, HCL_polling(SN_ID, final_HCL, CC_x), are broadcasted to neighboring SNs. SNs receiving the POLLING packets respond with JOIN packets to update their cluster membership information. At the end of the set-up phase, SNs are either tagged with HCM or final HCL status, while orphaned SNs establish a connection with the closest HCM to become an affiliate of the HC. This cluster arrangement ensures that SN-to-MS data transmission requires a maximum of three hops in the worst-case scenario of orphaned SNs. The cluster configuration algorithm is shown in Algorithm 1.

Algortihm 1 Cluster Configuration Algorithm for UBER.

```
for each SN_x received MS_CONNECT signal
    if (Z_{x,t} ← MS && Z_{x,t} ≠ NULL)
        status.connect(MS) ← TRUE
        compute CC_x ← find_peak(Z_{x,t})
    else
        status.connect(MS) ← FALSE
        CC_x ← −INF
    end if
    broadcast CCx to SNx.ADJ within CTR
    status.final_HCL ← FALSE
end for
    while (R ≠ R_max && HCL_PR ≠ 1)
        if (status.connect(MS) ← TRUE && status.trial_HCL ← TRUE)
            HHCL_PR ← rand(0,1)CL_PR ← rand(0,1)
            elect.MY_HCL ← trial_HCL.min(CC_x)
            if (MY_HCL = SN_ID && HCL_PR ≠ 1)
                broadcast HCL_polling(SN_ID, trial_HCL, CC_x)
                status.final_HCL ← FALSE
            end if
        else
            broadcast HCL_polling(SN_ID, final_HCL, CC_x)
        end if
        HCL_PR at t−1 ← HCL_PR
        HCL_PR ← min(2xHCL_PR,1)
    end while
  status.final_HCL ← TRUE
  update(trial_HCL) ← POLLING packet
  elect(trial_HCL) ← IDLE
  elect(final_HCL) ← ACTIVE
  broadcast POLLING packet within CTR
        for each ordinary SN_x received POLLING packet
                compute Euclidean distance cost
                multicast JOIN packet
        end for
        if SN_x did not receive POLLING packet
      compute euclidean distance cost to SN_x.ADJ within CTR
      construct EDGE using least distance cost
        end if
        for each HCL
                register HCM list
                construct EDGE with HCM set
        end for
```

3.5. Data Gathering Phase

In the steady-state phase, the radio of each HCM is triggered to WAKE state for monitoring and transmitting livestock data to their respective HCLs. It must be mentioned that SNs are in WAKE state only for active network operation time and are put into SLEEP state otherwise. The HCLs are assigned the network task of data aggregation and forwarding to the nearest MS. Multi-hop communication chain is relied upon for end-to-end (HCM-to-HCL, HCL-to-MS, and MS-to-LMS) data transmission. After the LMS receives all the desired livestock data, END_ROUND packet is broadcasted to the network by the HCLs based on the interrupting signal received from the MS.

4. Results and Discussions

This section discusses UBER's performance metrics, simulation parameters, algorithmic complexity, performance analysis to different MS altitudes, and comparative performance evaluation of UBER against HYBRID and MS-DVCR.

4.1. Performance Metrics

The measures employed for performance analysis are network coverage (COV), network stability (NST), energy consumption (ENC), received packets (RPK), topology fluctuation effect ratio (TFER), SN failures detected (SNFD), route failures detected (RFD), routing overhead (ROH), load balancing ratio (LBR), and end-to-end delay (ETE).

4.2. Simulation Parameters

Simulation experiments for this research work were conducted with OMNET++ and MATLAB. Selected key parameters employed for this simulation are provided in Table 1. From Table 1, energy tax threshold levels are obtained from Equation (3) and their significance is explained as step-wise programmable power levels adopted to prevent exponential E_{TAX} increase. Aggregation energy is the energy consumed to perform data aggregation by the HCL in order to reduce redundancy before onward transmission to the MS. Dual frequency (433 MHz for SN localization and handshaking and 2.4 GHz for data transmission) are used. The PHY/MAC layer characteristics are based on IEEE 802.15.4 protocol specifications while the NETWORK layer characteristics is based on ZigBee protocol specifications (as indicated in Figure 1). Command and control signaling is used to enable the sending of a command signal to the UAV from the LMS and receiving data traffic from the UAV payload. Due to the focus of this paper on a smart wireless livestock sensor network, the SNs deployment is by attachment to the neck region of the livestock (as shown in Figure 1) and the UAVs are deployed to follow the livestock herd within the LF network perimeter.

4.3. Algorithmic Complexity

Through simulation experiments, it was observed that when LIM_{LOW} is adjusted to 0.005, UBER converges at around 11 iterative rounds. Furthermore, it was also observed that if LIM_{UP} is adjusted to 0.05 (i.e., HCL of 5%), UBER converges at around seven iterative rounds, together with the simultaneous observation that HCL_{PR} converges to one after seven iterative rounds. Therefore, this shows that UBER successfully converges after a constant value of iterative rounds, and consequently, it has algorithmic complexity of Θ (1).

4.4. Performance Analysis

4.4.1. Analysis of UBER Performance

To examine the variations in MS altitude with respect to network stability, load balancing/distribution and topology fluctuation effect on connectivity on UBER performance, scenarios of low altitude (120 m), normal altitude (230 m), and high altitude (340 m) were experimented within the simulation. This choice of normal altitude (230 m) is duly guided by practical network design considerations. This normal altitude falls within the SN's transmission range of 250 m and lies above the close range of the SNs.

4.4.2. Effect of MS Altitude on Network Stability (NST)

NST is defined as the ratio of the number of stable HCL-to-MS connections to the total number of connections after R network operation rounds. From Figure 2, the red line represents UBER's network stability performance trend under low MS altitude of 120 m, the green line represents UBER's network stability performance trend under normal MS altitude of 230 m, and the blue line represents UBER's network stability performance trend under high MS altitude of 340 m. With reference to Figure 2, a higher NST value (NST $\geq$ 0.5) is desired as this means the MS can maintain a connection with the HCL for a desirably long period (after R network operation rounds) before losing the connection. The network recorded an average NST of 0.3004, 0.5391 and 0.3829 for the high-altitude (indicated as blue line), normal-altitude (indicated as green line) and low-altitude (indicated as red line) scenarios throughout network operation. This means that by operating the MS at normal altitude (indicated as green line), the network gains NST of 31.58% and 12.79% over a similar network configuration operating at high altitude (indicated as blue line) and low altitude (indicated as red line), respectively. The technical justification for this is that operating at high altitude (indicated with blue line) keeps the MS almost out of communication range; while operating at low altitude (indicated with red line) is not feasible due to the LF network terrain, interference, collision and congestion issues Results in Figure 2 emphasize the importance of adopting a suitable MS altitude on NST performance.

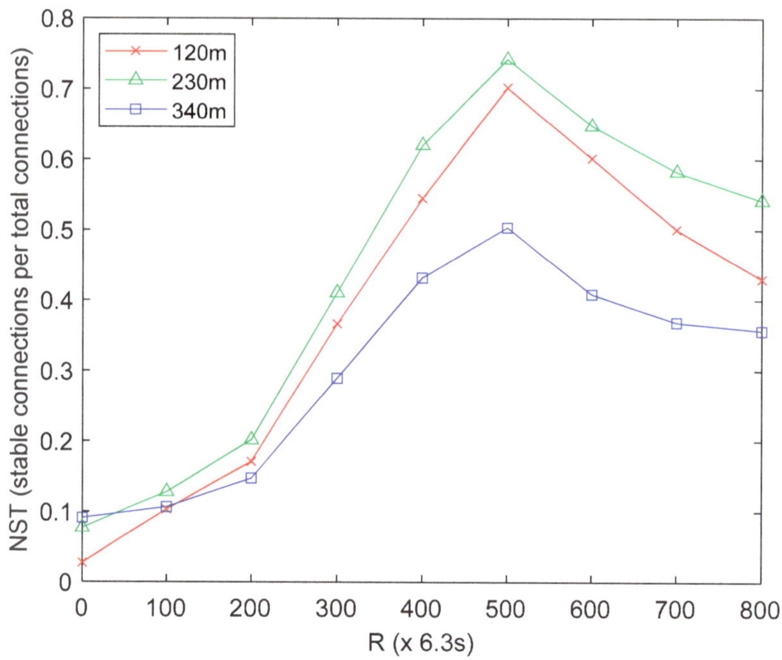

Figure 2. Effect of MS Altitude on Network Stability.

4.4.3. Effect of MS Altitude on Load Balancing Ratio (LBR)

The LBR is defined as the comparative ratio of net load successfully accepted by the MS to the total load offered by the HCLs, averaged for *R* network operation rounds. From Figure 3, the red line represents UBER's load balancing ratio performance trend under low MS altitude of 120 m, the green line represents UBER's load balancing ratio performance trend under normal MS altitude of 230 m, and the blue line represents UBER's load balancing ratio performance trend under high MS altitude of 340 m. With respect to Figure 3 a higher LBR value (LBR $\geq$ 0.5) is desired as this means data traffic coming from the HCLs is well distributed (or balanced) among the MS to avoid underloading and overloading scenarios. The network recorded an average LBR of 0.2478, 0.6466 and 0.3787 for the high altitude (indicated as blue line), normal-altitude (indicated as green line) and low-altitude (indicated as red line) scenarios over the period of network operation. This means that by operating the MS at normal altitude (indicated as green line), the network gains LBR of 61.67% and 41.42% over a similar network configuration operating at high altitude (indicated as blue line) and low altitude (indicated as red line), respectively. The reason for this is that operating at high altitude (indicated with blue line) results in underloading (as a result of long processing delays) due to weak strength of MS coverage; while operating at low altitude (indicated with red line) results in overloading (as a result of packet flooding frequent packet drops, reconnection and retransmissions) due to simultaneous detection of MS by multiple HCLs. Figure 3 results underscore the significance of utilizing suitable MS altitude on LBR performance.

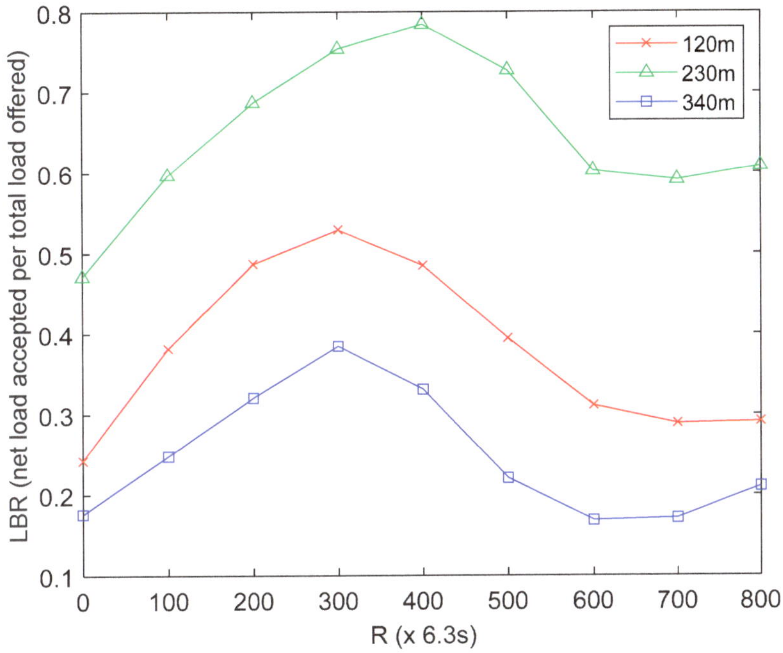

Figure 3. Effect of MS Altitude on Load Balancing Ratio.

4.4.4. Effect of MS Altitude on Topology Fluctuation Effect Ratio (TFER)

The TFER is defined as the frequency by which the HCLs detect and switch MS connections as a ratio of the total number of connections after *R* network operation rounds. From Figure 4, the red line represents UBER's topology fluctuation effect ratio performance trend under low MS altitude of 120 m, the green line represents UBER's topology fluctuation effect ratio performance trend under normal MS altitude of 230 m, and the blue line represents UBER's topology fluctuation effect ratio performance trend under high MS altitude of 340 m. With regard to Figure 4, a moderate TFER value ($0.2 \leq \text{TFER} \leq 0.4$) is desired as this measures the frequency of switching MS-to-HCL connectivity as a result of changing HC formation, location and re-assignment of HCLs. A moderate TFER value is desired to avoid under-sensitivity and oversensitivity scenarios. The network recorded an average TFER of 0.1291, 0.2925 and 0.5136 for the high-altitude (indicated as blue line), normal-altitude (indicated as green line) and low-altitude (indicated as red line) scenarios over the period of network operation. This means that by operating the MS at normal altitude (indicated as green line), the network gains TFER of 55.86% and 75.57% over a similar network configuration operating at high altitude (indicated as blue line) and low altitude (indicated as red line), respectively. The technical reason for this is that operating at high altitude (indicated with blue line) results in under-sensitivity due to near MS out-of-reach issues; while operating at low altitude (indicated with red line) results in oversensitivity to HC variations and re-configuration due to high MS proximity. Figure 4 results underline the influence of employing suitable MS altitude on TFER performance.

Table 2 summarizes UBER performance evaluation results for different MS altitudes.

4.5. Comparitive Performance Evaluation of UBER

To conduct a comparative performance evaluation of UBER, HYBRID and MS-DVCR are selected as baseline protocols for benchmarking in this research.

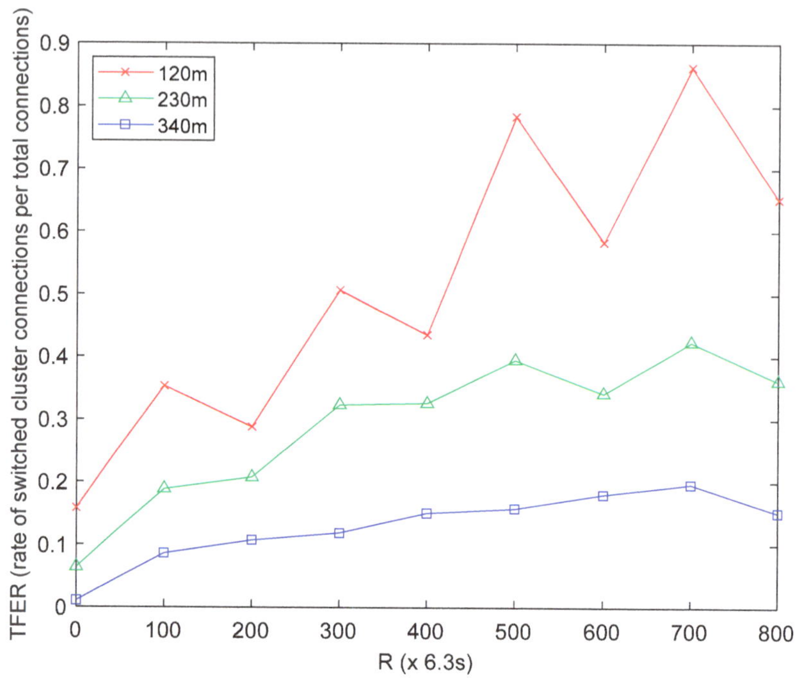

Figure 4. Effect of MS Altitude on Topology Fluctuation Effect Ratio.

Table 2. Summary of UBER Performance with MS Altitude Variations.

	% Gain of 230 M Over	
Metric	**340 M**	**120 M**
NST	31.58%	12.79%
LBR	61.67%	41.42%
TFER	55.86%	75.57%

4.5.1. Evaluation of Energy Consumption (ENC) Performance

ENC is defined as the aggregate energy tax (E_{TAX}) by the SNs after R network operation rounds. From Figure 5, the red line represents MS-DVCR's energy consumption performance trend, the green line represents HYBRID's energy consumption performance trend, and the blue line represents UBER's energy consumption performance trend under normal MS altitude of 230 m. Figure 5 shows the comparative plot of ENC for UBER (indicated as blue line) against HYBRID (indicated as green line) and MS-DVCR (indicated as red line). UBER (indicated with blue line) recorded lesser E_{TAX} with 25.59% and 46.48% improvements over HYBRID (indicated with green line) and MS-DVCR (indicated with red line), respectively. The technical justification for UBER's performance improvement is due to the energy conservation benefits from the dynamic residual energy thresholding technique, which ensures robust HCL-to-MS data transmission, and reduced energy consumption for HC setup and topology maintenance.

4.5.2. Evaluation of Network Coverage (COV) Performance

COV is defined as the percentage of successfully covered SNs to the total node density for the LF network field. From Figure 6, the red line represents HYBRID's network coverage performance trend, the green line represents MS-DVCR's network coverage performance trend, and the blue line represents UBER's network coverage performance trend under normal MS altitude of 230 m. Figure 6 provides the comparative plot of COV for UBER (indicated as blue line) with respect to HYBRID (indicated as red line) and MS-DVCR (indicated as green line). UBER (indicated with blue line) exhibited better network coverage by yielding improvements of 28.44% and 47.33% over HYBRID (indicated with red line) and MS-DVCR (indicated with green line), respectively. UBER's performance enhancements are due to the effective HCL-to-MS cluster-based connectivity chain and suitable selection of MS altitude for extending network coverage during network operation.

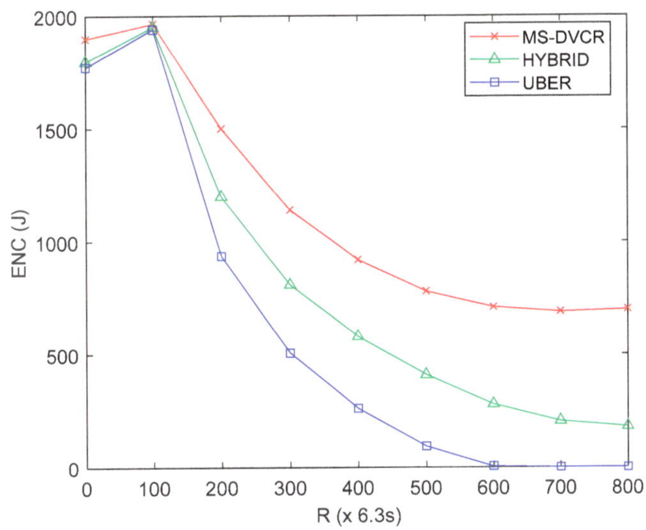

Figure 5. Energy Consumption Performance.

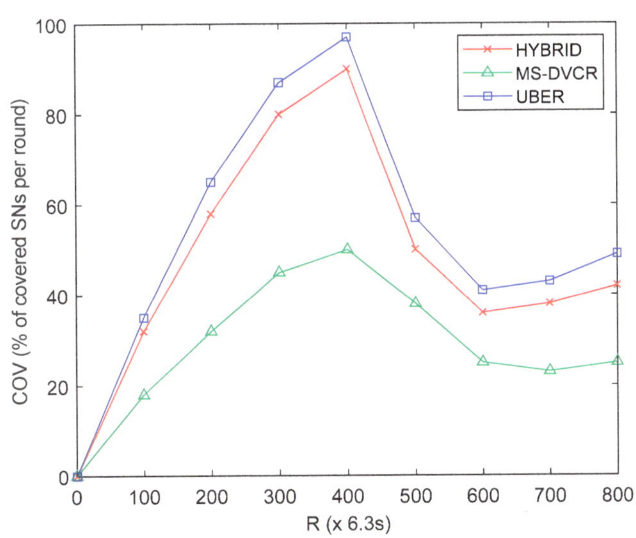

Figure 6. Network Coverage Performance.

4.5.3. Evaluation of Received Packets (RPK) Performance

RPK is defined as total received packets recorded via the MS-to-LMS transmission link after R network operation rounds. From Figure 7a, the red line represents MS-DVCR's received packets performance trend, the green line represents HYBRID's received packets performance trend, and the blue line represents UBER's received packets performance trend under normal MS altitude of 230 m. Figure 7a depicts the comparative plot of RPK for UBER (indicated as blue line) compared to HYBRID (indicated as green line) and MS-DVCR (indicated as red line). UBER (indicated with blue line) displayed higher RPK by giving improvements of 15.68% and 3.637% over HYBRID (indicated with green line) and MS-DVCR (indicated with red line), respectively. The technical justification for UBER's performance improvements is as a result of adopting an MS-assisted data-gathering strategy cluster resolution and assimilation for orphaned SNs, and minimal-hop SN-to-MS data transmission. The close RPK performance between UBER and MS-DVCR is simply a slight performance tradeoff and it is not as a result of high statistical error ranges or statistical dependency issues. The standard deviation for RPK performance is shown in Figure 7b to buttress this performance evaluation. The standard deviation plot demonstrates that UBER has a significantly lesser standard deviation values for RPK ($\leq$0.4) in all instances of network operation rounds compared to MS-DVCR and HYBRID. Irrespective of the close performance between UBER and MS-DVCR, the lesser standard deviation values recorded for UBER is a relatively strong indicator of stable packet reception and statistical significance of the obtained RPK results.

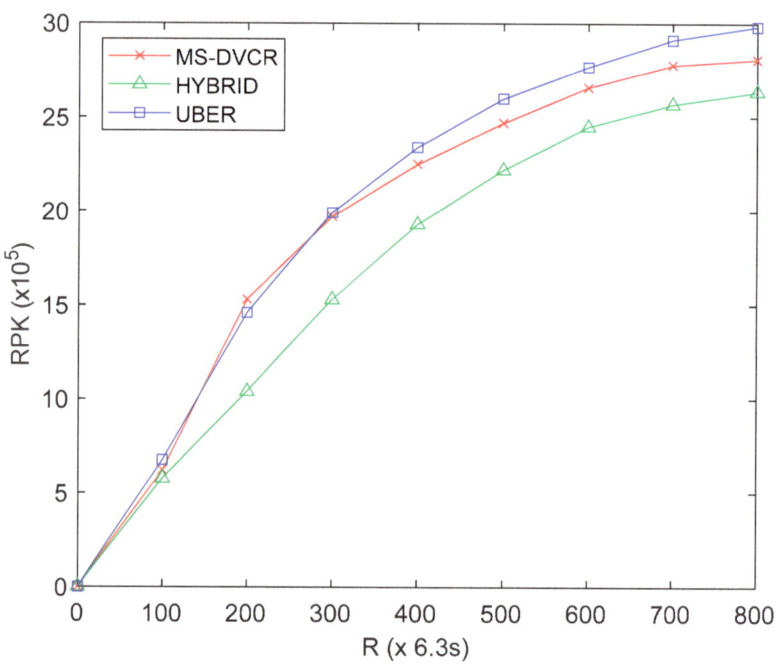

(a)

Figure 7. *Cont.*

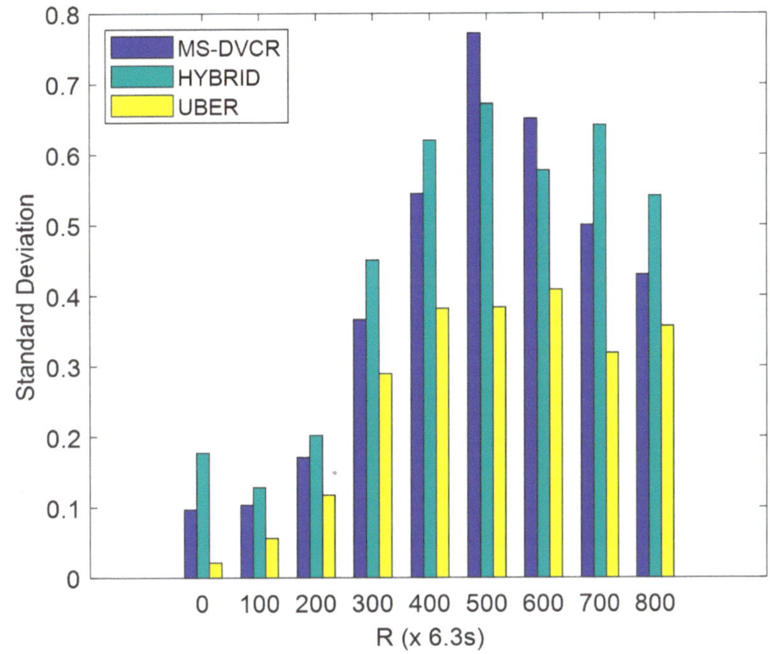

(**b**)

Figure 7. (**a**) Received Packets Performance. (**b**) Standard Deviation Performance Evaluation of SN Failures Detected (SNFD) Performance.

SNFD is defined as the percentage of SN failures detected at each specified round of network operation. From Figure 8, the blue bar represents MS-DVCR's SN failures detected performance trend, the green bar represents HYBRID's SN failures detected performance trend, and the yellow bar represents UBER's SN failures detected performance trend under normal MS altitude of 230 m. Figure 8 provides the comparative plot of SNFD for UBER (indicated as yellow bar) with respect to HYBRID (indicated as green bar) and MS-DVCR (indicated as blue bar). UBER (indicated with yellow bar) exhibited lower SNFD by recording gains of 19.78% and 11.35% over HYBRID (indicated with green bar) and MS-DVCR (indicated with blue bar), respectively. UBER's performance gains are due to the effective MS-assisted network coverage and recovery mechanism and effective HCL-to-HCM enlisting process, which makes it possible to achieve seamless end-to-end data transmission with reduced SN failures.

4.5.4. Evaluation of Route Failures Detected (RFD) Performance

RFD is defined as the percentage of route breakages detected at each specified round of network operation. From Figure 9, the blue bar represents MS-DVCR's route failures detected performance trend, the green bar represents HYBRID's route failures detected performance trend, and the yellow bar represents UBER's route failures detected performance trend under normal MS altitude of 230 m. Figure 9 shows the comparative plot of RFD for UBER (indicated as yellow bar) against HYBRID (indicated as green bar) and MS-DVCR (indicated as blue bar). UBER (indicated with yellow bar) recorded lesser RFD by showing improvements of 46.44% and 44.89% over HYBRID (indicated with green bar) and MS-DVCR (indicated with blue bar), respectively. The technical justification for UBER's performance improvement is due to the effective HCL-to-MS cluster-based connectivity chain and MS-based network monitoring and recovery mechanism that ensures robust HCL-to-MS data transmission with reduced route failures.

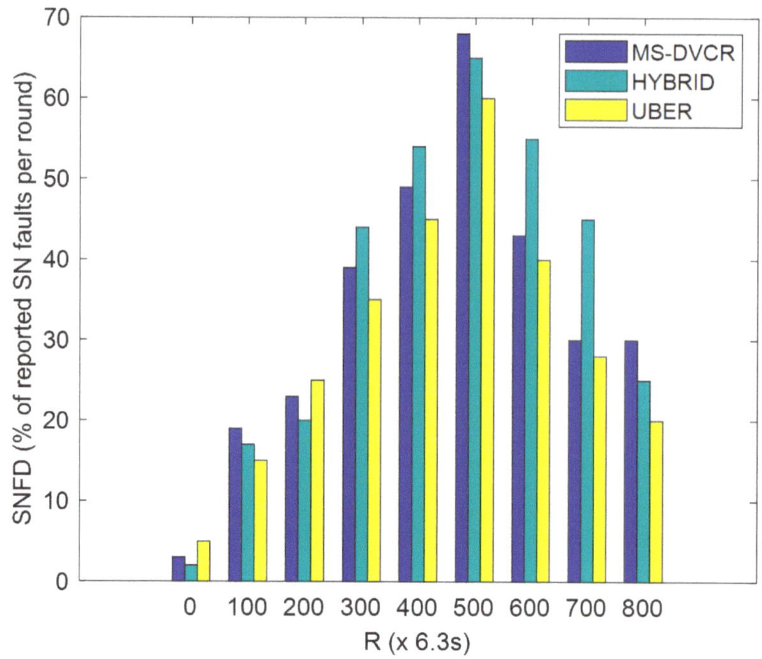

Figure 8. SN Failures Detected Performance.

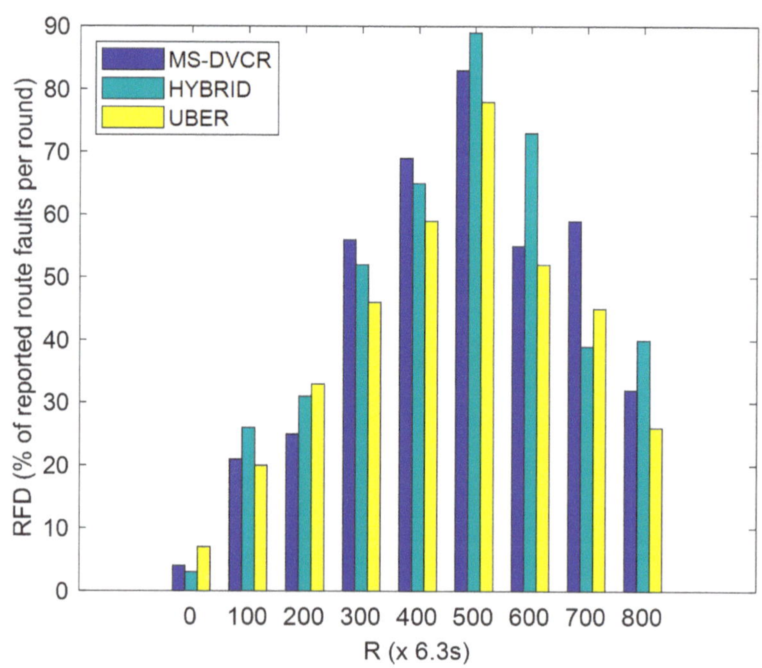

Figure 9. Route Failures Detected Performance.

4.5.5. Evaluation of Routing Overhead (ROH) Performance

ROH is defined as the ratio (in percentage) of packet processing (prior to actual data transmission) duration to network operation duration. From Figure 10, the blue bar segment represents MS-DVCR's routing overhead performance trend, the green bar segment represents HYBRID's routing overhead performance trend, and the yellow bar segment represents UBER's routing overhead performance trend under normal MS altitude of 230 m. Figure 10 provides the comparative plot of ROH for UBER (indicated as yellow bar segment) with respect to HYBRID (indicated as green bar segment) and MS-DVCR (indicated as blue bar segment). UBER (indicated with yellow bar segment) exhibited lower ROH by recording gains of 29.38% and 16.45% over HYBRID (indicated with green bar segment) and MS-DVCR (indicated with blue bar segment), respectively. UBER's performance gains are due to the reduced algorithmic complexity, which makes it possible for the network to construct routes and carry out routing operations with reduced computational costs.

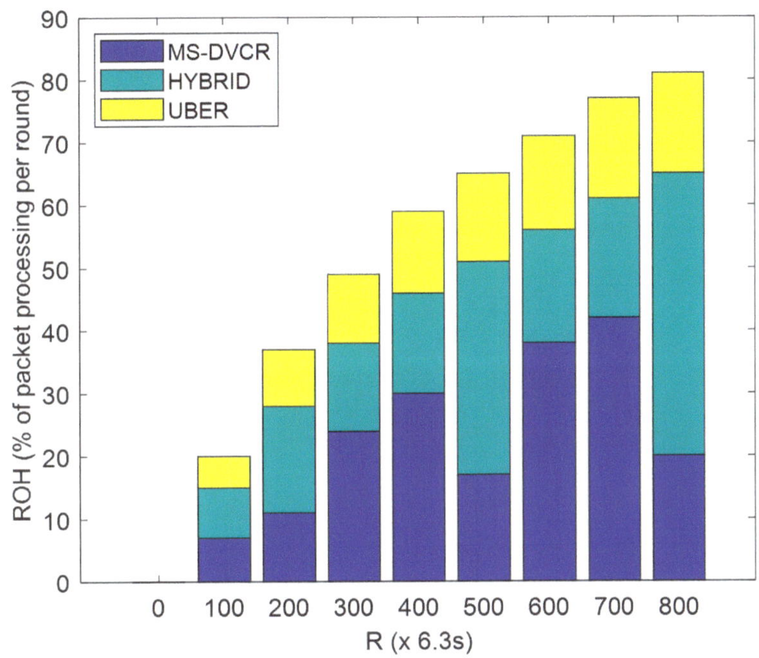

Figure 10. Routing Overhead Performance.

4.5.6. Evaluation of End-To-End Delay (ETE) Performance

ETE is defined as the total duration measured from initial packet generation at the HCM to eventual delivery via the MS-to-LMS transmission link. From Figure 11, the blue bar represents MS-DVCR's end-to-end delay performance trend, the green bar represents HYBRID's end-to-end delay performance trend, and the yellow bar represents UBER's end-to-end delay performance trend under normal MS altitude of 230 m. Figure 11 depicts the comparative plot of ETE for UBER (indicated as yellow bar) in comparison to HYBRID (indicated as green bar) and MS-DVCR (indicated as blue bar). UBER (indicated with yellow bar) displayed lesser ETE by 58.56% and 54.33% improvements over HYBRID (indicated with green bar) and MS-DVCR (indicated with blue bar), respectively. The technical justification for UBER's performance improvements is due to the adaptive MS-assisted maintenance of HCM-to-HCL, HCL-to-MS, and MS-to-LMS data forwarding chains. This adaptive nature makes it possible to preserve end-to-end data transmission links with lesser delays.

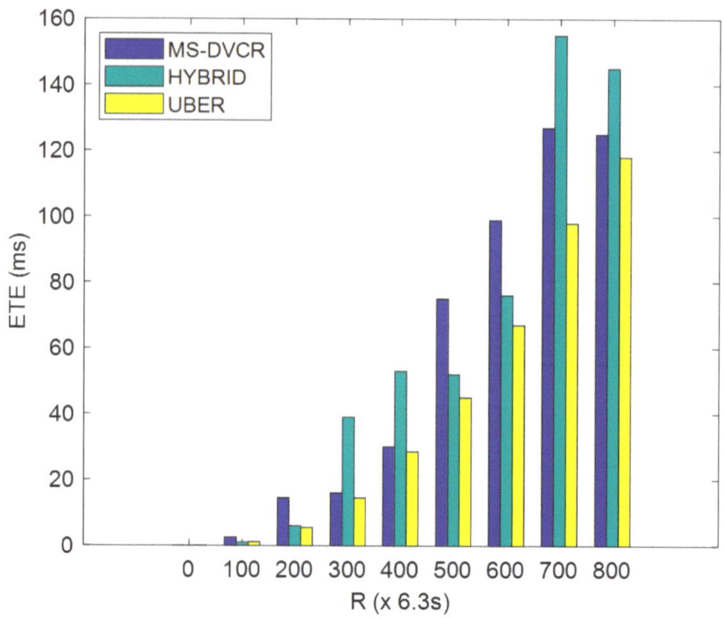

Figure 11. End-to-End Delay Performance.

Table 3 summarizes UBER's comparative performance evaluation against HYBRID and MS-DVCR.

Table 3. Summary of UBER Comparative Performance Results.

Metric	HYBRID	MS-DVCR
ENC	25.59%	46.48%
COV	28.44%	47.33%
RPK	15.68%	3.637%
SNFD	19.78%	11.35%
RFD	46.44%	44.89%
ROH	29.38%	16.45%
ETE	58.56%	54.33%

5. Conclusions

This paper treats the issues of loss of coverage and rapid energy drainage for wireless livestock sensor network applications by developing a UAV-based energy-efficient reconfigurable routing (UBER) scheme for smart wireless livestock sensor networking. This research contrived and incorporated a dynamic residual energy thresholding mathematical model, robust cluster-to-UAV link formation strategy, and UAV-assisted network coverage and recovery mechanisms into this scheme for the proposed integrated heterogenous network model. Experiments were carried out with OMNET++ and MATLAB. The performance of UBER was analyzed using low, normal, and high UAV altitude scenarios. Simulation results revealed that operating the network with a UAV altitude at normal range yielded performance gains with respect to network stability, load balancing ratio, and topology fluctuation effect ratio. A comparative performance analysis of UBER was performed with respect to HYBRID and MS-DVCR. With regard to the obtained results, UBER recorded significant performance improvements in terms of energy consumption, network coverage, received packets, SN failures detected, route failures detected, routing overhead, and end-to-end delay. The results obtained demonstrated that the UBER scheme is highly efficient with competitive performance against the benchmarked CBR schemes. Future research work will focus on developing a lightweight, reliable, and energy-efficient repair/recovery scheme for the identified SNFD and RFD issues.

Author Contributions: Conceptualization, A.F.S. and Y.A.S.; methodology, A.F.S., Y.A.S. and H.R.E.H.B.; software, A.F.S.; validation, Y.A.S., M.S.S. and H.R.E.H.B.; formal analysis, Y.A.S., M.K.S. and M.S.S.; investigation, A.F.S.; resources, A.F.S. and Y.A.S.; writing—original draft preparation, M.A.A., A.F.S., Y.A.S., H.R.E.H.B., M.S.S., M.K.S. and M.K.; writing—review and editing, M.A.A., A.F.S., Y.A.S., H.R.E.H.B., M.S.S. and M.K.; visualization, Y.A.S.; supervision, H.R.E.H.B. and M.A.A.; project administration, M.A.A.; funding acquisition, M.A.A. All authors have read and agreed to the published version of the manuscript.

Funding: This research was funded by Deputyship for Research and Innovation, Ministry of Education in Saudi Arabia, grant number IFP-A-201-2-1.

Institutional Review Board Statement: Not applicable.

Informed Consent Statement: Not applicable.

Acknowledgments: The authors extend their appreciation to the Deputyship for Research and Innovation, Ministry of Education in Saudi Arabia for funding this research work through the project No IFP-A-201-2-1.

Conflicts of Interest: The authors declare no conflict of interest.

References

1. Alanezi, M.A.; Shahriar, M.S.; Hasan, M.B.; Ahmed, S.; Sha'aban, Y.A.; Bouchekara, H.R.E.H. Livestock Management with Unmanned Aerial Vehicles: A Review. *IEEE Access* **2022**, *10*, 45001–45028. [CrossRef]
2. Barbedo, J.G.A.; Koenigkan, L.V. Perspectives on the use of unmanned aerial systems to monitor cattle. *Outlook Agric.* **2018**, *47*, 214–222. [CrossRef]
3. Awan, K.M.; Sherazi, H.H.R.; Ali, A.; Iqbal, R.; Khan, Z.A.; Mukherjee, M. Energy-aware cluster-based routing optimization for WSNs in the livestock industry. *Trans. Emerg. Telecommun. Technol.* **2019**, *33*, e3816. [CrossRef]
4. Maddikunta, P.R.; Hakak, S.; Alazab, M.; Bhattacharya, S.; Gadekallu, T.R.; Khan, W.Z.; Pham, Q.V. Unmanned Aerial Vehicles in Smart Agriculture: Applications, Requirements, and Challenges. *IEEE Sens. J.* **2021**, *21*, 17608–17619. [CrossRef]
5. Friha, O.; Ferrag, M.A.; Shu, L.; Maglaras, L.; Wang, X. Internet of Things for the Future of Smart Agriculture: A Comprehensive Survey of Emerging Technologies. *IEEE/CAA J. Autom. Sin.* **2021**, *8*, 718–752. [CrossRef]
6. Long, N.K.; Sammut, K.; Sgarioto, D.; Garratt, M.; Abbass, H.A. A Comprehensive Review of Shepherding as a Bio-Inspired Swarm-Robotics Guidance Approach. *IEEE Trans. Emerg. Top. Comput. Intell.* **2020**, *4*, 523–537. [CrossRef]
7. Xiang, T.-Z.; Xia, G.-S.; Zhang, L. Mini-Unmanned Aerial Vehicle-Based Remote Sensing: Techniques, applications, and prospects. *IEEE Geosci. Remote Sens. Mag.* **2019**, *7*, 29–63. [CrossRef]
8. Boursianis, A.D.; Papadopoulou, M.S.; Diamantoulakis, P.; Liopa-Tsakalidi, A.; Barouchas, P.; Salahas, G.; Karagiannidis, G.; Wan, S.; Goudos, S. KInternet of Things (IoT) and Agricultural Unmanned Aerial Vehicles (UAVs) in smart farming: A comprehensive review. *Internet Things* **2020**, *18*, 100187. [CrossRef]
9. Kakamoukas, G.; Sariciannidis, P.; Livanos, G.; Zervakis, M.; Ramnalis, D.; Polychronos, V.; Karamitsou, T.; Folinas, A.; Tsitsiokas, N. A Multi-collective, IoT-enabled, Adaptive Smart Farming Architecture. In Proceedings of the 2019 IEEE International Conference on Imaging Systems and Techniques (IST), Abu Dhabi, United Arab Emirates, 9–10 December 2019; pp. 1–6. [CrossRef]
10. Urdaneta, G.A.; Meyers, C.; Rogalski, L. How do drones facilitate human life? *Future Technol.* **2022**, *1*, 7–13. [CrossRef]
11. Mistry, C.; Ghosh, A.; Biswas, M.; Basak, B.B.A. Applications of Internet of Things and Unmanned Aerial Vehicle in Smart Agriculture: A Review. *OSF Prepr.* **2022**. [CrossRef]
12. Casas, R.; Hermosa, A.; Marco, Á.; Blanco, T.; Zarazaga-Soria, F.J. Real-Time Extensive Livestock Monitoring Using LPWAN Smart Wearable and Infrastructure. *Appl. Sci.* **2021**, *11*, 1240. [CrossRef]
13. Tahir, A.; Böling, J.; Haghbayan, M.-H.; Toivonen, H.T.; Plosila, J. Swarms of Unmanned Aerial Vehicles—A Survey. *J. Ind. Inf. Integr.* **2019**, *16*, 100106. [CrossRef]
14. Pajares, G. Overview and Current Status of Remote Sensing Applications Based on Unmanned Aerial Vehicles (UAVs). *Photogramm. Eng. Remote Sens.* **2015**, *81*, 281–330. [CrossRef]
15. Mukhamediev, R.I.; Symagulov, A.; Kuchin, Y.; Zaitseva, E.; Bekbotayeva, A.; Yakunin, K.; Assanov, I.; Levashenko, V.; Popova, Y.; Akzhalova, A.; et al. Review of Some Applications of Unmanned Aerial Vehicles Technology in the Resource-Rich Country. *Appl. Sci.* **2021**, *11*, 10171. [CrossRef]
16. Elmokadem, T.; Savkin, A.V. Towards Fully Autonomous UAVs: A Survey. *Sensors* **2021**, *21*, 6223. [CrossRef]
17. Sivakumar, M.; TYJ, N.M. A Literature Survey of Unmanned Aerial Vehicle Usage for Civil Applications. *J. Aerosp. Technol. Manag.* **2021**, *13*, 1–23. [CrossRef]

18. Petrova, T.; Petrov, Z. Analysis of Efficiency of the Unmanned Aerial Vehicles Use in Contemporary Agrotechnologies. *Int. J. Inf. Technol. Secur.* **2021**, *13*, 25–34. Available online: https://ijits-bg.com/contents/IJITS-2021-No4/2021-N4-03.pdf (accessed on 24 April 2022).
19. Freed, T.; Carson, V.C.; Doerr, K.H. Optimizing a RFID-UAV cattle search tour. *Int. J. RF Technol.* **2021**, *11*, 127–141. [CrossRef]
20. Chamoso, P.; Raveane, W.; Parra, V.; González, A. UAVs Applied to the Counting and Monitoring of Animals. In *Ambient Intelligence—Software and Applications*; Springer: Cham, Switzerland, 2014; pp. 71–80. [CrossRef]
21. Afrianto, I.; Wahjuni, S.; Djatna, T. Model of Ubiquitous Precision Livestock System 4.0: A Technological Review. *FoITIC* **2020**, 156–162. Available online: https://eproceeding.itenas.ac.id/index.php/foitic/article/view/74 (accessed on 25 April 2022).
22. Chabot, D.; Bird, D.M. Wildlife research and management methods in the 21st century: Where do unmanned aircraft fit in? *J. Unmanned Veh. Syst.* **2015**, *3*, 137–155. [CrossRef]
23. Rivas, A.; Chamoso, P.; González-Briones, A.; Corchado, J.M. Detection of Cattle Using Drones and Convolutional Neural Networks. *Sensors* **2018**, *18*, 2048. [CrossRef] [PubMed]
24. Behjati, M.; Noh, A.B.M.; Alobaidy, H.A.H.; Zulkifley, M.A.; Nordin, R.; Abdullah, N.F. LoRa Communications as an Enabler for Internet of Drones towards Large-Scale Livestock Monitoring in Rural Farms. *Sensors* **2021**, *21*, 5044. [CrossRef] [PubMed]
25. Xu, J.; Solmaz, G.; Rahmatizadeh, R.; Turgut, D.; Bölöni, L. Animal monitoring with unmanned aerial vehicle-aided wireless sensor networks. In Proceedings of the 2015 IEEE 40th Conference on Local Computer Networks (LCN), Clearwater Beach, FL, USA, 26–29 October 2015; pp. 125–132. [CrossRef]
26. Xu, J.; Solmaz, G.; Rahmatizadeh, R.; Turgut, D.; Boloni, L. Internet of Things Applications: Animal Monitoring with Unmanned Aerial Vehicle. *arXiv* **2016**, arXiv:1610.05287. Available online: https://arxiv.org/abs/1610.05287 (accessed on 25 April 2022).
27. Heinzelman, W.R.; Chandrakasan, A.; Balakrishnan, H. Energy-efficient communication protocol for wireless microsensor networks. In Proceedings of the 33rd Annual Hawaii International Conference on System Sciences, Maui, HI, USA, 7 January 2000; pp. 1–10. [CrossRef]
28. Heinzelman, W.B.; Chandrakasan, A.P.; Balakrishnan, H. An application-specific protocol architecture for wireless microsensor networks. *IEEE Trans. Wirel. Commun.* **2002**, *1*, 660–670. [CrossRef]
29. Bandyopadhyay, S.; Coyle, E.J. An energy efficient hierarchical clustering algorithm for wireless sensor networks. In Proceedings of the IEEE INFOCOM 2003—Twenty-Second Annual Joint Conference of the IEEE Computer and Communications Societies, San Francisco, CA, USA, 30 March–3 April 2003; pp. 1713–1723. [CrossRef]
30. Huang, H.; Wu, J. A probabilistic clustering algorithm in wireless sensor networks. In Proceedings of the VTC-2005-Fall—2005 IEEE 62nd Vehicular Technology Conference, Dallas, TX, USA, 28 September 2005; pp. 1796–1798. [CrossRef]
31. Sivakumar, R.; Sinha, P.; Bharghavan, V. CEDAR: A core-extraction distributed ad hoc routing algorithm. *IEEE J. Sel. Areas Commun.* **1999**, *17*, 1454–1465. [CrossRef]
32. Ding, P.; Holliday, J.; Celik, A. Distributed Energy-Efficient Hierarchical Clustering for Wireless Sensor Networks. In *Distributed Computing in Sensor Systems*; Springer: Berlin/Heidelberg, Germany, 2005; pp. 322–339. [CrossRef]
33. Neethirajan, S. Recent advances in wearable sensors for animal health management. *Sens. Bio-Sens. Res.* **2017**, *12*, 15–29. [CrossRef]
34. Lotfinezhad, M.; Liang, B. Energy efficient clustering in sensor networks with mobile agents. In Proceedings of the IEEE Wireless Communications and Networking Conference, New Orleans, LA, USA, 13–17 March 2005; pp. 1872–1877. [CrossRef]
35. Morsly, Y.; Aouf, N.; Djouadi, M.S. Dynamic decentralized/centralized free conflict UAV's team allocation. In Proceedings of the 2012 IEEE International Instrumentation and Measurement Technology Conference Proceedings, Graz, Austria, 13–16 May 2012 pp. 1–6. [CrossRef]
36. Behera, T.M.; Mohapatra, S.K.; Samal, U.C.; Khan, M.S. Hybrid heterogeneous routing scheme for improved network performance in WSNs for animal tracking. *Internet Things* **2019**, *6*, 100047. [CrossRef]
37. Rahmatizadeh, R.; Khan, S.A.; Jayasumana, A.P.; Turgut, D.; Bölöni, L. Routing towards a mobile sink using virtual coordinates in a wireless sensor network. In Proceedings of the 2014 IEEE International Conference on Communications (ICC), Sydney Australia, 10–14 June 2014; pp. 12–17. [CrossRef]
38. Rahman, G.M.E.; Wahid, K.A. LDCA: Lightweight Dynamic Clustering Algorithm for IoT-Connected Wide-Area WSN and Mobile Data Sink Using LoRa. *IEEE Internet Things J.* **2022**, *9*, 1313–1325. [CrossRef]
39. Wark, T.; Crossman, C.; Hu, W.; Guo, Y.; Valencia, P.; Sikka, P.; Corke, P.; Lee, C.; Henshall, J.; Prayaga, K.; et al. The Design and Evaluation of a Mobile Sensor/Actuator Network for Autonomous Animal Control. In Proceedings of the 2007 6th International Symposium on Information Processing in Sensor Networks, Cambridge, MA, USA, 25–27 April 2007; pp. 206–215. [CrossRef]
40. Gu, J.; Su, T.; Wang, Q.; Du, X.; Guizani, M. Multiple Moving Targets Surveillance Based on a Cooperative Network for Multi-UAV *IEEE Commun. Mag.* **2018**, *56*, 82–89. [CrossRef]
41. Hu, J.; Turgut, A.E.; Krajnik, T.; Lennox, B.; Arvin, F. Occlusion-Based Coordination Protocol Design for Autonomous Robotic Shepherding Tasks. *IEEE Trans. Cogn. Dev. Syst.* **2022**, *14*, 126–135. [CrossRef]
42. Lin, C.; Han, G.; Qi, X.; Du, J.; Xu, T.; Martínez-García, M. Energy-Optimal Data Collection for Unmanned Aerial Vehicle-Aided Industrial Wireless Sensor Network-Based Agricultural Monitoring System: A Clustering Compressed Sampling Approach. *IEEE Trans. Ind. Inform.* **2021**, *17*, 4411–4420. [CrossRef]
43. Gnanasekera, M.; Katupitiya, J.; Savkin, A.V.; de Silva, A.H.T.E. A Range-Based Algorithm for Autonomous Navigation of an Aerial Drone to Approach and Follow a Herd of Cattle. *Sensors* **2021**, *21*, 7218. [CrossRef] [PubMed]

44. Salami, A.F.; Anwar, F.; Priantoro, A.U. An investigation into clustering routing protocols for wireless sensor networks. *Sens. Transducers* **2009**, *106*, 48–61.
45. Salami, A.F.; Bari, S.M.S.; Anwar, F.; Khan, S. Feasibility analysis of clustering routing protocols for multipurpose sensor networking. In Proceedings of the 2nd International Conference on Multimedia and Computational Intelligence (ICMCI), Wuhan, China, 29–30 September 2010; pp. 432–435.
46. Astakhova, T. Research on the Energy Characteristics of Routing in Wireless Sensor Networks. In CEUR Workshop Proceedings 2020. Available online: http://ceur-ws.org/Vol-2590/short15.pdf (accessed on 20 April 2022).
47. Li, X.; Huang, H.; Savkin, A.V.; Zhang, J. Robotic Herding of Farm Animals Using a Network of Barking Aerial Drones. *Drones* **2022**, *6*, 29. [CrossRef]
48. Yaxley, K.J.; Joiner, K.F.; Abbass, H. Drone approach parameters leading to lower stress sheep flocking and movement: Sky shepherding. *Sci. Rep.* **2021**, *11*, 7803. [CrossRef]
49. Salami, A.F.; Anwar, F.; Aibinu, A.M.; Bello-Salau, H.; Abdalla, A.H. Investigative analysis of clustering routing protocols for scalable sensor networks. In Proceedings of the 4th IEEE International Conference on Mechatronics (ICOM), Kuala Lumpur, Malaysia, 17–19 May 2011; pp. 11–15.
50. Bello-Salau, H.; Salami, A.F.; Anwar, F.; Aibinu, A.M. Evaluation of Radio Propagation Techniques for Hierarchical Sensor Networks. In Proceedings of the 4th IEEE International Conference on Mechatronics (ICOM), Kuala Lumpur, Malaysia, 17–19 May 2011; pp. 001–005.
51. Bello-Salau, H.; Salami, A.F.; Anwar, F.; Islam, M.R. Analysis of radio model performance for clustering sensor networks. *Sens. Transducers* **2011**, *128*, 27–38.
52. Friis, H.T. A Note on a Simple Transmission Formula. *Proc. IRE* **1946**, *34*, 254–256. [CrossRef]
53. IEEE Std 802.15.4-2020 (Revision of IEEE Std 802.15.4-2015)*IEEE Standard for Low-Rate Wireless Networks*; IEEE Standards Department: Piscataway, NJ, USA, 2020; pp. 1–800. [CrossRef]
54. 4 GHz IEEE 802.15.4/ZigBee-ready RF Transceiver Applications. Available online: https://www.ti.com/product/CC2420 (accessed on 21 April 2022).

agriculture

MDPI

Article

Adaptive Sliding Mode Path Tracking Control of Unmanned Rice Transplanter

Jinyang Li [1,2,*], Zhijian Shang [1,2], Runfeng Li [1,2] and Bingbo Cui [1,2]

1 College of Agricultural Engineering, Jiangsu University, Zhenjiang 212013, China
2 Key Laboratory of Modern Agricultural Equipment and Technology, Ministry of Education, Zhenjiang 212013, China
* Correspondence: by0817136@163.com

Abstract: To decrease the impact of uncertainty disturbance such as sideslip from the field environment on the path tracking control accuracy of an unmanned rice transplanter, a path tracking method for an autonomous rice transplanter based on an adaptive sliding mode variable structure control was proposed. A radial basis function (RBF) neural network, which can precisely approximate arbitrary nonlinear function, was used for parameter auto-tuning on-line. The sliding surface was built by a combination of parameter auto-tuning and the power approach law, and thereafter an adaptive sliding controller was designed. Based on theoretical and simulation analysis, the performance of the proposed method was evaluated by field tests. After the appropriate hardware modification, the high-speed transplanter FLW 2ZG-6DM was adapted as a test platform in this study. The contribution of this study is providing an adaptive sliding mode path tracking control strategy in the face of the uncertainty influenced by the changeable slippery paddy soil environment in the actual operation process of the unmanned transplanter. The experimental results demonstrated that: compared to traditional sliding control methods, the maximum lateral deviation was degraded from 17.5 cm to 9.3 cm and the average of absolute lateral deviation was degraded from 9.1 cm to 3.2 cm. The maximum heading deviation was dropped from 46.7° to 3.1°, and the average absolute heading deviation from 10.7° to 1.3°. The proposed control method not only alleviated the system chattering caused by uncertain terms and environmental interference but also improved the path tracking performance of the autonomous rice transplanter. The results show that the designed control system provided good stability and reliability under the actual rice field conditions.

Keywords: autonomous rice transplanter; path tracking control; RBF neural network; automatic steering; navigation system

Citation: Li, J.; Shang, Z.; Li, R.; Cui, B. Adaptive Sliding Mode Path Tracking Control of Unmanned Rice Transplanter. *Agriculture* **2022**, *12*, 1225. https://doi.org/10.3390/agriculture12081225

Academic Editors: Monica Herrero-Huerta, Jose A. Jiménez-Berni, Shangpeng Sun, Ittai Herrmann and Diego González-Aguilera

Received: 28 May 2022
Accepted: 8 August 2022
Published: 15 August 2022

1. Introduction

Due to the high labor cost, the shortage of rural labor force, and the poor straightness accuracy of the seedling row, unmanned transplanters have gradually become a significantly sought-after in promoting the mechanized transplanting of rice in China. The path tracking control performance is the key to determining the straightness of transplanting [1–3].

Scholars, both domestically and abroad, have carried out relevant studies on path tracking control algorithms based on fuzzy control [4,5], the pure tracking model [2,6–8], sliding mode control [9,10], and other methods. A fuzzy adaptive pure tracking algorithm for agricultural machinery was proposed [11] which improved the tracking accuracy of the model path. However, the control laws were difficult to establish and thus increased the control difficulty. To enhance the adaptability of the navigation control system, a pure tracking control method for tractor navigation was developed based on the SVR reverse model; it was difficult to make real-time and dynamic adjustments for the tracking error because of complex model construction and sample training [12]. An improved Stanley controller was proposed to improve the path tracking the performance of autonomous tractors and the parameters of the controller were optimized based on a multiple-population genetic algorithm [13]. The effectiveness and advantages were evaluated by simulation. A control strategy based on nonsingular fast terminal sliding mode control with a finite time disturbance observer was presented to improve the trajectory tracking ability of the autonomous agricultural tractor [14]. However, the effectiveness of the designed control algorithm was not verified by field tests. Some of the state-of-the-art deep learning models have been used in agricultural applications. Deep learning methods were used in uncrewed vehicles to detect fruits [15–17]. A modified YOLOv3 model, called YOLO-Tomato models, was established to detect tomatoes in challenging environmental conditions [16]. The real-time growth stage detection model for a high degree of occultation was developed based on DenseNet-fused YOLOv4 [17]. The deep learning can be used for parameter optimization of control systems. In order to improve the trajectory tracking performance of autonomous orchard vehicles, a model-based control algorithm was presented for wheel slip compensation [18]. The model predictive control algorithm was designed to improve trajectory tracking performance, and nonlinear least squares frequency domain system identification was applied to identify the yaw dynamics [19].

Sliding mode control (SMC) has the advantages of fast response, insensitivity to parameter changes and disturbances, no online system identification, and simple physical implementation, which has widely concerned by scholars from all over the world. However, the chattering problem of sliding mode control in practical systems is a prominent obstacle. A robust controller based on backstepping sliding mode control for a 4WS/4WD agricultural robotic vehicle was designed against system uncertainties [20]. A layered multi-loop powerful control architecture is presented for an electro-hydraulic coupling system, and a sliding mode controller was applied to consider the parameter uncertainties and disturbances from the system [21]. Although an adaptive steering control strategy was proposed for varying yaw rate properties of a farm tractor, the performance is hardly validated by experimental tests. To weaken chattering and improve the dynamic performance of approaching movement by adopting various reaching laws, a significant number of studies was carried out. A path tracking control strategy for the tracked robot was proposed based on a sliding mode variable structure algorithm [22], and a good control effect on path tracking was demonstrated by simulation. A fuzzy sliding mode control algorithm was designed based on inverse control and good robustness was verified by simulation results [23]. However, these two sliding mode control methods have not taken into account the influence of uncertain disturbance such as sideslip caused by the variation of the working environment and have not been verified by actual field tests. Moreover, although widespread applications of sliding mode control algorithms including improved algorithms have been found, most research has been limited to computer simulations and constrained real-time in practical working processes.

Because of powerful approximation properties, neural networks (NN) and fuzzy systems were applied to modeling the uncertainties and external disturbances in the system or approaching the switching part of the controller by making discontinuous control signals continuous and thus effectively reducing chattering. A novel neural network sliding mode control method is proposed. The neural network is used to estimate the nonlinear part, the uncertain part, and the unknown disturbance of the linear system. An equivalent control based on a neural network is realized, and chattering is effectively eliminated [24]. Two neural networks are used to approximate the equivalent sliding mode control part and the switching sliding mode control part, respectively. The chattering of the controller is eliminated effectively without an object model [25]. Compared with BP-NN, RBF-NN has good generalization ability and a simple network structure, which can avoid unnecessary and lengthy calculations and approximate any nonlinear function in a compact set at arbitrary precision. Using the approximation ability of NN, a sliding mode controller based on an RBF-NN is proposed. The switching function s is used as the network input, and the controller is entirely realized by the continuous RBF (radius basis function) function, which eliminates the switching term and chattering [26].

To make up for the shortcomings of the above algorithms and realize the precise control of the expected rotation angle and the travel speed, an adaptive sliding mode variable structure path tracking control algorithm based on the power approach law and an RBF-NN was proposed in this study, aiming at overcoming the influence of uncertainty from the changeable slippery paddy soil environment in the actual operation process of the unmanned transplanter. The reliability of the model and the feasibility of the method are verified by the theoretical analysis, model simulation, and field tests.

2. System Hardware Modification

2.1. Steering Control

The hydraulic power steering device of the existing FLW 2ZG-6DM high-speed transplanter is installed as a whole, and its automatic steering was achieved by a parallel oil circuit [27,28]. The automatic steering control only can be realized by adding an electric steering wheel to the original steering mechanism and controlling the motor to simulate the manual operation of the steering wheel. The AF300 electric steering wheel, as shown in Figure 1, was adopted (Shanghai LIANSHI Navigation Co., Shanghai, China). The specific parameters of the electric steering wheel are shown in Table 1.

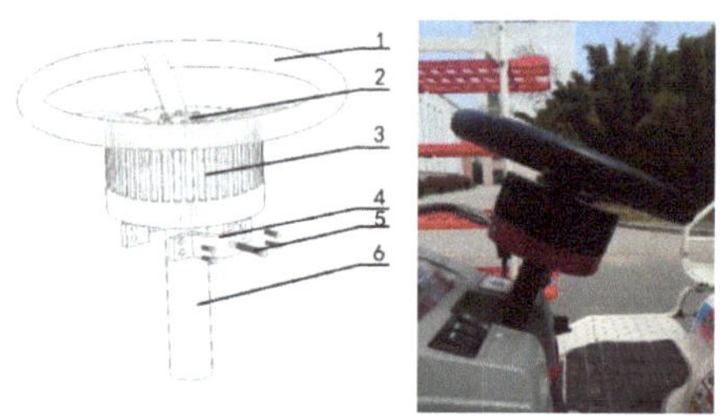

(a) (b)

Figure 1. Installation schematic diagram of steering wheel-like motor. (**a**) Structure diagram. (**b**) Installation drawing: 1. Steering wheel; 2. Tight set; 3. Barrel motor; 4. Fixed axle support; 5. Locking bolt; 6. Fixed shaft.

Table 1. Parameters of steering wheel-like motor.

Type	Parameter
Rated voltage/V	12
Rated torque/(N*m)	15
Output mode	RS232
Speed/rmp	100

To ensure accuracy, long-term operation stability, and reliable performance during driving operation, the fixed bracket (Figure 2) is designed and installed under the front axle arm of the rice transplanter. The connecting rod and the front wheel steering shaft constituted a parallel four-bar mechanism, and the angle sensor was installed at the wheel steering trapezoid, which provided accurate angle data for the steering control of the transplanter. Considering the cost performance, volume, and other factors, the DWQT RS485-G/J angle sensor (Beijing Tianhai Technology Co., Beijing, China) was selected, and the specific parameters are shown in Table 2.

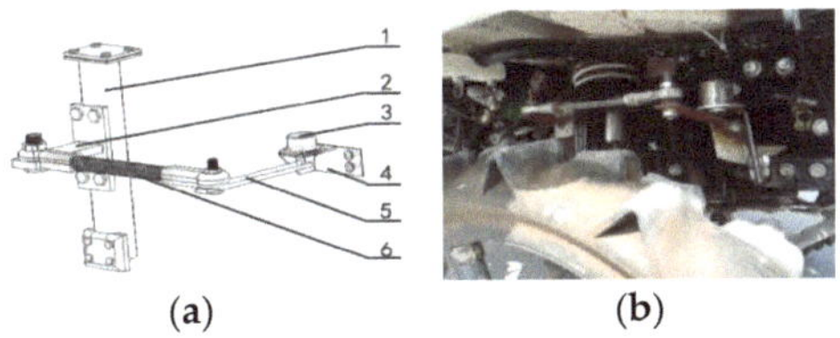

Figure 2. Installation schematic diagram of angle sensor. (**a**) Structure diagram; (**b**) installation drawing: 1. Steering shaft; 2. Trellis bar; 3. Angle sensor; 4. Fixed bracket; 5. Fixed link; 6. Ball tie rod.

Table 2. Parameters of angle sensor.

Type	Parameter
Rated voltage/V	8–24
Range/°	0–360
Output mode	RS485
Accuracy/°	0.1

2.2. Travel Speed Control

As one of the essential control units of the traveling speed control, the accelerator pedal mechanism is the key to ensuring the stability of the operating speed. The original mechanical accelerator pedal of the rice transplanter is transformed into an electric push rod pull accelerator pedal. The corresponding relationship between displacement and velocity is established through the real-time displacement information of the electric push rod. According to measurements, the actual pulling force of the accelerator pedal is about 50 N, and the displacement of the stay wire is 95 mm. Considering the water-resistance of the operating equipment under the paddy field environment, a DC electric push rod with position feedback (Taiwan, China, LD3-12-20-50-IP65-POT) was selected. Its structure and installation are shown in Figure 3, and the main performance parameters are shown in Table 3.

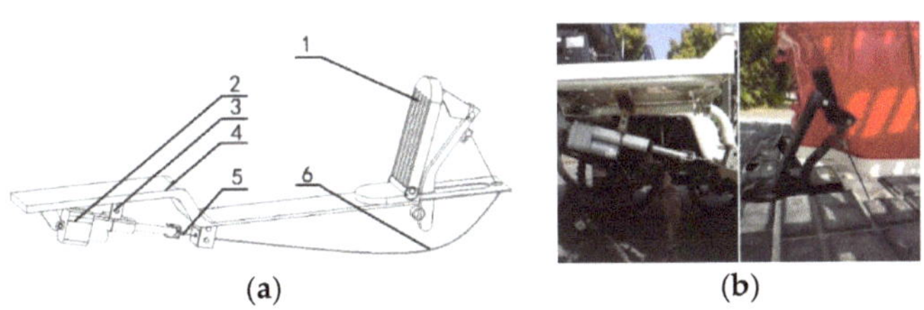

Figure 3. Installation schematic diagram of the electric push rod. (**a**) Structure diagram; (**b**) installation drawing: 1. Accelerator pedal; 2. Linear actuator; 3. Support plate; 4. Transplanter chassis; 5. Steel strand core; 6. Stay sleeve.

Table 3. Parameters of the electric push rod.

Type	Parameter
Rated voltage/V	12
Load/N	200
Trip/mm	100
Speed/(mm*s^{-1})	24

2.3. Location Information Collection

During transplanting operations, to achieve continuous and accurate navigation, reduce accumulated positioning error, and ensure the stability of the positioning system, the combined navigation of BEIDOU and INS (inertial navigation system) [29] was adopted.

A HUA XIN full band antenna, supporting C201 receiver, and a navigation reference station were selected as the navigation module. The RTK horizontal positioning accuracy is ±2 cm, and the data update rate is set at 5 Hz. The attitude sensor is a micro inertial sensor MTI-30 model manufactured by the XENS company of the Netherlands. The navigation reference station and attitude sensor are shown in Figure 4, and the performance parameters of the attitude sensor are shown in Table 4.

(**a**) (**b**)

Figure 4. Diagram of inertia sensor and navigation system. (**a**) Navigation reference station; (**b**) posture sensor.

Table 4. Parameters of inertia sensor.

Type	Parameter
Rated voltage/V	5
Sampling frequency/kHz	10
Output mode	RS485
Rated power/mW	480–570

2.4. Main Controller

The main board of the series STM32F103 is used as the central controller to realize the functions of position information collection, angle information collection, electric steering wheel control, and travel control. The controller has 64KB SRAM, 512KB FLASH, five serial ports, and 112 general I/O ports. It can meet the data processing requirements of the intelligent control system. The system block diagram is shown in Figure 5. The navigation module and angle transducer are attached to the master controller by two groups of the RS485 interface. The navigation module provides position and gesture data such as lateral deviation, heading deviation, and vehicle speed for the master controller at a 10 Hz frequency. An angle transducer is employed to sense the front steering wheel angle in real-time. The master controller is connected to the attitude sensor by RS485 interface, which manages the yaw angle, pitch, and roll angles from the attitude sensor at a 10 Hz frequency. The electric steering wheel is attached to the controller through the RS232 interface and is used to drive the front wheel of the rice transplanter. A linear actuator with displacement sensor, attached to the controller by A/D and digital output interfaces and assembled in the controller, is employed to adjust vehicle speed. Moreover, the touch screen, which connects to the controller by the RS232 interface, is applied to display data and set control parameters.

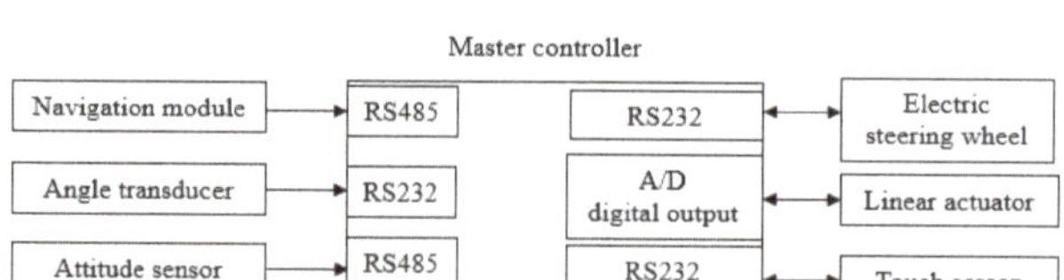

Figure 5. Construction diagram of the system.

3. Algorithm of Path Tracking Control System

3.1. Kinematic Modeling and Theoretical Analysis of Rice Transplanter

Take the transplanter as the controlled object and establish the relative kinematic model between the transplanter and the path as shown in Figure 6. The kinematic model can be expressed as [30].

$$\begin{cases} \dot{l} = \frac{vcos\rho}{1-c(l)d} \\ \dot{d} = vsin\rho \\ \dot{\rho} = v\left(\frac{tan\theta}{D_L} - \frac{c(l)cos\rho}{1-c(l)d}\right) \end{cases} \tag{1}$$

where d, D_L and l denote lateral deviation, wheel base of transplanter and path tracking curvature, respectively; $\rho = \varphi_c - \varphi$, θ denotes the heading deviation and steering wheel angle, respectively; $c(l)$ and v are the path arc length and speed of the rice transplanter, respectively.

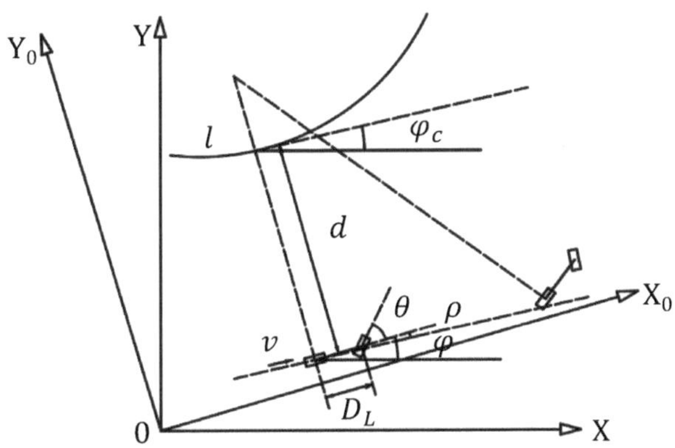

Figure 6. Kinematic model.

Let $z = [z_1, z_2]^T = [d, \rho]^T$ and calculate the derivative of z respect to l, the Equation (1) can be rewritten as

$$\dot{z} = f(z) + g(z)\frac{\tan\theta}{D_L} \tag{2}$$

where

$$f(z) = \begin{bmatrix} tanz_2(1 - z_1c(l)) \\ -c(l) \end{bmatrix}$$
$$g(z) = \begin{bmatrix} 0 \\ secz_2(1 - z_1c(l)) \end{bmatrix}$$

To ensure nonsingular of Equation (4), the following two conditions must be met:

$$1 - c(l)z_1 \neq 0,\ z_2 \neq k\pi + \frac{\pi}{2}(k = 0, 1, 2 \cdots)$$

The path curvature tracked by rice transplanters is generally not very large, $1/c(l) \gg z_1$; during operation, the rice transplanter travels along the direction of the path and thus $z_2 < \frac{\pi}{2}$. Therefore, the above conditions are easily met in the actual transplanting process.

Thus, Equation (2) can be expressed by the matrix form

$$\dot{z} = \begin{bmatrix} \dot{d} \\ \dot{\rho} \end{bmatrix} = \begin{bmatrix} v\sin\rho \\ -\dot{l}c(l) + v\frac{\tan\theta}{D_L} \end{bmatrix} \tag{3}$$

According to dynamic feedback linearization theory, Equation (4) can be obtained by applying state transformation and input transformation to Equation (3) [31],

$$\begin{cases} e_1 = z_1 \\ e_2 = \nabla e_1 f(z) = (1 - z_1 c(l))\tan z_2 \\ u = \varphi\left(z, \frac{\tan\theta}{D_L}\right) = \sec^3 e_2 (1 - z_1 c(l))^2 \frac{\tan\theta}{D_L} \\ +(z_1 c(l) - 1)c(l)\left(\tan^2 z_2 + \sec^2 z_2\right) \end{cases} \tag{4}$$

According to Equations (3) and (4), the controllable standard type can be expressed as

$$\begin{bmatrix} \dot{e}_1 \\ \dot{e}_2 \end{bmatrix} = \begin{bmatrix} 0 & 1 \\ 0 & 0 \end{bmatrix} \begin{bmatrix} e_1 \\ e_2 \end{bmatrix} + \begin{bmatrix} 0 \\ 1 \end{bmatrix} u \tag{5}$$

The complexity and spatial variability of the operating environment brings great uncertainty to the steady-state performance of the vehicle yaw rate [32]. The interference term $|\gamma| \leq T\ (T)$ is introduced to represent this uncertain disturbance. Thus, the Equation (5) can be rewritten as

$$\begin{bmatrix} \dot{e}_1 \\ \dot{e}_2 \end{bmatrix} = \begin{bmatrix} 0 & 1 \\ 0 & 0 \end{bmatrix} \begin{bmatrix} e_1 \\ e_2 \end{bmatrix} + \begin{bmatrix} 0 \\ 1 \end{bmatrix} u + \begin{bmatrix} 0 \\ 1 \end{bmatrix} \Delta\eta \tag{6}$$

where

$$\Delta\eta = \frac{(1 - dc(l))^3}{v\cos^3(\rho)}\gamma$$

It can be seen from Equations (5) and (6) that the path tracking ability of the model under the conditions of uncertain interference is effectively improved by eliminating the influence of time and speed on the system.

However, under the conditions of uncertain disturbance, the path tracking error of the controlled object will change with an amplitude of $\Delta\eta$. If the control of the influence of uncertain disturbances such as sideslip are realized, system chattering will occur in the process of path tracking, which will affect the control of the system. Therefore, to decrease or eliminate system chatting, sliding mode control is used to improve the convergence speed of the system, and, then, the RBF neural network is used to approach the disturbance adaptively and ensure the stability and convergence of the system.

3.2. Design and Analysis for Adaptive Sliding Mode Controller

The amplitude of the uncertain term $\Delta\eta$ in Equation (6) varies greatly with the path tracking error. If a large interference boundary is assumed in robust controller design, the chattering amount will increased and the control results will be too conservative. In order to decrease the lateral and heading deviations and improve the adaptive ability of the control system, an adaptive sliding mode path tracking control algorithm is presented based on the power approach rate (PAR) and RBF. The adaptive estimation of $\Delta\eta$ is achieved by RBF and PAR is used to reduce or eliminate the chattering. The PAR is expressed as

$$\dot{\sigma} = -\varepsilon|\sigma|_p sgn\sigma - k\sigma \tag{7}$$

where

$0 < p < 1, \varepsilon > 0, k > 0$

$$\sigma = \mu e_1 + e_2 \tag{8}$$

where

$$\mu > 0$$

By differentiation of l and combining Equations (6) and (7) as

$$\dot{\sigma} = \mu\dot{e}_1 + \dot{e}_2 = \mu\dot{e}_1 + u + \Delta\eta = -\varepsilon|\sigma|_p sgn\sigma - k\sigma \tag{9}$$

Let the control rate $\Delta\eta = 0$ be

$$u = -\mu\dot{e}_1 - \varepsilon|\sigma|_p sgn\sigma - k\sigma \tag{10}$$

The stability analysis is carried out by using the Lyapunov function $V = \sigma^2/2$ as

$$\dot{V} = \sigma\dot{\sigma} \le -\varepsilon|\sigma|^{p+1} - k\sigma^2 + |\sigma|\Phi \tag{11}$$

where

$$\Phi \ge |\Delta\eta|$$

According to Equation (11),

$$\dot{V} + k\sigma^2 = \dot{V} + 2kV \le -\varepsilon|\sigma|^{p+1} + |\sigma|T \tag{12}$$

If $|\sigma| \ge (T/\varepsilon)^{\frac{1}{p}}$, $\dot{V} + k\sigma^2 \le 0$ exists. Thus, there is an interference stability bound of $(T/\varepsilon)^{\frac{1}{p}}$, which can make the system converge to the region $|\sigma| \le (T/\varepsilon)^{\frac{1}{p}}$ stably in a limited distance. This implies that an interference stability bound $(T/\varepsilon)^{\frac{1}{p}}$ exists. In order to eliminate this interference stability bound, the online estimation of $\Delta\eta$ is performed by RBF (Figure 7) in this study. The corresponding input and output of RBF is $e = [e_1, e_2]^T$ and the control rate is $\Delta\eta$. The 2-5-1 neural network architecture is determined by trial and error. Mini-batch gradient descent is employed, and mini-batch size is set to 2048 by trial and error. The number of epochs is 10. To accelerate learning in the early stage of algorithm optimization and avoid large fluctuations in the later period, the learning rate is adjusted from 0.9 to 0.2. The detailed structure diagram is as follows.

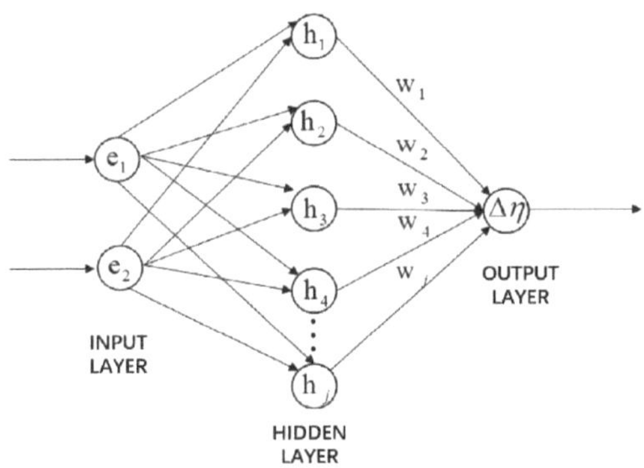

Figure 7. RBF neural network architecture diagram.

The network algorithm used to eliminate the interference stability bound can be described as

$$h_j = \exp\left(\frac{-||e - c_j||^2}{2b_j^2}\right) \tag{13}$$

where, h_j denotes the jth neuron output of the network hidden layer, j is the jth node of the network hidden layer ($j = 1, 2 \ldots 5$). c_j is the center vector value of the jth hidden layer and b_j is the base width parameter of the j hidden layer.

$$\Delta\eta = w^{*T}H \tag{14}$$

where w^* denotes the ideal weight, $H = [h_j]^T$ is the output of the Gaussian basis function. The estimation value of $\Delta\eta$ can be written as

$$\Delta\hat{\eta} = \hat{w}^T H \tag{15}$$

where $\hat{w}$ is the actual weight of the network.

$$\Delta\eta - \Delta\hat{\eta} = w^{*T} H - \hat{w}^T H = -\widetilde{w}^T H \tag{16}$$

where $\widetilde{w} = \hat{w} - w^*$. Substituting $\Delta\hat{\eta}$ for $\Delta\eta$, the developed control rate can be presented as

$$u = -\mu\dot{e}_1 - \varepsilon|\sigma|^p sgn\sigma - k\sigma - \hat{w}^T H \tag{17}$$

Lyapunov function is defined as $V = \frac{1}{2}\sigma^2 + \frac{1}{2\lambda}\widetilde{w}^T\widetilde{w}$ and λ is adjustment parameter and $\lambda > 0$.

$$\begin{aligned} \dot{V} &= \sigma\dot{\sigma} + \tfrac{1}{\lambda}\widetilde{w}^T\dot{\widetilde{w}} = \sigma(\Delta\eta - \Delta\hat{\eta} - \varepsilon|\sigma|_p sgn\sigma - k\sigma) + \tfrac{1}{\lambda}\widetilde{w}^T\dot{\widetilde{w}} \\ &= \sigma(-\widetilde{w}^T H - \varepsilon|\sigma|^p sgn\sigma - k\sigma) + \tfrac{1}{\lambda}\widetilde{w}^T\dot{\widetilde{w}} = -\varepsilon|\sigma|^{p+1} - k\sigma^2 + \widetilde{w}^T(\tfrac{1}{\lambda}\dot{\widetilde{w}} - \sigma H) \end{aligned} \tag{18}$$

The path tracking error converges to zero when the adaptive rate of RBF is selected as $\dot{\hat{w}} = \lambda\sigma H$ and $\dot{V} \leq 0$. Supposing β is the estimation error in the network, thus $\Delta\eta = w^{*T}H + \beta$ ($|\beta| \leq E$ and E is the upper bound of the estimation error). Finally, the tracking error of the system converges steadily to $(E/\varepsilon)^{\frac{1}{p}}$. From Equation (11), it can be seen that $(E/\varepsilon)^{\frac{1}{p}} \ll (T/\varepsilon)^{\frac{1}{p}}$, which can meet the tracking error requirements of the rice transplanter. From the above analysis, the RBF neural network has good robustness to approach the interference stability bound. In other words, when the system is disturbed by uncertainty, the tracking path of the transplanter can still move along the sliding surface towards the balance point, and the system is asymptotically stable.

3.3. Simulation Verification and Analysis

In order to evaluate the effectiveness of the proposed path tracking control strategy for the rice transplanter, the simulation was conducted by MATLAB software. The adaptive rate, neural network, wheelbase of the transplanter, and other relevant parameters are set. The specific information is as follows: wheelbase D_L = 1040 mm, approach law coefficient ″ = 1.2, k = 0.8, p = 0.5, linear sliding surface coefficient ¯ = 8, adaptive rate coefficient λ = 2000, neural network coefficient b_j = 2, c_j = (−0.1, −0.5, 0, 0.5, 1).

In the complex nonlinear system environment, the adaptive sliding mode path tracking control is used to simulate the lateral deviation and heading deviation in the range of 10 m. The simulation results are shown in Figure 8.

From Figure 8, under real field conditions with uncertain disturbance, it can be found that the lateral error converges from a large deviation value to the zero point after about 0.6 m. Meanwhile, the heading deviation converges to the zero point after approximately 0.5 m, and the system gradually enters the steady state. When the adaptive sliding mode path tracking control method is applied to approach uncertain disturbances such as vehicle sideslip, it can effectively improve the steady-state performance of the system and reduce the influence of chattering caused by the system track reaching the switching surface on the system. That is to say, the adaptive sliding mode path tracking control method can effectively eliminate the uncertain disturbances and chattering in the actual control process under the influence of interference, which ensures the accuracy of path tracking and the stability of long-term operation.

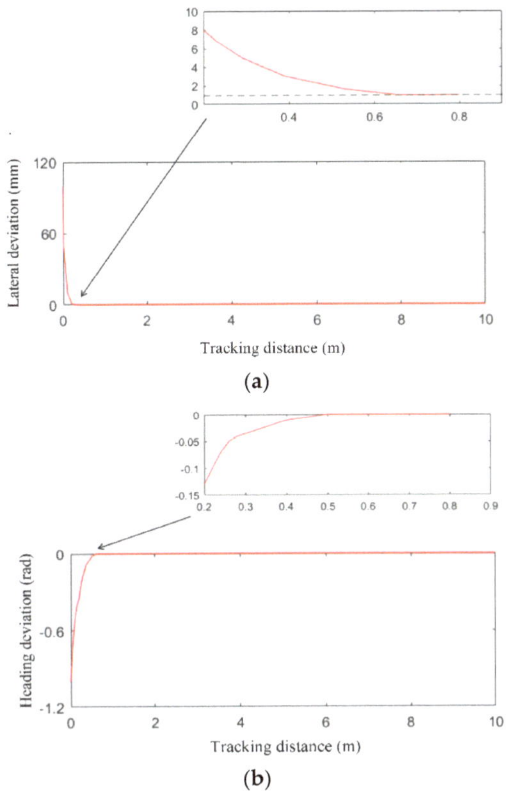

Figure 8. Path tracking error simulation results. (**a**) Lateral deviation; (**b**) heading deviation.

4. Field Test and Data Analysis

4.1. Field Experiment

The performance of the proposed adaptive sliding mode path tracking control algorithm is evaluated by comparing with a traditional sliding mode control algorithm, the paddy field tests are carried out in the grain industrial park of XINGHUA City, JIANGSU Province. The test platform and field test are shown in Figure 9.

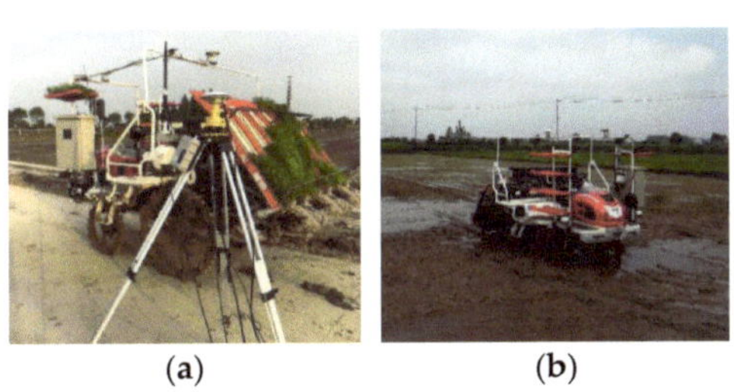

Figure 9. Experimental platform and Path tracking test. (**a**) Test platform; (**b**) field experiment.

A 100 m × 30 m rectangular field was selected for the experiment. The unevenness of paddy soil was less than 5 cm. The mud foot depth in the field is about 25 cm. During paddy field experiments, the vehicle track with depth about 25 cm was generated after the vehicle travelled across the planned paths. If the vehicle runs along the same path many times, the mud foot depth will become deeper and deeper. Due to the influence of sideslip and the restriction of the driving ability of the electric steering wheel, in the next round, it will be difficult for the vehicle to escape the track generated in the previous round of testing. Hence, four groups of field tests with different paths were carried out to validate effectiveness. Considering effect of the vehicle track of the previous trial on tracking performance of the current run, the traditional sliding control and proposed control algorithms were used alternately during each group of tests. For each group of tests, the vehicle followed the same planned paths. During the group 1 and 3 tests, the traditional sliding mode control was applied first and after the proposed control algorithms in this study were run. For groups 2 and 4, the running sequence of the two control algorithms was the opposite. After four sets of tests, the lateral and heading errors were averaged and the corresponding path tracking results from the two control algorithms were presented in Figures 10 and 11, respectively.

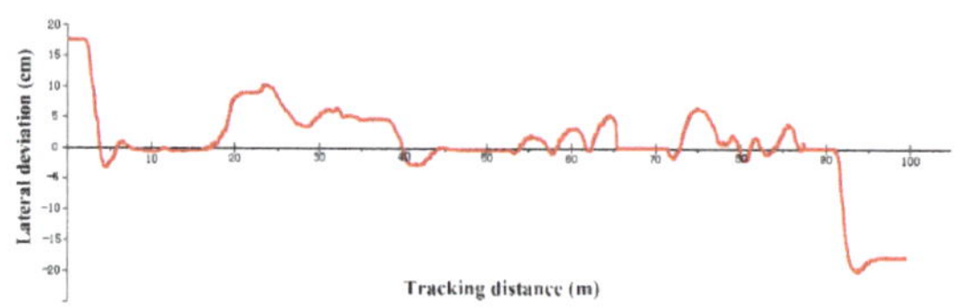

(a)

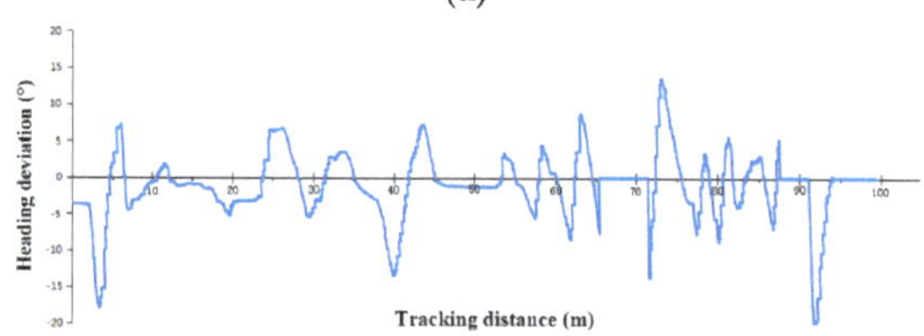

(b)

Figure 10. Path tracking error using traditional sliding mode. (**a**) Lateral deviation; (**b**) heading deviation.

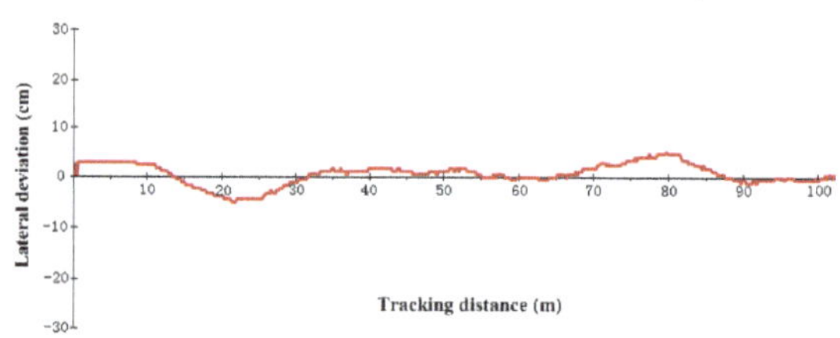

(a)

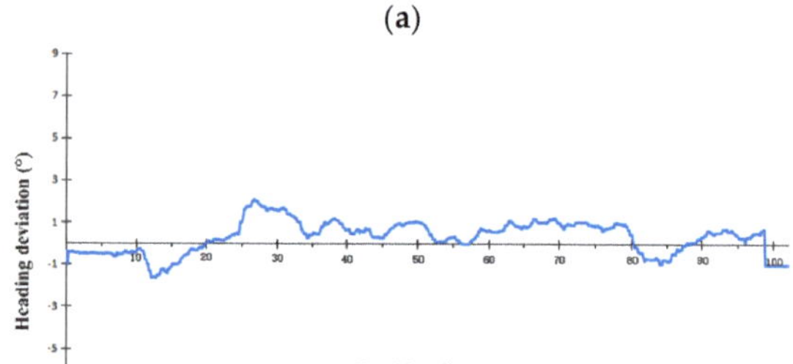

(b)

Figure 11. Path tracking error using proposed control algorithm. (**a**) Lateral deviation; (**b**) heading deviation.

The four longitude and latitude coordinates of the field vertices were collected by the navigation module and transformed into the current field coordinates. According to the operation width of the rice transplanter, the field operation path was independently planned and autonomously generated. Once the transplanter starts to work along the starting position of the long side, the real-time data of lateral deviation and heading deviation were collected through the RS485 serial port of the navigation module. It can be seen from Figure 10 that the lateral deviation obviously increases at about 93 m where the rice transplanter suffered a sudden side slip, which causes the electric steering wheel to rotate beyond the maximum range towards the planned path. In order to prevent the vehicle from rolling over and ensure driving safety, only a small turning angle can be used to approach the planned path step by step. This means that the robustness and adaptive ability of the traditional sliding mode control algorithm is poor when the rice transplanter suffers from uncertain interference such as sideslip or a bumpy road. Moreover, large fluctuations of lateral and heading deviations can be observed, which perhaps are caused by a large stability bound of the control algorithm.

The small fluctuation amplitude of lateral deviation and heading deviation is observed in Figure 11. This indicates that the proposed control algorithm is robust and has good adaptive ability to uncertain inference.

4.2. Data Analysis

In order to verify the performance of the adaptive sliding mode path tracking control method, it is compared with the traditional sliding mode path tracking control method. The compared results are shown in Table 5.

Table 5. Results of experiment.

Type	Traditional Sliding Mode	Adaptive Sliding Mode
Maximum lateral deviation/cm	17.5	7.9
Average of lateral deviation/cm	9.1	3.2
Maximum heading deviation/°	19.3	2.7
Mean of heading deviation/°	8.5	1.3

It can be seen from Table 5 and Figures 10 and 11 that system chattering is obvious when using the traditional sliding mode control; the maximum lateral deviation and the average lateral deviation of path tracking are 17.5 cm and 9.1 cm, respectively. However, the lateral deviation and the average lateral deviation from the adaptive sliding mode control reduce to 7.9 cm and 3.2 cm, respectively. Furthermore, the maximum heading deviation drops from 19.3° to 2.7°, and the mean value of the heading deviation decreases from 8.5° to 1.3°. This phenomenon implies that the control quantity from the controller changes smoothly by adopting the proposed control algorithm and chattering is effectively eliminated, which shows that the proposed path tracking control method is better than the traditional sliding mode path tracking control method.

5. Conclusions

In view of the chattering problem of the system, which is easily caused by external interference and uncertainty in the process of path tracking of the unmanned transplanter, an adaptive sliding mode path tracking control algorithm was proposed. The proposed control algorithm integrated an RBF neural network and sliding mode control which improved the adaptability of the transplanter to the complex and uncertain environment and reduced the lateral deviation and heading deviation.

Based on modified FLW 2ZG-6DM high-speed transplanter, the vehicle kinematic model was established. The simulation tests demonstrated that the position tracking distance approximated from a large deviation distance to near the zero point within about 0.6 m, the direction tracking angle converged to the zero point within 0.5 m, and the system gradually entered the steady state. The model can effectively improve the steady-state performance of the system, reduce the impact of chattering on the system when the system track reaches the switching surface, and ensure the path tracking accuracy and long-term operation stability in the process of driving operation.

The field test results demonstrated that the maximum value and the average value of the lateral deviation were 7.9 cm and 3.2 cm, respectively. The corresponding maximum angle and average value of the heading deviation were 2.7° and 1.3°, respectively. The experimental results implied that the proposed control strategy can meet the automatic driving requirements of rice transplanter in the field.

Author Contributions: Conceptualization, J.L. and R.L.; methodology, Z.S.; software, Z.S.; validation J.L.; formal analysis, Z.S.; investigation, J.L.; resources, J.L.; data curation, R.L.; writing—original draft preparation, R.L.; writing—review and editing, B.C.; visualization, B.C.; supervision, J.L.; project administration, J.L. All authors have read and agreed to the published version of the manuscript.

Funding: This research was funded by the National Key Research and Development Program (2016YFD0700104), and the Key Research and Development Program of Jiangsu Province (BE2018324).

Institutional Review Board Statement: Not applicable.

Informed Consent Statement: Not applicable.

Data Availability Statement: No new data were created or analyzed in this study. Data sharing is not applicable to this article.

Conflicts of Interest: The authors declare no conflict of interest.

References

1. Yang, F.X. Improved design of rice transplanter based on intelligent vision. *China South. Agric. Mach.* **2018**, *49*, 26–29.
2. Kondoh, T.; Nagasaka, Y.; Kato, A.; Shindo, H.; Yaji, Y. Reduction of nitrogen, phosphorus, and suspended solids effluent loads from paddy fields by transplanting into retained ponding water using a GNSS-controlled rice transplanter. *Paddy. Water Environ.* **2019**, *17*, 15–23. [CrossRef]
3. Wang, H.; Wang, G.M.; Luo, X.W.; Zhang, Z.G.; Gao, Y.; He, J.; Yue, B.B. Path tracking control method of agricultural machine navigation based on aiming pursuit mode. *Trans. Chin. Soc. Agric. Eng.* **2019**, *35*, 11–19.
4. Aldair, A.A.; Rashid, A.T.; Rashid, M.T.; Alsaedee, E.B. Adaptive fuzzy control applied to seven-link biped robot using ant colony optimization algorithm. *Iran. J. Sci. Technol. Trans. Electr. Eng.* **2019**, *43*, 797–811. [CrossRef]
5. Li, S.C.; Xu, H.Z.; Ji, Y.H.; Cao, R.Y.; Zhang, M.; Li, H. Development of a following agricultural machinery automatic navigation system. *Comput. Electron. Agric.* **2019**, *158*, 335–344. [CrossRef]
6. Li, T.C.; Hu, J.T.; Gao, L.; Liu, X.G.; Bai, X.P. Agricultural machine path tracking method based on fuzzy adaptive pure pursuit model. *Trans. Chin. Soc. Agric. Mach.* **2013**, *44*, 205–210.
7. Ye, Y.; He, L.; Zhang, Q. Steering control strategies for a four-wheel-independent-steering bin managing robot. *IFAC-PapersOnLine* **2016**, *49*, 39–44. [CrossRef]
8. Xiong, Z.H.; Hak, J.K.; Joon, Y.K.; Sang, Y.Y.; Hee, C.M.; Jung, H.K.; Young, J.K. Path-tracking simulation and field tests for an auto-guidance tillage tractor for a paddy field. *Comput. Electron. Agric.* **2015**, *112*, 161–171.
9. Huang, C.; Naghdy, F.; Du, H. Fault tolerant sliding mode predictive control for uncertain steer-by-wire system. *IEEE. Trans. Cybern.* **2019**, *49*, 261–272. [CrossRef]
10. Wu, X.; Jin, P.; Zou, T.; Qi, Z.; Xiao, H.; Lou, P. Backstepping trajectory tracking based on fuzzy sliding mode control for differential mobile robots. *J. Intell. Robot. Syst.* **2019**, *96*, 109–121. [CrossRef]
11. Wu, H.M.; Karkoub, M.; Hwang, C.L. Mixed fuzzy sliding-mode tracking with backstepping formation control for multi-nonholonomic mobile robots subject to uncertainties. *J. Intell. Robot. Syst.* **2015**, *79*, 73–86. [CrossRef]
12. Zhang, W.Y.; Ding, Y.C.; Wang, X.L.; Zhang, X.; Cai, X.; Liao, Q.X. Pure pursuit control method based on SVR inverse-model for tractor navigation. *Trans. Chin. Soc. Agric. Mach.* **2016**, *47*, 29–36.
13. Wang, L.; Zhai, Z.Q.; Zhu, Z.X.; Mao, E.R. Path racking control of an autonomous tractor using improved Stanley controller optimized with multiple-population genetic algorithm. *Actuators* **2022**, *11*, 22. [CrossRef]
14. Zhang, T.; Jiao, X.H.; Lin, Z.M. Finite time trajectory tracking control of autonomous agricultural tractor integrated nonsingular fast terminal sliding mode and disturbance observer. *Biosyst. Eng.* **2022**, *219*, 153–164. [CrossRef]
15. Xiong, J.T.; Liu, Z.; Chen, S.M.; Liu, B.L.; Zheng, Z.H.; Zhong, Z.; Yang, Z.G.; Peng, H.X. Visual detection of green mangoes by an unmanned aerial vehicle in orchards based on a deep learning method. *Biosyst. Eng.* **2020**, *194*, 261–272. [CrossRef]
16. Mubashiru, O.L. Tomato detection based on modifed YOLOv3 framework. *Sci. Rep.* **2021**, *11*, 1447.

17. Arunabha, M.R.; Jayabrata, B. Real-time growth stage detection model for high degree of occultation using DenseNet-fused YOLOv4. *Comput. Electron. Agric.* **2022**, *193*, 106694.
18. Gokhan, B.; Marcel, B.; Erhan, I.K.; Ahmet, B.K. Improving the trajectory tracking performance of autonomous orchard vehicles using wheel slip compensation. *Biosyst. Eng.* **2016**, *146*, 149–164.
19. Erkan, K.; Erdal, K.; Herman, R.; Wouter, S. Towards agrobots: Identification of the yaw dynamics and trajectory tracking of an autonomous tractor. *Comput. Electron. Agric.* **2015**, *115*, 78–87.
20. Tu, X.Y.; Gai, J.Y.; Tang, L. Robust navigation control of a 4WD/4WS agricultural robotic vehicle. *Comput. Electron. Agric.* **2019**, *164*, 104892. [CrossRef]
21. Xu, G.F.; Chen, M.Z.; He, X.K.; Pang, H.X.; Miao, H.Q.; Cui, P.D.; Wang, W.J.; Diao, P.S. Path following control of tractor with an electro-hydraulic coupling steering system: Layered multi-loop robust control architecture. *Biosyst. Eng.* **2021**, *209*, 282–299. [CrossRef]
22. Li, Z.Q.; Chen, L.Q.; Zheng, Q.; Dou, X.Y.; Yang, L. Control of a path following caterpillar robot based on a sliding mode variable structure algorithm. *Biosystems. Eng.* **2019**, *186*, 80–87. [CrossRef]
23. Deng, H.; Sun, F.C.; Sun, Z.Q. Robust control of robotic manipulators using fuzzy inverse model. *Acta Autom. Sin.* **2001**, *27*, 521–529.
24. Morioka, H.; Wada, K.; Sabanovic, A.; Jezernik, K. Neural network based chattering free sliding mode control. In Proceeding of the 34th SICE Annual Conference, Hokkaido, Japan, 26–28 July 1995; pp. 1303–1308.
25. Ertugrul, M.; Kaynak, O. Neuro sliding mode control of robotic manipulators. *Mechatronics* **2000**, *10*, 239–263. [CrossRef]
26. Huang, S.J.; Huang, K.C.; Chiou, K.C. Development and applications of a novel radial basis function sliding mode controller. *Mechatronics* **2003**, *13*, 313–329. [CrossRef]
27. Zhu, G.Y. *Study on Automatic Steering Control System of High Speed Rice Transplanter*; Engineering College of Anhui Agricultural University: Hefei, China, 2018.
28. He, J.; Zhu, J.G.; Luo, X.W.; Zhang, Z.G.; Hu, L.; Gao, Y. Design of steering control system for rice transplanter equipped with steering wheel-like motor. *Trans. Chin. Soc. Agric. Eng.* **2019**, *35*, 10–17.
29. Hu, J.T.; Gao, L.; Bai, X.P.; Li, T.C.; Liu, X.G. Review of research on automatic guidance of agricultural vehicles. *Trans. Chin. Soc. Agric. Eng.* **2015**, *31*, 1–10.
30. Mohammad, R.R.; Zahra, T.S. A novel adaptive sliding mode controller design for tracking problem of an AUV in the horizontal plane. *Int. J. Dyn. Control.* **2019**, *7*, 679–689.
31. Li, T.C. Adaptive sliding mode path tracking control of agricultural wheeled mobile robots. *China Mech. Eng.* **2018**, *29*, 579–584.
32. Derrick, J.B.; Bevly, D.M. Adaptive steering control of a farm tractor with varying yaw rate properties. *J. Field Robot.* **2009**, *26*, 519–536. [CrossRef]

sensors

MDPI

Article

Vision-Based Module for Herding with a Sheepdog Robot

Virginia Riego del Castillo [1], Lidia Sánchez-González [1,*], Adrián Campazas-Vega [1] and Nicola Strisciuglio [2]

[1] Departamento de Ingenierías Mecánica, Informática y Aeroespacial, Universidad de León, 24071 León, Spain; vriec@unileon.es (V.R.d.C.); acamv@unileon.es (A.C.-V.)
[2] Faculty of Electrical Engineering, Mathematics and Computer Science, University of Twente, 7522 NB Enschede, The Netherlands; n.strisciuglio@utwente.nl
* Correspondence: lidia.sanchez@unileon.es

Abstract: Livestock farming is assisted more and more by technological solutions, such as robots. One of the main problems for shepherds is the control and care of livestock in areas difficult to access where grazing animals are attacked by predators such as the Iberian wolf in the northwest of the Iberian Peninsula. In this paper, we propose a system to automatically generate benchmarks of animal images of different species from iNaturalist API, which is coupled with a vision-based module that allows us to automatically detect predators and distinguish them from other animals. We tested multiple existing object detection models to determine the best one in terms of efficiency and speed, as it is conceived for real-time environments. YOLOv5m achieves the best performance as it can process 64 FPS, achieving an mAP (with IoU of 50%) of 99.49% for a dataset where wolves (predator) or dogs (prey) have to be detected and distinguished. This result meets the requirements of pasture-based livestock farms.

Keywords: computer vision; threat identification; wolf recognition; herding; sheepdog robots; precision livestock farming

Citation: Riego del Castillo, V.; Sánchez-González, L.; Campazas, A.; Strisciuglio, N. Vision-Based Module for Herding with a Sheepdog Robot. *Sensors* **2022**, *22*, 5321. https://doi.org/10.3390/s22145321

Academic Editors: Diego González-Aguilera, Jose A. Jiménez-Berni, Ittai Herrmann, Shangpeng Sun and Monica Herrero-Huerta

Received: 30 May 2022
Accepted: 12 July 2022
Published: 16 July 2022

1. Introduction

Nowadays, people are more concerned about sustainability and biodiversity, and they demand eco-products and healthier food [1]. This trend also helps to enhance regions that have experienced a huge depopulation in recent years. For example, traditional jobs such as sheep herding have become again popular, initiating extensive livestock farming in rural areas as a form of entrepreneurship. Industrial livestock production systems are mainly indoors and have a higher profitability [2]. On the contrary, pasture-based livestock farming increases animal welfare as they can behave naturally and move freely, and it is also positive for biodiversity [3,4]. However, grazing and monitoring the welfare of cattle and sheep is an arduous and time-consuming task, as the animals are often scattered over large areas and require year-round attention [5]. For this reason, assisting the required tasks by deploying autonomous systems is of utmost timeliness, although it presents challenges.

Precision Livestock Farming (PLF) offers technological tools to assist farmers in livestock management [6]. Through the use of sensors and data-driven systems, the herdsmen are able to manage and control several stages of the production flow [7]. In this sense, the benefits of adding new technologies to improve herder productivity are being explored, such as in the case of pasture-based livestock farming. Several different approaches in the literature use certain sensors such as accelerometers, cameras or GPS collars among others to obtain data useful to understand the animal behaviour and monitor them in order to detect diseases, supervise feeding and weight gain control [8].

Pasture-based livestock production remains an important sector in the European Mediterranean basin. It contributes to preserving large agricultural areas of high nature value, which are often located in less industrialised regions with low productive capacity, such as mountainous regions, e.g., in southern Europe [9]. These grazing systems present environmental advantages, facilitating biodiversity and encouraging cultural and landscape diversity [9]. Furthermore the shepherd's working conditions have personal benefits, such as working in a natural environment. On the downside, they may suffer from inclement weather.

However, pasture-based livestock farming has several trade-offs [9]. Exclusively pasture-feeding animals ensures they maintain nutrient cycles, but it implies landscape modification. Another issue to consider is the prevention of environmental risks, such as wolf attacks. They might indeed harm the herd, causing a reduction in profits. Therefore, the introduction of a mobile robot capable of monitoring the herd, performing grazing tasks and sending information to the farmers when they cannot cover the whole herd would help to improve current PLF solutions. It also helps biodiversity as it can locate wolf packs and send their position to the herder to avoid encountering them.

Robots have applications in many fields as they have a multitude of functionalities such as the ability to swim [10], navigation in dark underground mines [11], assistance in subterranean search-and-rescue [12], and others. In livestock production, many commercial solutions help farmers in their daily routines. One example is robotic milking farms, which allow experts to analyse herd behaviour during the summer, enabling them to control how temperature affects milk production as well as contributing to animal welfare [13]. Other tasks have also been automated to assist farmers, such as feeding robots (https://www.lely.com/solutions/feeding/vector/, accessed on 28 May 2022), forage pushers (https://milkingparlour.co.uk/portfolio/joz-moov-robotic-silage-pusher/, accessed on 28 May 2022), scraping robots (https://www.lely.com/gb/centers/eglish/farm-business-improvement-scheme/, accessed on 28 May 2022) or herd-monitoring robots (https://www.gea.com/en/articles/data-for-better-fresh-cow-management/index.jsp, accessed on 28 May 2022), which provide data to monitor animal welfare, control feeding and prevent disease. However, most of them are used indoors, as they require a specific infrastructure.

Although the use of a four-legged robot as a sheepdog has been reported in the news, it is still an unsolved problem [14]. The reason is basically that the challenges of deploying a robot into the wild (perception of the environment, herd control or communication problems, among others) still requires investigation to allow its use in real-world scenarios. In this work, we focus on the development of a perception system that can build automatically a dataset with the desired predators using the iNaturalist API and determine the presence of such potential threats in order to prevent damages and respecting biodiversity. Otherwise, the proposed system can be deployed on an autonomous robot operating as a sheepdog in an outdoor farm.

We present a vision-based system that provides herders with valuable information from on-site sheepdog robots in real time. This information helps to increase the profitability of the sheep farm by avoiding some threats and helping the shepherd in their daily tasks. The proposed system identifies threats such as the presence of wolves, which helps make decisions about which grazing areas to drive the flock to. Moreover, we have developed a method to automatically build datasets of images of certain potential predators of a region by using the iNaturalist API [15]. The proposed system can be thus adapted to any region, including species of animals specific of a certain area or part of the world.

This paper is organised as follows. In Section 2, the related works are discussed. Section 3 presents the proposed system. The dataset considered to conduct the experiments and all the experimental setup is given in Section 4. Section 5 shows the obtained results. Finally, Section 6 includes a comparison with existing techniques, and Section 7 gathers the achieved conclusions.

2. Related Works

PLF involves data-driven systems to control animals, supervising all the related aspects such as their welfare or health, improving the production process. As sensor technology has advanced enormously in recent years, many measures related to physiological, behavioural and productivity aspects can be acquired [16]. Most of them are focused on monitoring and tracking animal behaviour [17,18] as it is related to livestock diseases and also allows for the analysis of extensive and hilly pastures [19].

Different sensors are employed to develop these tasks such as GPS, cameras, accelerometers or thermographic devices [20,21]. Generally, these sensors are classified in two groups sensors worn by animals such as ear tags, collars, leg straps or implants [5], and sensors placed in the animal's surroundings as in the case of cameras. Sensors placed on animals have several drawbacks. For example, the use of GPS collars can harm animals or even become stuck in forest. This is solved with sensors placed in the environment, which have more advantages by allowing the tracking of many animals simultaneously instead of a single one.

Among wearable sensors, there are approaches that use accelerometers or gyroscopes. These sensors are attached to the ear or collar to classify grazing and ruminant behaviour in sheep, providing information about their health or even detecting the lack of pasture quality [22].

With non-wearable sensors, in [23], a system for tracking sheep that detects if they are standing or lying with infrared radiation cameras and computer vision techniques is proposed. Using video cameras and deep learning, wild animals can be successfully identified, counted and described [24,25] as well as other particular species such as Holstein Friesian cattle [26]. A quadcopter with a Mask R-CNN architecture has been used to detect and count cattle in both extensive production pastures and in feedlots with an accuracy of 94% [27,28]. A complete review of the use of images from satellites, manned aircraft and unmanned aerial systems to detect wild animals is given in [29].

Regarding herding, most of the existing approaches are based on monitoring animals from a distance, since it results in a performance increase of the livestock farm, improves animal welfare and reduces the environmental impact. This is called Virtual Fencing (VF) and it is used for handling free-ranging livestock, as it optimises the rangelands and acts like a virtual shepherd [30]. By combining GPS information from the collars and remote sensing, it is possible to monitor animal interactions with the landscape to predict animal behaviour and manage rangelands, protecting those regions that are sensitive or likely to suffer degradation due to overgrazing [31]. Other approaches use drones to herd and move animals, especially in areas dangerous for herders [5,32]. Regarding the impact of drones to animals, there are studies that conclude that terrestrial mammals are more tolerant to unmanned aircraft systems [33], becoming accustomed to them [34], but other species, such as penguins, react differently [35]. Moreover, while unmanned ground vehicles (UGVs) can operate in adverse weather conditions, drones cannot and have problems in forested areas [36].

As the proposed system is developed outdoors, wireless coverage is not guaranteed. In [37,38], a review of the existing wireless sensor networks that can be employed in Precision Agriculture is gathered. Among them, the Long-Range Radio (LoRa) Protocol is discussed as well as its use in different approaches such as smart irrigation systems or the monitoring of agricultural production [38,39].

The proposed method goes a step further, focusing not only on the herd but also on potential threats such as the presence of predators such as wolves, which is a major challenge as it involves both livestock safety and wolf conservation [40,41]. In other regions, predators are jaguars [42] or bears [43,44]. Thus, an adaptive method should be automatically configured to work with different species of predators. In Section 4.1, we show how to configure the method to detect other predators. In addition to this, contact between wildlife and livestock can also be studied as it can potentially transmit zoonotic diseases [45].

In order to detect the presence of wolves, computer vision techniques are employed. Traditionally, features such as Scale-Invariant Feature Transform (SIFT), Speeded Up Robust Features (SURF) or Histogram of Oriented Gradients (HOG) were extracted from images and then were classified using Support Vector Machines, among other classifiers, in order to detect objects in an image [46]. These HOG features have been employed, i.e., for human detection [47], animal detection to avoid animal–vehicle collisions [48], and also in real time [49].

Modern approaches use deep learning techniques such as Convolutional Neural Networks (CNN) that extract features from images, which were subsequently classified using several dense layers. There are also solutions that combine traditional features such as HoG with CNN [50,51]. An extensive review of the object detection techniques using deep learning is provided in [52]. Different CNN architectures are used for object detection and classification, which are trained for large datasets—a time-consuming task [53]. As it is not possible to handle certain problems if the available dataset is not so large or the hardware requirements do not allow obtaining results in a short period of time, a technique named transfer learning is used to take advantage of those existing models that have been trained to detect certain objects and adapt them to a different domain [54].

Through transfer learning, pre-trained models can be adapted to new domains in a way that takes advantage of the knowledge extracted from the set of millions of images on which the network was trained and adapts it to the new problem [54]. They also reduce training time as only the added layers, which introduce the information of the particular problem, are trained. The model layers are frozen or fine-tuned, so the existing knowledge is maintained but adapted to the new problem. Most of the popular solutions for object detection are: YOLO (You Only Look Once), Single Shot Detector (SSD) and R-CNN. The existing architectures with the pre-trained models have been applied to animal classification with an accuracy over 90%. Some examples are fish classification [55], bird classification [56], camera trap images classification [57] or wild animal image classification [39].

3. System Architecture

The imitation of sheepdog behaviours with a shepherd robot requires a decision-making system that is able to manage information collected by its sensors and to generate actions through its actuators, taking into account the changing characteristics of the environment. As this system will be deployed in a real environment, some conditions, such as lighting and obstacles that must be detected for a proper navigation, are not fixed, since they vary over time. Moreover, these decisions have to offer long and short-term opportunities to be reactive to any expected and unexpected behaviour in the scenario.

This research uses MERLIN [58], which combines deliberative and behavioural capacities. Deliberative capacities define the characteristics of planning to infer long-term tasks. Behavioural capabilities provide the set of actions capable of responding to changes in the environment more quickly. Thus, Merlin allows us to set a certain goal but is also able to react if an unexpected event occurs.

Once the robot is deployed in a real context, it is crucial to handle unexpected behaviours, and for this, the use of different sensors is required. Traditionally, the literature proposes the use of camera sensors, such as the one proposed in this research. This sensor provides information about the context, such as dangerous species, identification of certain animals or predator attacks. The information provided by the proposed method in the robot's closed environment will promote alternative navigation routes, keep herd control updated or trigger an alarm process to alert the herder remotely if a wolf is detected. The proposed vision module can also be deployed in fixed cameras near grazing areas.

The proposed vision module is included along with the other reactive components. It assumes a background approach, feeding a knowledge database and keeping the deliberative monitoring system that would interrupt the active task up to date. The architecture proposed in [58] requires to be updated by including the new vision-based module proposed in this paper in the reactive layer. As this reactive layer gathers all the sensors of the robot, with the proposed vision module, the images acquired by the robot camera are processed. The obtained information is sent to the rest of the layers of the architecture. The useful provided knowledge about what the robot has seen helps the rest of the architecture to make decisions about the subsequent actions to carry out (see Figure 1).

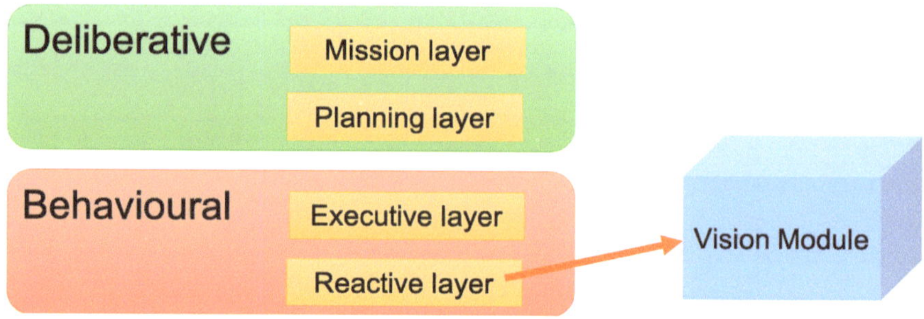

Figure 1. MERLIN architecture and proposed vision module.

In some scenarios, the autonomous features of the robot are not sufficient for the continuous monitoring of the herd, and there are some specific events where the herder needs to access the robot in real time to monitor dangerous external contexts or improve productivity. The point of view of the sheperd and of the sheepdog are not always the same, so the robot can respond to the commands of the shepherd, who has previous experience and knows how to deal with situations such as weather conditions or grazing specific areas Thus, images from the robot can be sent to the shepherd's mobile phone on demand.

The acquired images are processed in the vision module in order to obtain information about the environment. This information is sent to the shepherd by using the LoRaWan® networking protocol. It is widely used in those areas where there are not wireless connections satisfying Internet of Things requirements. The LoRaWan® network architecture allows bi-directional communication, and messages are sent to a gateway that functions as a bridge to the Internet network [59]. Those gateways can be located at different points of the region, making it possible to cover a huge area (≈15 km).

4. Vision Module

The vision module belongs to the perception system, which plays an important role in the behaviour of a robot. The use of a Unitree A1, which is a robot dog, has been proposed to detect victim and pedestrians in emergency rescue situations by using thermal and color cameras [60]. These systems usually use a Robot Operating System (ROS) with a vision module to acquire images and detect the existing objects, i.e., smart glasses [61] or cameras on drones [62]. Cameras can also be used to help in robot navigation, exploration and mapping as in [63,64].

We propose a vision module that can be used in fixed cameras near pastures and villages or in cameras built into SheepDog robot systems. Figure 2 shows the complete pipeline of the process, where images are acquired and labelled in order to train the object detection models.

There are datasets that are very frequently used in object detection problems [65]. PASCAL Visual Object Classes (VOC) [66] includes 20 categories with animals such as birds, cats, cows, dogs, horses and sheep. ImageNet [67] provides an important source of data for object detection with 200 labelled categories in the 2017 version. Common Objects in Context (COCO) [68] includes the localisation, label and semantic information for each object in one of the 80 categories. In this work, the proposed system can automatically generate species-specific animal datasets through an API.

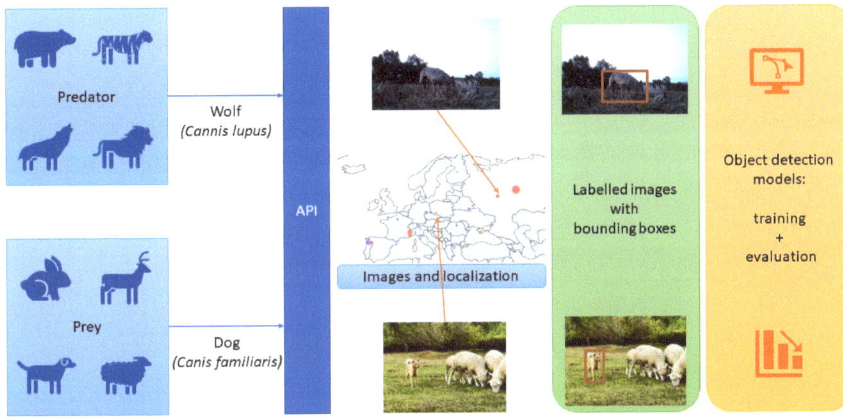

Figure 2. Pipeline of the vision module. First, images and their location are acquired using an API; then, images are labelled to train object detection models.

4.1. Data Acquisition And Labelling

Shepherds have to deal with predator attacks on the herd, so it is necessary to anticipate this situation by evading the threat and distinguishing if there is a potential risk [69]. We consider the presence of predatory animals as a potential risk, which can be a bear, tiger, lion or wolf. In the proposed system, prey animals are distinguished from predators to determine suitable grazing areas to maintain distances from predator locations.

We focus on a predatory species of the northwest of the Iberian Peninsula: the Iberian wolf. In [70], a study of the diet of the Iberian wolf shows that it tends to eat goats, cattle, sheep and rabbits, which are some of the animals that farmers raise in the area. Iberian wolves have phylogenetic proximity to other European wolf populations (*Canis lupus*), being considered as a sub-specie of it (*Canis lupus signatus*). Otherwise, dogs are the domesticated descendants of the wolf, presenting similarities as specie (*Canis familiaris*). In this paper, we have created a dataset to differentiate a predator (wolf) from a prey (dog) that can be implemented for more species diversity.

We have used the iNaturalist API [15] to create the dataset, obtaining images from two species: *Canis lupus* (wolf) and *Canis familiaris* (dog). As the code is available in [71], the vision module can be adapted to other predators of other regions by using the notebook `get_inaturalist` and choosing the desired species. Localisation has allowed us to divide the images into two groups: Europe and Outside Europe. Then, images were labelled manually by experts, removing images with low quality. Figure 3 shows how 925 images and 1137 detections are split by species and location.

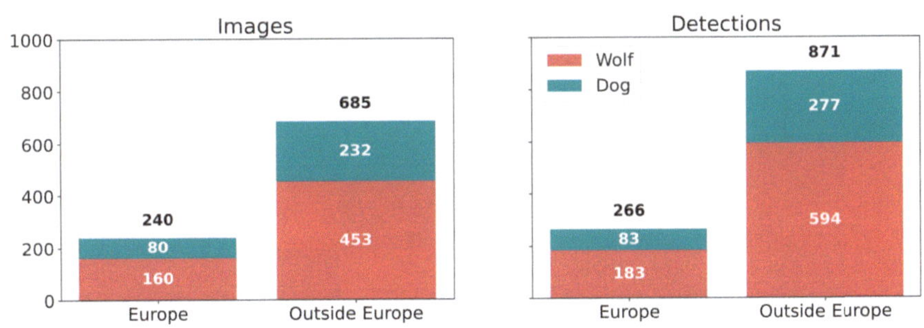

Figure 3. Information of the dataset disaggregated by species and location in number of images (**left side**) and number of detections (**right side**).

Images present diversity, as different animals can appear on the images (Figure 4a), lying (Figure 4b), looking to the camera (Figure 4c), partially occluded (Figure 4d), with different lighting conditions (Figure 4e), with multiple detections (Figure 4f) or just one (Figure 4g), feeding (Figure 4h) with different illuminations and different distances from the camera. Due to this information, in Europe, we can observe wolves in couples or alone, as can be shown in Figure 5.

Figure 4. Samples of dogs (**upper row**) and wolves (**bottom row**) in Europe (**left**) and the rest of the world (**right**).

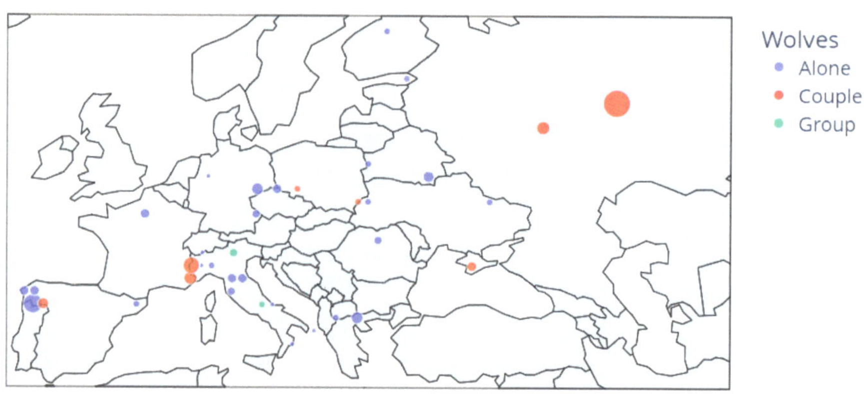

Figure 5. Iberian wolves detected in Europe. Bubble size depends on the number of images.

4.2. Object Detection Architectures

Object detectors are evaluated based on accuracy, speed and complexity. Two-stage detectors have two steps: extract features from the input image (feature extractor) and recognise the features (classifier). Meanwhile, one-stage detectors combine the feature extractor and classifier into one, reducing complexity and improving speed, but accuracy may be reduced. As the proposed module is deployed in real-time environments, it is based on one-stage detectors. The considered state-of-the-art algorithms [72] based on one stage are You Only Look Once (YOLO) in different versions (YOLOv1, YOLOv2, YOLOv3, YOLOv4, YOLOv5) and Single-Shot MultiBox Detector (SSD). SSD has improved versions such as Deconvolutional SSD (DSSD) that includes large-scale context in object detection, Rainbow SSD (RSSD) that concatenates different feature maps using deconvolution and batch normalisation [73], and Feature-fusion SSD (FSSD) that balances semantic and positional information using bilinear interpolation to resize feature maps to the same size to be subsequently concatenated [74]. The comparison of different architectures for real-time applications presented in [75] also mentions RetinaNet because it has higher accuracy, but it is not recommended for real-time applications, as it has a frame rate lower than 25 frames per second (FPS). EdgeEye [76] proposes an edge computing framework to analyse real-time video with a mean of 55 FPS as inference speed.

For evaluation, different metrics have been considered. First, Intersection over Union (IoU) measures the overlapping area between the predicted bounding box and the ground-truth divided by the union of the areas. IoU is fixed by a threshold (t) generating the confusion matrix as:

- True Positives (TP) are those objects detected by the model with an IoU greater than the considered threshold ($IoU \geq t$);
- False Positives (FP) are the detected objects whose IoU is less than the fixed threshold ($IoU < t$);
- False Negatives (FN) stand for those objects that are not detected;
- True Negatives (TN) are the number of objects detected by the model when actually the image does not have such objects.

Detector models use performance metrics computed from the confusion matrix mentioned above as follows:

- *Precision*: measures how many of the predicted outputs labelled as true predictions are correctly predicted:

$$Precision = \frac{TP}{TP + FP} \tag{1}$$

- *Recall*: measures how many of the real true predictions are correctly predicted:

$$Recall = \frac{TP}{TP + FN} \tag{2}$$

In order to compare the results of different authors, there are well-established metrics based on mean Average Precision (mAP), which is the average of the accuracy obtained in the object detection over all the dataset categories. Specifically, metrics are related to datasets mentioned previously [77]:

- COCO metric (mAP_{COCO} or $mAP@50:5:95$): evaluates 10 $IoUs$ between 50% and 95% with steps of 5% of mean Average Precision as

$$mAP_{COCO} = \frac{mAP_{0.50} + mAP_{0.55} + mAP_{0.60} + \cdots + mAP_{0.95}}{10} \tag{3}$$

- PASCAL VOC metric (mAP_{VOC} or mAP_{50}): evaluates IoU at 50%.

4.3. SSD

SSD is composed by two components: a backbone model (in this case, a pre-trained VGG16) and an SSD head with convolutional layers to obtain the bounding boxes and categories of the detected objects. From an image of 300 by 300 pixels, SSD achieves 72.1% mAP_{VOC} on a VOC2007 test at 58 FPS on a Nvidia Titan X [78]. The model has been trained from the PAZ (Perception for Autonomous Systems) library with the object detection module [79].

4.4. YOLO

YOLO is also based on a classification backbone with new headers to obtain the bounding box and the assigned class of the object. There are multiple implementations of the architecture, for example, YOLOv3 [80], which uses a DarkNet framework and makes predictions in three different scales. It achieves 57.9% mAP_{VOC} in 51 ms (around 20 FPS). An improvement was made with YOLOv4 [81] obtaining 65.7% of mAP_{VOC} and a speed of 65 FPS. After this, a tiny version of YOLOv3 [82] was proposed with higher speed but lower precision. The newest version, which is YOLOv5 [83], can achieve 68.9% of mAP_{VOC} with more than 80 FPS. Moreover, it includes different versions of complexity, obtaining more precision and more inference time when the model is more complex. The YOLOv5 models are: Nano (YOLOv5n), Small (YOLOv5s), Medium (YOLOv5m), Large (YOLOv5l) and Extra-Large (YOLOv5x).

5. Results

The goal of the following experiment is to establish which of the existing architectures is better to detect predators as wolves in images taking into account performance and speed due to it being a real-time vision module.

On the one hand, an SSD model was trained with pre-trained weights from VGG and Stochastic Gradient Descent (SGD) optimizer (as in [78]) with a learning rate of 0.0001. Training was completed during 100 epochs with a batch size of 16. The model achieves 92.90% of mAP_{VOC} in the training set and 85.49% in the test set. Inference takes 80 ms on average, which corresponds with 12.5 FPS (in an NVIDIA GeForce RTX 2060).

On the other hand, multiple YOLO models were trained with an SGD optimizer configured with a learning rate of 0.01 (as in [80]) and a batch size of 4 during 50 epochs. Table 1 shows the obtained results of the different models. As it can be observed, YOLOv3 achieved the best results in mAP_{COCO} with 88.63%, while the tiny model is the fastest one with 64 FPS also with an NVIDIA GeForce RTX 2060. With the newest version of YOLO, the extra-large model YOLOv5x yielded the highest mAP_{COCO} with 88.24% but with the lower frame rate, whereas the nano, small and medium architectures are faster, achieving 64 FPS and keeping a mAP sligthly lower. Small architectures (nano YOLOv5n, small YOLOv5s and medium YOLOv5m) are lighter in weight, and therefore, the execution time is lower. Heavier architectures (large YOLOv5l and extra-large YOLOv5x) take longer to run but have higher precision in the results.

Table 1. Results of the models over the test set, where mAP_{VOC} is the PASCAL VOC metric and mAP_{COCO} corresponds with the COCO metric. Inference time is measured in ms, and YOLOv3t is YOLOv3-tiny. Best results are highlighted in bold.

Model	Precision	Recall	mAP_{VOC}	mAP_{COCO}	Inference	FPS
SSD	85.49	93.33	85.49	47.87	80	12.5
YOLOv3	99.35	99.56	99.49	**88.63**	31.24	32.01
YOLOv3t	94.86	94.97	98.63	71.98	**15.62**	64.01
YOLOv5n	95.88	96.13	99.07	75.86	**15.62**	**64.02**
YOLOv5s	98.54	98.54	99.47	82.78	**15.62**	64.01
YOLOv5m	99.17	99.52	99.49	85.53	**15.62**	**64.02**
YOLOv5l	**99.47**	99.61	**99.50**	87.57	31.24	32.01
YOLOv5x	99.38	**99.73**	**99.50**	88.24	46.86	21.34

Figure 6 shows the scheme of the final module. Images are divided into training and validation using localisation: species from outside Europe are used for training and validation is performed with images of European species. First, the YOLO model is trained and then, metrics are calculated using the validation set (2nd step). Once the model is trained and evaluated as it is proposed in this paper, further steps involve integrating the model in an autonomous system such as a robot or in any device with a fixed camera which acquires images (3rd step). The acquired images are processed by the model (4th step). Finally, identified threats such as the presence of wolves can be used to raise an alarm and inform the shepherd to avoid the area where they are located (5th step).

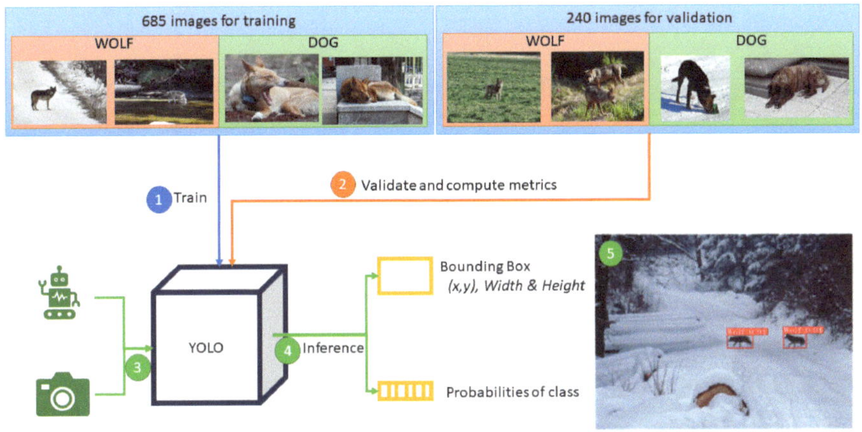

Figure 6. Complete scheme of the proposed vision module.

Further study of the results indicates that YOLO performs best in terms of accuracy and speed; that is, it is able to recognise wolves or wolf packs and distinguish them from dogs with high accuracy and in real time. Figure 7 shows a graph with the COCO metric and speed (FPS), where the best results are in the top right. We can determine that YOLOv5m (medium) is the best model in terms of speed and mAP, as it can process 64 FPS with 99.49% of mAP_{VOC} and 85.53% of mAP_{COCO}. Observing these results, YOLOv5m is chosen to form the vision module.

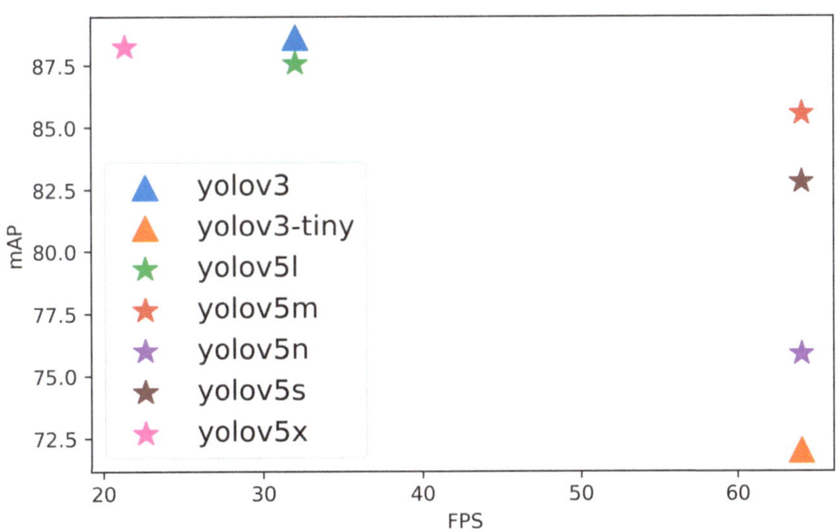

Figure 7. Results of the mAP_{COCO} over the speed for YOLO models.

YOLOv5 employs a loss function composed by three losses: the bounding box loss, which uses a regression loss for object location (Mean Squared Error, MSE), the object loss, which obtains the confidence of object presence (Binary Cross-Entropy) and classification loss, which determines that the classification is correct (Cross-Entropy). Figure 8 displays the evolution of the loss functions during the training that shows the good behaviour of the model. Moreover, an evolution of the obtained metrics is shown in Figure 9, achieving a high performance after 30 epochs.

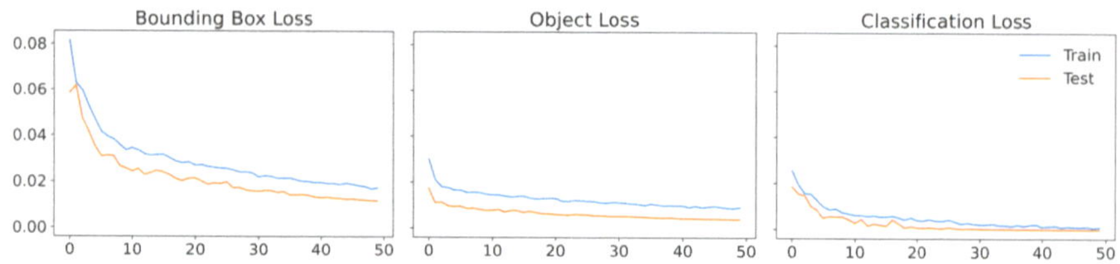

Figure 8. Loss functions of YOLOv5m during the training: bounding box loss with Mean Squared Error (**first graph**), object loss with Binary Cross-Entropy (**second graph**) and classification loss with Cross-Entropy (**third graph**).

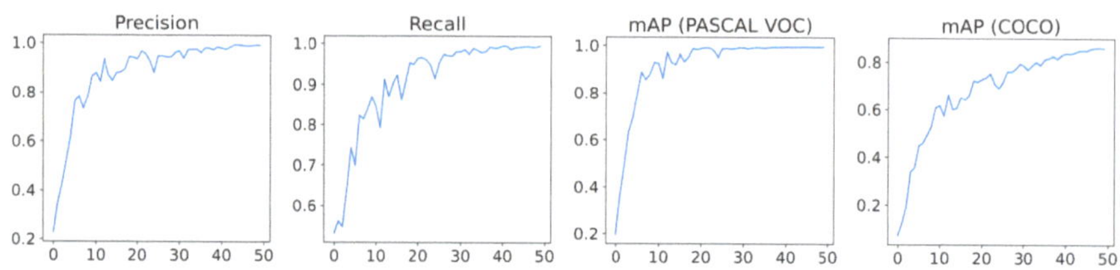

Figure 9. Metrics of YOLOv5m: precision (**first graph**), recall (**second graph**), mAP_{VOC} which corresponds with PASCAL VOC metric (**third graph**) and mAP_{COCO} which corresponds with COCO metric (**fourth graph**).

Finally, Figure 10 gathers some samples of the dataset with the objects detected by the YOLOv5m model. As it can be pointed out, there are images of the considered categories in which the objects to be detected, wolves and dogs, are at different distances to the camera and also deal with occlusions. Figure 11 shows the confusion matrix for training and validation data with an IoU greater than 0.5.

Figure 10. Samples of dogs (**upper row**) and wolves (**bottom row**) in Europe (**left**) and the rest of the world (**right**) with the detections yielded by YOLOv5m.

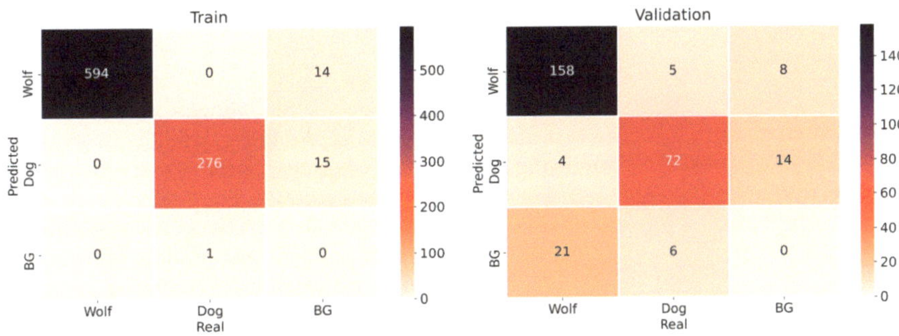

Figure 11. Confusion matrix for training (**left**) and validation (**right**) of YOLOv5m.

6. Discussion

There are approaches in the literature where researchers address similar problems of animal detection and classification. For instance, in [27], cattle are counted by analysing the images with a Mask R-CNN, obtaining 92% accuracy and AP for the detection of 91%, which is outperformed by the proposed method. In addition, using image processing and Mask R-CNN for counting animals, in [28], livestock, sheep and cattle are counted and classified, achieving a precision of 95.5% and mAP_{40} values of 95.2%, 95% and 95.4% for livestock, sheep and cattle, respectively. Using thermal images, animals such as dogs, cats, deer, rhinos, horses and elephants are detected and classified with an mAP_{VOC} of 75.98% with YOLOv3, 84.52% with YOLOv4 and 98.54% with Fast-RCNN [25]. In [24], the problem is turned into a classification as follows. First, a binary classification is performed to decide whether or not there are animals in the image, with an accuracy of 96.8%. If animals are detected, then a classification of the number of animals in the image is carried out, considering 1, 2, 3, 4, 5, 6, 7, 8, 9, 10, 11–50, or +51 individuals and achieving 63.1% top-1 accuracy.

According to the previously discussed methods, the proposed vision module outperforms the state-of-the-art results not only in precision and accuracy but also in speed to be able to couple to real-time systems. Table 2 summarises the comparison between the state-of-art methods explained above and the proposed method.

Table 2. Comparison of the state-of-the-art results with the proposed method for animal detection and count of different species.

Problem	Results
Holstein Friesian cattle detection and count [24]	Animal detection: accuracy of 96.8% Counting animals: 63.1% top-1 accuracy
Cattle count [27]	Counting animals: accuracy of 92% Bounding box prediction (localisation): AP of 91%
Livestock, Sheep, Cattle detection [28]	Precision rate: 95.5%, 96% and 95% Recall with IoU of 0.4: 95,2%, 95% and 95.4%
Animal detection in thermal images [25]	mAP_{VOC} with YOLOv4: 75.98% mAP_{VOC} with YOLOv3: 84.52% mAP_{VOC} with Faster-RCNN: 98.54%
Wolf and Dog detection (Vision Module)	mAP_{VOC} with YOLOv3: 99.49% (FPS: 32) mAP_{VOC} with YOLOv5m: 99.49% (FPS: 64)

7. Conclusions

In this paper, a vision-based module to detect predators in pasture-based livestock farming and distinguish them from other species is proposed. This module can be deployed within on-site sheepdog robots and fixed cameras to assist shepherds in threat detection.

First, we propose a system that can automatically generate datasets of different species through the iNaturalist API in order to obtain a module that can be used in any region depending on the existing predators. Focusing on a predator specie of the northwest of the Iberian Peninsula, namely the Iberian wolf, a particular dataset is obtained. The generated benchmark has the aim of providing data and an evaluation framework to test different algorithms to detect wolves, as a predator specie, and differentiate from other animals such as dogs, which have a similar physical appearance. These data can be automatically extended with the new predator and prey species of the region.

Then, multiple object detection models have been trained to establish which one achieves better results in a real-time module. According to the obtained results, the best results are achieved with YOLOv5m yielding an inference time of 15.62 ms, which allows 64 FPS. This model achieves a precision of 99.17% and a recall of 99.52% on the considered benchmark, outperforming other existing approaches, with an mAP_{VOC} of 99.49% and mAP_{COCO} of 85.53%. These results fulfil the requirements of a real-time detection module and improve state-of-the-art methods.

Future development lines involve integration into autonomous systems and data collection in the field. Information about potential threats will enable early warning alerts to be managed for herders in difficult-to-access terrain.

Author Contributions: Conceptualization, N.S. and L.S.-G.; software, V.R.d.C.; validation, V.R.d.C., A.C.-V. and L.S.-G.; investigation, V.R.d.C. and L.S.-G.; resources, V.R.d.C. and A.C.-V.; data curation, V.R.d.C. and A.C.-V.; writing, V.R.d.C., L.S.-G., A.C.-V. and N.S.; supervision, N.S. and L.S.-G. All authors have read and agreed to the published version of the manuscript.

Funding: The APC was funded by Universidad de León.

Data Availability Statement: All experiments and data are available in [71]. Images can be downloaded from iNaturalist API with the proposed system.

Acknowledgments: Virginia Riego would like to thank Universidad de León for its funding support for her doctoral studies. We also express our grateful to Centro de Supercomputación de Castilla y León (SCAYLE) for its infrastructure support.

Conflicts of Interest: The authors declare no conflict of interest.

Abbreviations

The following abbreviations are used in this manuscript:

CNN	Convolutional Neural Network
COCO	Common Objects in Context
FPS	Frame per second
IoU	Intersection over Union
LoRa	Long-Range Radio
mAP	mean Average Precision
PAZ	Perception for Autonomous Systems
PLF	Precision Livestock Farming
ROS	Robot Operating System
SGD	Stochastic Gradient Descent
SSD	Single-Shot MultiBox Detector
UGVs	Unmanned Ground Vehicles
VF	Virtual Fencing
VOC	Visual Object Classes
YOLO	You Only Look Once

References

1. Delaby, L.; Finn, J.A.; Grange, G.; Horan, B. Pasture-Based Dairy Systems in Temperate Lowlands: Challenges and Opportunities for the Future. *Front. Sustain. Food Syst.* **2020**, *4*, 543587. [CrossRef]
2. Campos, P.; Mesa, B.; Álvarez, A. Pasture-Based Livestock Economics under Joint Production of Commodities and Private Amenity Self-Consumption: Testing in Large Nonindustrial Privately Owned Dehesa Case Studies in Andalusia, Spain. *Agriculture* **2021**, *11*, 214. [CrossRef]
3. Lessire, F.; Moula, N.; Hornick, J.L.; Dufrasne, I. Systematic Review and Meta-Analysis: Identification of Factors Influencing Milking Frequency of Cows in Automatic Milking Systems Combined with Grazing. *Animals* **2020**, *10*, 913. [CrossRef] [PubMed]
4. Yu, H.; Li, Y.; Oshunsanya, S.O.; Are, K.S.; Geng, Y.; Saggar, S.; Liu, W. Re-introduction of light grazing reduces soil erosion and soil respiration in a converted grassland on the Loess Plateau, China. *Agric. Ecosyst. Environ.* **2019**, *280*, 43–52. [CrossRef]
5. Herlin, A.; Brunberg, E.; Hultgren, J.; Högberg, N.; Rydberg, A.; Skarin, A. Animal Welfare Implications of Digital Tools for Monitoring and Management of Cattle and Sheep on Pasture. *Animals* **2021**, *11*, 829. [CrossRef]
6. Schillings, J.; Bennett, R.; Rose, D.C. Exploring the Potential of Precision Livestock Farming Technologies to Help Address Farm Animal Welfare. *Front. Anim. Sci.* **2021**, *2*, 639678. [CrossRef]
7. Niloofar, P.; Francis, D.P.; Lazarova-Molnar, S.; Vulpe, A.; Vochin, M.C.; Suciu, G.; Balanescu, M.; Anestis, V.; Bartzanas, T. Data-driven decision support in livestock farming for improved animal health, welfare and greenhouse gas emissions: Overview and challenges. *Comput. Electron. Agric.* **2021**, *190*, 106406. [CrossRef]
8. Samperio, E.; Lidón, I.; Rebollar, R.; Castejón-Limas, M.; Álvarez-Aparicio, C. Lambs' live weight estimation using 3D images. *Animal* **2021**, *15*, 100212. [CrossRef]
9. Bernués, A.; Ruiz, R.; Olaizola, A.; Villalba, D.; Casasús, I. Sustainability of pasture-based livestock farming systems in the European Mediterranean context: Synergies and trade-offs. *Livest. Sci.* **2011**, *139*, 44–57. [CrossRef]
10. Chen, G.; Lu, Y.; Yang, X.; Hu, H. Reinforcement learning control for the swimming motions of a beaver-like, single-legged robot based on biological inspiration. *Robot. Auton. Syst.* **2022**, *154*, 104116. [CrossRef]
11. Mansouri, S.S.; Kanellakis, C.; Kominiak, D.; Nikolakopoulos, G. Deploying MAVs for autonomous navigation in dark underground mine environments. *Robot. Auton. Syst.* **2020**, *126*, 103472. [CrossRef]
12. Lindqvist, B.; Karlsson, S.; Koval, A.; Tevetzidis, I.; Haluška, J.; Kanellakis, C.; akbar Agha-mohammadi, A.; Nikolakopoulos, G. Multimodality robotic systems: Integrated combined legged-aerial mobility for subterranean search-and-rescue. *Robot. Auton. Syst.* **2022**, *154*, 104134. [CrossRef]
13. Osei-Amponsah, R.; Dunshea, F.R.; Leury, B.J.; Cheng, L.; Cullen, B.; Joy, A.; Abhijith, A.; Zhang, M.H.; Chauhan, S.S. Heat Stress Impacts on Lactating Cows Grazing Australian Summer Pastures on an Automatic Robotic Dairy. *Animals* **2020**, *10*, 869. [CrossRef] [PubMed]
14. Vincent, J. A Robot Sheepdog? 'No One Wants This,' Says One Shepherd. 2020. Available online: https://www.theverge.com/2020/5/22/21267379/robot-dog-rocos-boston-dynamics-video-spot-shepherd-reaction (accessed on 12 January 2022).
15. Matheson, C.A. iNaturalist. *Ref. Rev.* **2014**, *28*, 36–38. [CrossRef]
16. Tedeschi, L.O.; Greenwood, P.L.; Halachmi, I. Advancements in sensor technology and decision support intelligent tools to assist smart livestock farming. *J. Anim. Sci.* **2021**, *99*, skab038. [CrossRef]
17. Porto, S.; Arcidiacono, C.; Giummarra, A.; Anguzza, U.; Cascone, G. Localisation and identification performances of a real-time location system based on ultra wide band technology for monitoring and tracking dairy cow behaviour in a semi-open free-stall barn. *Comput. Electron. Agric.* **2014**, *108*, 221–229. [CrossRef]
18. Spedener, M.; Tofastrud, M.; Devineau, O.; Zimmermann, B. Microhabitat selection of free-ranging beef cattle in south-boreal forest. *Appl. Anim. Behav. Sci.* **2019**, *213*, 33–39. [CrossRef]
19. Bailey, D.W.; Trotter, M.G.; Knight, C.W.; Thomas, M.G. Use of GPS tracking collars and accelerometers for rangeland livestock production research1. *Transl. Anim. Sci.* **2018**, *2*, 81–88. [CrossRef]
20. Stygar, A.H.; Gómez, Y.; Berteselli, G.V.; Costa, E.D.; Canali, E.; Niemi, J.K.; Llonch, P.; Pastell, M. A Systematic Review on Commercially Available and Validated Sensor Technologies for Welfare Assessment of Dairy Cattle. *Front. Vet. Sci.* **2021**, *8*, 634338. [CrossRef]
21. Aquilani, C.; Confessore, A.; Bozzi, R.; Sirtori, F.; Pugliese, C. Review: Precision Livestock Farming technologies in pasture-based livestock systems. *Animal* **2022**, *16*, 100429. [CrossRef]
22. Mansbridge, N.; Mitsch, J.; Bollard, N.; Ellis, K.; Miguel-Pacheco, G.G.; Dottorini, T.; Kaler, J. Feature Selection and Comparison of Machine Learning Algorithms in Classification of Grazing and Rumination Behaviour in Sheep. *Sensors* **2018**, *18*, 3532. [CrossRef] [PubMed]
23. Ren, K.; Karlsson, J.; Liuska, M.; Hartikainen, M.; Hansen, I.; Jørgensen, G.H. A sensor-fusion-system for tracking sheep location and behaviour. *Int. J. Distrib. Sens. Netw.* **2020**, *16*. [CrossRef]
24. Norouzzadeh, M.S.; Nguyen, A.; Kosmala, M.; Swanson, A.; Palmer, M.S.; Packer, C.; Clune, J. Automatically identifying, counting, and describing wild animals in camera-trap images with deep learning. *Proc. Natl. Acad. Sci. USA* **2018**, *115*, E5716–E5725. [CrossRef]
25. Khatri, K.; Asha, C.C.; D'Souza, J.M. Detection of Animals in Thermal Imagery for Surveillance using GAN and Object Detection Framework. In Proceedings of the 2022 International Conference for Advancement in Technology (ICONAT), Goa, India, 21–22 January 2022 ; pp. 1–6. [CrossRef]

26. Andrew, W.; Greatwood, C.; Burghardt, T. Visual Localisation and Individual Identification of Holstein Friesian Cattle via Deep Learning. In Proceedings of the 2017 IEEE International Conference on Computer Vision Workshops (ICCVW), Venice, Italy, 22–29 October 2017; pp. 2850–2859. [CrossRef]
27. Xu, B.; Wang, W.; Falzon, G.; Kwan, P.; Guo, L.; Chen, G.; Tait, A.; Schneider, D. Automated cattle counting using Mask R-CNN in quadcopter vision system. *Comput. Electron. Agric.* **2020**, *171*, 105300. [CrossRef]
28. Xu, B.; Wang, W.; Falzon, G.; Kwan, P.; Guo, L.; Sun, Z.; Li, C. Livestock classification and counting in quadcopter aerial images using Mask R-CNN. *Int. J. Remote Sens.* **2020**, *41*, 8121–8142. [CrossRef]
29. Wang, D.; Shao, Q.; Yue, H. Surveying wild animals from satellites, manned aircraft and unmanned aerial systems (UASs): A review. *Remote Sens.* **2019**, *11*, 1308. [CrossRef]
30. Anderson, D.M.; Estell, R.E.; Holechek, J.L.; Ivey, S.; Smith, G.B. Virtual herding for flexible livestock management –a review *Rangel. J.* **2014**, *36*, 205–221. [CrossRef]
31. Handcock, R.N.; Swain, D.L.; Bishop-Hurley, G.J.; Patison, K.P.; Wark, T.; Valencia, P.; Corke, P.; O'Neill, C.J. Monitoring Animal Behaviour and Environmental Interactions Using Wireless Sensor Networks, GPS Collars and Satellite Remote Sensing. *Sensors* **2009**, *9*, 3586–3603. [CrossRef]
32. Rivas, A.; Chamoso, P.; González-Briones, A.; Corchado, J.M. Detection of Cattle Using Drones and Convolutional Neural Networks. *Sensors* **2018**, *18*, 2048. [CrossRef]
33. Schroeder, N.M.; Panebianco, A.; Musso, R.G.; Carmanchahi, P. An experimental approach to evaluate the potential of drones in terrestrial mammal research: A gregarious ungulate as a study model. *R. Soc. Open Sci.* **2020**, *7*, 191482. [CrossRef]
34. Ditmer, M.A.; Werden, L.K.; Tanner, J.C.; Vincent, J.B.; Callahan, P.; Iaizzo, P.A.; Laske, T.G.; Garshelis, D.L. Bears habituate to the repeated exposure of a novel stimulus, unmanned aircraft systems. *Conservation Physiology* **2019**, *7*, coy067. [CrossRef] [PubMed]
35. Rümmler, M.C.; Mustafa, O.; Maercker, J.; Peter, H.U.; Esefeld, J. Sensitivity of Adélie and Gentoo penguins to various flight activities of a micro UAV. *Polar Biol.* **2018**, *41*, 2481–2493. [CrossRef]
36. Meena, S.D.; Agilandeeswari, L. Smart Animal Detection and Counting Framework for Monitoring Livestock in an Autonomous Unmanned Ground Vehicle Using Restricted Supervised Learning and Image Fusion. *Neural Process. Lett.* **2021**, *53*, 1253–1285 [CrossRef]
37. Ruiz-Garcia, L.; Lunadei, L.; Barreiro, P.; Robla, I. A Review of Wireless Sensor Technologies and Applications in Agriculture and Food Industry: State of the Art and Current Trends. *Sensors* **2009**, *9*, 4728–4750. [CrossRef] [PubMed]
38. Jawad, H.M.; Nordin, R.; Gharghan, S.K.; Jawad, A.M.; Ismail, M. Energy-Efficient Wireless Sensor Networks for Precision Agriculture: A Review. *Sensors* **2017**, *17*, 1781. [CrossRef]
39. Gresl., J.; Fazackerley., S.; Lawrence., R. Practical Precision Agriculture with LoRa based Wireless Sensor Networks. In Proceedings of the Proceedings of the 10th International Conference on Sensor Networks—WSN4PA, Vienna, Austria, 9–10 February 2021 INSTICC, SciTePress: Setúbal, Portugal, 2021; pp. 131–140. [CrossRef]
40. Axelsson Linkowski, W.; Kvarnström, M.; Westin, A.; Moen, J.; Östlund, L. Wolf and Bear Depredation on Livestock in Northern Sweden 1827–2014: Combining History, Ecology and Interviews. *Land* **2017**, *6*, 63. [CrossRef]
41. Laporte, I.; Muhly, T.B.; Pitt, J.A.; Alexander, M.; Musiani, M. Effects of wolves on elk and cattle behaviors: Implications for livestock production and wolf conservation. *PLoS ONE* **2010**, *5*, e11954. [CrossRef]
42. Cavalcanti, S.M.C.; Gese, E.M. Kill rates and predation patterns of jaguars (Panthera onca) in the southern Pantanal, Brazil. *J. Mammal.* **2010**, *91*, 722–736. [CrossRef]
43. Steyaert, S.M.J.G.; Søten, O.G.; Elfström, M.; Karlsson, J.; Lammeren, R.V.; Bokdam, J.; Zedrosser, A.; Brunberg, S.; Swenson, J.E. Resource selection by sympatric free-ranging dairy cattle and brown bears Ursus arctos. *Wildl. Biol.* **2011**, *17*, 389–403. [CrossRef]
44. Wells, S.L.; McNew, L.B.; Tyers, D.B.; Van Manen, F.T.; Thompson, D.J. Grizzly bear depredation on grazing allotments in the Yellowstone Ecosystem. *J. Wildl. Manag.* **2019**, *83*, 556–566. [CrossRef]
45. Bacigalupo, S.A.; Dixon, L.K.; Gubbins, S.; Kucharski, A.J.; Drewe, J.A. Towards a unified generic framework to define and observe contacts between livestock and wildlife: A systematic review. *PeerJ* **2020**, *8*, e10221. [CrossRef] [PubMed]
46. Lowe, D.G. Distinctive Image Features from Scale-Invariant Keypoints. *Int. J. Comput. Vis.* **2004**, *60*, 91–110. [CrossRef]
47. Dalal, N.; Triggs, B. Histograms of oriented gradients for human detection. In Proceedings of the 2005 IEEE Computer Society Conference on Computer Vision and Pattern Recognition (CVPR'05), San Diego, CA, USA, 20–25 June 2005; Volume 1, pp. 886–893 [CrossRef]
48. Mammeri, A.; Zhou, D.; Boukerche, A.; Almulla, M. An efficient animal detection system for smart cars using cascaded classifiers. In Proceedings of the 2014 IEEE International Conference on Communications (ICC), Sydney, Australia, 10–14 June 2014; pp. 1854–1859. [CrossRef]
49. Komorkiewicz, M.; Kluczewski, M.; Gorgon, M. Floating point HOG implementation for real-time multiple object detection. In Proceedings of the 22nd International Conference on Field Programmable Logic and Applications (FPL), Oslo, Norway, 29–31 August 2012; pp. 711–714. [CrossRef]
50. Munian, Y.; Martinez-Molina, A.; Alamaniotis, M. Intelligent System for Detection of Wild Animals Using HOG and CNN in Automobile Applications. In Proceedings of the 2020 11th International Conference on Information, Intelligence, Systems and Applications (IISA), Piraeus, Greece, 15–17 July, 2020; pp. 1–8. [CrossRef]

51. Munian, Y.; Martinez-Molina, A.; Alamaniotis, M. Comparison of Image segmentation, HOG and CNN Techniques for the Animal Detection using Thermography Images in Automobile Applications. In Proceedings of the 2021 12th International Conference on Information, Intelligence, Systems Applications (IISA), Chania Crete, Greece, 12–14 July 2021; pp. 1–8. [CrossRef]
52. Sharma, V.; Mir, R.N. A comprehensive and systematic look up into deep learning based object detection techniques: A review. *Comput. Sci. Rev.* **2020**, *38*, 100301. [CrossRef]
53. Ren, J.; Wang, Y. Overview of Object Detection Algorithms Using Convolutional Neural Networks. *J. Comput. Commun.* **2022**, *10*, 115–132.
54. Elgendy, M. *Deep Learning for Vision Systems*; Manning: Shelter Island, NY, USA, 2020.
55. Allken, V.; Handegard, N.O.; Rosen, S.; Schreyeck, T.; Mahiout, T.; Malde, K. Fish species identification using a convolutional neural network trained on synthetic data. *ICES J. Mar. Sci.* **2019**, *76*, 342–349. [CrossRef]
56. Huang, Y.P.; Basanta, H. Bird image retrieval and recognition using a deep learning platform. *IEEE Access* **2019**, *7*, 66980–66989. [CrossRef]
57. Zualkernan, I.; Dhou, S.; Judas, J.; Sajun, A.R.; Gomez, B.R.; Hussain, L.A. An IoT System Using Deep Learning to Classify Camera Trap Images on the Edge. *Computers* **2022**, *11*, 13. [CrossRef]
58. González-Santamarta, M.A.; Rodríguez-Lera, F.J.; Álvarez-Aparicio, C.; Guerrero-Higueras, A.M.; Fernández-Llamas, C. MERLIN a Cognitive Architecture for Service Robots. *Appl. Sci.* **2020**, *10*, 5989. [CrossRef]
59. Alliance, L. LoRaWAN Specification. 2022. Available online: https://lora-alliance.org/about-lorawan/ (accessed on 24 April 2022).
60. Cruz Ulloa, C.; Prieto Sánchez, G.; Barrientos, A.; Del Cerro, J. Autonomous thermal vision robotic system for victims recognition in search and rescue missions. *Sensors* **2021**, *21*, 7346. [CrossRef]
61. Suresh, A.; Arora, C.; Laha, D.; Gaba, D.; Bhambri, S. Intelligent Smart Glass for Visually Impaired Using Deep Learning Machine Vision Techniques and Robot Operating System (ROS). In *Robot Intelligence Technology and Applications 5*; Kim, J.H., Myung, H., Kim, J., Xu, W., Matson, E.T., Jung, J.W., Choi, H.L., Eds.; Springer International Publishing: Cham, Switzerland, 2019; pp. 99–112.
62. Lee, J.; Wang, J.; Crandall, D.; Šabanović, S.; Fox, G. Real-Time, Cloud-Based Object Detection for Unmanned Aerial Vehicles. In Proceedings of the 2017 First IEEE International Conference on Robotic Computing (IRC), Taichung, Taiwan, 10–12 April 2017; pp. 36–43. [CrossRef]
63. Puthussery, A.R.; Haradi, K.P.; Erol, B.A.; Benavidez, P.; Rad, P.; Jamshidi, M. A deep vision landmark framework for robot navigation. In Proceedings of the 2017 12th System of Systems Engineering Conference (SoSE), Waikoloa, HI, USA, 18–21 June 2017; pp. 1–6. [CrossRef]
64. Reid, R.; Cann, A.; Meiklejohn, C.; Poli, L.; Boeing, A.; Braunl, T. Cooperative multi-robot navigation, exploration, mapping and object detection with ROS. In Proceedings of the 2013 IEEE Intelligent Vehicles Symposium (IV), Gold Coast City, Australia, 23 June 2013; pp. 1083–1088. [CrossRef]
65. Zhiqiang, W.; Jun, L. A review of object detection based on convolutional neural network. In Proceedings of the 2017 36th Chinese Control Conference (CCC), Dalian, China, 26–28 July 2017; pp. 11104–11109. [CrossRef]
66. Everingham, M.; Van Gool, L.; Williams, C.K.; Winn, J.; Zisserman, A. The Pascal Visual Object Classes (VOC) Challenge. *Int. J. Comput. Vis.* **2009**, *88*, 303–338. [CrossRef]
67. Russakovsky, O.; Deng, J.; Su, H.; Krause, J.; Satheesh, S.; Ma, S.; Huang, Z.; Karpathy, A.; Khosla, A.; Bernstein, M.; et al. ImageNet Large Scale Visual Recognition Challenge. *Int. J. Comput. Vis. (IJCV)* **2015**, *115*, 211–252. [CrossRef]
68. Lin, T.Y.; Maire, M.; Belongie, S.; Hays, J.; Perona, P.; Ramanan, D.; Dollár, P.; Zitnick, C.L. Microsoft COCO: Common Objects in Context. In Proceedings of the Computer Vision—ECCV 2014, Zurich, Switzerland, 6–12 September 2014; Fleet, D., Pajdla, T., Schiele, B., Tuytelaars, T., Eds.; Springer International Publishing: Cham, Switzerland, 2014; pp. 740–755.
69. Panda, P.K.; Kumar, C.S.; Vivek, B.S.; Balachandra, M.; Dargar, S.K. Implementation of a Wild Animal Intrusion Detection Model Based on Internet of Things. In Proceedings of the 2022 Second International Conference on Artificial Intelligence and Smart Energy (ICAIS), Coimbatore, India, 23–25 February 2022; pp. 1256–1261. [CrossRef]
70. Figueiredo, A.M.; Valente, A.M.; Barros, T.; Carvalho, J.; Silva, D.A.; Fonseca, C.; de Carvalho, L.M.; Torres, R.T. What does the wolf eat? Assessing the diet of the endangered Iberian wolf (Canis lupus signatus) in northeast Portugal. *PLoS ONE* **2020**, *15*, e0230433. [CrossRef] [PubMed]
71. Github: VISORED. Available online: https://github.com/uleroboticsgroup/VISORED (accessed on 28 May 2022).
72. Li, D.; Wang, R.; Chen, P.; Xie, C.; Zhou, Q.; Jia, X. Visual Feature Learning on Video Object and Human Action Detection: A Systematic Review. *Micromachines* **2022**, *13*, 72. [CrossRef] [PubMed]
73. Jeong, J.; Park, H.; Kwak, N. Enhancement of SSD by concatenating feature maps for object detection. *arXiv* **2017**, arXiv:1705.09587.
74. Li, Z.; Zhou, F. FSSD: Feature Fusion Single Shot Multibox Detector. *arXiv* **2017**. arXiv:1712.00960.
75. Tan, L.; Huangfu, T.; Wu, L.; Chen, W. Comparison of RetinaNet, SSD, and YOLO v3 for real-time pill identification. *BMC Med. Inform. Decis. Mak.* **2021**, *21*, 324. [CrossRef] [PubMed]
76. Liu, P.; Qi, B.; Banerjee, S. Edgeeye: An edge service framework for real-time intelligent video analytics. In Proceedings of the 1st International Workshop on Edge Systems, Analytics and Networking, Munich, Germany, 10–15 June 2018; pp. 1–6.
77. Padilla, R.; Netto, S.L.; da Silva, E.A.B. A Survey on Performance Metrics for Object-Detection Algorithms. In Proceedings of the 2020 International Conference on Systems, Signals and Image Processing (IWSSIP), Niteroi, Brazil, 1–3 July 2020; pp. 237–242. [CrossRef]

78. Liu, W.; Anguelov, D.; Erhan, D.; Szegedy, C.; Reed, S.E.; Fu, C.; Berg, A.C. SSD: Single Shot MultiBox Detector. *arXiv* **2015**, arXiv:1512.02325v5.
79. Arriaga, O.; Valdenegro-Toro, M.; Muthuraja, M.; Devaramani, S.; Kirchner, F. Perception for Autonomous Systems (PAZ). *arXiv* **2020**, arXiv:cs.CV/2010.14541.
80. Redmon, J.; Farhadi, A. YOLOv3: An Incremental Improvement. *arXiv* **2018**, arXiv:1804.02767v1.
81. Bochkovskiy, A.; Wang, C.; Liao, H.M. YOLOv4: Optimal Speed and Accuracy of Object Detection. *arXiv* **2020**, arXiv:2004.10934v1
82. Adarsh, P.; Rathi, P.; Kumar, M. YOLO v3-Tiny: Object Detection and Recognition using one stage improved model. In Proceedings of the 2020 6th International Conference on Advanced Computing and Communication Systems (ICACCS), Coimbatore, India, 6–7 March 2020; pp. 687–694. [CrossRef]
83. Github: Ultralytics. yolov5. Available online: https://github.com/ultralytics/yolov5 (accessed on 28 May 2022).

remote sensing

MDPI

Article

Similarity Analysis between Contour Lines by Remotely Piloted Aircraft and Topography Using Hausdorff Distance: Application on Contour Planting

Alexandre Araujo Ribeiro Freire [1,*], Mauro Antonio Homem Antunes [2], Murilo Machado de Barros [2], Wagner Dias de Souza [2], Wesley de Sousa da Silva [2] and Thaís Machado de Souza [2]

1 Technical School, Federal Rural University of Rio de Janeiro, Seropédica 23890-000, RJ, Brazil
2 Engineering Department, Federal Rural University of Rio de Janeiro, Seropédica 23890-000, RJ, Brazil; mauroantunes@ufrrj.br (M.A.H.A.); barrosmm@ufrrj.br (M.M.d.B.); wagners@ufrrj.br (W.D.d.S.); wesleyss@ufrrj.br (W.d.S.d.S.); thaaiismds@ufrrj.br (T.M.d.S.)
* Correspondence: alexandrefreire@ufrrj.br

Abstract: Contour planting minimizes soil degradation, making agricultural production more sustainable. Currently, geotechnologies can provide more precise and fast data from relief than rudimentary data acquisition for agricultural management. Thus, the objective of this work was to analyze the similarities between contour lines from topography and Remotely Piloted Aircraft, using the Hausdorff distance algorithm. This study was carried out in the period between January 2020 and November 2021 in four localities in the State of Rio de Janeiro, Brazil: two areas located in the municipality of Bom Jardim and two areas in the municipality of Seropédica. Data were acquired through a conventional topographic survey and an aerial photogrammetric survey by Remotely Piloted Aircraft. From the acquired field data for the studied areas, the Digital Elevation Models were generated with a spatial resolution of 0.20 m and the contour lines with an equidistance of one meter. The contour lines obtained by both techniques were superimposed and their similarity was verified using the Hausdorff distance. The results show that there was a better similarity among the contour lines in areas with a very rugged relief than in a smooth relief. Also, the lowest altimetric differences observed in the Digital Elevation Models were associated with the smallest Hausdorff distance. These adjustments correspond, respectively, to the segments between the contour lines with the best and the worst individual similarity for each area. We observed that the similarity between the contour lines from topography and RPA yielded slope differences lower than 6.1% for at least 95% of all studied areas. The Hausdorff distance analysis allowed us to conclude that contour planting can be performed from data obtained via Remotely Piloted Aircraft, provided that vertical accuracy analysis controls the quality of the Digital Elevation Models.

Keywords: aerial photogrammetry by Remotely Piloted Aircraft; contour planting; Digital Elevation Model

Citation: Freire, A.A.R.; Antunes, M.A.H.; de Barros, M.M.; de Souza, W.D.; de Sousa da Silva, W.; de Souza, T.M. Similarity Analysis between Contour Lines by Remotely Piloted Aircraft and Topography Using Hausdorff Distance: Application on Contour Planting. *Remote Sens.* **2022**, *14*, 3269. https://doi.org/10.3390/rs14143269

Academic Editors: Diego González-Aguilera, Jose A. Jiménez-Berni, Ittai Herrmann, Shangpeng Sun and Monica Herrero-Huerta

Received: 27 May 2022
Accepted: 4 July 2022
Published: 7 July 2022

1. Introduction

Food demand is expected to increase, and the rational use of natural resources available on Earth becomes an incontestable premise from any perspective in the near future. Soil is one of the most precious and necessary resources for agricultural activities, and because of it, its use must be subjected to processes that increasingly generate less damage to the environment.

Conservation agricultural practices play a key role in a sustainable agricultural production scenario [1]. Agricultural productivity could expand without an increase in environmental degradation if the amount of impact per unit of product or activity were reduced [2].

Among the usable conservation practices, contour planting is one of the viable options to preserve soil erosion processes, according to Santos et al. [3], Griebeler et al. [4], Leite et al. [5], and Xavier et al. [6]. Contour planting contributes to sustainability by controlling the loss of soil particles so that the soil becomes more resistant to the erosive process [7]. In this context, techniques that result in more efficient agricultural production [8], such as implementing Intelligent Agriculture practices will become relevant for preserving natural resources [9].

Presenting different technologies to farmers will allow them to acquire skills to achieve socially, economically, and environmentally advanced agriculture technology [1]. From this premise, geotechnologies contribute to decision making, providing promising tools to perform different analyzes in rural space which can be used as a means of control and knowledge concerning land use [10,11].

Hunt and Daughtry [12] and Maes and Steppe [13] reviewed the progress of remote sensing with Remotely Piloted Aircrafts (RPAs) in the scope of Precision Agriculture (PA). The authors emphasized that the focus of the most recently discussed applications in the scientific environment has been the detection of water stress, pathogens, and pesticide applications, monitoring and detection of weeds, and assessment of the nutritional status growth, biomass, and yield forecast of crops. Literature reviews on this subject are available in Bendig et al. [14], Pérez-Ortiz et al. [15], Chang et al. [16], Schut et al. [17], Tsouros, Bibi and Sarigianni-dis [18], Santos et al. [19] and Delavarpour et al. [20].

RPAs generate digital cartographic products, such as Digital Elevation Models (DEMs) [21,22] and slope maps [23,24], which are necessary for the discretization of the relief, where the farming will be installed. Such products are considered indispensable for several farmland management, e.g., coffee [25]. Coffee is generally growing on sloping lands and contour planting provides a well-distributed crop, increasing operational efficiency, mainly those arising from mechanized activity [25,26].

For digital cartographic products via RPA it is necessary to proceed with the quality data analysis referring to the DEMs and their derived attributes, e.g., the contour lines for the correct application in PA. This work presented a new perspective regarding the use of RPAs in the data acquired, i.e., similarity analysis by the Hausdorff distance algorithm among the contour lines compared to a more accurate elevation data source, such as conventional topography. The objective of this comparison is based on the use of the contour lines from aerial photogrammetry by RPAs, to carry out contour planting, optimizing the PA cycle.

2. Study Area

We selected four rural localities in the State of Rio de Janeiro, Brazil: two properties located in the municipality of Bom Jardim, in the Mountain Region of Rio de Janeiro, and two areas located at the Federal Rural University of Rio de Janeiro (UFRRJ), in the lowlands of Rio de Janeiro, municipality of Seropédica.

Areas 1 and 2 are located in the municipality of Bom Jardim. The first area of approximately 0.344 ha, between latitude 22°17′23.987″S–22°17′21.494″S, longitude 42°20′6.930″W–42°20′3.157″W, the slope average of 38.4%. The second area of approximately 0.183 ha, between latitude 22°17′30.056″S–22°17′28.826″S, longitude 42°19′54.851″W–42°19′52.368″W, the average slope 41.2%. Both sites at the time of data acquisition (January 2020) had collard (*Brassica oleracea*, Acephala group), plantations carried out in beds, with an average height of the crop around 0.25 m, planted intermittently, without weeds and the interference of trees or buildings in the polygon of interest. Bare soil was predominant in these two areas.

Areas 3 and 4 are located on the UFRRJ campus. Area 3 of approximately 1044 ha, between latitude 22°47′11.965″S–22°47′6.392″S, longitude 43°40′49.271″W–43°40′44.625″W and an average slope 6.5%. Area 4 of approximately 0.756 ha, between latitude 22°46′47.015″S–22°46′40.282″, longitude 43°41′7.111″W–43°41′2.119″W and slope average of 6.3%. Both areas had grass cover at a height close to 0.05 m with patches of bare soil and without the interference of trees or buildings in the polygon of interest at the time of image acquisition in November 2021.

According to the Brazilian soil classification system [27], areas 1 and 2 have topography characterized as very rugged relief. Areas 3 and 4 have topography characterized as smooth relief. Figures 1 and 2 present the four study areas separated by region and the spatial distribution of the control and checkpoints to perform the external orientation of the images and the quality control.

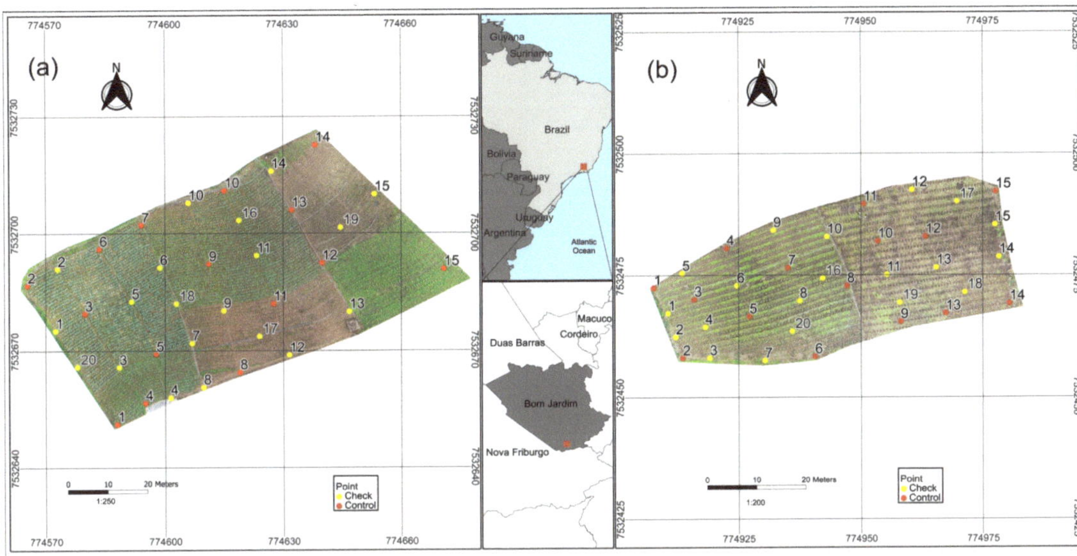

Figure 1. Orthophotomosaic of area 1 (**a**) and area 2 (**b**), with their respective control and checkpoints.

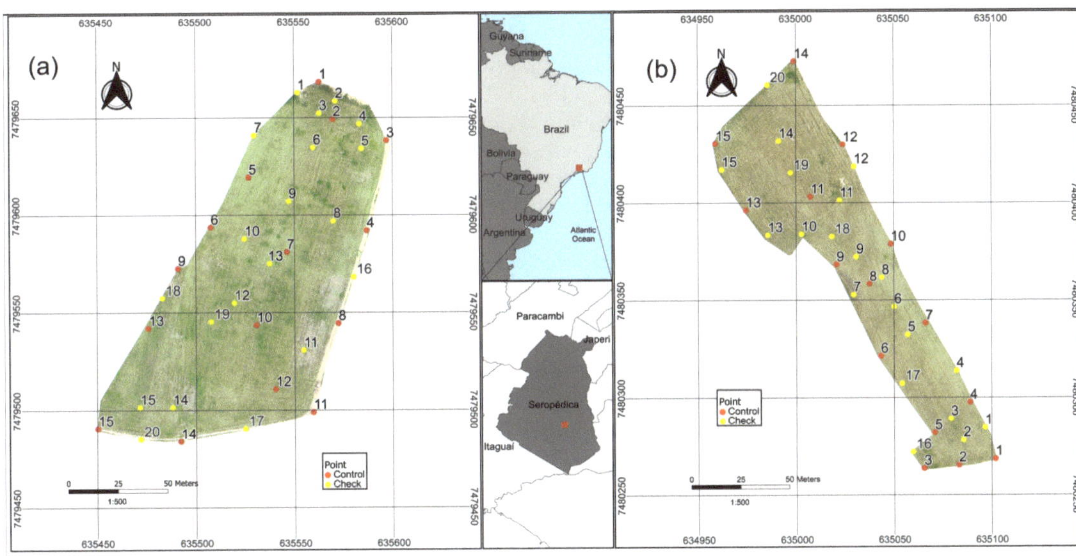

Figure 2. Orthophotomosaic of areas 3 (**a**) and 4 (**b**), with their respective control and checkpoints.

3. Data and Methods

3.1. Data Acquisition and Analysis

In this section, the procedures for data acquisition and analysis were presented according to the steps shown in the flowchart (Figure 3).

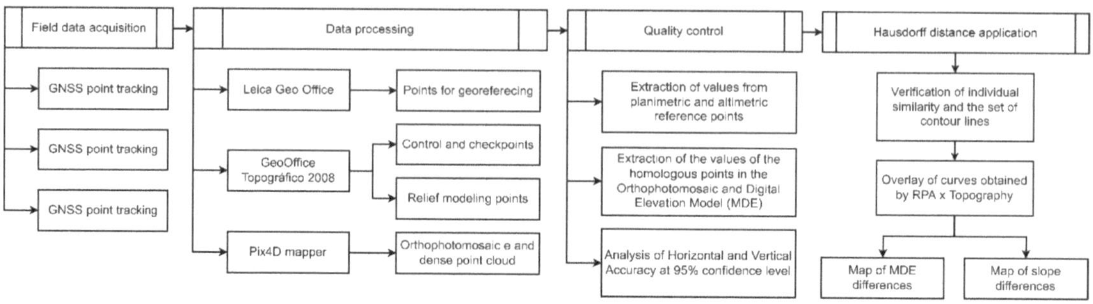

Figure 3. Flowchart for data acquisition and analysis.

The following sections presented the detailed steps in Figure 3.

3.1.1. Tracking Points by GNSS

We chose a pair of GNSS points implanted in the field in each of the four studied areas for georeferencing topographical points and to carry out the external orientation of the images obtained by the RPA. The tracking was performed using the fast static relative type method, with three hours of occupancy at each vertex. The equipment used was a dual-frequency (L1, L2) Spectra Precision EPOCH 25 GNSS receiver, performing the data post-processing using the Leica Geo Office software.

The processing was made with the coordinates (E, N, and h) of the control stations originating from the Brazilian Network for Continuous Monitoring of GPS (RBMC), located in the municipalities of Niterói (RJNI) and Vassouras (RJVA), State of Rio de Janeiro, Brazil. The coordinates were extracted from the descriptions of the databases provided by the Brazilian Institute of Geography and Statistics (IBGE), responsible for regulating the Brazilian Geodetic System (SGB). The geoid height component (N) was used to convert the ellipsoidal height (h) into orthometric height (H), where N was calculated from the MAPGEO 2015 software provided by the IBGE.

We defined the reference ellipsoid and the projection system for the elaboration of the products of interest. For this purpose, the elliptical Geodetic Reference System 1980 (GRS80) was used in the Geocentric Reference System for the Americas (SIRGAS 2000). The GRS80 ellipsoid was used due it is the basis of SIRGAS 2000, the current Geodetic Reference System of the SGB. The projection system was the Universal Transverse Mercator (UTM), Zone 23 South, central meridian equal to 45°W.

3.1.2. Conventional Topographic Survey

The topographic planialtimetric survey was carried out using a total station, Spectra Precision, model Focus 2 by the irradiation method. The equipment has a nominal angular precision of 2 s and a nominal linear precision of 2 mm + 2 ppm, classified as a high precision device according to the Brazilian Standard (NBR) 13.133 (Brazilian Association of Technical Standards—ABNT, 1994) which regulates topographic surveys.

The topographic survey aimed to register the relief variations of the areas for the production of DEMs, contour lines, and topographic slope maps, as shown in Figures 4 and 5. We surveyed the 15 control points and 20 checkpoints implemented in the field as reference for the processes inherent to the aerial photogrammetric survey by RPA and subsequent validation in the quality control stage. The polygonal and topographic irradiation data were processed using the GeoOffice Topographic 2008 software v. 2.8.4.0.

exporting the attributes spreadsheet in text format and the ".txt" extension to ArcGis V. 10.8. In the GIS environment, Digital Elevation Models were generated and then contour lines and slope maps were derived. To perform these operations, the "Topo to Raster" interpolator was used. This interpolation algorithm is based on iterative finite differences to generate a regular grid from elevation points and/or contour lines ESRI [28]. According to Hutchinson [29], the "Topo to Raster" uses known information from surface elevation, such as elevation points (such as in this work), contour lines, and water body delimitations, among others. Also, according to the author, the insertion of this information optimizes the resolution of the DEM, improving the quality of the generated product.

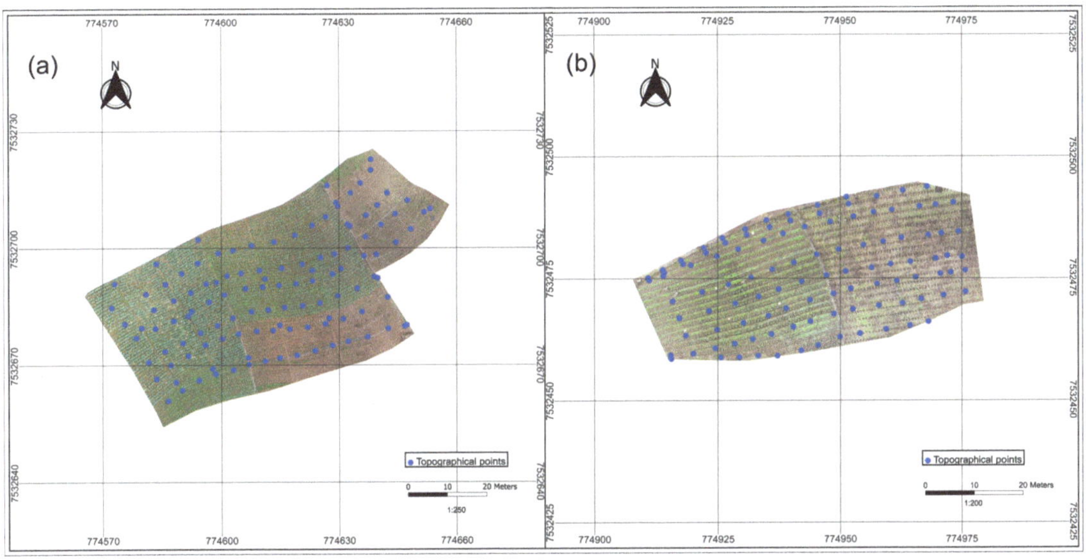

Figure 4. Surveyed points from the areas of Bom Jardim municipality: (**a**) location of points registered in area 1 by conventional topographic survey and (**b**) location of points registered in area 2 by the conventional topographic survey.

3.1.3. Remotely Piloted Aircraft Survey

The RPA Phantom 4 Pro equipment was used to carry out the aerial photogrammetric survey (Figure 6). The 20-megapixel CMOS camera was programmed to obtain aerial images at previously stipulated time intervals and according to the flight plan prepared for the area.

The aerial photogrammetric survey was conducted following the steps: flight planning, field implementation of control points and checkpoints, image acquisition, project configuration, and the digital processing of the images.

Initially, the prior planning was prepared through the Drone Deploy application, with the information relevant to the flight lines configured for the imaging of the areas. Then, we defined the flight height of 60 m, flight direction in the longitudinal direction, and lateral and longitudinal overlaps of 80% and 80%, respectively, for the four areas.

Before the flight, we checked the propellers' fixation, the energy load of the remote control and the aircraft, the amount of satellite signal the RPA was receiving and enabled the starting point system, checked the compass, radio, IMU, and gimbal calibration using DJI GO v.4 software (DJI, G.O.4, 2017).

After the flight, the digital processing of the images was performed using the software Pix4D mapper v. 4.5.6. This software automatically calculated the positions and orientations of the original images obtained using the RPA through aerial triangulation, finally performing the bundle block adjustment. The interior and exterior orientation parameters were provided and recognized by the software automatically, as they were in the RPA camera library.

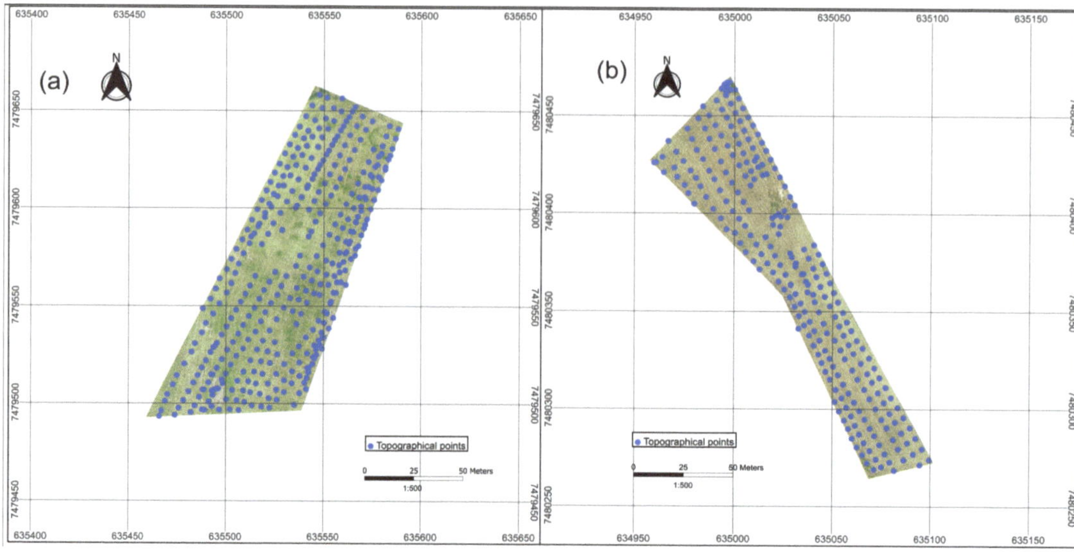

Figure 5. Surveyed points from the areas of Seropédica municipality: (**a**) location of points registered in area 3 by conventional topographic survey and (**b**) location of points registered in area 4 by the conventional topographic survey.

Figure 6. Phantom 4 Pro RPA for field survey. (**a**) Above view, and (**b**) Side view.

In the Pix4D mapper environment, all three processing steps were performed automatically [30]. These steps followed interior orientation, automatic correspondence among images imported into the software, and with some overlap, the simultaneous bundle block adjustment, and the generation of final products, such as the orthophotomosaic and the dense point cloud generation. However, in the external orientation stage in the image orthorectification process, the control points in the images were identified manually with the insertion of their respective coordinates.

The Structure from Motion (SfM) algorithm, which uses the combined approach to know the pattern of correspondence in the images, was used to derive dense point clouds in three dimensions (3D) from the images obtained by the camera coupled to RPA [31]. Initial processing was performed by selecting the Pre-Set Standard 3D Maps option, using the original image scale. Digital Surface Models (DSM) were generated, performing the automatic filtering procedure for the dense cloud of points generation. The densification of the point cloud was performed based on the original scale of the image. The density of points of the dense cloud was elaborated with the optimized parameter (Pix4D default).

After performing the steps of digital image processing, orthophotomosaics and dense point clouds were generated for all areas in the Pix4D mapper software, which were exported in LAS format to the ArcGIS software. In the GIS environment, the data were processed to obtain the DEMs, derive the contour lines, and prepare the slope maps, using the "Topo to Raster" interpolator.

In a GIS environment, the coordinates E (m), N (m), and H (m) of homologous points to the checkpoints registered by topography were extracted from the orthophotomosaics and the DEMs of areas 1, 2, 3, and 4. This procedure was necessary to carry out the planimetric and altimetric quality control of the generated digital cartographic products.

3.2. Quality Control

The product accuracy measured by comparing its information with those observed in the field is more reliable and of better quality. The verification of the accuracy of the product was determined based on statistical parameters linked to a certain level of confidence adopted, following the standardization recommended for each country [32]. Therefore, those obtained by conventional topography were considered reference data, as one of the most accurate methods for obtaining data [33]. Therefore, the digital cartographic products obtained through RPA were considered data to be validated.

The assessment of the positional accuracy of the orthophotomosaics and the altimetric accuracy of the DEMs was based on the Positional Accuracy Standards for Digital Geospatial Data (PASDGD) (ASPRS 2014), considering open terrain and areas without vegetation (NVA classification) [34]. This was possible because the vegetation had low heights in the four areas (0.25 m in areas 1 and 2 with mostly bare soil and 0.05 m in areas 3 and 4 with patches of bare soil). GeoPEC software v. 3.5.2 was used to evaluate the horizontal and vertical accuracy, where the coordinate data of the 20 checkpoints (reference and test) in each area were inserted through a file (.txt). The ASPRS PASDGD standard was used for the analysis of geospatial datasets based on the Root Mean Square Error (RMSE) statistic of the checkpoints. The RMSE was obtained as a function of the difference between the reference coordinates and the observed coordinates, for the three axes (x, y, and z), as presented in Equations (1)–(3), respectively [35].

$$RMSE_x = \sqrt{\frac{1}{n}\sum_{i=1}^{n}(X_{reference} - X_{test})^2} \tag{1}$$

where:

n is the number of samples;
$X_{reference}$ as coordinate value (x) in the reference product (m);
X_{test} as coordinate value (x) in the product tested (m).

$$RMSE_y = \sqrt{\frac{1}{n}\sum_{i=1}^{n}(Y_{reference} - Y_{test})^2} \tag{2}$$

where:

n is the number of samples;
$Y_{reference}$ as coordinate value (y) in the product reference (m);
Y_{test} as coordinate value (y) in the reference product (m).

$$RMSE_z = \sqrt{\frac{1}{n}\sum_{i=1}^{n}(Z_{reference} - Z_{test})^2} \quad (3)$$

where:

n is the number of samples;
$Z_{reference}$ as coordinate value (z) in the reference product (m);
Z_{test} as coordinate value (z) in the reference product (m).

From the results found in the RMSEx and RMSEy, the value of the RMSEr (radial direction) was determined using Equation (4) [36].

$$RMSE_r = \sqrt{(RMSE_x^2) + RMSE_y^2)} \quad (4)$$

After RMSEr and RMSEz results, horizontal (Equation (5)) and vertical (Equation (6)) accuracy values were calculated for each of the areas, according to the PASDGD standard (ASPRS 2014).

$$Accuracy_r = 1.7308 \times RMSE_r \quad (5)$$

$$Accuracy_v = 1.960 \times RMSE_z \quad (6)$$

3.3. *Hausdorff Distance Application*

The Hausdorff distance (dH) algorithm was used to quantify the similarity between the homologous contour lines referring to the digital products generated by two different data acquisition techniques. According to Gregoire and Bouillot [37], the algorithm had its conception developed by Felix Hausdorff, according to Equation (7):

$$dH(A,B) = sup\{h(A,B), h(B,A)\} \quad (7)$$

where:

dH(A,B): as Hausdorff distance.
sup: as the highest value between two data set.
h(A,B): as the highest distance between the minimum data set from A to B.
h(B,A): as the highest distance between the minimum data set from B to A.

h(A,B) and h(A,B) were given by Equations (8) and (9):

$$h(A,B) = sup\{min_{b \in B}\{d\{(a,b)\}\} \quad (8)$$

$$h(B,A) = sup_{b \in B}\{min_{a \in A}\{d\{(b,a)\}\} \quad (9)$$

According to data sets (A and B), which are the linear features that describe the contour lines, the Euclidean distance from each point a ϵ A to all points b ϵ B was determined. The data set (A) is composed of contour lines from conventional topography (reference line), whereas the data set (B) is formed by contours obtained via RPAs (test line). Figure 7 represents the distance from the data set A to B (d1) and the distance from data set B to A (d2) between the reference line (LR) and the test line (LT).

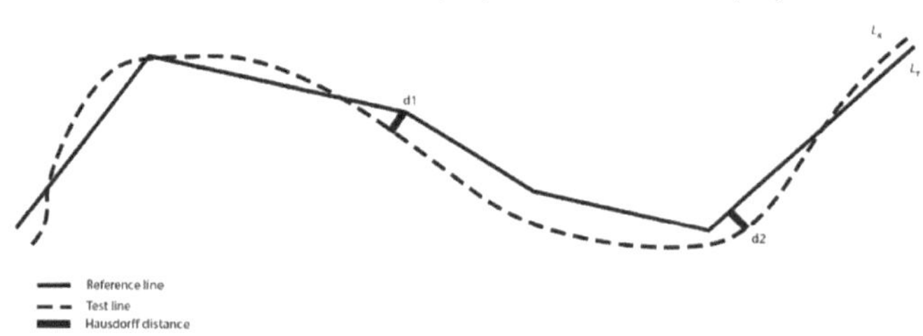

Figure 7. Hausdorff distance survey. Adapted with permission from Santos et al. [38].

The smallest distances between the points in the data set A to B were determined, and the supreme value of the smallest ones was taken, that is, the greatest distance between the smallest measured distances was fixed. This same process was repeated, starting from data set B to A. At the end of the process, two maximum values h(A,B) and h(B,A) were determined, therefore, the dH was the biggest value.

The analysis of the dH algorithm between the contour lines was performed with the aid of the similarity checker software, developed for this experiment. For such, the user inserted the data in the format (GeoJSON) of the georeferenced files containing the contour lines of both representations to be compared.

From the insertion of the data, the software applied the algorithm based on the recognition of the elevation of the homologous contour lines and their georeferenced positioning, returning with the dH for each pair of curves. The similarity of a cartographic representation obtained by different techniques can be individually and jointly analyzed, determining statistical parameters such as mean ($\overline{x}$), standard deviation (σ), maximum (max), and minimum (min) distance from a set of contour lines in the same representation.

Finally, the contour lines were superimposed on the maps that represent the altimetric differences and the resulting slope differences between both techniques to associate them with the results found by the dH.

3.4. Database System Application of Hausdorff Algorithm

A Database System (DBS) was created for the application of the Hausdorff algorithm, which is composed of, at least, an application program, a Database Management System (DBMS), and a Database [39]. The flowchart for using this system is shown in Figure 8. The application program for this system was available on a website divided into front-end and back-end. Apache Lounge 2.4 was used on the Windows 10 Pro operating system to provide this application on a local network.

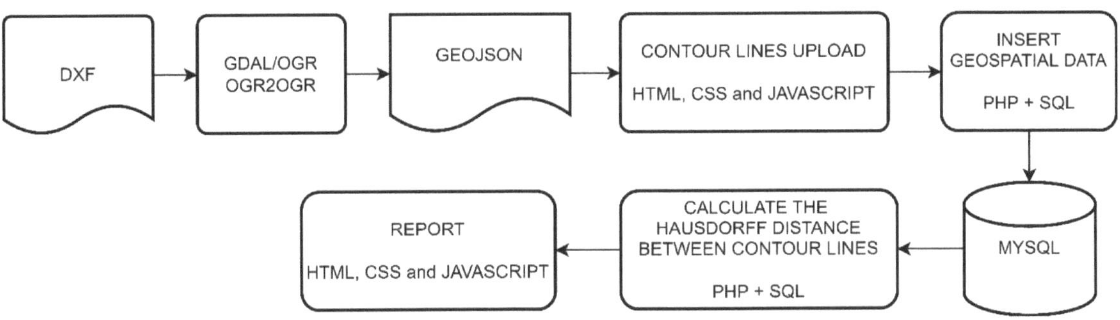

Figure 8. Flowchart of the Database System.

The HTML, the CSS style, and the Javascript programming languages were used to develop the front end. The Hypertext Preprocessor (PHP) version 8.1 and Standard Query Language (SQL) were used for the back-end implementation, while the MySQL Community Server DBMS version 8.0 was used for geospatial data storage.

Thus, the application consisted of a Web system with a Geographic Database. MySQL was chosen due to its fast processing speed and spatial function with the implementation of the Hausdorff distance algorithm between two linear features. This function is called "ST_HausdorffDistance" and can calculate the similarity of two different geometries inserted in the WEB system.

Data entry in this system was performed by uploading a file in GeoJSON format [40,41]. The choice of GeoJSON was based on their fast processing speed, easy data readability by humans and machines, reduced file size, and full compatibility between all software technologies used in this work. Therefore, all contour lines were created or converted to a file in GeoJSON format. The European Petroleum Survey Group (EPSG) Geodetic Parameter

Dataset of the geospatial data used corresponds to the MySQL-controlled geographic database. EPSG 31983 was configured in the SIRGAS 2000 Reference System, zone 23 S, Central Meridian 45°W, of the UTM Projection System.

The "ogr2ogr" is a Geospatial Data Abstraction Library (GDAL/OGR) command-line tool to convert one OGR Simple Features Library into another. In this work procedure, the software "ogr2ogr" was used to convert files in DXF format, which contained the contour lines, to files in GeoJSON format. However, the file in GeoJSON format was constructed with features of the type LineString (shown in the blue rectangle of Figure 9) and each point of the LineString had the three axes x, y, and z.

```
{
"type": "FeatureCollection",
"name": "entities",
"crs": { "type": "name", "properties": { "name": "urn:ogc:def:crs:EPSG::31983" } },
"features": [
{
"type": "Feature",
"properties": {
    "Layer": "AREA1_ARP_REC2",
    "SubClasses": "AcDbEntity:AcDb3dPolyline",
    "EntityHandle": "69" },
    "geometry": {
        "type": "LineString",
        "coordinates": [ [ 774644.394759248942137, 7532719.995867665857077, 877.0 ], ... ] } },
{
"type": "Feature",
"properties": {
    "Layer": "AREA1_ARP_REC2",
    "SubClasses": "AcDbEntity:AcDb3dPolyline",
    "EntityHandle": "E4" },
    "geometry": {
        "type": "LineString",
        "coordinates": [ [ 774659.094501582672819, 7532707.735314433462918, 871.0 ], ... ] } },
...,
...,
...
]
}
```

Figure 9. Example of a system input GeoJSON file.

Thus, the z coordinate consisted of the elevation of the contour lines (shown in the red rectangle in Figure 9). The z coordinate value is used in the software to sort and organize the contour lines in this system.

The contour lines were uploaded by the user via the website's graphical interface. After performing this procedure, the system's back-end received and sent the geospatial data to be stored in a MySQL database. After correctly populating the geographic database, spatial queries were created using SQL. A new webpage was developed to present the results of these queries. On this webpage, the results obtained according to the dH algorithm between the linear features, which represent the contour lines, were presented in table format. In Figure 10, the PHP and SQL code of the query is analyzed to calculate the Hausdorff distance between curves of different geospatial objects in MySQL. Figure 11 shows the PHP and SQL implementation of the query to calculate the Hausdorff distance between contour lines in a single MySQL geospatial object.

```
$query="";
for($i=1; $i<=$total_curves; ++$i){
    $query .= "
        SELECT
            ROUND(
                ST_HausdorffDistance(
                    ST_GeometryN(ct.curves, $i),
                    ST_GeometryN(cd.curves, $i),
                    'metre'
                ),
                3
            ) as hausdorff_distance
        FROM curves ct, curves cd
        WHERE ct.id < cd.id
        UNION
    ";
}
$query = trim(preg_replace('/\s+/S', " ", $query));
$query = substr($query, 0, -6);
$query .= ";";
$result_query = mysqli_query($connection, $query);
```

Figure 10. Implementation of the query in PHP and SQL to obtain the Hausdorff distance among features of different geospatial objects.

```
$query="";
for($i=1; $i<$total_curves; ++$i){
    $query .= "
        SELECT
            ct.id as method_id,
            $i as id_a,
            $i+1 as id_b,
            ROUND(
                ST_HausdorffDistance(
                    ST_GeometryN(ct.curves, $i),
                    ST_GeometryN(ct.curves, $i+1),
                    'metre'
                ),
                3
            ) as hausdorff_distance,
            FROM curves ct, curves cd
        WHERE ct.id = cd.id
        UNION
    ";
}
$query = trim(preg_replace('/\s+/S', " ", $query));
$query = substr($query, 0, -6);
$query .= " ORDER BY id_a;";
$result_query = mysqli_query($connection, $query);
```

Figure 11. Implementation of the query in PHP and SQL to obtain the Hausdorff distance among features of the same geospatial object.

4. Results and Discussion

4.1. Differences in Relief Modeling

The DEMs were generated for areas 1, 2, 3, and 4 from the topographic and aerial photogrammetric surveys. Subsequently, the map algebra operation among their respective products was carried out to obtain the altimetric difference between both techniques (Figures 12 and 13).

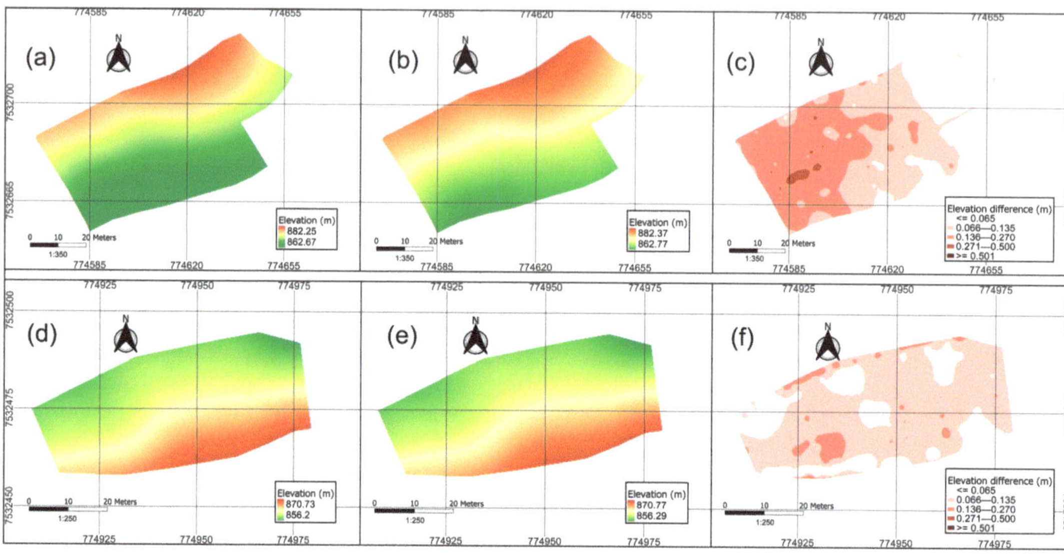

Figure 12. DEMs for areas 1 and 2 and the differences of DEM from both methodologies: (**a**) DEM of area 1, obtained by conventional topographic survey; (**b**) DEM of area 1, obtained using RPA; (**c**) altimetric difference between DEMs (**a**,**b**). (**d**) DEM of area 2, obtained by conventional topographic survey; (**e**) DEM of area 2, obtained using RPA; (**f**) altimetric difference between DEMs (**d**,**e**).

The maps represented in Figure 12c,f and Figure 13c,f show the differences among the DEMs for area 1, area 2, area 3, and area 4. Based on the area occupied by the classifications indicated on the maps (≤0.065 m; 0.066–0.135 m; 0.136–0.270 m; 0.271–0.500 m; ≥0.501 m) the percentages of differences obtained in the total area measured (Table 1).

Table 1. Percentages of the differences among the DEMs in the total area.

Intervals of Difference among DEMs (m)	Area 1 (%)	Area 2 (%)	Area 3 (%)	Area 4 (%)
≤0.065	23.50	25.23	55.76	84.99
0.066–0.135	36.50	69.43	38.66	14.03
0.136–0.270	38.86	5.33	5.58	0.84
0.271–0.500	1.14	0.01	0.00	0.14
≥0.501	0.00	0.00	0.00	0.00

After generating the DEMs, slope maps were derived for all areas, repeating the map algebra operation (Figures 14 and 15).

The difference between the slope maps of area 1, area 2, area 3, and area 4 are represented sequentially by Figure 14c,f and Figure 15c,f. From the classifications presented in the maps (≤2.0%; 2.1–4.0%; 4.1–6.0%; ≥6.1%), the percentage of occurrence of these intervals of differences in the total area was verified (Table 2).

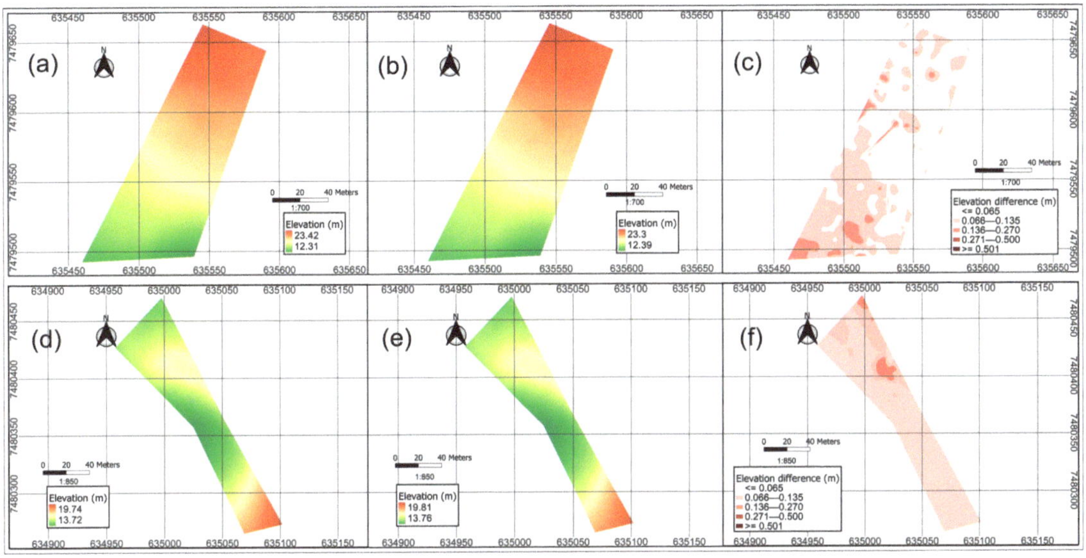

Figure 13. DEMs for areas 3 and 4 and the differences of DEM from both methodologies: (**a**) DEM of area 3, obtained by conventional topographic survey; (**b**) DEM of area 3, obtained using RPA; (**c**) altimetric difference between DEMs (**a**,**b**). (**d**) DEM of area 4, obtained by conventional topographic survey; (**e**) DEM of area 4, obtained using RPA; (**f**) altimetric difference between DEMs (**d**,**e**).

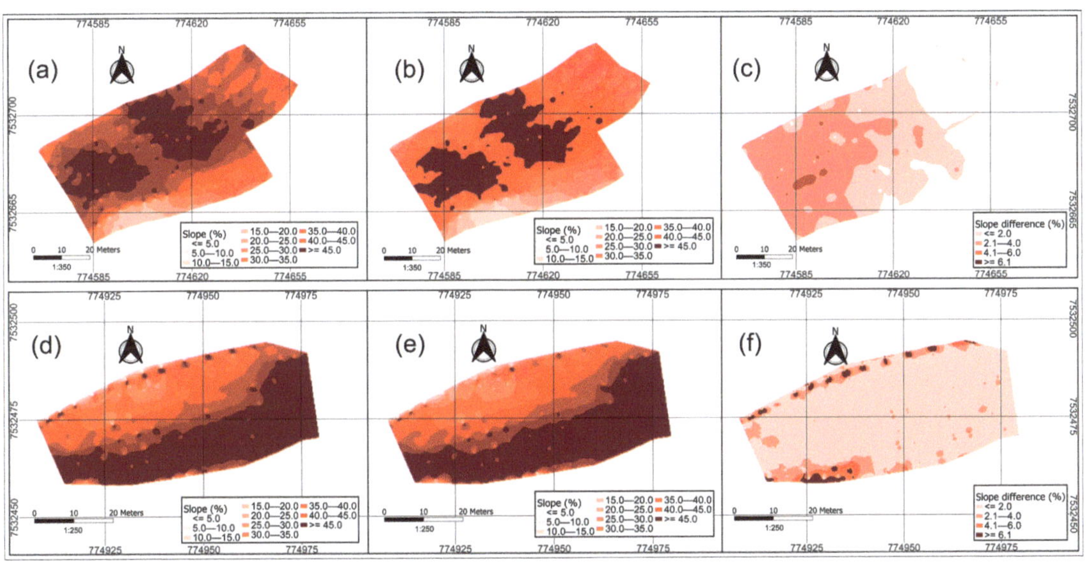

Figure 14. Slope maps for areas 1 and 2 and the differences in slope from both methodologies: (**a**) slope map of area 1, obtained by conventional topographic survey; (**b**) slope map of area 1 obtained using RPA; (**c**) difference between slope maps (**a**,**b**). (**d**) slope map of area 2 obtained by conventional topographic survey; (**e**) slope map of area 2 obtained using RPA; (**f**) difference between slope maps (**d**,**e**).

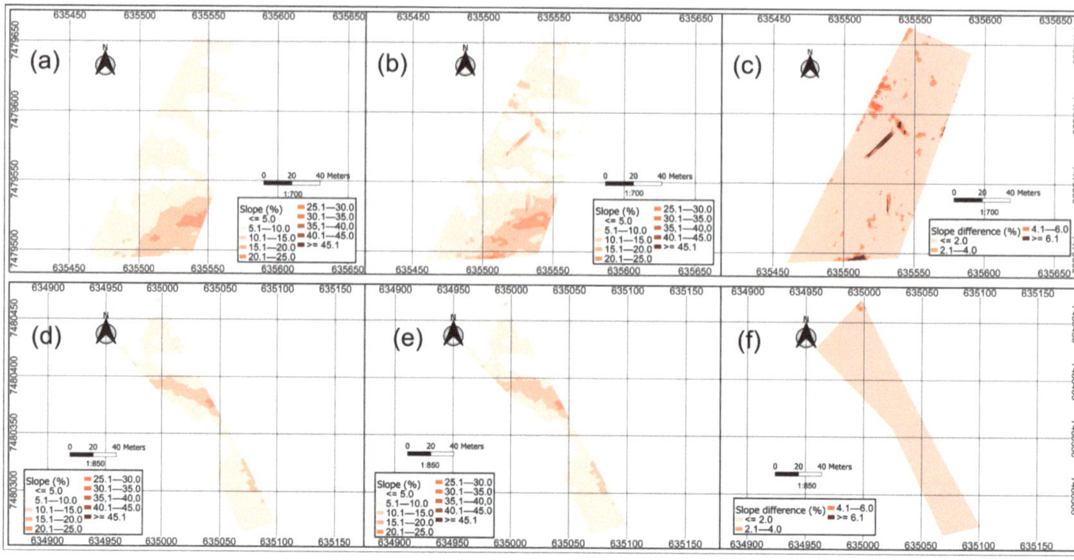

Figure 15. Slope maps for areas 3 and 4 and the differences in slope from both methodologies: (**a**) slope map of area 3, obtained by conventional topographic survey; (**b**) slope map of area 3 obtained using RPA; (**c**) difference in slope between maps (**a**,**b**). (**d**) slope map of area 4, obtained by conventional topographic survey; (**e**) slope map, of area 4, obtained using RPA; (**f**) difference in slope between maps (**d**,**e**).

Table 2. Occurrence percentages of the slope difference in the total area.

Slope Difference Intervals (%)	Area 1 (%)	Area 2 (%)	Area 3 (%)	Area 4 (%)
≤2.0	93.15	84.69	92.73	99.59
2.1–4.0	6.25	9.78	4.92	0.31
4.1–6.0	0.51	3.29	1.39	0.10
6.1	0.09	2.24	0.96	0.00

4.2. Validation and Accuracy Analysis

Validation was performed after extracting the planimetric and altimetric coordinates of the 20 checkpoints determined by the topography (reference). The points were compared to their respective counterparts extracted from orthophotomosaics (planimetry) and DEMs (altimetry), generated after digital image processing obtained through RPA (test). Tables 3 and 4 sequentially present the results of the statistical parameters of the planimetric validations of the orthophotomosaics and altimetric of the DEMs.

Table 3. Result of the statistical parameters of planimetric validations of orthophotomosaics.

Parameters	Area 1	Area 2	Area 3	Area 4
$\bar{x}$ (m)	0.032	0.043	0.029	0.018
σ (m)	0.019	0.037	0.019	0.008
max. (m)	0.078	0.112	0.071	0.036
min. (m)	0.010	0.000	0.000	0.010
$RMSE_x$ (m)	0.027	0.035	0.029	0.014
$RMSE_y$ (m)	0.025	0.044	0.018	0.014
$RMSE_r$ (m)	0.037	0.057	0.035	0.020

Table 4. Result of statistical parameters of DEM altimetric validations.

Parameters	Area 1	Area 2	Area 3	Area 4
$\bar{x}$ (m)	0.046	−0.003	0.039	0.022
σ (m)	0.056	0.065	0.048	0.033
max. (m)	0.210	0.140	0.150	0.090
min. (m)	−0.050	−0.100	−0.050	−0.040
$RMSE_z$ (m)	0.071	0.063	0.061	0.039

The highest RMSEr values occurred in areas 1 and 2, whose average slopes are higher than areas 3 and 4, as they are areas classified as very rugged relief (Table 3). Similarly to what was perceived in the determination of the planimetric statistical parameters of the orthophotomosaics (Table 3), the highest values of RMSEz occurred in areas 1 and 2, whose behavior was already expected because they are areas with greater slopes when compared with areas 3 and 4 (Table 4).

The horizontal and vertical absolute accuracy values were independently determined at the 95% confidence level for the four study areas (Table 5).

Table 5. Horizontal and vertical absolute accuracy of the areas.

Parameters	Area 1	Area 2	Area 3	Area 4
Horizontal accuracy (m)	0.064	0.099	0.061	0.035
Vertical accuracy (m)	0.139	0.123	0.120	0.076

The results found in Table 5 for horizontal accuracy indicate that the worst value occurred for area 2, in the order of 0.099 m, whose average relief slope is higher than the others. While the best value occurred for area 4, in the order of 0.035 m, whose average relief slope is the smallest among all areas.

For vertical accuracy, area 1 presented the worst value (0.139 m) when compared to the other areas and area 4 presented the best overall result, in the order of 0.076 m. However, it is noteworthy that all areas presented a vertical accuracy smaller than the size of a pixel of the DEMs (0.20 m). The results found in the analysis of DEM vertical accuracy for areas 1 and 2 (very rugged relief) were similar to those found by Pedreira et al. [42]. The authors used 20 checkpoints to determine accuracy and performed an aerophotogrammetric survey in a single study area, dividing it into relief classes according to the slope. The altimetric accuracy results found by the authors for the relief class which is comparable to areas 1 and 2 of this study, were in the order of 0.125 m. In our study area, 1 presented a lower vertical accuracy value (0.139 m), however, area 2 was slightly higher (0.123 m). They found a vertical accuracy in the order of 0.156 m for the relief class similar to areas 3 and 4 of this work (smooth relief). Areas 3 and 4 presented vertical accuracy in the order of 0.120 and 0.076 m, respectively, therefore both were higher than those found by the authors [42].

4.3. Simiarity of Hausdorff Distance

After validating the aerophotogrammetric DEMs, contour lines were derived for each area. Concomitantly, the contour lines were generated using the conventional topographic survey. The contour lines were superimposed and the similarity evaluator software returned with the dH for each pair of homologous contour lines in each area, as shown in Table 6.

The results show that the best fit represented by the dH in area 1, occurred in the contour line of 881 m of elevation, with the measure of similarity in the order of 0.295 m. While in the contour line at 863 m of elevation, the measure of similarity found was 1.275 m, which is considered the worst fit in this representation.

The best fit represented by the dH in area 2 was found in the contour line at 869 m of elevation, with the similarity measure in the order of 0.218 m. The worst fit occurred in the contour line at 857 m of elevation, with a similarity measure of 0.573 m.

Table 6. Determination of dH for contour lines in areas 1 to 4.

Contour Line Elevation (m)	Hausdorff Distance (m)		Contour Line Elevation (m)	Hausdorff Distance (m)	
	Area 1	Area 2		Area 3	Area 4
857	-	0.573	13	1.046	-
858	-	0.382	14	2.005	0.983
859	-	0.448	15	1.570	1.400
860	-	0.448	16	2.247	0.290
861	-	0.424	17	3.102	0.238
862	-	0.354	18	1.372	0.187
863	1.275	0.416	19	2.808	0.223
864	1.096	0.334	20	2.083	-
865	0.535	0.388	21	2.861	-
866	0.663	0.262	22	1.677	-
867	0.545	0.352	23	3.011	-
868	0.765	0.258	-	-	-
869	0.600	0.218	-	-	-
870	0.547	0.249	-	-	-
871	0.637	-	-	-	-
872	0.560	-	-	-	-
873	0.551	-	-	-	-
874	0.491	-	-	-	-
875	0.572	-	-	-	-
876	0.841	-	-	-	-
877	0.581	-	-	-	-
878	0.592	-	-	-	-
879	0.331	-	-	-	-
880	0.437	-	-	-	-
881	0.295	-	-	-	-

The best fit indicated by the dH for area 3 occurred in the contour line at 13 m of elevation, in the order of 1.046 m. The worst fit was identified in the contour line at 17 m of elevation, in the order of 3.102 m.

The best fit by the dH in area 4 was in the contour line at 18 m of elevation, in the order of 0.187 m. The worst adjustment was found in the contour line at 15 m of elevation in the order of 1.400 m.

From the dH data set presented in Table 6, the statistical parameters (mean, standard deviation, maximum and minimum) of the data set that represent all the contour lines of the same area were determined, as shown in Table 7.

Table 7. Statistical parameters of the dH analysis of the set of contour lines.

Parameters	Area 1	Area 2	Area 3	Area 4
$\bar{x}$ (m)	0.627	0.365	2.162	0.554
σ (m)	0.235	0.097	0.707	0.512
max. (m)	1.275	0.573	3.102	1.400
min. (m)	0.295	0.218	1.046	0.187

The highest value of the mean and standard deviation of contour lines using dH was credited to area 3 (smooth relief), indicating the worst similarity. Additionally, area 4 (smooth relief) also presented a dispersion of data associated with dH higher than areas 1 and 2, even reporting a maximum value above these areas.

The lowest value observed for the mean and standard deviation of the contour lines associated with dH occurred in area 2 (very rugged relief). Area 1 (very rugged relief) despite having an average associated with dH higher than area 4, has better similarity than area 4, it has a higher number of contour lines than those arranged in area 4.

However, the dH must be interpreted in association with the correlated effects of the differences found between the DEMs and their respective slopes that occurred between the data acquisition techniques. Figures 16 and 17 present the sets of contour lines derived from the two techniques, superimposed respectively on maps that represent differences in elevation and slope for the areas.

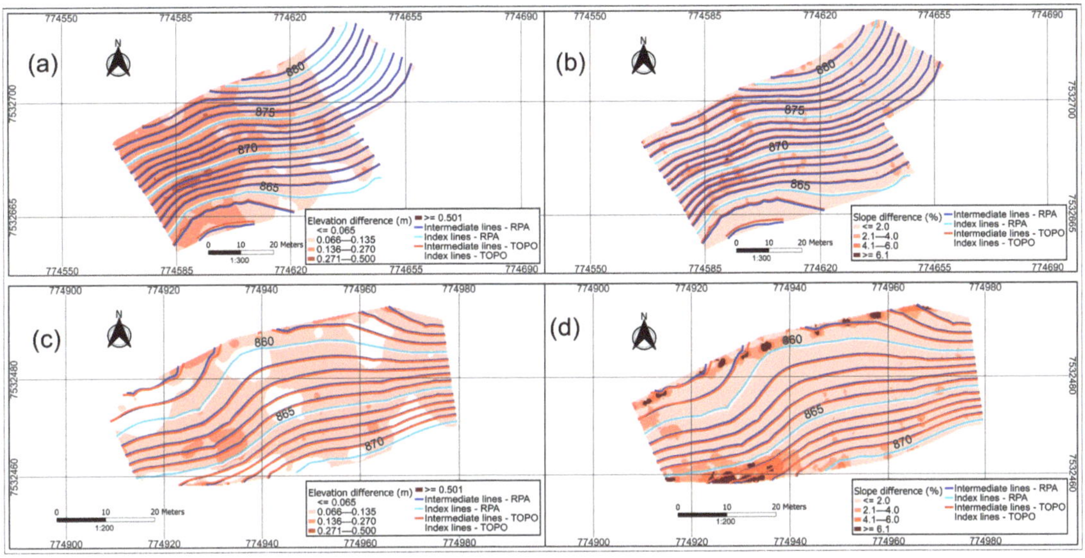

Figure 16. Superposition of homologous contour lines from topography and RPA of areas 1 and 2 over elevation and slope difference maps: (**a**) superposition of contour lines over DEM differences of area 1, (**b**) superposition of contour lines over slope differences of area 1, (**c**) superposition of contour lines over DEM differences of area 2, (**d**) superposition of contour lines over slope differences of area 2.

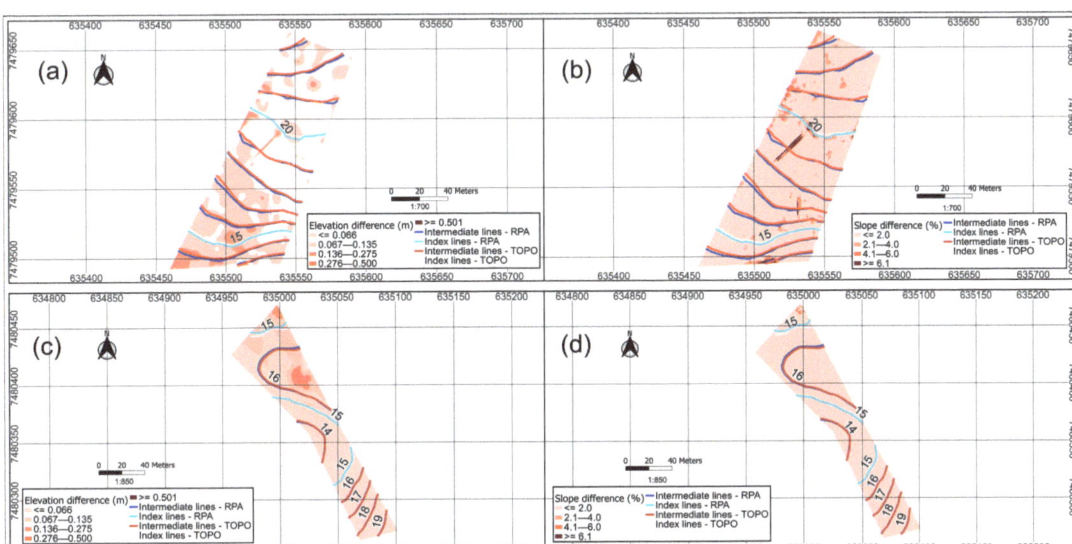

Figure 17. Superposition of homologous contour lines from topography and RPA of areas 3 and 4 over elevation and slope difference maps: (**a**) superposition of contour lines over DEM differences of area 3, (**b**) superposition of contour lines over slope differences of area 3, (**c**) superposition of contour lines over DEM differences of area 4, (**d**) superposition of contour lines over slope differences of area 4.

Considering the overlap in the differences of DEMs of the contour lines in area 1 (Figure 16a), we observed that the contour line at 881 m of elevation (best similarity) is inserted in a place with less altimetric divergence compared to the contour line at 863 m of elevation (worst similarity). However, the slope differences (Figure 16b) remain in a range of 0 to 2% where these curves are located.

In area 2, the contour line at the 869 m of elevation (best similarity) is located in lower altimetric divergences, when compared to the trajectory of the contour line at the elevation of 857 m (worst similarity), as shown in Figure 16c. In areas with these contour lines, the slope differences remained at 0 to 2% in almost the entire path of the contour lines, however presenting variations in small intervals of 2.1 to 4% for the contour line at the elevation of 869 m (Figure 16d). As for the contour line at the elevation of 857 m, the range from 0 to 2% of slope difference was predominant, but there were small stretches with values above 6.1% (Figure 16d).

In area 3 (Figure 17a), the contour lines at the elevation of 13 m (best similarity) and 17 m (worst similarity) were inserted in places with the same altimetric divergences. When checking the slope differences on these curves (Figure 17b), discrete stretches with variations in this parameter were observed. In the places where these curves are located the slope differences remained in a range of 0 to 2% in most of their entire path, however they present different intervals above 6.1% for both contour lines.

We found for area 4 (Figure 17c) that the contour lines at an elevation of 18 m (best similarity) passed through places with lower classes of altimetric differences when compared to the contour lines at an elevation of 15 m in the upper part of Figure 17c (worst similarity). No differences were observed in slope associated with dH in both curves (Figure 17d), as contour lines at 18 m and 15 m of elevation remained in a slope range difference from 0 to 2%.

The results found by the analysis of Hausdorff distance between homologous contour lines allows us to infer that this algorithm can be used to evaluate the accuracy of vector geometries. Gonçalves and Mitishita [43] also used Hausdorff distance for reviewing and updating urban maps by analyzing the similarity between vector geometries. After determining the quality and verifying the similarity of the features described by the contour lines obtained via RPA, a new perspective on using the RPAs for Precision Agriculture was introduced here. Martos et al. [44] describe the need to ensure sustainable agriculture, including the use of RPA technology to achieve this goal. Therefore, this work applies remote sensing with a robust methodology to achieve the goal described by Martos et al. [44], opening this segment to new perspectives of research and applications in Precision Agriculture.

5. Conclusions

In general, area 2 had the best similarity by the average of the data set provided by dH while area 3 had the worst similarity. By comparing areas with similar reliefs, area 2 had smaller differences between the DEMs (topographic x RPA) and better vertical accuracy when compared to area 1. Both areas have a very rugged relief. This performance is repeated in the data found by similarity analysis when observing the general parameters of the dataset provided by dH. All indicators (average, standard deviation, maximum and minimum value) in area 2 were better than those found in area 1, according to Table 7.

For areas 3 and 4 classified as smooth relief, we found that area 4 presented the smallest differences between the DEMs (topographic x RPA), also associated with better vertical accuracy when compared to area 3. Area 3 and area 4 had the same performance when the similarity was evaluated by dH. All parameters evaluated (mean, standard deviation, maximum and minimum value) were better than those found in area 3, according to data in Table 7.

Although areas 3 and 4 have respectively presented the highest values associated with dH (3.102 m and 1.400 m), when compared to areas 1 and 2, whose maximum values were 1.275 m and 0.573 m, respectively, the differences between altimetry and slope were lower than those found in area 2 (best similarity of the set). The average slope rate of area 2 is much higher than in areas 3 and 4. Therefore, the best general similarity found by applying the dH algorithm in area 2 did not mean a smaller difference between the DEMs, nor a better difference in slope when compared to areas with flatter reliefs.

The similarity by dH associated with vertical accuracy can be used as a quality indicator of an aerophotogrammetric product obtained by RPA when compared to the same product obtained by a source of superior precision. Therefore, the similarity algorithm used concomitantly with vertical accuracy is likely to be used for decision-making on issues involving application in precision agriculture.

Based on the results, the application of RPAs for the generation of contour lines proved to be a sufficiently effective technology capable of providing adequate contour planting. This finding is especially valid when analyzing the quantitative differences in slope, with the similarity obtained by the dH. It was noticed that the similarity obtained between the curves generated by topography and by RPA did not cause differences in slope greater than 6% in at least 95% of all study areas. This result indicates a high possibility of applying the contour lines derived from the RPA technology, for the safe application of contour planting.

In conclusion, the dH similarity algorithm can analyze the feasibility of contour planting with the products obtained by RPA according to the situations addressed in this study. The determination of vertical accuracy of Digital Elevation Models is of paramount importance and the Hausdorff similarity algorithm is a predictor of this quality by comparing different data acquisition techniques. Therefore, RPAs can generate contour lines for contour planting in Precision Agriculture, provided that vertical accuracy analysis controls the quality of the Digital Elevation Models.

Author Contributions: Conceptualization, A.A.R.F., M.A.H.A. and M.M.d.B.; methodology, A.A.R.F., M.A.H.A., M.M.d.B. and W.D.d.S.; software, W.D.d.S. and A.A.R.F.; validation, A.A.R.F., W.d.S.d.S. and T.M.d.S.; formal analysis, A.A.R.F. and M.A.H.A.; investigation, A.A.R.F. and T.M.d.S.; data curation, A.A.R.F. and W.d.S.d.S.; writing—original draft preparation, A.A.R.F. and T.M.d.S.; writing—review and editing, M.A.H.A., M.M.d.B. and W.D.d.S.; visualization, A.A.R.F., T.M.d.S., W.d.S.d.S. and W.D.d.S.; supervision, M.A.H.A. and M.M.d.B.; project administrator, A.A.R.F. All authors have read and agreed to the published version of the manuscript.

Funding: This research was financed by the Higher Education Personnel Improvement Coordination (CAPES) and by the Technical School of the Rural University (CTUR), both linked to the Federal Rural University of Rio de Janeiro (UFRRJ).

Data Availability Statement: Not applicable.

Acknowledgments: We thank the Coordination for the Improvement of Higher Education Personnel (CAPES), Technical School of the Rural University (CTUR), Federal Rural University of Rio de Janeiro (UFRRJ), and the Graduate Program in Science, Technology, and Innovation in Agriculture (PPGCTIA).

Conflicts of Interest: The authors declare no conflict of interest.

References

1. Silva, L.L.; Baptista, F.; Cruz, V.F.; da Silva, J.R.M. Aumentar as competências dos agricultores para a prática de uma agricultura sustentável. *Rev. Ciências Agrárias* **2020**, *43*, 240–252. [CrossRef]
2. Neumann, P.S.; Loch, C. Legislação ambiental, desenvolvimento rural e práticas agrícolas. *Ciência Rural* **2002**, *32*, 243–249. [CrossRef]
3. Santos, F.S.; Amaral Sobrinho, N.M.B.; Mazur, N. Influência de diferentes manejos agrícolas na distribuição de metais pesados no solo e em plantas de tomate. *Rev. Bras. Ciência Solo* **2002**, *26*, 535–543. [CrossRef]
4. Griebeler, N.P.; Pruski, F.F.; Teixeira, A.F.; de Oliveira, L.F. Software to planning the use of level terracing systems in more rational ways. *Eng. Agrícola* **2005**, *25*, 841–851. [CrossRef]
5. Leite, E.F.; Rosa, R. Análise do uso, ocupação e cobertura da terra na bacia hidrográfica do Rio Formiga, Tocantins. *Rev. Eletrônica Geogr.* **2012**, *4*, 90–106.
6. Xavier, M.V.B.; Santos, L.L.; Fonseca, A.P.M.; de Almeida, E.S.; Almeida, L.V.O.; Aguiar, R.M.A.S.; Moreira, C.D.D.; Semensato, B.D.; Ferreira, J.M.; de Oliveira, P.V.A. Capacidade de uso e manejo conservacionista do solo de um fragmento de cerrado. *Res. Soc. Dev.* **2021**, *10*, e41410716697. [CrossRef]
7. Fortini, R.M.; Braga, M.J.; Freitas, C.O. Impacto das práticas agrícolas conservacionistas na produtividade da terra e no lucro dos estabelecimentos agropecuários brasileiros. *Rev. Econ. Sociol. Rural* **2020**, *58*, e199479. [CrossRef]

8. Paccioretti, P.; Córdoba, M.; Balzarini, M. FastMapping: Software to create field maps and identify management zones in precision agriculture. *Comput. Electron. Agric.* **2020**, *175*, 105–556. [CrossRef]
9. Yost, M.A.; Kitchen, N.R.; Sudduth, K.A.; Massey, R.E.; Sadler, E.J.; Drummond, S.T.; Volkmann, M.R. A long-term precision agriculture system sustains grain profitability. *Precis. Agric.* **2019**, *20*, 1177–1198. [CrossRef]
10. Nascimento, H.R.; de Abreu, Y.V. Geração de informações sobre a agricultura de energia por meio das geotecnologias. *Interações* **2012**, *13*, 181–189. [CrossRef]
11. Francisco, H.R.; Corrêia, A.F.; Feiden, A. Classification of areas suitable for fish farming using geotechnology and multi-criteria analysis. *ISPRS Int. J. Geo-Inf.* **2019**, *8*, 394. [CrossRef]
12. Hunt Jr, E.R.; Daughtry, C.S.T. What Good Are Unmanned Aircraft Systems for Agricultural Remote Sensing and Precision Agriculture. *Int. J. Remote Sens.* **2018**, *39*, 5345–5376. [CrossRef]
13. Maes, W.H.; Steppe, K. Perspectives for remote sensing with unmanned aerial vehicles in precision agriculture. *Trends Plant Sci.* **2019**, *24*, 152–164. [CrossRef] [PubMed]
14. Bendig, J.; Yu, K.; Aasen, H.; Bolten, A.; Bennertz, S.; Broscheit, J.; Gnyp, M.L.; Bareth, G. Combining UAV-Based plant height from crop surface models, visible, and near-infrared vegetation indices for biomass monitoring in barley. *Int. J. Appl. Earth Obs. Geoinf.* **2015**, *39*, 79–87. [CrossRef]
15. Pérez-Ortiz, M.; Peña, J.M.; Gutiérrez, P.A.; Torres-Sánchez, J.; Hervás-Martínez, C.; López-Granados, F. Selecting patterns and features for between- and within- crop-row weed mapping using UAV-imagery. *Expert Syst. Appl.* **2016**, *47*, 85–94. [CrossRef]
16. Chang, A.; Jung, J.; Maeda, M.M.; Landivar, J. Crop height monitoring with digital imagery from Unmanned Aerial System (UAS). *Comput. Electron. Agric.* **2017**, *141*, 232–237. [CrossRef]
17. Schut, A.G.T.; Traore, P.C.S.; Blaes, X.; By, R.A. Assessing yield and fertilizer response in heterogeneous smallholder fields with UAVs and satellites. *Field Crops Res.* **2018**, *221*, 98–107. [CrossRef]
18. Tsouros, D.C.; Bibi, S.; Sarigiannidis, P.G. A review on UAV-based applications for precision agriculture. *Information* **2019**, *10*, 349 [CrossRef]
19. Santos, L.M.; Ferraz, G.A.S.; Barbosa, B.D.S.; Andrade, A.D. Use of remotely piloted aircraft in precision agriculture: A review. *Dyna* **2019**, *86*, 284–291. [CrossRef]
20. Delavarpour, N.; Koparan, C.; Nowatzki, J.; Bajwa, S.; Sun, X. A technical study on UAV characteristics for precision agriculture applications and associated practical challenges. *Remote Sens.* **2021**, *13*, 1204. [CrossRef]
21. Nemmaoui, A.; Aguilar, F.J.; Aguilar, M.A.; Qin, R. DSM and DTM generation from VHR satellite stereo imagery over plastic covered greenhouse areas. *Comput. Electron. Agric.* **2019**, *164*, 104903. [CrossRef]
22. Akturk, E.; Altunel, A.O. Accuracy assessment of a low-cost UAV derived digital elevation model (DEM) in a highly broken and vegetated terrain. *Meas. J. Int. Meas. Confed.* **2019**, *136*, 382–386. [CrossRef]
23. Mukherjee, A.; Misra, S.; Raghuwanshi, N.S. A survey of unmanned aerial sensing solutions in precision agriculture. *J. Netw Comput. Appl.* **2019**, *148*, 102461. [CrossRef]
24. Iticha, B.; Takele, C. Digital soil mapping for site-specific management of soils. *Geoderma* **2019**, *351*, 85–91. [CrossRef]
25. Santana, L.S.; Ferraz, G.A.e.S.; Marin, D.B.; Faria, R.d.O.; Santana, M.S.; Rossi, G.; Palchetti, E. Digital Terrain Modelling by Remotely Piloted Aircraft: Optimization and Geometric Uncertainties in Precision Coffee Growing Projects. *Remote Sens.* **2022**, *14*, 911. [CrossRef]
26. Marchi, E.C.S.; Campos, K.P.; Corrêa, J.B.D.; Guimarães, R.J.; Souza, C.A.S. Sobrevivência de mudas de cafeeiro produzidas em sacos plásticos e tubetes no sistema convencional e plantio direto, em duas classes de solo. *Ceres* **2003**, *50*, 407–416.
27. Solos, E. *Sistema Brasileiro de Classificação de Solos*, 2nd ed.; Centro Nacional de Pesquisa de Solos: Rio de Janeiro, Brazil, 2006; p. 306.
28. Esri, R. *ArcGIS Desktop*; Environmental Systems Research Institute: Redlands, CA, USA, 2016.
29. Hutchinson, M.F. A new procedure for gridding elevation and stream line data with automatic removal of spurious pits. *J. Hydrol* **1989**, *106*, 211–232. [CrossRef]
30. Oikonomou, C.; Stathopoulou, E.K.; Georgopoulos, A. Contemporary data acquisition technologies for large scale mapping. In Proceedings of the 35th EARSeL Symposium–European Remote Sensing: Progress, Challenges and Opportunities, Stockholm Sweden, 15–18 June 2015.
31. Westoby, M.J.; Brasington, J.; Glasser, N.F.; Hambrey, M.J.; Reynolds, J.M. 'Structure-from-Motion' photogrammetry: A low-cost effective tool for geoscience applications. *Geomorphology* **2012**, *179*, 300–314. [CrossRef]
32. Zanetti, J.; Braga, F.L.S.; Dos Santos, A.d.P. Comparativo das normas de controle de qualidade posicional de produtos cartográficos do Brasil, da ASPRS e da OTAN. *Rev. Bras. Cartogr.* **2018**, *70*, 359–390. [CrossRef]
33. Mora, O.E.; Chen, J.; Stoiber, P.; Koppanyi, Z.; Pluta, D.; Josenhans, R.; Okubo, M. Accuracy of stockpile estimates using low-costsUAS photogrammetry. *Int. J. Remote Sens.* **2020**, *41*, 4512–4529. [CrossRef]
34. American Society for Photogrammetry and Remote Sensing (ASPRS). ASPRS Positional Accuracy Standards for Digital Geospatial Data. Photogramm. *Eng. Remote Sens.* **2015**, *81*, A1–A26.
35. Ghilani, C.D.; Wolf, P.R. *Adjustment Computations: Spatial Data Analysis*, 4th ed.; John Wiley & Sons, Inc.: Hoboken, NJ, USA, 2006; pp. 326–327. [CrossRef]
36. López, F.J.A. Calidad en la producción cartográfica. *Mapping* **2003**, *84*, 96–98.
37. Gregoire, N.; Bouillot, M. Hausdorff Distance between convex polygons. *Comput. Geom. Web Proj.* **1998**, *8*, 2008.

38. Santos, A.D.P.D.; Medeiros, N.D.G.; Santos, G.R.D.; Rodrigues, D.D. Controle de qualidade posicional em dados espaciais utilizando feições lineares. *Bol. De Ciências Geodésicas.* **2015**, *21*, 233–250. [CrossRef]
39. Elmasri, R.; Navathe, S.B. *Fundamentals of Database Systems*, 7th ed.; University of Texas at Arlington Georgia Institute of Technology: Arlington, TX, USA, 2016; p. 1280.
40. Butler, H.; Daly, M.; Doyle, A.; Gillies, S.; Hagen, S.; Schaub, T. The geojson format. *Internet Eng. Task Force (IETF)* **2016**, *RFC 7946*, 1–28.
41. Butler, H.; Daly, M.; Doyle, A.; Gillies, S.; Schaub, T.; Schmidt, C. GeoJSON. Electronic. 2014. Available online: http://geojson.org (accessed on 26 May 2022).
42. Pedreira, W.J.P.; de Andrade Oliveira, J.; Santos, P.S. Avaliação da Acurácia Altimétrica usando a Tecnologia VANT. *Rev. Caminhos Geogr.* **2020**, *21*, 209–222. [CrossRef]
43. Gonçalves, G.A.; Mitishita, E.A. O Uso da Distância de Hausdorff como Medida de Similaridade em Sistemas Automáticos de Atualização Cartográfica. *Bol. Ciências Geodésicas.* **2016**, *22*, 719–735. [CrossRef]
44. Martos, V.; Ahmad, A.; Cartujo, P.; Ordoñez, J. Ensuring agricultural sustainability through remote sensing in the era of agriculture 5.0. *Appl. Sci.* **2021**, *11*, 5911. [CrossRef]

agriculture

MDPI

Article

Classification of Maize Lodging Extents Using Deep Learning Algorithms by UAV-Based RGB and Multispectral Images

Xin Yang [1,2], Shichen Gao [2], Qian Sun [1], Xiaohe Gu [1,*], Tianen Chen [3,*], Jingping Zhou [1] and Yuchun Pan [1]

[1] Research Center of Information Technology, Beijing Academy of Agriculture and Forestry Sciences, Beijing 100089, China; 15803304826@163.com (X.Y.); sunq817@163.com (Q.S.); zhoujp@nercita.org.cn (J.Z.); panyc@nercita.org.cn (Y.P.)

[2] School of Science, China University of Geosciences, Beijing 100089, China; gsc2039@163.com

[3] National Engineering Research Center for Information Technology in Agriculture, Beijing 100089, China

* Correspondence: guxh@nercita.org.cn (X.G.); chente@nercita.org.cn (T.C.)

Abstract: Lodging depresses the grain yield and quality of maize crop. Previous machine learning methods are used to classify crop lodging extents through visual interpretation and sensitive features extraction manually, which are cost-intensive, subjective and inefficient. The analysis on the accuracy of subdivision categories is insufficient for multi-grade crop lodging. In this study, a classification method of maize lodging extents was proposed based on deep learning algorithms and unmanned aerial vehicle (UAV) RGB and multispectral images. The characteristic variation of three lodging extents in RGB and multispectral images were analyzed. The VGG-16, Inception-V3 and ResNet-50 algorithms were trained and compared depending on classification accuracy and Kappa coefficient. The results showed that the more severe the lodging, the higher the intensity value and spectral reflectance of RGB and multispectral image. The reflectance variation in red edge band were more evident than that in visible band with different lodging extents. The classification performance using multispectral images was better than that of RGB images in various lodging extents. The test accuracies of three deep learning algorithms in non-lodging based on RGB images were high, i.e., over 90%, but the classification performance between moderate lodging and severe lodging needed to be improved. The test accuracy of ResNet-50 was 96.32% with Kappa coefficients of 0.9551 by using multispectral images, which was superior to VGG-16 and Inception-V3, and the accuracies of ResNet-50 on each lodging subdivision category all reached 96%. The ResNet-50 algorithm of deep learning combined with multispectral images can realize accurate lodging classification to promote post-stress field management and production assessment.

Keywords: lodging classification; unmanned aerial vehicle (UAV); sensitive band; ResNet algorithm

Citation: Yang, X.; Gao, S.; Sun, Q.; Gu, X.; Chen, T.; Zhou, J.; Pan, Y. Classification of Maize Lodging Extents Using Deep Learning Algorithms by UAV-Based RGB and Multispectral Images. *Agriculture* **2022**, *12*, 970. https://doi.org/10.3390/agriculture12070970

Academic Editors: Monica Herrero-Huerta, Jose A. Jiménez-Berni, Shangpeng Sun and Diego González-Aguilera

Received: 16 June 2022
Accepted: 5 July 2022
Published: 6 July 2022

1. Introduction

According to the data released by China's National Bureau of Statistics, the planting area of maize reached 43.324 million hectares in 2021, increasing by 2.059 million hectares compared with 2020. The total output of maize achieved 272 million tons, which made it the most productive of China's major crops. The yield variance of maize has an important impact on national food security and agricultural economic development. However, crop lodging is one of the major negative elements to affect maize output. It is stated as the displacement of the above-ground stems from their upright position or failure of root-soil attachment [1]. Lodging is generally caused by rainstorms, loose soil, high planting density and unreasonable fertilization [2–4]. Lodging hinders the growth of maize [5], reduces grain quality [6] and affects mechanized harvesting [7], which is becoming an important restricting issue to increase maize yield [8]. Therefore, precise and efficient classification with different maize lodging extents can help agricultural departments to investigate the influence of maize growth, guide farmers to implement post-stress field management and facilitate insurance firms to settle disputes properly [9,10].

The traditional lodging assessment methods widely used are mainly visual inspection and artificial measurement [11], which are inefficient, time-consuming and environment-constraining [12]. The inaccuracy and subjectivity of them may lead to compensation disputes between farmers and insurance companies, which cannot meet the needs of precision agriculture. Remote sensing technology, as a new approach, has greatly promoted the development of crop lodging detection [13]. The lodging incidence in wheat, rice and barley were detected using visible and thermal infrared images based on the ground-based and space-borne platforms [14–17]. In recent decades, the unmanned aerial vehicle (UAV) has been increasingly applied for lodging monitoring due to its advantages of convenient, flexible, low cost and high resolution [18,19]. It can timely and accurately obtain centimeter-level images with multiple sensors, which plays a powerful role in lodging detection [20]. Many studies detected crop lodging based on a UAV system equipped with a digital camera. They discriminated lodging from non-lodging and evaluated the extents of crop lodging by analyzing color and texture features [21–23]. However, compared to RGB images with only three visible bands, multispectral images with red edge and infrared bands reflecting the growth capacity of crops can offer more information in crop lodging [12,24,25]. Both spatial and spectral information of ground targets are obtained in the meantime. Therefore, the information richness of lodging features between these two types of images are different. It is worth studying to verify the performance of discriminating lodging severity extents using RGB and multispectral images.

The appropriate classification methods for crop lodging extents are significant as well as the selection of data source. Traditionally, machine learning algorithms consisting of a support vector machine (SVM) [8], decision tree [26] and nearest neighbor [5] were used to classify lodging by extracting crop morphology and spectral characteristics [21,23,27]. However, these manual approaches for extracting features often required empirical knowledge and were typically suboptimal in the results [28]. With the development of machine learning, the convolutional neural network (CNN) of deep learning has gradually become the mainstream. CNN algorithms can automatically extract image features, and depict rich intrinsic information with strong nonlinear modeling ability. Xia hao et al. [29] proposed a classification model named GL-CNN on account of convolutional neural networks to determine the optimal growth stage of leafy vegetables. Ananda et al. [30] used the Visual Geometry Group (VGG) model to achieve the disease detection and classification of grapes and tomatoes. CNN has been proved to be superior to existing traditional machine classification algorithms [31]. The Inception and ResNet algorithms were proposed with better performance, which could automatically extract target features from images more accurately. They have been widely used in disease detection and crop classification in intelligent agriculture [32,33]. However, there are few studies on maize lodging classification based on deep learning algorithms. The maize lodging characteristics of multiple data types need to be analyzed. The performance difference using RGB and multispectral images will be compared. Previous studies have often focused on the overall classification accuracy of crop lodging, which were unable to fully embody the quality of the model. The classification effects of algorithms under subdivision categories are also worthy of attention.

The purpose of this study is to use deep learning algorithms to monitor the lodging extents of maize based on RGB or multispectral images. The lodging extents are discriminated as non-lodging, moderate lodging and severe lodging by lodging angle. The specific objectives are as follows: (1) to analyze the characteristics of obtained images with different lodging extents, (2) classify lodging extents of maize based on RGB and multispectral images through VGG-16, Inception-V3 and ResNet-50 algorithms and (3) evaluate classification performance in different lodging extents to determine the optimal algorithm.

2. Materials and Methods

The processes of classifying maize lodging extents in the study were showed in Figure 1. The RGB and multispectral images were acquired via UAV, which were respectively cropped, augmented and labeled to build the datasets. The difference of each band of RGB and multispectral images caused by maize lodging extents were analyzed. The classification results of maize lodging extents using three deep learning algorithms were compared and validated.

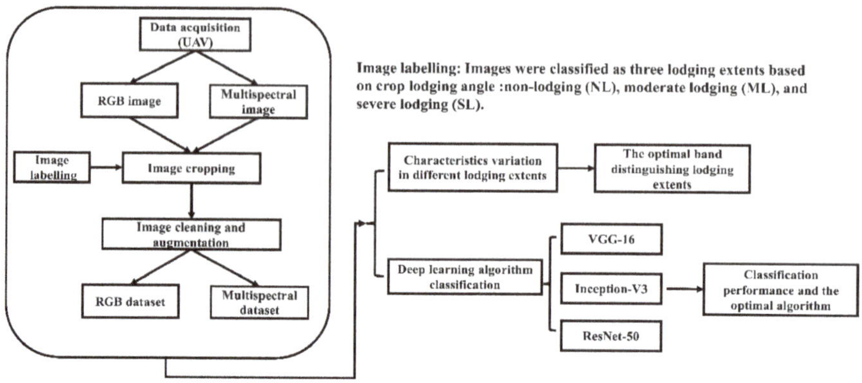

Figure 1. Flowchart of RGB and multispectral dataset acquisition and classification of different lodging extents using deep learning algorithms.

2.1. Study Area

The study area is located in Lishu County, Siping City, southwest Jilin Province, China (Figure 2). The geographical coordinates of it are 43°02′ N–43°46′ N, 123°45′ E–124°53′ E Lishu is in the hinterland of Songliao Plain and the major grain producing county with a maize planting area of 213,300 hectares. During the maize growth period, sunshine and precipitation are sufficient, which can fully meet the growth needs of one ripe a year. From late August to early September in 2020, strong winds and heavy rain caused crop lodging

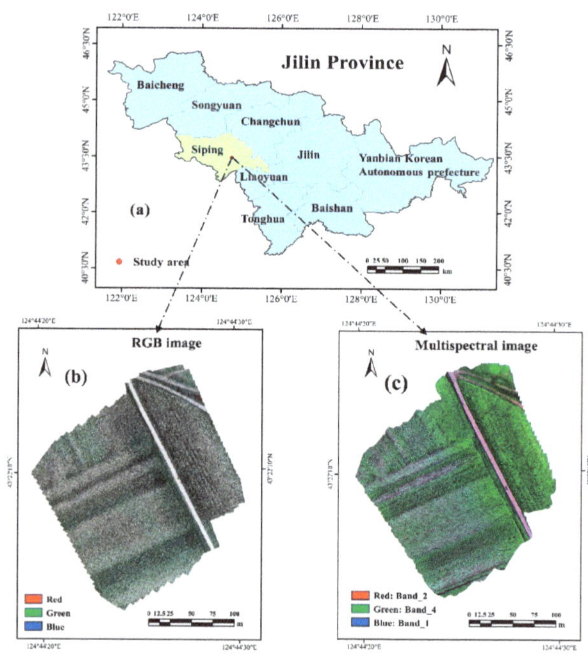

Figure 2. Overview of the study area (**a**) Geographical location of the study area. (**b**) The UAV RGB image. (**c**) The UAV multispectral image (false color composite, R: Red, G: NIR, B: Green).

2.2. Data Acquisition

The data collection of maize lodging canopy images in this study was performed with a DJI Phantom 4 Pro (DJI-Innovations, Inc., Shenzhen, China) at 12:00 am on 12 September 2020. The weather was cloudless and windless. The overall weight of UAV system is 1388 g, and the duration of flight is about 30 min. In this study, the flight altitude was 30 m above the ground. The forward and lateral overlap was 80%. The digital camera had three color channels of red, green and blue with a resolution of 1 cm/pixel. The multispectral images were collected by a Parrot Sequoia camera (MicaSense, Inc., Seattle, DC, USA). It consisted of four multispectral channels of green (550 nm), red (660 nm), red edge (735 nm) and near-infrared (790 nm) with a resolution of 2 cm/pixel. The global positioning system (GPS) and irradiance sensors were equipped at the same time. Before and after each flight, radiometric calibration images were obtained by a calibrated reflectance panel. The field inspection was taken after UAV data acquisition. Lodging has a huge impact on both yield and grain quality. Lodging caused a maize yield loss of approximately 0–50% at different lodging angles [3]. In general, the smaller the lodging angle, the smaller the yield loss. Lodging classification can provide a basis for predicting future harvest yield. According to the investigation of maize lodging in the study area, we categorically defined three lodging extents based on crop lodging angle: non-lodging (NL) maize with a crop angle <10°, moderate lodging (ML) maize with a crop angle between 10–50° and severe lodging (SL) maize with a crop angle >50° (Figure 3).

Figure 3. Maize lodging data collection (**left**), aerial imagery collection; (**right**), classification of three lodging extents based on crop lodging angle.

2.3. Data Cleaning and Augmentation

RGB and multispectral images of the entire study area were obtained by Agisoft Photoscan software. RGB images were resampled to 2 cm/pixel to match the resolution of multispectral images. Then, the images of the entire study area were cropped into small images with a resolution of 300 × 300 pixels. The actual spatial size of each image was 6 m, achieving a precise classification of maize lodging. Considering the partial areas of the images were not related to maize lodging, the original dataset of 1326 images was acquired by deleting the cropped images containing roads and weeds. Then, each sample was labeled as non-lodging, moderate lodging and severe lodging by an expert through visual interpretation (Figure 4).

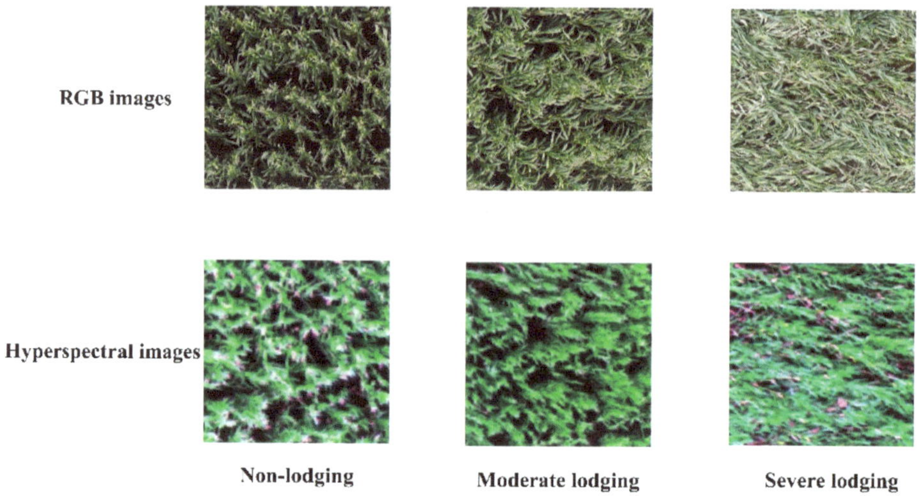

Figure 4. Maize lodging samples after data cleaning.

For the purpose of improving the overall generalization ability of the model, abundant training images are needed in the deep learning algorithms to avoid over-fitting. Data augmentation undertakes a more crucial improvement upon the classification accuracy in the dataset [28]. Therefore, we performed data augmentation on the obtained dataset to expand the number of samples. In this study, we enhanced image numbers by random rotation, horizontal inversion and vertical inversion. A dataset of 5000 RGB images and 5000 multispectral images was generated by data augmentation without introducing extra labeling costs. The dataset included 1616 non-lodging samples, 1684 moderate lodging samples and 1700 severe lodging samples. The results of image augmentation taking an RGB image as an example are shown in Figure 5.

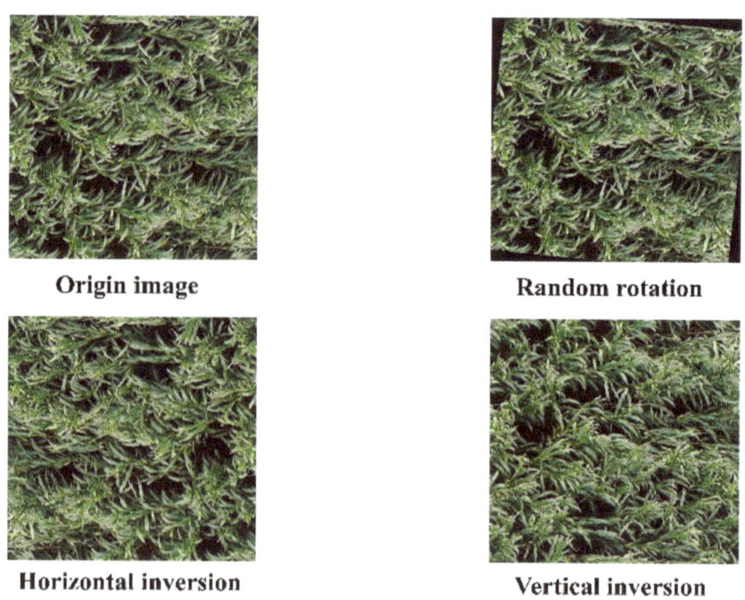

Figure 5. Original UAV image and three augmented images.

2.4. Deep Learning

2.4.1. Convolutional Neural Networks

Convolutional neural networks (CNN) have been essential to the development of deep learning. Remarkable advancements have been made on image classification [34]. CNN architecture is mainly divided into convolution layer, pooling layer and fully connected layer (Figure 6). The various aspects in the whole image are assigned importance for establishing a distinction between different objects in convolution layer. The weights of convolution kernels (not directly accessible to users) are constantly updated during algorithm iterations. After the convolution, the pooling operation can reduce the spatial size of the convolved features. It can help reduce the computing power requirements of data processing. We generally use two pooling methods, including maximum pooling and average pooling. Maximum pooling was superior in this study, because it could suppress noise while reducing dimension. Convolution and pooling layers were combined to extract image features of different levels. The last layer is the fully connected layer, which identifies the extracted features and provides the predicted label by using Softmax regression classifier eventually.

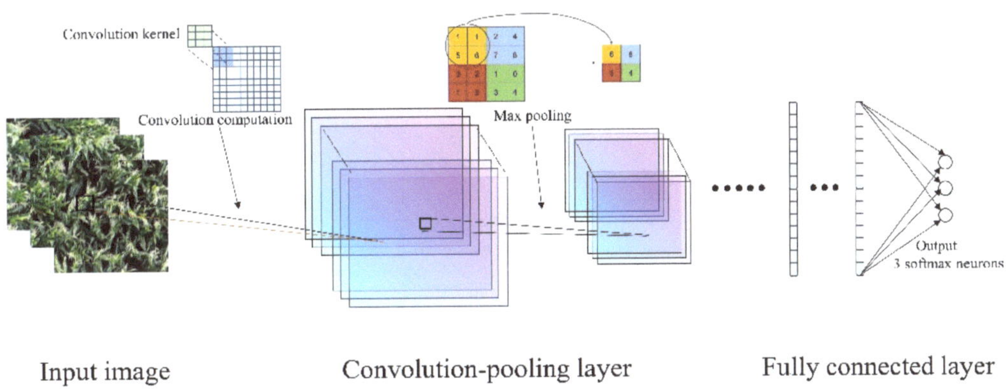

Figure 6. Structure diagram of the convolutional neural network.

2.4.2. VGG-16

VGG-16 is a CNN algorithm proposed by the Visual Geometry Group of Oxford University [35]. It consists of thirteen convolution layers (extracting image features), five maximum pooling layers (reducing image spatial size) and three fully connected layers (classifying images into labels) (Figure 7). Compared with traditional convolutional neural networks, this algorithm uses a 3×3 convolution kernel to replace the larger one (e.g., 5×5, 7×7). This optimization effectively reduces the number of model parameters and extracts the detail features of the images more accurately. Hence, it can improve the computing speed and has good generalization performance.

Figure 7. Structure diagram of VGG-16.

2.4.3. Inception-V3

Inception-V3 is the most representative algorithm among inception algorithms [36]. It uses the Inception module, which performs multiple convolution and max pooling operations in parallel to obtain a deeper feature map. The Inception-V2 references VGG net using small convolution kernels (e.g., 1×1, 3×3) to reduce the computational cost effectively. On the basis of that, Inception-V3 decomposes the 3×3 convolution kernel into 1×3 and 3×1 convolution kernels (Figure 8). The depth and nonlinearity of the network increase, which makes the network classification ability stronger.

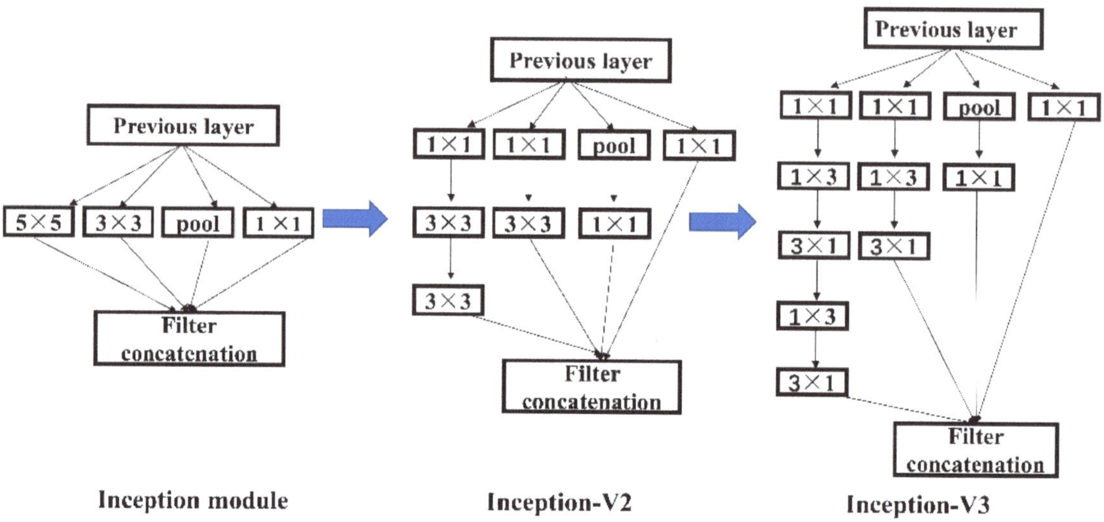

Figure 8. The optimization procedure of Inception module: (**left**), architecture of the initial Inception module; (**middle**), module architecture in Inception-V2; (**right**), module architecture in Inceptiom-V3.

2.4.4. ResNet-50

ResNet-50 is proposed to solve the degradation problem in neural network training which means the performance of the algorithm decreases with the deepening of network layers [37]. Residual block is the core of ResNet network (Figure 9), which mainly connects the convolution layer across layer by jumping connection and short circuit methods. It can transfer the input x as the initial result directly to the output, ensuring the integrity of the information. The output result is H(x) = F(x) + x, where F(x) is the residual function which helps to transmit information to deeper neural networks and improve the accuracy of the algorithm.

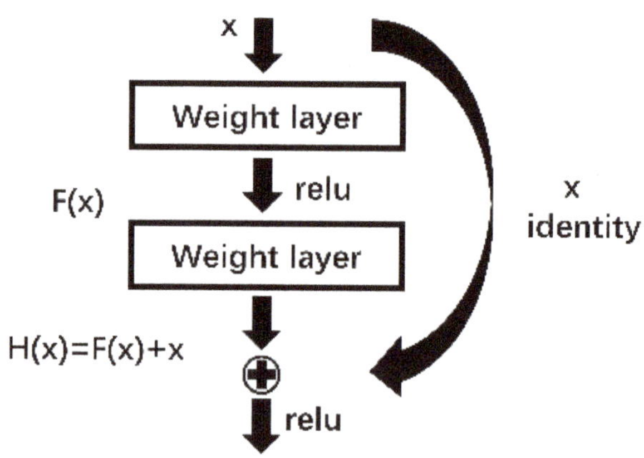

Figure 9. Architecture of the residual block to solve the degradation problem.

These three algorithms were used to classify different lodging extents and test their accuracy performance. The ReLU function was used as the activation function, and the dropout layer was imported to prevent the algorithms from overfitting (dropout_ratio = 0.5). The last layer (fully connected layer) was replaced by three classification categories to adapt to the dataset of this study.

To demonstrate the algorithms' validity and reliability, 70% of the samples (without substitution) were randomly selected as the training set and the remaining 30% of samples were the test set.

2.5. Validation

The image classification results of the dataset are evaluated by the confusion matrix, test accuracy and Kappa coefficient. The test accuracy is figured by the ratio between the number of correctly classified samples and the total number of samples in the test set. Kappa coefficient is a robust measure of the extents of agreement. In order to evaluate these indicators more persuasively, we repeated the experiments 10 times. The test accuracy and Kappa coefficients were calculated by the following formula and recorded as the average of ten repetitions:

$$\text{Test Accuracy} = \frac{\sum_{i=1}^{n} x_{ii}}{N} \tag{1}$$

$$\text{Kappa} = \frac{\sum_{i=1}^{n} x_{ii}/N - \sum_{i=1}^{n}(\sum_{j=1}^{n} x_{ij} \sum_{j=1}^{n} x_{ji})/N^2}{1 - \sum_{i=1}^{n}(\sum_{j=1}^{n} x_{ij} \sum_{j=1}^{n} x_{ji})/N^2} \tag{2}$$

where x_{ii} refers to the correctly predicted samples, x_{ij} refers to the elements of the i-th row and j-th column of the confusion matrix, n is the number of classifications and N is the total number of samples in test set.

3. Results

3.1. Research Images Analysis

In order to obtain an understanding of the lodging features under different types of images better, all the samples were used to observe the characteristics variation of maize canopy in the different lodging extents. The intensity values of RGB images and the reflectance of multispectral images were extracted directly by the statistical function of the ENVI 5.3 software.

3.1.1. RGB Images Analysis

RGB images contain the intensity values in red, green and blue color channels ranging from 0 to 255. Different intensity values of the three channels are combined into different colors. The means and standard deviations of three channels with different lodging extents were calculated in Figure 10. The intensity values of lodging (moderate lodging and severe lodging) were all significantly higher than that of non-lodging in three bands, but those of moderate and severe lodging were close relatively. In the non-lodging area, there were interspaces between maize plants along with shadows, and the soil was exposed to aerial photography, which made the intensity values low. After lodging, the plants tilted and piled each other, causing the soil to be covered. The intensity values increased with the decrease of soil bareness and the increase of plant density. Meanwhile, the changes of intensity values with different lodging extents were consistent in three bands, which were the lowest in blue and highest in green band. In Table 1, compared with the intensity values of non-lodging maize in three bands, those of the moderate lodging increased by 37.64%, 21.68% and 27.73%, and those of the severe lodging increased by 53.81%, 32.89% and 40.81%, respectively. It showed that the intensity values increased rapidly after lodging, and the increase rate of the values in the blue band was the highest.

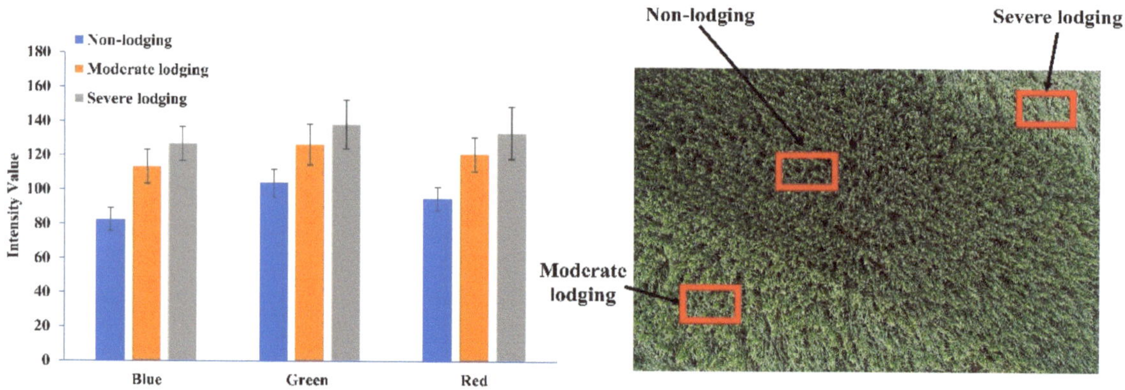

Figure 10. The intensity values variation (**left**) and the maize canopy under the UAV aerial with different lodging extents (**right**).

Table 1. The average intensity values with three lodging extents in different bands.

Extents \ Bands	Blue	Green	Red
Non-lodging	82.49	103.99	94.84
Moderate lodging	113.54	126.54	121.14
Severe lodging	126.88	138.20	133.55

3.1.2. Multispectral Images Analysis

Multispectral images show the reflectance in green, red, red edge and near-infrared bands with different lodging extents. The reflectance ranges from 0 to 1. The means and standard deviations of four channels with different lodging extents were calculated in Figure 11. The spectral reflectance increased following the enhancement of lodging extents in four bands. The reason was that lodging has changed the morphological structure of the maize population. The original maize canopy was damaged with the stems exposed. As the severity of maize lodging increased, more stems were exposed in aerial images taken by the UAV. Furthermore, the reflectance of the leaf was lower than that of the stem [24]. In Table 2, the reflectance of the red edge and near-infrared bands was significantly higher than that of the green and red bands. Compared with the reflectance of non-lodging maize in four bands, that of the moderate lodging increased by 6.45%, 12.50%, 19.51% and 13.20% and that of the severe lodging increased by 19.35%, 25%, 36.58% and 20.75%, respectively. It indicated that in different lodging extents, the variation of reflectance in the red edge band was more evident than that in the visible band, which meant the increase rate of reflectance in the red edge band was the largest as well.

Table 2. The average reflectance with three lodging extents in different bands.

Extents \ Bands	Green	Red	Red Edge	Near-Infrared
Non-lodging	0.31	0.24	0.41	0.53
Moderate lodging	0.33	0.27	0.49	0.60
Severe lodging	0.37	0.30	0.56	0.64

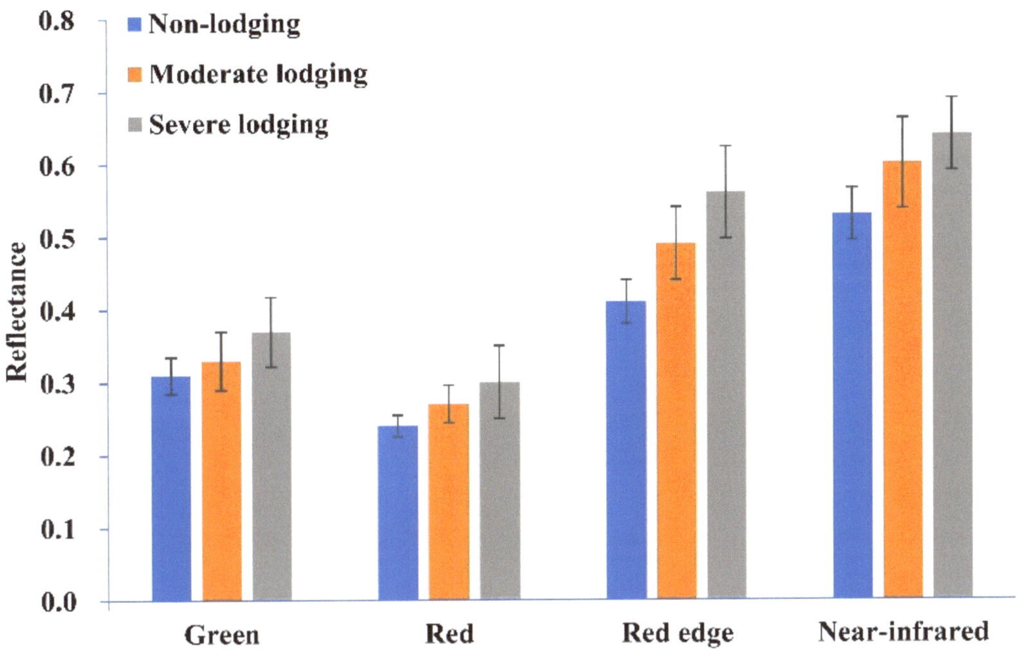

Figure 11. Spectral reflectance of four bands under different lodging extents.

3.2. Lodging Classification Using RGB Images

VGG-16, Inception-V3 and ResNet-50 had pre-trained CNN models to deal with RGB images. Their weight parameters were trained and identified based on a huge number of RGB images from the ImageNet Dataset (http://image-net.org/index, accessed on 4 March 2022). The transfer-learning method could achieve sharing of model features through the hyperparameter transfer. Therefore, the backbone parameters of three CNN algorithms were initialized using the pre-trained weights, which could save algorithm training time and obtain accurate results. The PyTorch framework with Python 3.6 was used to support all experiments, and GTX 1070 6G GPU was employed to accelerate the overall process. The learning rate, batch size and the number of iterations of the three algorithms were set to 0.0001, 20 and 100, respectively. The networks were trained with the Adam optimizer and cross-entropy loss function to optimize the objectives.

The changes of classification accuracy and loss in three algorithms during 100 iterations were shown in Figure 12. Due to the use of pre-training models, the initial training accuracies were all more than 0.6. With the continuous optimization of the algorithms, the classification accuracy improved rapidly. Eventually, the training accuracy of the three algorithms reached 86.16%, 91.89% and 94.16%, respectively. In addition, we chose cross-entropy as the loss function, and loss gradually decreased following the opposite overall trend to accuracy curves. Both of them began to maintain stability after approximately 20 iterations. The convergence rate of ResNet-50 algorithm was obviously faster than the other two algorithms. In Table 3, the test accuracies of the three algorithms were 83.55%, 87.32% and 90.08% with Kappa coefficients of 0.7421, 0.8040 and 0.8599, respectively. The overfitting phenomenon did not occur in the training process. ResNet-50 obtained the optimal performance in three algorithms, whose test accuracy was 7.81% and 3.16% higher than VGG-16 and Inception-V3. However, in addition to discussing the overall classification accuracies of the three algorithms, the classification performance of different categories was also worth further analysis.

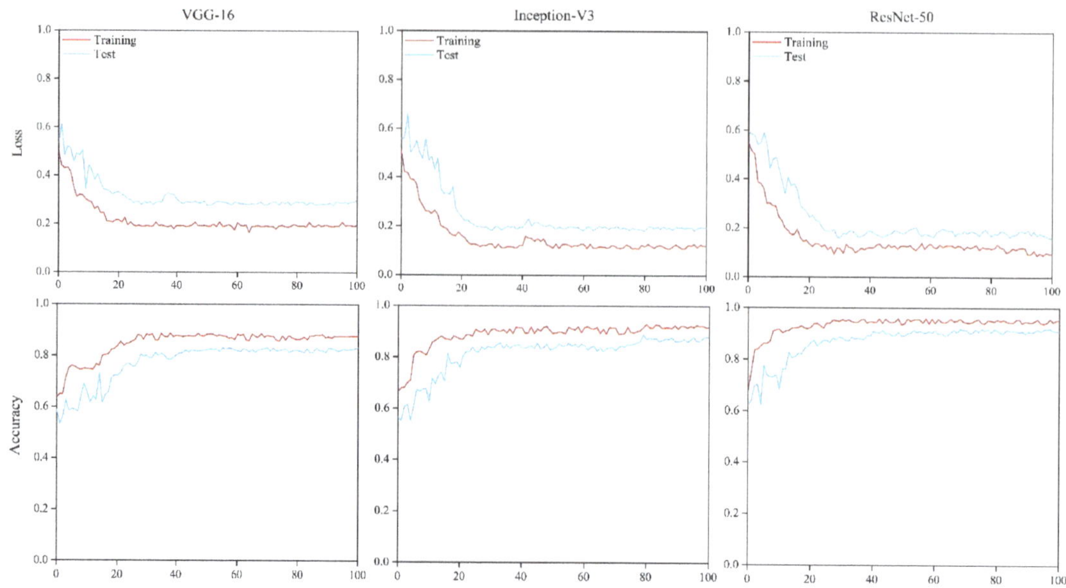

Figure 12. Accuracy and loss for the training and test sets of the three algorithms through RGB images.

Table 3. Performance of the three algorithms for the test sets of RGB images.

Algorithms	Test Accuracy	Kappa
VGG-16	83.55%	0.7421
Inception-V3	87.32%	0.8040
ResNet-50	90.08%	0.8599

The confusion matrices of the three algorithms are shown in Figure 13. It indicated that the performance varied for different lodging severity extents in three algorithms. The identifications of non-lodging all achieved good results, whose classification accuracies were more than 90%. The accuracies of Inception-V3 and ResNet-50 in moderate lodging were improved over 10% compared to that of VGG-16. The three algorithms had no distinct differences in severe lodging. However, the classification error of the three algorithms between moderate lodging and severe lodging was high with almost over 10%, especially VGG-16, which made it difficult to identify the subdivision of lodging effectively.

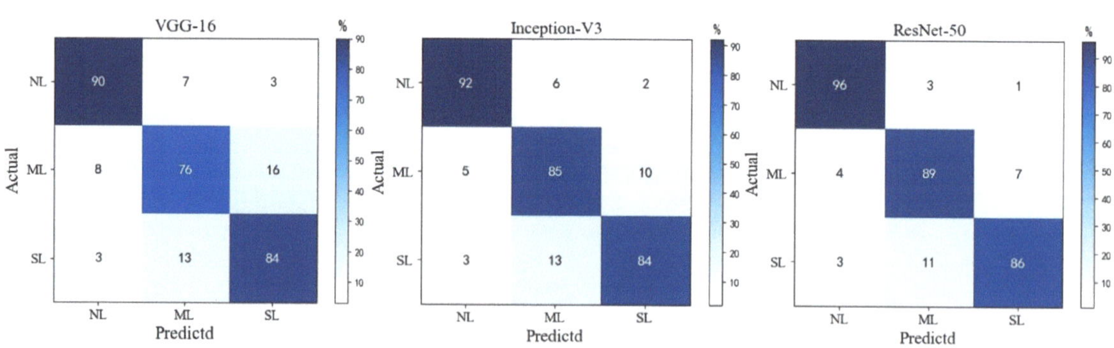

Figure 13. The confusion matrices of the three algorithms through RGB images. Types of lodging extents: NL is non-lodging, ML is moderate lodging, SL is severe lodging.

3.3. Lodging Classification Using Multispectral Images

For multispectral images, the backbone parameters of the three algorithms needed to be randomly initialized to retrain models by the Xavier initialization method [38]. The last layer (fully connected layer) was replaced by three classification categories as well. The software environment and hyperparameter settings were the same as the operation on the RGB images.

The fluctuations of classification accuracy and loss of the three algorithms during 100 iterations using multispectral images were represented in Figure 14. In the early stage of algorithm optimization, the accuracy and loss curves showed an oscillating trend. That was because the algorithms were quickly adjusting the parameters to meet the classification requirements at the beginning. Then, the training accuracy of the three algorithms gradually increased and converged after 60 iterations with 92.34%, 94.70% and 98.55%, respectively. With the continuous optimization, the loss decreased quickly and ResNet-50 was the first to realize convergence. In Table 4, the test accuracies of the three algorithms were 89.91%, 92.36% and 96.32% with Kappa coefficients of 0.8318, 0.8935 and 0.9551, respectively. There was no over-fitting phenomenon in the training process as well. The test accuracy of ResNet-50 was 7.12% and 4.28% higher than VGG-16 and Inception-V3, respectively.

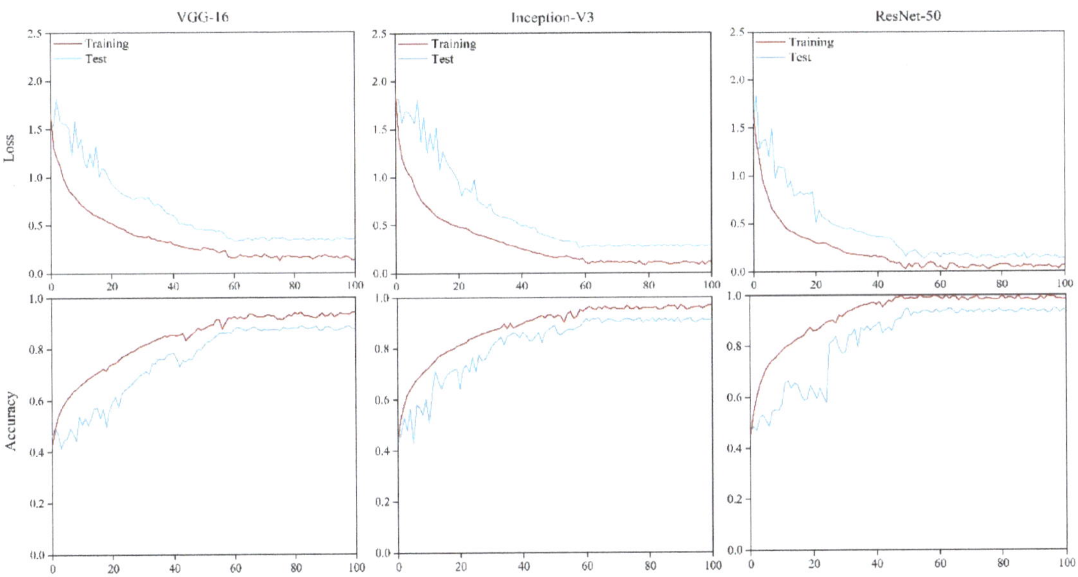

Figure 14. Accuracy and loss for the training and test sets of the three algorithms through multispectral images.

Table 4. Performance of the three algorithms for the test sets of multispectral images.

Algorithms	Test Accuracy	Kappa
VGG-16	88.91%	0.8318
Inception-V3	92.36%	0.8935
ResNet-50	96.32%	0.9551

The confusion matrices of the three algorithms through multispectral images are shown in Figure 15. The three algorithms still performed well in the classification of non-lodging, which was higher than 92%. Compared with the RGB images, the classification of moderate lodging and severe lodging was significantly improved by multispectral images, and the accuracies of Inception-V3 and ResNet-50 were more than 90%. The accuracy error of ResNet-50 between moderate lodging and severe lodging was less than 5%, which can better classify the three extents of maize lodging.

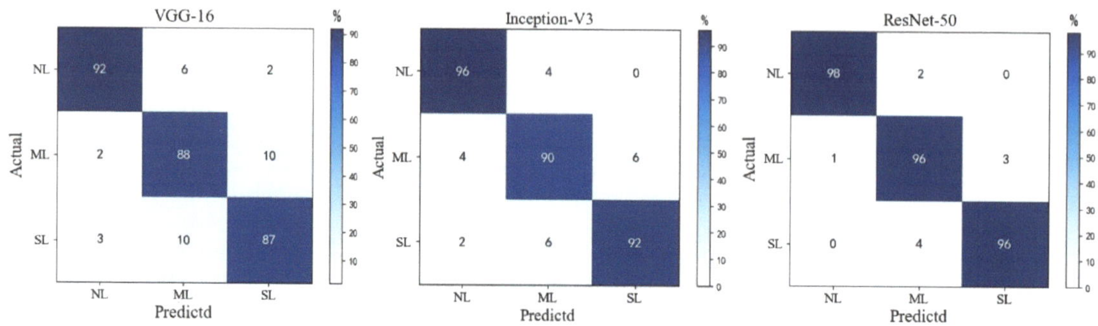

Figure 15. Confusion matrix for the three algorithms through multispectral images. NL is non-lodging, ML is moderate lodging, SL is severe lodging.

3.4. Classification Results

In this study, the experiment results indicated that the overall performance of three deep learning algorithms using multispectral images in classification of different maize lodging extents was better than that of RGB images with an increase of 6.42%, 5.77% and 6.93% (Figure 16). The maize lodging classification based on RGB images using three algorithms realized high accuracy in non-lodging, which was suitable for the binary classification of lodging and non-lodging. Among the three deep learning algorithms, ResNet-50 was efficient and robust to classify the different lodging extents with the fastest convergence rate and highest classification accuracy during algorithm training. ResNet-50 also had the highest improvement in classification accuracy of multispectral images compared with RGB images, which could extract the lodging features more effectively. Therefore, ResNet-50 was the optimal algorithm to realize the classification of maize lodging extents.

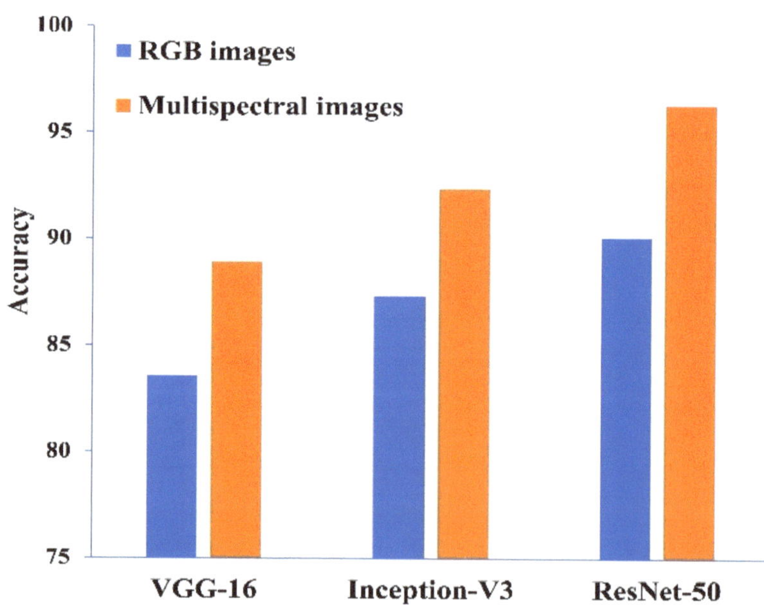

Figure 16. Classification accuracy of the three algorithms with two image types.

4. Discussion

Lodging is a major factor in decreasing the crop yields worldwide. Accurate classification of lodging extents is beneficial to monitoring crop production and conducting reasonable decision-making. Timely and effectively obtaining experimental data plays a crucial role in it. Some researchers used satellite data to conduct crop lodging studies [14,39]. However, it is susceptible to clouds and the revisiting time is long with low spatial resolution. With the development of UAV technology, remote sensing research based on UAVs platform has been highly valued and become a hotspot [40]. The wide application of UAVs has indeed facilitated the monitoring of crop lodging. Tan et al. [23] used RGB images for grading lodging severity with the accuracy of 79.1%. Sun et al. [25] realized the detection of maize lodging with the overall accuracy of 86.61% and the Kappa coefficient of 0.8327 using maximum likelihood classification (MLC) by multispectral images. Furthermore, through applying machine learning methods, such as nearest neighborhood classification and Support Vector Machine (SVM), Chauhan et al. [24] and Rajapaksa et al. [41] reported the wheat lodging classification using multispectral images with 90% and 92.6% accuracies, respectively. Multispectral images had more potential to explore the characteristics of crop lodging. The lodging feature extraction is also of great significance for the classification result. Canopy texture, crop height, spectral reflectance and vegetation indices were extracted separately to research in the above study. The extraction process was both time-consuming and subjective. The features extracted for different crops were also different. It created difficult problems for further research of crop lodging.

We further realized the maize lodging classification based on deep learning algorithms. Deep learning algorithms can automatically extract intrinsic features from massive data through supervised learning to classify different lodging extents. Among the three deep learning algorithms in this study, the ResNet-50 algorithm performed best, with a test accuracy of 96.32% and Kappa coefficients of 0.9551, which was significantly better than traditional machine learning algorithms. On the type of images used above, although the lodging classification using multispectral images was more accurate, the low cost of RGB images acquisition and more than 80% test accuracy made it more beneficial for smallholders to detect crop lodging. The application of transfer-learning method can greatly shorten the training time of the models, which can facilitate more timely agricultural disaster assessment and management. In addition, through using multispectral images, the reflectance variation in red edge band was more evident than that in visible band with the increase of lodging severity extents, which may be an important factor for better lodging classification using multispectral images. Using red edge band to extract sensitive features for classifying lodging extents is worth further study.

There are still some deficiencies that need to be improved. We divided the experimental plots into three lodging extents. Further detailed classification of the lodging extents is necessary, which meets the requirements of precision agriculture. Moreover, the models presented in this study need to be tested and validated in other crop lodging classifications. The solution of them can serve crop yield prediction and precise agricultural insurance claim.

5. Conclusions

In this study, unmanned aerial vehicles (UAVs) provided convenience for multiple types of data acquisition. The RGB and multispectral images of maize lodging canopy were tested to classify different lodging extents. The images were preprocessed by cropping, cleaning and enhancing to generate the dataset containing 5000 subimages. The experimental results indicated that the spectral reflectance increased with the increase of lodging severity on the multispectral images of maize lodging. The red edge band was the most sensitive to the change of lodging severity extents. The classification performance of the three algorithms using RGB images, although good for non-lodging with over 90% accuracy, was unsatisfactory for moderate and severe lodging. The test accuracies of VGG-16, Inception-V3 and ResNet-50 were 89.91%, 92.36% and 96.32% with Kappa coefficients of

0.8318, 0.8935 and 0.9551, respectively, by using multispectral images. The accuracy of ResNet-50 on each lodging subdivision category all reached 96%. Therefore, ResNet-50 outperformed the Inception-V3 and VGG-16 algorithms, and multispectral images were more suitable for crop lodging classification than RGB images. This study provides a more accurate and effective method for the classification of crop lodging extents. Further detailed lodging classification and the general applicability of the method will be the focus of subsequent research.

Author Contributions: X.G. and T.C. designed and initiated the experiments; X.Y. wrote the article; Q.S. collected the data; J.Z. processed the data and prepared the figures; S.G. and Y.P. helped in revising the manuscript. All authors have read and agreed to the published version of the manuscript.

Funding: This research was funded by the National Key Research and Development Program of China (2021YFD1500203) and Beijing Talents Project (2020A58).

Institutional Review Board Statement: Not applicable.

Informed Consent Statement: Not applicable.

Data Availability Statement: The data presented in this study are available on request from the corresponding author. The data are not publicly available due to the use of subsequent studies.

Conflicts of Interest: The authors declare no conflict of interest.

References

1. Wu, W.; Ma, B. A new method for assessing plant lodging and the impact of management options on lodging in canola crop production. *Sci. Rep.* **2016**, *6*, 31890. [CrossRef]
2. Ma, D.; Xie, R.; Liu, X.; Niu, X.; Hou, P.; Wang, K.; Lu, Y.; Li, S. Lodging-related stalk characteristics of maize varieties in China since the 1950s. *Crop Sci.* **2014**, *54*, 2805–2814. [CrossRef]
3. Sun, Q.; Gu, X.; Chen, L.; Xu, X.; Wei, Z.; Pan, Y.; Gao, Y. Monitoring maize canopy chlorophyll density under lodging stress based on UAV hyperspectral imagery. *Comput. Electron. Agric.* **2022**, *193*, 106671. [CrossRef]
4. Shu, M.; Zhou, L.; Gu, X.; Ma, Y.; Sun, Q.; Yang, G.; Zhou, C. Monitoring of maize lodging using multi-temporal Sentinel-1 SAR data. *Adv. Space Res.* **2020**, *65*, 470–480. [CrossRef]
5. Han, L.; Yang, G.; Yang, X.; Song, X.; Xu, B.; Li, Z.; Wu, J.; Yang, H.; Wu, J. An explainable XGBoost model improved by SMOTE-ENN technique for maize lodging detection based on multi-source unmanned aerial vehicle images. *Comput. Electron Agric.* **2022**, *194*, 106804. [CrossRef]
6. Islam, M.S.; Peng, S.; Visperas, R.M.; Ereful, N.; Bhuiya, M.S.U.; Julfiquar, A.W. Lodging-related morphological traits of hybrid rice in a tropical irrigated ecosystem. *Field Crop. Res.* **2007**, *101*, 240–248. [CrossRef]
7. Guo, Y.; Hu, Y.; Chen, H.; Yan, P.; Du, Q.; Wang, Y.; Wang, H.; Wang, Z.; Kang, D.; Li, W.-X. Identification of traits and genes associated with lodging resistance in maize. *Crop J.* **2021**, *9*, 1408–1417. [CrossRef]
8. Liu, T.; Li, R.; Zhong, X.; Jiang, M.; Jin, X.; Zhou, P.; Liu, S.; Sun, C.; Guo, W. Estimates of rice lodging using indices derived from UAV visible and thermal infrared images. *Agric. For. Meteorol.* **2018**, *252*, 144–154. [CrossRef]
9. Sposaro, M.M.; Berry, P.M.; Sterling, M.; Hall, A.J.; Chimenti, C.A. Modelling root and stem lodging in sunflower. *Field Crop. Res.* **2010**, *119*, 125–134. [CrossRef]
10. Zhang, P.; Gu, S.; Wang, Y.; Yang, R.; Yan, Y.; Zhang, S.; Sheng, D.; Cui, T.; Huang, S.; Wang, P. Morphological and mechanical variables associated with lodging in maize (*Zea mays* L.). *Field Crop. Res.* **2021**, *269*, 108178. [CrossRef]
11. Bock, C.H.; Poole, G.H.; Parker, P.E.; Gottwald, T.R. Plant disease severity estimated visually, by digital photography and image analysis, and by hyperspectral imaging. *Crit. Rev. Plant Sci.* **2010**, *29*, 59–107. [CrossRef]
12. Chu, T.; Starek, M.; Brewer, M.; Murray, S.; Pruter, L. Assessing lodging severity over an experimental maize (*Zea mays* L.) field Using UAS images. *Remote Sens.* **2017**, *9*, 923. [CrossRef]
13. Jay, S.; Maupas, F.; Bendoula, R.; Gorretta, N. Retrieving LAI, chlorophyll and nitrogen contents in sugar beet crops from multi-angular optical remote sensing: Comparison of vegetation indices and PROSAIL inversion for field phenotyping. *Field Crop. Res.* **2017**, *210*, 33–46. [CrossRef]
14. Chauhan, S.; Darvishzadeh, R.; Lu, Y.; Boschetti, M.; Nelson, A. Understanding wheat lodging using multi-temporal Sentinel-1 and Sentinel-2 data. *Remote Sens. Environ.* **2020**, *243*, 111804. [CrossRef]
15. Sakamoto, T.; Shibayama, M.; Takada, E.; Inoue, A.; Morita, K.; Takahashi, W.; Miura, S.; Kimura, A. Detecting seasonal changes in crop community structure using day and night digital images. *Photogramm. Eng. Remote Sens.* **2010**, *76*, 713–726. [CrossRef]
16. Liu, Z.; Li, C.; Wang, Y.; Huang, W.; Ding, X.; Zhou, B.; Wu, H.; Wang, D.; Shi, J. Comparison of spectral indices and principal component analysis for differentiating lodged rice crop from normal ones. In *Computer and Computing Technologies in Agriculture V. IFIP Advances in Information and Communication Technology*; Li, D., Chen, Y., Eds.; Springer: Berlin/Heidelberg, Germany, 2012; Volume 369, pp. 84–92.

17. Murakami, T.; Yui, M.; Amaha, K. Canopy height measurement by photogrammetric analysis of aerial images: Application to buckwheat (*Fagopyrum esculentum* Moench) lodging evaluation. *Comput. Electron. Agric.* **2012**, *89*, 70–75. [CrossRef]
18. Xiang, H.; Tian, L. Development of a low-cost agricultural remote sensing system based on an autonomous unmanned aerial vehicle (UAV). *Biosyst. Eng.* **2011**, *108*, 174–190. [CrossRef]
19. Chapman, S.; Merz, T.; Chan, A.; Jackway, P.; Hrabar, S.; Dreccer, M.; Holland, E.; Zheng, B.; Ling, T.; Jimenez-Berni, J. Pheno-copter: A Low-altitude, autonomous remote-sensing robotic helicopter for high-throughput field-based phenotyping. *Agronomy* **2014**, *4*, 279–301. [CrossRef]
20. Burkart, A.; Aasen, H.; Alonso, L.; Menz, G.; Bareth, G.; Rascher, U. Angular dependency of hyperspectral measurements over wheat characterized by a novel UAV based goniometer. *Remote Sens.* **2015**, *7*, 725–746. [CrossRef]
21. Wang, J.-J.; Ge, H.; Dai, Q.; Ahmad, I.; Dai, Q.; Zhou, G.; Qin, M.; Gu, C. Unsupervised discrimination between lodged and non-lodged winter wheat: A case study using a low-cost unmanned aerial vehicle. *Int. J. Remote Sens.* **2018**, *39*, 2079–2088. [CrossRef]
22. Molaei, B.; Chandel, A.; Peters, R.T.; Khot, L.R.; Vargas, J.Q. Investigating lodging in spearmint with overhead sprinklers compared to drag hoses using entropy values from low altitude RGB-imagery. *Inf. Process. Agric.* **2021**, *9*, 335–341. [CrossRef]
23. Tan, S.; Mortensen, A.K.; Ma, X.; Boelt, B.; Gislum, R. Assessment of grass lodging using texture and canopy height distribution features derived from UAV visual-band images. *Agric. For. Meteorol.* **2021**, *308*, 108541. [CrossRef]
24. Chauhan, S.; Darvishzadeh, R.; Lu, Y.; Stroppiana, D.; Boschetti, M.; Pepe, M.; Nelson, A. Wheat Lodging Assessment Using Multispectral UAV Data. *Int. Arch. Photogramm. Remote Sens. Spat. Inf. Sci.* **2019**, *XLII-2/W13*, 235–240. [CrossRef]
25. Sun, Q.; Sun, L.; Shu, M.; Gu, X.; Yang, G.; Zhou, L. Monitoring maize lodging grades via unmanned aerial vehicle multispectral Image. *Plant Phenomics* **2019**, *2019*, 1–16. [CrossRef] [PubMed]
26. Yang, M.-D.; Huang, K.-S.; Kuo, Y.-H.; Tsai, H.; Lin, L.-M. Spatial and spectral hybrid image classification for rice lodging assessment through UAV imagery. *Remote Sens.* **2017**, *9*, 583. [CrossRef]
27. Wilke, N.; Siegmann, B.; Klingbeil, L.; Burkart, A.; Kraska, T.; Muller, O.; van Doorn, A.; Heinemann, S.; Rascher, U. Quantifying lodging percentage and lodging severity using a UAV-based canopy height model combined with an objective threshold approach. *Remote Sens.* **2019**, *11*, 515. [CrossRef]
28. Zhang, H.; Li, Y.; Zhang, Y.; Shen, Q. Spectral-spatial classification of hyperspectral imagery using a dual-channel convolutional neural network. *Remote Sens. Lett.* **2017**, *8*, 438–447. [CrossRef]
29. Hao, X.; Jia, J.; Khattak, A.M.; Zhang, L.; Guo, X.; Gao, W.; Wang, M. Growing period classification of Gynura bicolor DC using GL-CNN. *Comput. Electron. Agric.* **2020**, *174*, 105497. [CrossRef]
30. Paymode, A.S.; Malode, V.B. Transfer learning for multi-crop leaf disease image classification using convolutional neural network VGG. *Artif. Intell. Agric.* **2022**, *6*, 23–33. [CrossRef]
31. Zhang, Z.; Flores, P.; Igathinathane, C.; Naik, D.L.; Kiran, R.; Ransom, J.K. Wheat lodging detection from UAS imagery using machine learning algorithms. *Remote Sens.* **2020**, *12*, 1838. [CrossRef]
32. Subetha, T.; Khilar, R.; Christo, M.S. A comparative analysis on plant pathology classification using deep learning architecture–Resnet and VGG19. *Mater. Today Proc.* **2021**. [CrossRef]
33. Zhao, Y.; Sun, C.; Xu, X.; Chen, J. RIC-Net: A plant disease classification model based on the fusion of inception and residual structure and embedded attention mechanism. *Comput. Electron. Agric.* **2022**, *193*, 106644. [CrossRef]
34. Zhang, H.; Feng, L.; Zhang, X.; Yang, Y.; Li, J. Necessary conditions for convergence of CNNs and initialization of convolution kernels. *Digit. Signal Process.* **2022**, *123*, 103397. [CrossRef]
35. Simonyan, K.; Zisserman, A. Very deep convolutional networks for large-scale image recognition. *arXiv* **2015**, arXiv:1409.1556.
36. Szegedy, C.; Vanhoucke, V.; Ioffe, S.; Shlens, J.; Wojna, Z. Rethinking the inception architecture for computer vision. In Proceedings of the 2016 IEEE Conference on Computer Vision and Pattern Recognition (CVPR), Las Vegas, NV, USA, 27–30 June 2016; pp. 2818–2826.
37. Lu, Z.; Bai, Y.; Chen, Y.; Su, C.; Lu, S.; Zhan, T.; Hong, X.; Wang, S. The classification of gliomas based on a pyramid dilated convolution resnet model. *Pattern Recognit. Lett.* **2020**, *133*, 173–179. [CrossRef]
38. Glorot, X.; Bengio, Y. Understanding the difficulty of training deep feedforward neural networks. In Proceedings of the Thirteenth International Conference on Artificial Intelligence and Statistics, Sardinia, Italy, 13–15 May 2010; pp. 249–256.
39. Kumpumaki, T.; Linna, P.; Lipping, T. Crop lodging analysis from UAS orthophoto mosaic, sentinel-2 image and crop yield monitor data. In Proceedings of the IGARSS 2018: IEEE International Geoscience and Remote Sensing Symposium, Valencia, Spain, 23–27 July 2018.
40. Ampatzidis, Y.; Partel, V. UAV-based high throughput phenotyping in citrus utilizing multispectral imaging and artificial intelligence. *Remote Sens.* **2019**, *11*, 410. [CrossRef]
41. Rajapaksa, S. Classification of crop lodging with gray level co-occurrence matrix. In Proceedings of the 2018 IEEE Winter Conference on Applications of Computer Vision (WACV), Lake Tahoe, NV, USA, 12–15 March 2018.

applied sciences

MDPI

Article

Centralized Task Allocation and Alignment Based on Constraint Table and Alignment Rules

Nam Eung Hwang *, Hyung Jun Kim and Jae Gwan Kim

Hanwha Systems Co., Gyeonggi-do, Seongnam-si 13524, Korea; hyungjun02.kim@hanwha.com (H.J.K.); jg07.kim@hanwha.com (J.G.K.)
* Correspondence: skadnd144@hanwha.com

Abstract: In this paper, we propose a centralized task allocation and an alignment technique based on constraint table and alignment rules. For task allocation, a scoring scheme has to be set. The existing time-discounted scoring scheme has two problems; if the score is calculated based on arrival time the agent who arrives in a task point first may finish the task late, and if the score is calculated based on end-time of the task, agents who have the same score may appear because of temporal constraints. Therefore, a modified time-discounted reward scheme based on both arrival and end-time is proposed. Additionally, an accumulated distance cost scheme is proposed for minimum fuel consumption. The constraint table made by tasks that are already aligned is also considered in scoring. For centralized task alignment based on the constraint table and alignment rules, a technique based on sequential greedy algorithm is proposed. Resolving conflicts on alignment is described in detail using constraint table and alignment rules, which are composed of four basic principles. We demonstrate simulations about task allocation and alignment for multi-agent with coupled constraints. Simple and complicated cases are used to verify the scoring schemes and the proposed techniques. Additionally, a huge case is used to show computational efficiency. The results are feasibly good when the constraints are properly set.

Keywords: task allocation; task alignment; mission planning; task planning; multi-agent systems; multi-agent planning and scheduling

Citation: Hwang, N.E.; Kim, H.J.; Kim, J.G. Centralized Task Allocation and Alignment Based on Constraint Table and Alignment Rules. *Appl. Sci.* **2022**, *12*, 6780. https://doi.org/10.3390/app12136780

Academic Editors: Diego González-Aguilera, Jose A. Jiménez-Berni, Shangpeng Sun and Monica Herrero-Huerta

Received: 2 June 2022
Accepted: 28 June 2022
Published: 4 July 2022

1. Introduction

Automation techniques of systems are among some highly interesting research themes currently. According to this trend, concepts of swarming with a lot of the Unmanned Vehicles (UxVs) or teaming with manned and unmanned vehicles occur to overcome hard tasks with manned or single vehicle using the UxVs. To coordinate several tasks with the UxVs as described above, the sequence of tasks for each UxV has to be issued to them based on their situations and capabilities. In addition, the purpose of task allocation, for example minimizing total mission time, minimizing fuel consumption, etc., and spatial or temporal constraints have to be considered for successful task allocation and alignment. Hence management techniques for multi-agent systems have been widely studied with various approaches and algorithms [1–11]. One of them is developed to recommend tasks to MOD drivers by analyzing and modeling the Mobility-on-Demand Vehicular Crowdsensing (MOD-CS) and MOD-Human-Crowdsensing (MOMAN-CS) market [10]. It is for manned vehicles, so they have freedom to refuse the recommendation. Another is developed for an urban delivery drone system considering energy consumption [11]. Previous research supposed that the energy consumption of robots, especially drones, is proportional to their moving distance but it did not match for delivery drone. Therefore, the weight of the drone including the consignment is also considered when calculating energy consumption.

One of the general task allocation algorithms is Consensus-Based Bundle Algorithm (CBBA), which is based on the auction process [12]. Several agents calculate the task rewards or costs which differ by the purpose, and the task is allocated to the agent that has the highest reward in CBBA. To calculate task rewards, the time-discounted reward scheme, one of the general scoring themes in score-based task allocation algorithm, is used in CBBA. This scheme calculates scores as decreasing time-invariant maximum rewards as time passes. This algorithm is fast to obtain the local optimum solution but CBBA cannot consider the constraints as depicted above. Therefore, Coupled-constraint CBBA (CCBBA), further improved CBBA, is developed [13,14]. In CCBBA, some spatial and temporal constraints can be considered in the task allocation and alignment. Each agent decides the bidding strategy for each task before bidding. According to the decided strategy, task rewards for each agent are calculated for task allocation. For task alignment, 'number of iterations in constraint violation', and some counts are considered. In the optimistic strategy, if the agent violates some constraints when performing a task, the number of iterations in constraint violation increases but the task can be aligned to the agent if it is lower than some thresholds. CCBBA has the ability to handle most multi-agent management problems, but there is a problem concerning the convergence of solution in certain scenarios. Additionally, there is a minor problem in setting temporal constraints; the user must set the maximum permitted time for each task unless they cannot predict it exactly.

CBBA and CCBBA are fast task allocation and alignment algorithms, but computational cost rapidly increases as the number of agents and tasks increases. The reason in CBBA is in the bundle construction phase. When creating a bundle for each agent, the score is calculated not only after the last task of the agent but also before and after all allocated tasks of the agent. For example, an agent has been allocated tasks 1 and 2, and there is an unallocated task 3. Then, scores about sequence of tasks (3-1-2), (1-3-2), and (1-2-3) are calculated to allocate the task 3. Note that task allocation and alignment with coupled constraints are NP-hard problems, and the solution of CBBA is just the local optimum solution. It is a time-consuming process to obtain this kind of solution. The reason in CCBBA is the process about whether an agent can bid to tasks or not. First, the groups of tasks are made based on coupled constraints, called activities. Dependency matrices and temporal constraint matrices are made for each activity to obtain a solution applying constraints. Then, each agent decides optimistic or pessimistic strategies for bidding on each task. For bidding, each agent counts violation iterations, number of bidding alone to task, and number of bidding of any task. This process is also time-consuming.

There is another way to allocate and align tasks from a different point of view, game theory [15]. In this method, the individual agent has freedom for its actions and the agent decides its action by sampling from its possible action set stochastically. When the agent violates the constraints for collective action, penalties that are larger than other rewards are given to it. These penalties are added to the score according to time-discounted reward scheme, so each agent is constrained to obey the constraints. This method solved a minor problem of CCBBA as setting the temporal constraints by just the relationship, such as 'before' or 'after' etc., so there is no need to set permitted time, such as 'do in 30 min after task 1'. While CBBA and CCBBA did not permit the task to be allocated to multi-agents, task allocation based on game theory permits collective actions for one task. In this case, duration time of the task decreases because several agents work together. However, this method results in acceptable and stable solutions only when the probability distributions of actions in action set for each agent are set based on large experiments. It means that a lot of experiments for several cases are needed to obtain the suitable probability distributions. Additionally, computation time is reduced compared to CCBBA, but it is still slow so the update interval of task allocation and alignment results is long. According to the results in [15], 18.086 s (mean time) to obtain reasonable central task allocation solution with six agents and 27 tasks.

In this paper, we consider an algorithm for coupled-constraint task allocation and alignment problems using a constraint table and alignment rules. This is based on CBBA and CCBBA, but a new method is proposed alternating some time-consuming procedures, so local optimum solutions are obtained fast enough. For the task allocation, two scoring schemes are proposed; modified time-discounted scoring scheme and accumulated distance scoring scheme. Additionally, a spatial constraint in CCBBA is considered, called MUTually EXclusive (MUTEX). In addition, a new spatial constraint, called local MUTEX, is considered for essential cases; cooperative work, and etc. For classification, MUTEX in CCBBA is called global MUTEX. Unilateral dependency in CCBBA is considered by temporal constraints instead. For the task alignment, some temporal constraints in CCBBA are considered afterwards, called simultaneous. Temporal constraints 'before' and 'between' are replaced to 'after'. Additionally, constraints 'during' is separated to two constraints; 'start during' and 'end during'. Constraint 'not during' in CCBBA is not used. These temporal constraints are considered using the constraint table which is made when alignment of a task is completed. To apply these constraints and scoring schemes to the task allocation and alignment, the algorithm called centralized Task Allocation and alignment based on Constraint Table and Alignment Rules (TACTAR) is proposed.

2. Preliminaries

2.1. Task Allocation and Alignment Problems

The task allocation problem is to find matching of tasks to agents that maximizes global reward as below [12,16]:

$$\max \sum_{i=1}^{N_u}\left(\sum_{j=1}^{N_t} c_{ij}(\boldsymbol{x}_i, \boldsymbol{p}_i)x_{ij}\right), \tag{1}$$

subject to

$$\sum_{i=1}^{N_u} x_{ij} \leq 1 \ \forall j \in J, \tag{2}$$

$$\sum_{j=1}^{N_t} x_{ij} \leq L_t \ \forall i \in I, \tag{3}$$

$$\sum_{i=1}^{N_u}\sum_{j=1}^{N_t} x_{ij} = N_{min}, \tag{4}$$

$$x_{ij} \in \{0,1\}, \tag{5}$$

where N_u is number of agents, N_t is number of tasks, and L_t is maximum number of tasks each agent can be allocated. x_{ij} is the flag that is 1 if task j is allocated to agent i and is 0 otherwise, $\boldsymbol{x}_i$ is the flag vector whose jth element is x_{ij}, and $\boldsymbol{p}_i$ is the ordered sequence of tasks for agent i. N_{min} is the lowest number between N_t and $N_u L_t$. c_{ij} is scoring function and $c_{ij}(\boldsymbol{x}_i, \boldsymbol{p}_i)$ means reward of task j if agent i completes tasks along $\boldsymbol{p}_i$ based on $\boldsymbol{x}_i$. I and J are agent group and task group, respectively, so number of elements for I and J are N_u and N_t, respectively. As in Equation (2), one task is allocated to only one agent.

The task alignment problem is to calculate the arrival time, start-time, and end-time of each task satisfying temporal constraints and allocated sequence of tasks for each agent. If the temporal constraints have logical error, for example a task1 has to be started after finishing task2, and task2 has to be started after finishing task1, the task alignment result cannot be calculated. To prevent this, we assume that temporal constraints have no logical error so there must be at least one task alignment solution satisfying temporal constraints.

2.2. Consensus-Based Bundle Algorithm (CBBA)

CBBA is a multi-agent task allocation algorithm based on auction concept. CBBA is organized into two phases, bundle construction phase and consensus phase. In bundle construction phase, each agent makes its own ordered sequence of tasks and bundle by calculating the rewards of each task. The rewards are obtained by arriving or completing tasks so the sequence of tasks for each agent is changed according to reward function to obtain maximum global reward. Algorithm of the bundle construction phase is as below [12,17]:

In lines 1~4 of Algorithm 1, vectors about agent i for bundle construction are initialized to previous iteration results, where t is number of iteration, $\boldsymbol{y}_i$ is winning bids list that the agent i knows, $\boldsymbol{z}_i$ is winning agent list that the agent i knows, $\boldsymbol{b}_i$ is bundle of the agent i, and $\boldsymbol{p}_i$ is sequence of tasks for agent i. Then, the bundle and the sequence of tasks are constructed until the size of the bundle is same with L_t in lines 5~14 of Algorithm 1. In line 6 of Algorithm 1, it founds agent i's maximum reward when a task is added to a previous sequence of tasks of agent i, where $S_i^{\boldsymbol{p}_i}$ is reward of agent i along $\boldsymbol{p}_i$ and $\boldsymbol{p}_i(+)_n\{j\}$ means that task j is added after nth component of $\boldsymbol{p}_i$. The rewards are then multiplied to the flag whether the reward is higher than winning bid or not. Finally, the task and the sequence that maximize the agent i's reward are found in lines 8~9 of Algorithm 1 and the task is added to iteration results in lines 10~13 of Algorithm 1.

Algorithm 1 CBBA Phase 1: Bundle construction

1: $\boldsymbol{y}_i(t) = \boldsymbol{y}_i(t-1)$
2: $\boldsymbol{z}_i(t) = \boldsymbol{z}_i(t-1)$
3: $\boldsymbol{b}_i(t) = \boldsymbol{b}_i(t-1)$
4: $\boldsymbol{p}_i(t) = \boldsymbol{p}_i(t-1)$
5: **while** $|\boldsymbol{b}_i| < L_t$, **do**
6: $c_{ij} = max_{n \le |\boldsymbol{p}_i|} S_i^{\boldsymbol{p}_i(+)_n\{j\}} - S_i^{\boldsymbol{p}_i}, \forall j \in J \backslash b_i$
7: $h_{ij} = II\left(c_{ij} > y_{ij}\right), \forall j \in J$
8: $JJ_i = argmax_j c_{ij} \cdot h_{ij}$
9: $n_{i,JJ_i} = argmax_n\ S_i^{\boldsymbol{p}_i(+)_n\{JJ_i\}}$
10: $\boldsymbol{b}_i = \boldsymbol{b}_i(+)_{end}\{JJ_i\}$
11: $\boldsymbol{p}_i = \boldsymbol{p}_i(+)_{n_{i,JJ_i}}\{JJ_i\}$
12: $y_{i,JJ_i}(t) = c_{i,JJ_i}$
13: $z_{i,JJ_i}(t) = i$
14: **end while**

In the consensus phase, agents confirm tasks to perform based on the bundles made in the bundle construction phase. In the decentralized system, the agents have to communicate data with each other for deciding the winning agent. However, in a centralized system, the agents send data only to the central computer. In the central computer, the data from the agents are collected and the consensus algorithm is processed with the data. The general centralized consensus algorithm is Sequential Greedy Algorithm (SGA) as shown in Algorithm 2 [12]. In lines 1~3 of Algorithm 2, variables for SGA are initialized, where I and J are groups of agents and tasks, respectively, η_i is number of confirmed tasks for agent i, and $c_{ij}\left[\boldsymbol{b}_i^{(n)}\right]$ is a reward that is obtained by finishing task j by agent i based on the bundle of agent i, $\boldsymbol{b}_i^{(n)}$. Superscript means the number of iterations. In lines 4~18 of Algorithm 2, agent i_n^* and task j_n^*, which are row and column of maximum value of c^n, respectively, are found, and the agent i_n^* is confirmed to do the task j_n^*. For the next iteration, the task j_n^* is excluded from the group of tasks J because of the assumption 'one task is allocated to only one agent' in Equation (2). If the number of confirmed tasks $\eta_{i_n^*}$ for the agent i_n^* is equal to the limit number of task L_t, the agent i_n^* is also excluded from the group of agents I. This procedure is repeated for N_{min} times to allocate all tasks to the agents.

Algorithm 2 Sequential greedy algorithm

$I_1 = I,\ J_1 = J$
$\eta_i = 0,\ \forall i \in I$
$c_{ij}^{(1)} = c_{ij}[\{\varnothing\}],\ (i,j) \in I \times J$
for $n = 1$ to N_{min} **do**
$(i_n^*, j_n^*) = argmax_{(i,j) \in I \times J} c_{ij}^{n}$
$\eta_{i_n^*} = \eta_{i_n^*} + 1$
$J_{n+1} = J_n \backslash \{j_n^*\}$
$\boldsymbol{b}_{i_n^*}^{(n)} = \boldsymbol{b}_{i_n^*}^{(n-1)} (+)_{end} \{j_n^*\}$
$\boldsymbol{b}_{i_n}^{(n)} = \boldsymbol{b}_{i_n}^{(n-1)},\ \forall i \neq i_n^*$
if $\eta_{i_n^*} = L_t$ **then**
$I_{n+1} = I_n \backslash \{i_n^*\}$
$c_{i_n^*, j}^{(n+1)} = 0,\ \forall j \in J$
else
$I_{n+1} = I_n$
end if
$c_{i, j_n^*}^{(n+1)} = 0,\ \forall i \in I_{n+1}$
$c_{ij}^{(n+1)} = c_{ij}\left[\boldsymbol{b}_i^{(n)}\right],\ \forall (i,j) \in I_{n+1} \times J_{n+1}$
end for

2.3. *Coupled-Constraint Consensus-Based Bundle Algorithm (CCBBA)*

CCBBA is an extension version of CBBA. CCBBA is designed to supplement the weakness of CBBA which is incapacity to resolve task allocation and alignment problems with coupled constraints. The coupled constraints which are considered in CCBBA are described in Tables 1 and 2 [13–15].

Table 1. Spatial coupled constraints in CCBBA.

Name	Description
Unilateral dependency	Task A can be allocated if task B is allocated
Mutual dependency	Task A and B must either be all allocated or not at all
Mutual Exclusive	Only either task A or B can be allocated at each time

Table 2. Temporal coupled constraints in CCBBA.

Name	Description
Simultaneous	Task A and B must start at the same time
Before	Task A must end before task B starts
After	Task A must start after task B ends
During	Task A must start between task B starts and ends
Not during	Task A must either end before task B starts or start after task B ends
Between	Task A must start after task B ends and end before task C starts

For applying these coupled constraints to CCBBA, tasks are partitioned into subgroups called activity that share coupled constraints. The activity does not share constraints with other activities. For each activity, two matrices called dependency matrix and temporal constraint matrix, which are defined as shown in Tables 3 and 4, are used [13–15]. Using these matrices, each agent decides bidding strategies, optimistic and pessimistic, for each task before bundle construction. The agent decides to bid or not according to these strategies in bundle construction and consensus phase. In this process, the concept of 'number of iterations in constraint violation' is applied. These are additional processes compared to CBBA. Although CCBBA can resolve most coupled constraint task allocation and alignment cases, there are some problems in CCBBA. First of all, it faces convergence problems in some cases due to latency. Additionally, the user must set maximum amount of time to the temporal constraint matrix. It means that the user should calculate the time for task alignment in a defined order. However, expected arrival time, start-time, and end-time of tasks are changed frequently based on multi-agent operation environment, so the user finds it hard to expect the maximum amount of time.

Table 3. Entry codes for dependency matrix in CCBBA.

Entry Code ($D(q, p)$)	Description
0	Task p can be allocated independently of task q
−1	Task p and q are mutually exclusive
1	Task p depends on task q
$a > 1$	Task p requires either q or another element with the same code, a *

* Entries are used sequentially: 3 is not used unless 2 is used.

Table 4. Entry for temporal constraint matrix in CCBBA.

Entry ($T(q, p)$)	Description
0	Diagonal component (if task p and task q are same)
$a \neq \infty, a \neq 0$	Maximum amount of time at which task q can start after task p starts
∞	No temporal constraint exists between task p and q

3. Scoring Scheme

A method that is similar to the bundle construction phase of CBBA is used in centralized TACTAR for task allocation. As we can see in Algorithm 1, rewarding function has to decide on making $\boldsymbol{p}_i$. In this paper, two scoring schemes are used: modified time-discounted reward and accumulated distance cost.

3.1. Modified Time-Discounted Reward

The basic time-discounted reward function calculates the reward of agent i as the maximum reward of each task decreases over time as below [12,17]:

$$S_i^{\boldsymbol{p}_i} = \sum_{j \in \boldsymbol{p}_i} \bar{c}_j \rho^{\tau_i^j(\boldsymbol{p}_i)} \tag{6}$$

where $\bar{c}_j$ is time-invariant maximum reward of task j, ρ is decreasing ratio, and $\tau_i^j(\boldsymbol{p}_i)$ is arrival time of task j for agent i along the sequence of task $\boldsymbol{p}_i$. If the reward can be obtained when the agent finishes the task, $\tau_i^j(\boldsymbol{p}_i)$ can be substituted to the end-time of task j. This method can be used in various ways. For example, if the importance of each task is different, a reasonable sequence of tasks can be made by setting $\bar{c}_j$ to the importance level of task j. Additionally, it can be used in food delivery problems if $\bar{c}_j$ is the same for all tasks and ρ is different for each task. In that case, delivering the food to appropriate places would be tasks, and ρ differs by the kind of food. If the reward of task is considered as safety level, it can be used in finding the safest sequence of tasks for agent i.

However, this rewarding scheme needs to be fixed because of two cases: one is that an agent is confirmed to perform a task because they arrive at the task point early, but finishes the task later than other agents; the other is that the agent cannot be confirmed to perform a task based on the rewards when several agents finish the task simultaneously. The second case occurs usually because of temporal constraints. Therefore, a modified time-discounted rewarding scheme has to be applied in TACTAR.

$$S_i^{\boldsymbol{p}_i} = \sum_{j \in \boldsymbol{p}_i} \bar{c}_j \left(w_{arr} \rho^{\tau_{i,arr}^j(\boldsymbol{p}_i)} + w_e \rho^{\tau_{i,end}^j(\boldsymbol{p}_i)} \right) \tag{7}$$

where w_{arr} and w_e are weights for arrival and end rewards, respectively, and $\tau_{i,arr}^j(\boldsymbol{p}_i)$ and $\tau_{i,end}^j(\boldsymbol{p}_i)$ are arrival time and end-time of task j for agent i based on sequence of tasks $\boldsymbol{p}_i$.

3.2. Accumulated Distance Cost

This cost function is used for finding the sequence of tasks for agents that minimizes total distance. To apply this cost function, bundle construction phase of CBBA should be changed as in Algorithm 3.

Algorithm 3 CBBA Phase 1: Bundle construction for Accumulated Distance Cost

1: $\boldsymbol{y}_i(t) = \boldsymbol{y}_i(t-1)$
2: $\boldsymbol{z}_i(t) = \boldsymbol{z}_i(t-1)$
3: $\boldsymbol{b}_i(t) = \boldsymbol{b}_i(t-1)$
4: $\boldsymbol{p}_i(t) = \boldsymbol{p}_i(t-1)$
5: **while** $|\boldsymbol{b}_i| < L_t$, **do**
6: $c_{ij} = d_{accum} + d^j_{p_i^{end}}, \ \forall j \in J \backslash \boldsymbol{b}_i$
7: $h_{ij} = \begin{cases} 1 & (c_{ij} < y_{ij}) \\ \infty & (otherwise) \end{cases}, \ \forall j \in J$
8: $JJ_i = argmin_j c_{ij} \cdot h_{ij}$
9: $\boldsymbol{b}_i = \boldsymbol{b}_i (+)_{end} \{JJ_i\}$
10: $\boldsymbol{p}_i = \boldsymbol{p}_i (+)_{end} \{JJ_i\}$
11: $y_{i,JJ_i}(t) = c_{i,JJ_i}$
12: $z_{i,JJ_i}(t) = i$
13: **end while**

In line 6 of Algorithm 3, d_{accum} is total accumulated distance and $d^j_{p_i^{end}}$ is distance between agent i and point of task in $\boldsymbol{p}_i(end)$. Note that the inequality sign and argument of the maxima in lines 7~8 of Algorithm 3 are changed for finding the sequence of tasks that minimizes total accumulated distance. To allocate tasks to agents, sequential greedy algorithm in Algorithm 2 also should be changed as described in Algorithm 4.

Algorithm 4 Sequential greedy algorithm for Accumulated Distance Cost

1: $I_1 = I, \ J_1 = J$
2: $\eta_i = 0, \ \forall i \in I$
3: $c_{ij}^{(1)} = c_{ij}[\{\varnothing\}], \ (i,j) \in I \times J$
4: **for** n = 1 to N_{min} **do**
5: $(i_n^*, j_n^*) = argmin_{(i,j) \in I \times J} c_{ij}^n$
6: $\eta_{i_n^*} = \eta_{i_n^*} + 1$
7: $J_{n+1} = J_n \backslash \{j_n^*\}$
8: $\boldsymbol{b}_{i_n^*}^{(n)} = \boldsymbol{b}_{i_n^*}^{(n-1)} (+)_{end} \{j_n^*\}$
9: $\boldsymbol{b}_{i_n}^{(n)} = \boldsymbol{b}_{i_n}^{(n-1)}, \ \forall i \neq i_n^*$
10: **if** $\eta_{i_n^*} = L_t$ **then**
11: $I_{n+1} = I_n \backslash \{i_n^*\}$
12: $c_{i_n^*,j}^{(n+1)} = 0, \ \forall j \in J$
13: **else**
14: $I_{n+1} = I_n$
15: **end if**
16: $c_{i,j_n^*}^{(n+1)} = 0, \ \forall i \in I_{n+1}$
17: $c_{ij}^{(n+1)} = c_{ij}\left[\boldsymbol{b}_i^{(n)}\right], \ \forall (i,j) \in I_{n+1} \times J_{n+1}$
18: **end for**

In line 5 of Algorithm 4, argument of maxima is changed to argument of minimum as line 8 of Algorithm 3. Note that d_{accum} should be replaced to $c_{i_n^*,j_n^*}^{(n)}$ when calculating the next iteration costs, $c_{ij}^{(n+1)}$ in line 17 of Algorithm 4. Additionally, note that accumulated distance cost can be only applied to centralized task allocation. This method can be used in the situation; a minimum number of agents are used for completing all tasks with minimum fuel consumption.

4. Scheduling Scheme

After task allocation, task alignment should be processed to meet temporal constraints. Assume that every agent has a speed limit and moves to target task points as soon as possible. Additionally, the duration time of each task is not changed by scheduling, such as shrinking or stretching; it is changed only by agents. Then, four basic rules can be made as in Table 5. If the task B is aligned, temporal constraints about the task B are generated. These constraints have to be considered for the next alignment of tasks, the task A in Table 5. If the task B is dropped by rules in Table 5, these constraints cannot be considered any more. Therefore, a constraint table is used in centralized TACTAR to manage generated constraints by aligned tasks. This table has number-of-constraints rows and six columns that contain that shown in Table 6.

Table 5. Basic rules for scheduling.

No.	Description *
1	If task A should start at or after t and start-time of task A is before t, start-time of task A is set to t and recalculate rewards/costs of candidate tasks.
2	If task A should start at or before t and start-time of task A is after t, task B and tasks aligned after task B are dropped and re-alignment proceeds.
3	If task A should end at or after t and end-time of task A is before t, end-time of task A is set to t and recalculate rewards/costs of candidate tasks.
4	If task A should end at or before t and end-time of task A is after t, task B and tasks aligned after task B are dropped and re-alignment proceeds.

* Task A is a candidate task of alignment and task B is an aligned task already. Additionally, temporal constraint at time t is made by task B.

Table 6. Contents of constraint table.

Column	Description
1	Target task ID
2	Allocated agent of task that makes constraint
3	ID of task that makes constraint
4	Target time(0: start-time, 1: end-time)
5	Constraint time
6	Constraint type(0: before, 1: after, 2: simultaneous)

For example, task 2 must start after task 1 ends and task 2 is aligned to agent 3. Then, a row of the constraint table is created. Column 1 of row 1 would be 1 which means task 1. Columns 2 and 3 would be 3 and 2 which mean agent 3 and task 2, respectively. Note that columns 2 and 3 are needed for deleting constraints when tasks are dropped. Column 4 would be 1 which means end-time of task 1. Columns 5 and 6 would be start-time of tasks 2 and 0, respectively. The zero in column 6 means that task 1 should end before the time in column 5. If temporal constraints on task 2 are more than one, other rows of the constraint table will be made.

Following the rules in Table 5, one of the temporal constraints 'Not During' in Table 2 cannot be applied because it cannot decide which rule should be applied when violating the constraint. For example, if task A must either end before task B starts or start after task B ends but task A starts after task B starts and task A ends before task B ends, it corresponds to rule 2 and 4. In this situation, task A cannot be aligned without changing the duration time of task A. Additionally, the temporal constraint 'Between' in Table 2 can be separated to 'Before' and 'After', so 'Between' is no longer needed in temporal constraints. The temporal constraint 'Before' can be replaced to 'After'. For example, 'Task A must end before task B starts' can be 'Task B must start after task A ends'.

The spatial constraint 'Mutual exclusive' in Table 1 means that only either task A or B can be allocated to all agents at each time. For example, if task A is allocated to agent 1, task B cannot be allocated to all agents. However, it needs to allocate task B to agents except agent 1. Hence, the spatial constraint 'Mutual exclusive' should be separated to two constraints: global and local. Additionally, 'Unilateral dependency' in Table 1 is useless because tasks which have unilateral dependency relationship have at least one temporal constraint so checking only temporal constraints violation is enough. Therefore, spatial and temporal constraints in centralized TACTAR are as in Tables 7 and 8.

Table 7. Spatial coupled constraints in centralized TACTAR.

Name	Description
Local Mutual Exclusive	Only either task A or B can be allocated to each agent at each time
Global Mutual Exclusive	Only either task A or B can be allocated to all agents at each time

Table 8. Temporal coupled constraints in centralized TACTAR.

Name	Description
Simultaneous	Task A and B must start at the same time
After	Task A must start after task B ends
Start during	Task A must start between task B starts and ends
End during	Task A must end between task B starts and ends

For applying these coupled constraints to centralized TACTAR, two matrices are defined as in Tables 9 and 10. Note that the components of temporal constraint matrix are just codes in Table 9, not maximum amount of time in Table 4.

Table 9. Entry codes for dependency matrix in centralized TACTAR.

Entry Code ($D(q, p)$)	Description
0	Task p can be allocated independently of task q
1	Task p and q are locally mutual exclusive
2	Task p and q are globally mutual exclusive

Table 10. Entry for temporal constraint matrix in centralized TACTAR.

Entry ($T(q, p)$)	Description
0	No temporal constraint exist between task p and q
1	Task p and q must start at the same time
2	Task p must start after task q ends
3	Task p must start between task q starts and ends
4	Task p must end between task q starts and ends

5. Algorithm

The algorithm of centralized TACTAR including scoring and scheduling schemes as described in Sections 3 and 4, respectively, can be summarized as in Algorithm 5. In line 1 of Algorithm 5, variables for centralized TACTAR are set where $\boldsymbol{X}_a$ is positions of agents, $\boldsymbol{X}_t$ is task points, and $\boldsymbol{D}_t$ and $\boldsymbol{T}_t$ are spatial and temporal constraint matrix, respectively.

In line 2 of Algorithm 5, constants for centralized TACTAR are set where $\boldsymbol{T}_{t,dur}$ is duration time of tasks, $\boldsymbol{V}_{max,a}$ is maximum velocity of each agent, $\boldsymbol{F}_{bid,a}$ is bidding flags whether each agent can bid to tasks or not, and $\boldsymbol{T}_{min,dur}$ is minimum overlapping time between working time of tasks in 'start during' or 'end during' relationship. $\bar{c}$ is time invariant maximum reward for each task, $\boldsymbol{\rho}$ is decreasing ratio of it for each task as described in Equation (7), w_{arr} and w_{end} are weight of arrival and end-reward (cost) for Equation (7), respectively.

Algorithm 5 Centralized TACTAR: main

1: **set** variables $X_a^{(0)}$, $X_t^{(0)}$, D_t, T_t
2: **set** constants $T_{t,dur}$, $V_{max,a}$, $F_{bid,a}$, $T_{min,dur}$, $\bar{c}$, ρ, w_{arr}, w_{end}
3: **While** (at least one capable task is remained incompletion)
4: **initialize** uncompleted tasks
5: **make** sequence of tasks
6: **end while**

In lines 3~7 of Algorithm 5, task allocation and alignment are performed until all tasks are completed. At the first iteration, variables for uncompleted tasks need to be initialized. The initialization for iterations is as below:

In Algorithm 6, lines 2~15 are performed for each agent. In lines 3~9 of Algorithm 6, initialization is performed based on condition of each task where p_m is sequence of tasks for agent m. If the agent has not arrived at the task point, initialization in line 4 of Algorithm 6 is performed. In the case that the agent has arrived at the task point or started the task, initialization in line 7 of Algorithm 6 is performed. t_{arr}, t_s, and t_e are arrival time, start-time, and end-time of tasks, respectively, z is the agent in charge of each task. If the user chooses accumulated distance cost scheme, lines 11~13 of Algorithm 6 are performed for recalculating accumulated distance, d_{accum}. After all, constraints made by deleted tasks in lines 4 or 7 of Algorithm 6 are deleted from the constraint table in line 14 of Algorithm 6. Note that if the agent in charge of the task differs from the agent in column 3 of the constraint table, these constraints also need to be deleted.

Algorithm 6 Centralized TACTAR: line 4 of main

1: **for** $m = 1{:}N_u$
2: **for** $n = 1{:}\mathrm{size}(p_m)$
3: **if** agent m is not arrived $p_m(n)$ point
4: **initialize** t_{arr}, t_s, t_e, z, p_m for tasks $p_m(n)$ and after
5: **Break**
6: **else if** agent m is arrived $p_m(n)$ point or start $p_m(n)$ task last
7: **initialize** t_{arr}, t_s, t_e, z, p_m for after $p_m(n)$
8: **Break**
9: **end if**
10: **end for**
11: **if** accumulated distance cost scheme
12: **recalculate** d_{accum} for p_m
13: **end if**
14: **delete** constraints made by deleted tasks
15: **end for**

After initialization is completed in line 4 of Algorithm 5, making a sequence of tasks, i.e., task allocation and alignment, is performed in line 5 of Algorithm 5 as follows:

In Algorithm 7, start-time, end-time, and reward(cost) of each task need to be calculated in line 1 before making the sequence of tasks if the task can be allocated. The conditions for the prohibition of task allocation are as in Table 11. Note that start-time and end-time of each task is calculated by assuming that agents move at maximum velocity. If at least one condition in Table 11 fails, the reward(cost), start-time, and end-time of the task are not calculated. In that case, the reward will be 0 if the user selects a time-discounted reward scheme and the cost will be infinity if the user selects an accumulated distance cost scheme.

Algorithm 7 Centralized TACTAR: line 5 of main

calculate t_s, t_e, and reward(cost) of each task
while (at least one task is not allocated)
if time-discounted reward scheme
find maximum reward
if maximum reward == 0
Break
end if
else if accumulated distance cost scheme
find minimum cost
if minimum cost == infinity
break
end if
end if
find (agent, task) that have maximum reward(minimum cost)
align (agent, task)
renew t_{arr}, t_s, t_e, z, p_m for (agent, task)
add constraints to constraint table
if accumulated distance cost scheme
recalculate d_{accum} for p_m
end if
calculate reward(cost) of each task
end while

Table 11. Conditions for prohibition of allocating task p to agent A in TACTAR.

No.	Contents	Cannot Be Allocated
1	Bidding flags, $F_{bid,a}(p)$	(1.1) 0
2	Condition of task p	(2.1) After arriving point of task p
3	Local MUTEX with task q	(3.1) Task q is allocated to agent A
4	Global MUTEX with task q	(4.1) Task q is allocated already
5	Simultaneous with task q	(5.1) Task q is allocated to agent A
6	After task q	(6.1) Task q is not allocated yet
7	Start during task q	(7.1) Task q is not allocated yet or (7.2) Task q is allocated to agent A or (7.3) Task q has relationship (except 'After') with task r, task r is allocated to agent A and agent A cannot arrive point of task p after finishing task r
8	End during task q	(8.1) Task q is not allocated yet or (8.2) Task q is allocated to agent A or (8.3) Task q has relationship (except 'After') with task r, task r is allocated to agent A and agent A cannot finish task p after finishing task r
9	Task r starts simultaneous with tasks q	(9.1 *) Some tasks in $\boldsymbol{q}$, called $\boldsymbol{q}_{sub}$, are allocated and agent A cannot arrive point of task r after finishing task p
10	Task r starts during task q	(10.1 **) Only task q is allocated and agent A cannot start task r after finishing task p before end-time of task q ***
11	Task r ends during task q	(11.1 **) Only task q is allocated and agent A cannot end task r after finishing task p before end-time of task q

* There is no task p in tasks $(\boldsymbol{q} - \boldsymbol{q}_{sub})$. ** The task p is not equal to task r. *** If the $\boldsymbol{T}_{min,dur}$ is not 0, the 'end-time of task q' is changed to '(end-time of task q)—$\boldsymbol{T}_{min,dur}$'.

If 11 conditions in Table 11 are all passed, start-time and end-time of each task are checked by the constraints table. If temporal violation can be solved by delaying these times, time correction proceeds. If not, these times are not corrected and reward(cost) calculation is processed. It is for time correction of already aligned tasks. Note that this procedure is not equal to bundle construction in Algorithm 1 or 3. In bundle construction, the sequence of whole tasks of each agent is made by calculating rewards(costs). However, in line 1 of Algorithm 7, reward(cost) of each task is calculated just for the next step. For example, there is an agent that is assigned task 1 and 2 and there are three remaining tasks. Then the rewards(costs), calculated in line 1 of Algorithm 7, are three rewards(costs) of remaining tasks for three sequences, (1-2-3), (1-2-4), and (1-2-5). The cost based on accumulated distance cost scheme is calculated as described in lines 6~7 of Algorithm 3.

The lines 2~22 in Algorithm 7 are repeated until all tasks are allocated. It will find (agent, task) set that has maximum reward(minimum cost) in line 14 of Algorithm 7. The alignment of this set is tried in line 15 of Algorithm 7 as described in Algorithm 8. In line 1 of Algorithm 8, constraints associated with the task in line 14 of Algorithm 7 are found in the constraint table. The lines 4~8 and lines 10~14 of Algorithm 8 are violation checked for start-time and end-time of the task, respectively. Note that only violations for 'before' and 'simultaneous' in Table 6 are checked because start-time and end-time are delayed based on the constraint table in line 1 of Algorithm 7 if they violate 'after' in Table 6. In lines 5 and 11 of Algorithm 8, aligned tasks correction proceeds. First, a task which made violated constraints afterwards are removed from tasks of the agent who made violated constraint. Then, constraints made by deleted tasks are removed from the constraint table. After that, constraints associated with the task in renewed constraint table are found in line 1 again. If the user choses accumulated distance cost scheme, d_{accum} is recalculated based on $\boldsymbol{p}_m$ renewed in line 16 of Algorithm 7. Then, the calculation of rewards(costs) for remaining tasks proceeds for next iteration of lines 2~21 of Algorithm 7.

After the process in line 5 of Algorithm 5, task allocation and alignment for one iteration is performed. Lines 4 and 5 in Algorithm 5 are repeated every iteration until all capable tasks are completed, so real-time task allocation and alignment are performed by TACTAR.

Algorithm 8 Centralized TACTAR: line 15 of Algorithm 7

1: **find** constraints associated the task in constraint table
2: **for** a = 1:number of constraints in line 1
3: **if** column 4 is 0 //start-time of task
4: **if** t_s > column 5 and column 6 is 0 or 2 //violate before or simultaneous
5: **delete** the tasks constraint(a,3) and after all from constraint(a,2)
6: **delete** constraints made by deleted tasks in line 5
7: **goto** line 1
8: **end if**
9: **else if** column 4 is 1 //end-time of task
10: **if** t_e > column 5 and column 6 is 0 or 2 //violate before or simultaneous
11: **delete** the tasks constraint(a,3) and after all from constraint(a,3)
12: **delete** constraints made by deleted tasks in line 5
13: **goto** line 1
14: **end if**
15: **end if**
16: **end for**

6. Results and Discussion

In this section, simulation results are provided to assess TACTAR. First, two scoring schemes as described in Section 3 are applied to a simple case for verification of scoring schemes because it is hard to verify the schemes in complicated cases. This case is intuitive enough to prove the performance of scoring schemes. Then, these schemes are applied to complicated cases to check that TACTAR can derive a solution satisfying spatial and temporal constraints. Note that the complicated case must contain almost constraints in Tables 7 and 8, and there are one or more solutions satisfying constraints for verification. Finally, a huge case which is composed of six agents, 30 tasks, and coupled constraints similar to the complicated case is used to emphasize the computational efficiency of TACTAR. These simulations are run by MATLAB 2018b, and specifications of the computer are in Table 12. Additionally, the settings for three cases are as shown in Table 13. To apply constraints in Table 13 to TACTAR, spatial and temporal matrices for the simple and complicated cases, respectively, should be made as described in Table 14.

Table 12. Specifications of computer.

Contents	Specification
CPU	Intel(R) Core(TM) i5-10500 3.10 GHz
GPU	Intel(R) UHD Graphics 630
RAM	16GB
Storage	256GB SSD
Program	MATALB 2018b

Table 13. Settings for simulations.

Contents	Simple Case	Complicated Case	Huge Case
N_u	3	3	6
N_t	8	8	30
$\mathbf{X}_a^{(0)}$	(1,0), (2,0), (3,0)	(1,0), (2,0), (3,0)	(1,0), (1,1), (2,0), (2,1), (3,0), (3,1)
$\mathbf{X}_t^{(0)}$ *	(−2,3), (−4,3), (−4,5.1), (−2,5.2), (7,7), (9,7), (9,9.1), (7,9.2)	(−2,3), (−4,3), (−4,5.1), (−2,5.2), (7,7), (9,7), (9,9.1), (7,9.2)	20 rand (30,1) −10
$T_{t,dur}$	0.5 [s] for all tasks		
$V_{max,a}$	2 [m/s] for all agents		
$T_{min,dur}$	0.3 [s] for all tasks		
$F_{bid,a}$	1 for all tasks and agents		
$\bar{c}$	100 for all tasks		
ρ	0.8 for all tasks		
w_{arr}	0.1 for all tasks		
w_{end}	1 for all tasks		
Spatial constraints	None	Local MUTEX (1-3) Global MUTEX (4-8)	Local MUTEX (1-3) Global MUTEX (4-8)
Temporal constraints	Task sequence (1-2-3-4) Task sequence (5-6-7-8)	Task sequence (1-2-3-4) Task simultaneously (1-5) Task 6 starts during task 7	Task sequence (1-2-3-4) Task simultaneously (1-5) Task 6 starts during task 7

* If arrival time, start-time, end-time, or distances between tasks are same for some agents, scoring schemes may not operate appropriately. In real world, these are hard to be same so the task points are adjusted slightly. This adjustment is reasonable enough.

6.1. *Time-Discounted Rewarding for Simple Case*

Results for the simple case using the time-discounted scoring scheme are as in Figures 1–3. In Figure 1, 'A' and 'T' mean agent and task, respectively, and numbers after these are the indices for agents or tasks. The red lines mean sequence of tasks for each agent. The number of tasks are larger than the number of agents so all agents are allocated (and aligned) tasks to get high total reward. Task 1 is allocated to agent 1 because agent 1 is the first to arrive at point of the task 1. Task 2 is allocated to agent 2 because the arrival time of task 2 for agent 2 is faster than agent 1. Figure 1 is the reasonable result from this point of view. Note that agent 2 is allocated to near tasks, task 2 and task 4, instead of far tasks, tasks 5~8, to obtain a high reward. It means that the modified time-discounted scoring scheme (and its original version as described in Equation (6)) is not used for minimizing total time for finishing all tasks. Figure 2 shows the computational time in this case. Most iterations are under 0.02 s so it is fast enough. Figure 3 describes the timetable of this case. Each row of Figure 2 shows the schedule of each agent. Blue bars mean working time of each task and the labels of y-axis mean agent in charge of tasks. 'T' and indices after that have same meaning as described in Figure 1. Note that the temporal constraints described in Table 13 are satisfied in the results.

Table 14. Settings for simulations.

Contents	Simple Case	Complicated Case
D	$\begin{bmatrix} 0&0&0&0&0&0&0&0\\ 0&0&0&0&0&0&0&0\\ 0&0&0&0&0&0&0&0\\ 0&0&0&0&0&0&0&0\\ 0&0&0&0&0&0&0&0\\ 0&0&0&0&0&0&0&0\\ 0&0&0&0&0&0&0&0\\ 0&0&0&0&0&0&0&0 \end{bmatrix}$	$\begin{bmatrix} 0&0&1&0&0&0&0&0\\ 0&0&0&0&0&0&0&0\\ 1&0&0&0&0&0&0&0\\ 0&0&0&0&0&0&0&2\\ 0&0&0&0&0&0&0&0\\ 0&0&0&0&0&0&0&0\\ 0&0&0&0&0&0&0&0\\ 0&0&0&2&0&0&0&0 \end{bmatrix}$
T	$\begin{bmatrix} 0&2&0&0&0&0&0&0\\ 0&0&2&0&0&0&0&0\\ 0&0&0&2&0&0&0&0\\ 0&0&0&0&0&0&0&0\\ 0&0&0&0&0&2&0&0\\ 0&0&0&0&0&0&2&0\\ 0&0&0&0&0&0&0&2\\ 0&0&0&0&0&0&0&0 \end{bmatrix}$	$\begin{bmatrix} 0&2&0&0&1&0&0&0\\ 0&0&2&0&0&0&0&0\\ 0&0&0&2&0&0&0&0\\ 0&0&0&0&0&0&0&0\\ 1&0&0&0&0&0&0&0\\ 0&0&0&0&0&0&0&0\\ 0&0&0&0&0&3&0&0\\ 0&0&0&0&0&0&0&0 \end{bmatrix}$

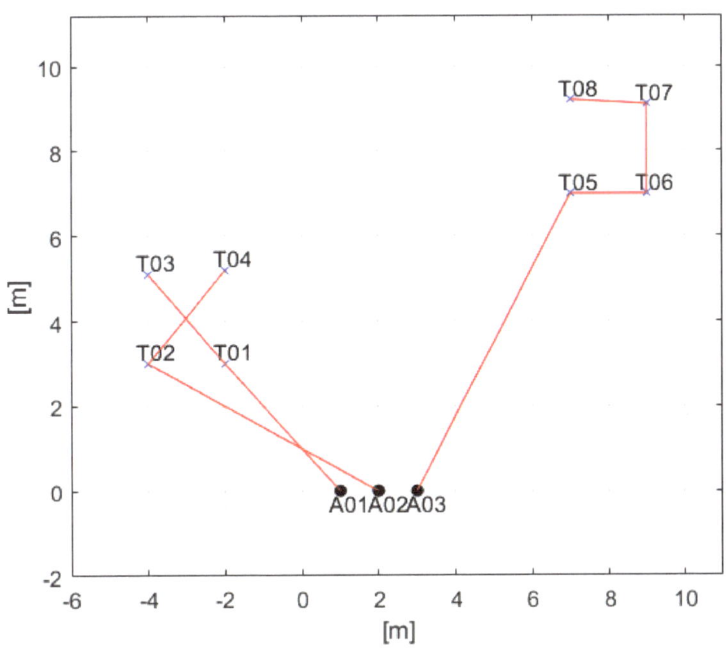

Figure 1. 2-D result of TACTAR for a simple case applying a modified time-discounted reward scheme.

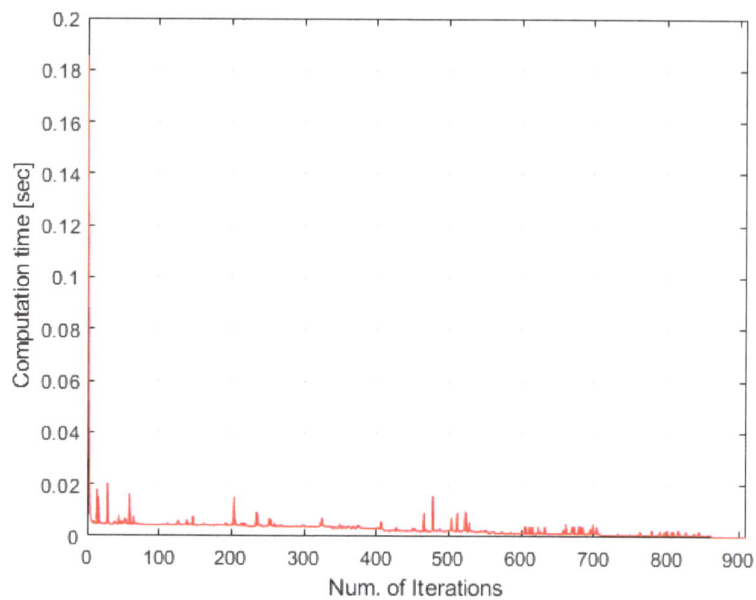

Figure 2. Computational time of TACTAR for a simple case applying a modified time-discounted reward scheme.

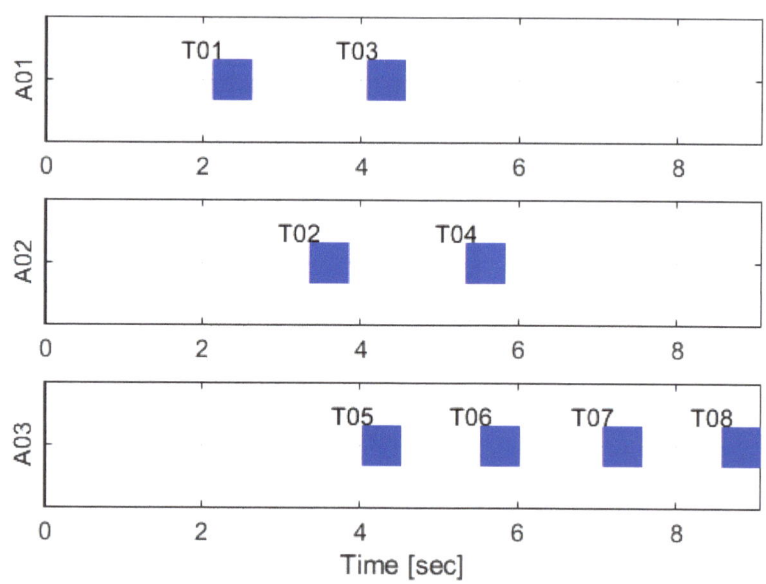

Figure 3. Timetable result of TACTAR for a simple case applying a modified time-discounted reward scheme.

6.2. Accumulated Distance Cost for Simple Case

The results for the simple case using the accumulated distance scoring scheme are as in Figures 4–6. Unlike the results in Section 6.1, only two agents, except agent two, are allocated tasks, although the number of tasks are larger than the number of agents. Task 1 is allocated to agent 1 first because distance between the point of the task 1 and the agent 1 is the shortest. Next, task 2 is allocated to agent 1 because increased total travel distance after finishing the task 1 is the smallest in this case. Figure 4 is an intuitive and reasonable result from this point of view. Figure 5 shows the computational time in this case. Most iterations are also under 0.02 s as seen in Figure 2 so it is fast enough. Figure 6 describes the timetable of this case. Agent 2 is not allocated to any task, so there is no timetable in row 2 of Figure 6. Note that the temporal constraints described in Table 13 are satisfied in the results.

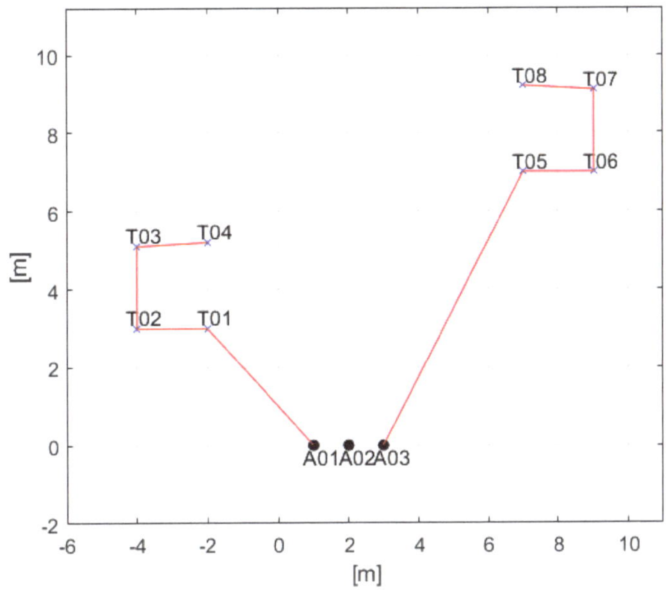

Figure 4. 2-D result of TACTAR for a simple case applying an accumulated distance cost scheme.

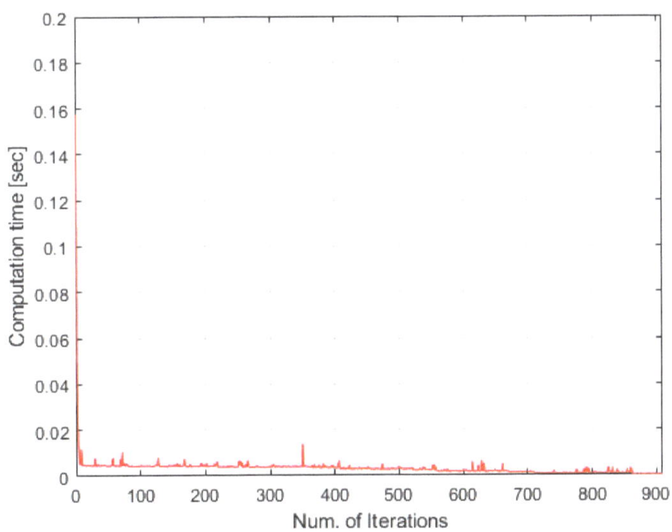

Figure 5. Computational time of TACTAR for a simple case applying an accumulated distance cost scheme.

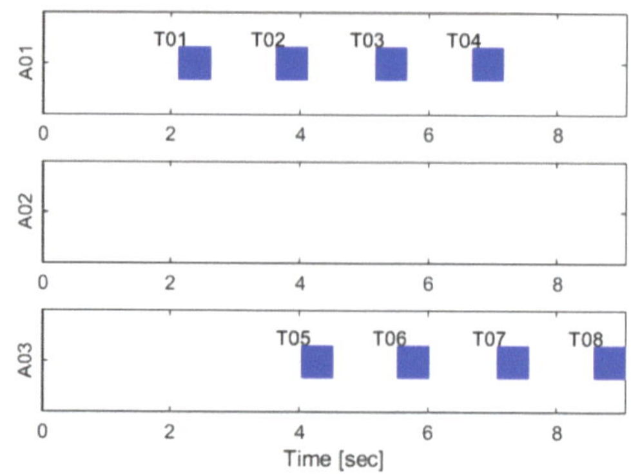

Figure 6. Timetable result of TACTAR for a simple case applying an accumulated distance cost scheme.

6.3. Time-Discounted Rewarding for Complicated Case

The time-discounted scoring scheme is verified in simple cases as in Section 6.1. Therefore, this scheme is applied to the complicated case and the results are as in Figures 7–9. These results are different from the results in Section 6.1 because the spatial and temporal constraints are changed. In Figure 7, agent 1 is not allocated task 3 because task 1 is in local MUTEX relationship with the task 3. Additionally, the start-time of task 8 is faster than task 4 which is in global MUTEX relationship with task 8 so task 4 is not allocated to all agents. Therefore, all spatial constraints in Table 13 are satisfied in the results. Figure 8 shows the computational time in this case. Most iterations are under 0.05 s so it is fast enough. In Figure 9, agent 1 arrives at the point of task 1 early but task 1 starts late because task 1 must start with task 5 simultaneously. The red bar means hold time at point of each task. Note that the hold time is from arrival time to start time of task. Task 2 must start after task 1 ends so agent 2 waits until the task 1 ends. The third temporal constraint is that task 6 must start during work at task 7. To satisfy this constraint, agent 3 waits until 0.2 s before the start-time of the task 6 because $T_{min,dur}$ is set to 0.3 s and $T_{t,dur}$ is set to 0.5 s. Note that TACTAR is iterated frequently so if the start-time of task 6 is delayed, the start-time of task 7 will be also delayed reflecting the situation of task 6. All temporal constraints in Table 13 are also satisfied in the results.

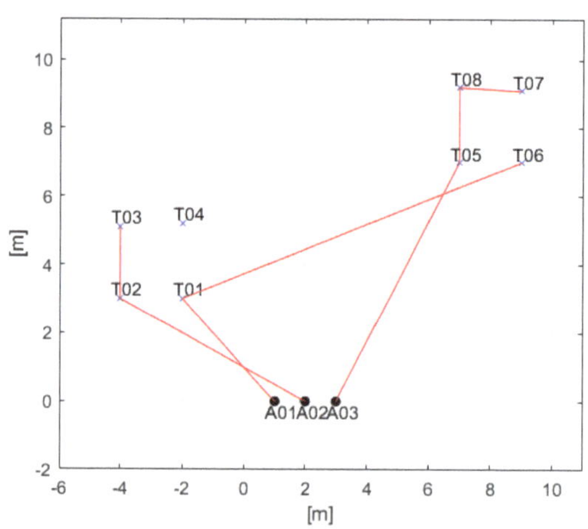

Figure 7. 2-D result of TACTAR for a complicated case applying a modified time-discounted reward scheme.

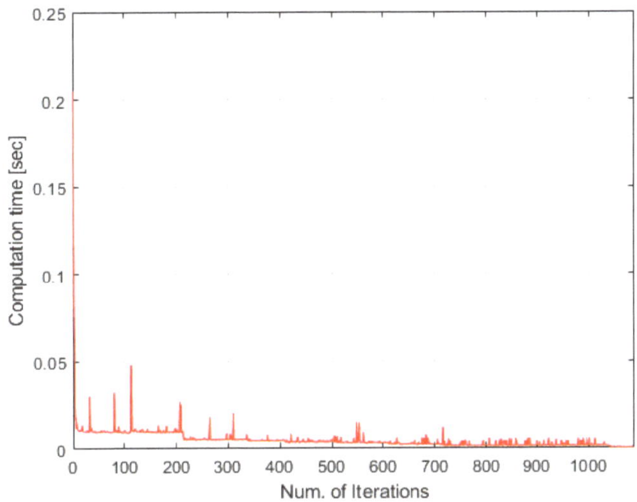

Figure 8. Computational time of TACTAR for a complicated case applying a modified time-discounted reward scheme.

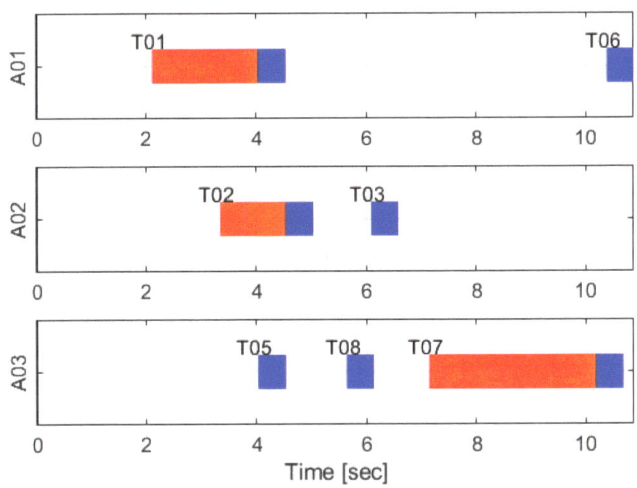

Figure 9. Timetable result of TACTAR for a complicated case applying a modified time-discounted reward scheme.

6.4. Accumulated Distance Cost for Complicated Case

The accumulated distance scoring scheme is also verified in the simple case as in Section 6.2, so this scheme is applied to the complicated case. The results are as in Figures 10–12. These results are also different to the results in Section 6.2 because of constraints. In Figure 10, agent 1 is not allocated task 3 because task 1 is in local MUTEX relationship with the task 3. Additionally, start-time of task 4 is faster than task 8 which is in global MUTEX relationship with the task 4, so the task 8 is not allocated to all agents. Therefore, all spatial constraints in Table 13 are satisfied in the results. Figure 11 shows the computational time in this case. Most iterations are also under 0.05 s as seen in Figure 8 so it is fast enough. In Figure 12, agent 1 arrives at the point of task 1 early but task 1 starts late because the task 1 must start with task 5 simultaneously. Task 3 also starts late because task 3 must start after the end of task 2. Task 7 is paused to start so long because of the third temporal constraint in Table 13. When the current time is 0.2 s before the start-time of task 6, the task is started by agent 3. Therefore, all temporal constraints in Table 13 are also satisfied in the results.

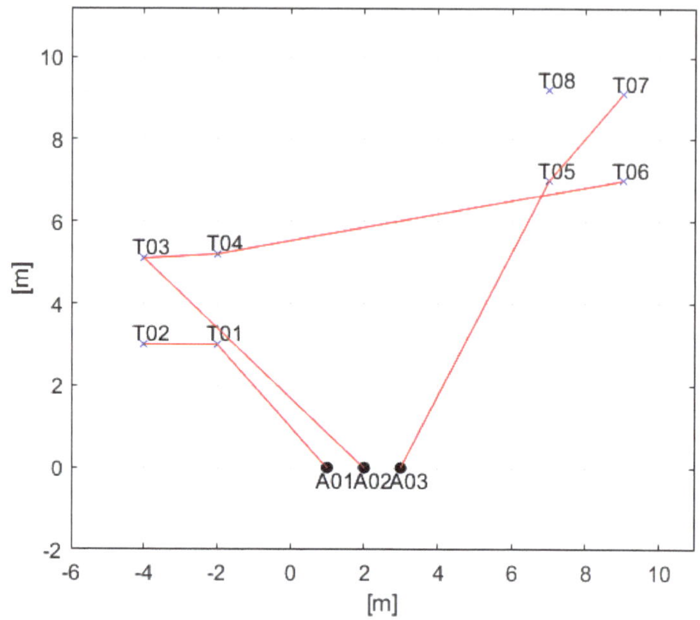

Figure 10. 2-D result of TACTAR for a complicated case applying an accumulated distance cost scheme.

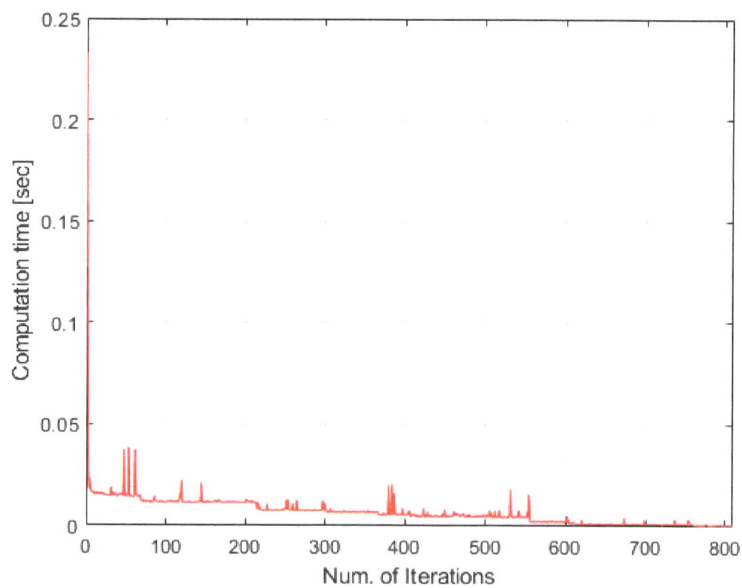

Figure 11. Computational time of TACTAR for a complicated case applying an accumulated distance cost scheme.

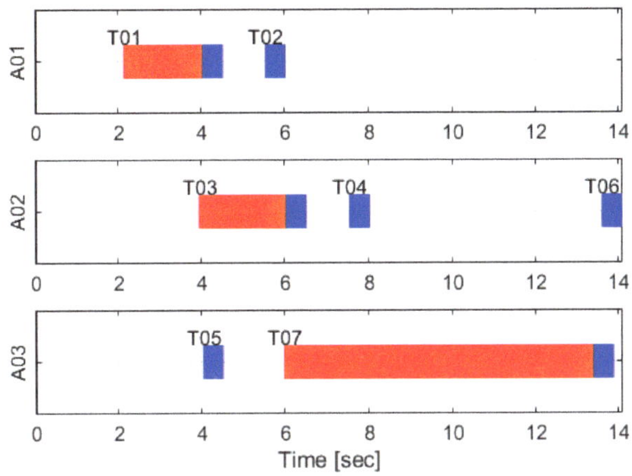

Figure 12. Timetable result of TACTAR for a complicated case applying an accumulated distance cost scheme.

6.5. Time-Discounted Rewarding for Huge Case

The time-discounted scoring scheme is verified in simple and complicated cases as in Sections 6.1 and 6.3, so this scheme is applied to the huge case for showing computational efficiency. The results are as in Figures 13–15. As we can see, all spatial and temporal constraints in Table 13 are satisfied in Figures 13 and 15. Figure 14 describes computational time of huge cases. The computational time is under 0.15 s for all iterations except the first one. Additionally, it decreases as the number of iterations increases because tasks are finished by agents. Note that these are the results using MATLAB which uses only a single CPU core. If the simulations are run by another compiler such as Visual Studio etc., the computational time will be much lower.

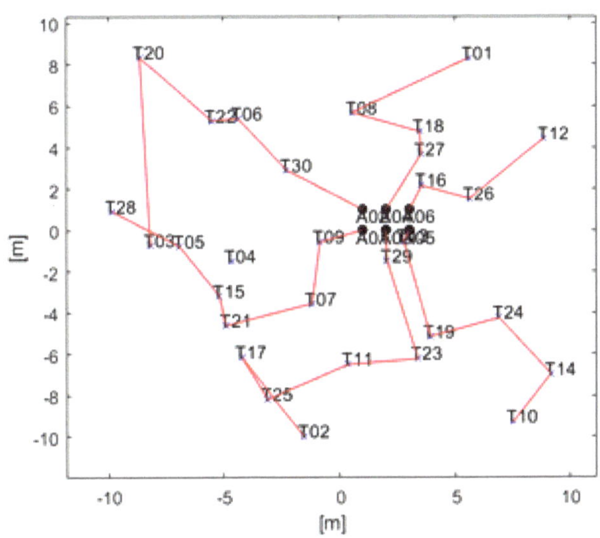

Figure 13. 2-D result of TACTAR for a huge case applying a modified time-discounted reward scheme.

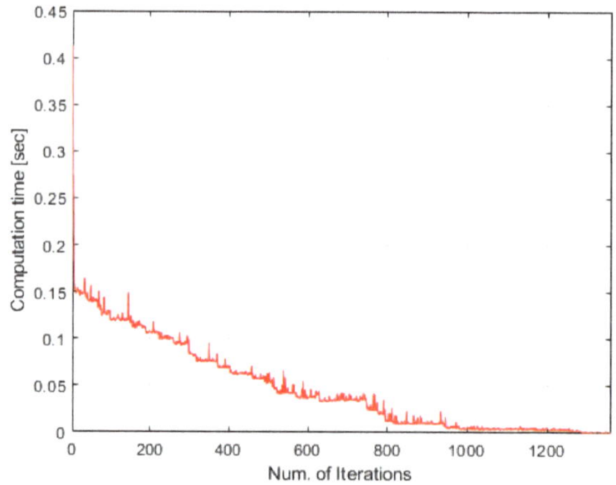

Figure 14. Computational time of TACTAR for a huge case applying a modified time-discounted reward scheme.

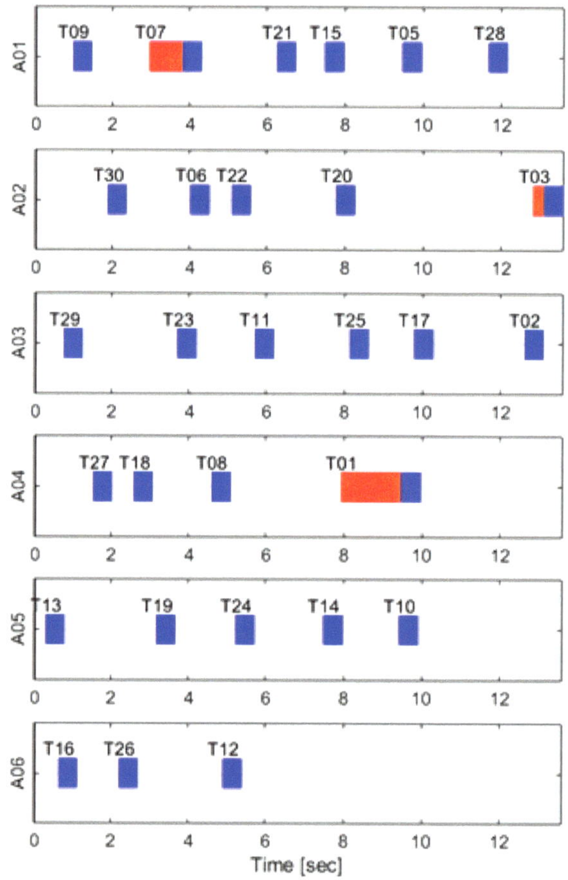

Figure 15. Timetable result of TACTAR for a huge case applying a modified time-discounted reward scheme.

7. Conclusions

In this paper, we introduced an algorithm for coupled-constraint task allocation and alignment problems using a constraint table and alignment rules called TACTAR. The two scoring schemes are used for task allocation. One is time-discounted reward scheme which is used in CBBA. For preventing the situation that two or more rewards of tasks are the same and calculating rewards based on both arrival time and end-time of tasks, a modified version of the reward scheme is used in TACTAR. To obtain reasonable allocation results, we need to set weights of arrival and end-rewards appropriately. The other is accumulated distance cost scheme which is proposed first in this paper. This scoring scheme is used for allocation with minimum fuel consumption.

The constraint table is used for task allocation and alignment. The constraint table has six columns which contain constrained task, who and which task makes constraints, and the contents of constraints. It is used in task allocation when arrival, start, and end-time of each task are used for calculating reward(cost). Additionally, it is used in task alignment, which is based on alignment rules. The alignment rules are about methods between candidate task of alignment and tasks already aligned. Finally, pseudo-codes of TACTAR and simulations for simple and complicated cases are described. The constraints used in TACTAR are different from those in CCBBA because some constraints are added for needs. Additionally, some constraints are deleted because it can be replaced with another one. However, the temporal constraint 'Not During' in CCBBA cannot be handled in this paper, which is a minor limitation of TACTAR. Some limitations including this have to be solved in future works.

Author Contributions: Conceptualization, N.E.H.; methodology, N.E.H.; software, N.E.H.; validation, N.E.H.; formal analysis, N.E.H.; investigation, N.E.H.; resources, N.E.H.; data curation, N.E.H.; writing—original draft preparation, N.E.H.; writing—review and editing, N.E.H.; visualization, N.E.H.; supervision, H.J.K. and J.G.K.; project administration, H.J.K. and J.G.K.; funding acquisition, H.J.K. and J.G.K. All authors have read and agreed to the published version of the manuscript.

Funding: This work was supported by Korea Research Institute for defense Technology planning and advancement (KRIT) grant funded by the Korea government (DAPA (Defense Acquisition Program Administration)) (No. KRIT-CT-21-009, Development of Realtime Automatic Mission Execution and Correction Technology based on Battlefield Information, 2022).

Institutional Review Board Statement: Not applicable.

Informed Consent Statement: Not applicable.

Data Availability Statement: The data presented in this study might be available on reasonable request from the corresponding author.

Conflicts of Interest: The authors declare no conflict of interest.

References

1. Jeong, B.M.; Ha, J.S.; Choi, H.L. MDP-Based Mission Planning for Multi-UAV Persistent Surveillance. In Proceedings of the 2014 14th International Conference on Control, Automation and Systems, Gyeonggi-do, Korea, 22–25 October 2014.
2. Billionnet, A.; Costa, M.C.; Sutter, A. An Efficient Algorithm for a Task Allocation Problem. *J. ACM* **1992**, *39*, 502–518. [CrossRef]
3. Bellingham, J.; Tillerson, M.; Richards, A.; How, J.P. Multi-Task Allocation and Path Planning for Cooperating UAVs. In *Cooperative Control: Models, Applications and Algorithms*, 1st ed.; Butenko, S., Murphey, R., Pardalos, P.M., Eds.; Springer: Boston, MA, USA, 2003; Volume 1, pp. 23–41.
4. Liu, C.; Kroll, A. A Centralized Multi-Robot Task Allocation for Industrial Plant Inspection by Using A* and Genetic Algorithms. In Proceedings of the International Conference on Artificial Intelligence and Soft Computing, Zakopane, Poland, 29 April–3 May 2012.
5. Jin, Y.; Minai, A.A.; Polycarpou, M.M. Cooperative Real-Time Search and Task Allocation in UAV Teams. In Proceedings of the 42nd IEEE International Conference on Decision and Control, Hawaii, HI, USA, 9–12 December 2003.
6. Lagoudakis, M.G.; Berhault, M.; Koenig, S.; Keskinocak, P.; Kleywegt, A.J. Simple Auctions with Performance Guarantees for Multi-Robot Task Allocation. In Proceedings of the 2004 IEEE/RSJ International Conference on Intelligent Robots and Systems, Sendai, Japan, 28 September–2 October 2004.

7. Marcarthur, K.S.; Stranders, R.; Ramchurn, S.D.; Jennings, N.R. A Distributed Anytime Algorithm for Dynamic Task Allocation in Multi-Agent Systems. In Proceedings of the 25th AAAI Conference on Artificial Intelligence, San Francisco, CA, USA, 7–11 August 2011.
8. Oh, K.T.; Kim, W.D. Task Assignment Algorithm for Rendezvous of Multiple UAVs. In Proceedings of the 2012 Korean Society for Aeronautical and Space Sciences Fall Conference, Jeju, Korea, 7–11 November 2012.
9. Wei, H.; Lv, Q.; Duo, N.; Wang, G.S.; Liang, B. Consensus Algorithms Based Multi-Robot Formation Control under Noise and Time Delay Conditions. *Appl. Sci.* **2019**, *9*, 229–244. [CrossRef]
10. Xiang, C.; Li, Y.; Zhou, Y.; He, S.; Qu, Y.; Li, Z.; Gong, L.; Chen, C. A Comparative Approach to Resurrecting the Market of MOD Vehicular Crowdsensing. In Proceedings of the IEEE International Conference on Computer Communications, Virtual Conference, 2–5 May 2022.
11. Xiang, C.; Zhou, Y.; Dai, H.; Qu, Y.; He, S.; Chen, C.; Yang, P. Reusing Delivery Drones for Urban Crowdsensing. *IEEE Trans. Mob. Compt.* 2021; *preprint*. [CrossRef]
12. Choi, H.L.; Brunet, L.; How, J.P. Consensus-Based Decentralized Auctions for Robust Task Allocation. *IEEE Trans. Robot.* **2009**, *25*, 912–926. [CrossRef]
13. Whitten, A.K.; Choi, H.L.; Johnson, L.B.; How, J.P. Decentralized Task Allocation with Coupled Constraints in Complex Missions. In Proceedings of the 2011 American Control Conference, San Francisco, CA, USA, 29 June–1 July 2011.
14. Whitten, A. Decentralized Planning for Autonomous Agents Cooperating in Complex Missions. Master's Thesis, Massachusetts Institute of Technology, Cambridge, MA, USA, 2010.
15. Lim, M.C.; Choi, H.L. Improving Computational Efficiency in Crowded Task Allocation Games with Coupled Constraints. *Appl. Sci.* **2019**, *9*, 29–52. [CrossRef]
16. Lee, C.H.; Moon, G.H.; Yoo, D.W.; Tahk, M.J.; Lee, I.S. Distributed Task Assignment Algorithm for SEAD Mission of Heterogeneous UAVs Based on CBBA Algorithm. *J. Korean Soc. Aeronaut. Space Sci.* **2012**, *40*, 988–996.
17. Moon, S.; Kim, H.J. Cooperation with Ground and Arieal Vehicles for Multiple Tasks: Decentralized Task Assignment and Graph Connectivity Control. *J. Inst. Contr. Robot Syst.* **2012**, *18*, 218–223. [CrossRef]

MDPI AG
Grosspeteranlage 5
4052 Basel
Switzerland
Tel.: +41 61 683 77 34

MDPI Books Editorial Office
E-mail: books@mdpi.com
www.mdpi.com/books

www.ingramcontent.com/pod-product-compliance
Lightning Source LLC
Chambersburg PA
CBHW041424210926
51850CB00013B/3352
* 9 7 8 3 7 2 5 8 2 0 9 5 5 *